Mechanical Behavior of Materials under Dynamic Loads

Mechanical Behavior of Materials under Dynamic Loads

SYMPOSIUM

HELD IN SAN ANTONIO, TEXAS, SEPTEMBER 6–8, 1967
SPONSORED BY THE ARMY RESEARCH OFFICE, DURHAM
AND THE SOUTHWEST RESEARCH INSTITUTE

EDITOR
Ulric S. Lindholm
SOUTHWEST RESEARCH INSTITUTE
SAN ANTONIO, TEXAS

Springer-Verlag New York Inc. • 1968

ISBN-13: 978-3-642-87447-5 e-ISBN-13: 978-3-642-87445-1

DOI: 10.1007/ 978-3-642-87445-1

Softcover reprint of the hardcover 1st edition 1968
Library of Congress Catalog Card Number 68-22687

Title Number 1501

PREFACE

An adequate physical and mathematical description of material behavior is basic to all engineering applications. Fortunately, many problems may be treated entirely within the framework of elastic material response. While even these problems may become quite complex because of geometrical and loading conditions, the linearity, reversibility, and rate independence generally applicable to elastic material description certainly eases the task of the analyst. Today, however, we are increasingly confronted with practical problems which involve material response which is inelastic, hysteretic and rate dependent combined with loading which is transient in nature. These problems include, for instance, structural response to moving or impulsive loads, all the areas of ballistics (internal, external and terminal), contact stresses under high speed bearings, high speed machining, rolling and other metal working processes, explosive and impact forming, shock attenuation structures, seismic wave propagation, and many others of equal importance. As these problems were encountered, it became increasingly evident that we did not have at hand the physical or mathematical description of the behavior of materials necessary to produce realistic solutions. Thus, during the last ten years particularly, there has been considerable effort expended toward the generation of both experimental data on the dynamic mechanical response of materials as well as the formulation of realistic constitutive theories. It was the purpose of the Symposium at which the articles in this book were presented to discuss and review recent developments in this field.

Dynamic loading in the present context was taken to include deformation rates above those achieved on the standard laboratory testing machine (commonly designated as static or quasi-static). While slow tests may encounter time-dependent effects, such as creep and stress-relaxation, and therefore are in a sense dynamic, these topics were not explicitly included in the Symposium. Rather, the emphasis was on rates of loading encountered in impact problems where inertia forces as well as viscous type forces may play a dominant role. The reader will find papers related to the interpretation of results from high-speed testing machines, impact or impulsive loading of bars, plates, cylinders and spheres, and the propagation of cracks; all with the purpose of defining the intrinsic dynamic mechanical properties of the materials under study. There are, in addition, papers dealing with the fundamentals of the development of a basic constitutive theory for solids. This theory is developed in different papers from the standpoints of classical continuum mechanics, from the fundamental laws of thermodynamics, and from the microscopic point of view through consideration of dynamic dislocation mechanisms for crystalline solids. It was the intention of the organizers of the Symposium to bring together those applying these apparently divergent approaches to the same basic end objective, since it was felt that future developments in this area will depend upon a synthesis of the results from such areas as mechanics, thermodynamics, the materials sciences, and certain aspects of solid state physics.

While the majority of the papers are concerned with the plastic flow of metals, other materials treated include those which exhibit viscoelastic behavior, rocks, and a specific foam composition. This distribution is not in proportion to importance, but rather reflects the relative level of effort in the technical community on dynamic properties of the several classes of materials. The future will undoubtedly see an increased interest in the non-metallics as their applications expand.

The Symposium was held on September 6–8, 1967, in San Antonio, Texas, and was attended by approximately 150 persons. There were five technical sessions with invited technical papers being presented by distinguished researchers from this country and abroad. At the conclusion of the technical presentations, the five session chairmen were given the unenviable task of summarizing in a few (10) minutes the papers presented and the state-of-the-art in their areas of interest. These closing comments are included at the end of this volume and are well worth the readers' close attention.

The program was as follows:

WEDNESDAY, SEPTEMBER 6

Session Ia 9:00 am

CHAIRMAN: U. S. Lindholm, Southwest Research Institute

Martin Goland, President, Southwest Research Institute: *Welcoming Remarks*

Sudhir Kumar, Army Research Office—Durham, Durham, N.C.: *Mechanics/Materials Aspects of Dynamic Loading*

Session Ib 10:00 am

CHAIRMAN: W. Prager, University of California at San Diego

M. Reiner, Technion—Israel Institute of Technology, Haifa, Israel: *Dynamical Strength of an Ideal Solid With Definite Constitutive Equation* (Presented by Dr. Amnon Foux)

J. F. Bell, The Johns Hopkins University, Baltimore, Maryland: *An Ex-*

perimental Study of Instability Phenomena in the Initiation of Plastic Waves in Long Rods

O. W. Dillon, Jr., University of Kentucky, Lexington, Kentucky: *Plastic Deformation Waves and Heat Generated Near the Yield Point of Annealed Aluminum*

Session II 2:00 pm

CHAIRMAN: D. C. Drucker, Brown University, Providence, R.I.

P. Perzyna, Polish Academy of Sciences, Warsaw, Poland: *On Thermodynamic Foundations of Viscoplasticity*

U. S. Lindholm, Southwest Research Institute: *Some Experiments in Dynamic Plasticity Under Combined Stress*

M. F. Kanninen, A. K. Mukherjee, A. R. Rosenfield, and G. T. Hahn, Battelle Memorial Institute, Columbus, Ohio: *The Speed of Ductile-Crack Propagation and the Dynamics of Flow in Metals*

J. M. Krafft, Naval Research Laboratory, Washington, D.C.: *Dynamic Mechanical Behavior of Metal at the Tip of a Plane Strain Crack*

THURSDAY, SEPTEMBER 7

Session III 9:00 am

CHAIRMAN: J. D. Campbell, University of Oxford, Oxford, England

J. J. Gilman, University of Illinois, Urbana, Illinois: *Dynamical Behavior of Dislocations*

S. R. Bodner, Technion – Israel Institute of Technology, Haifa, Israel: *Constitutive Equations for Dynamic Material Behavior*

J. W. Edington, Battelle Memorial Institute, Columbus, Ohio: *Effect of Strain Rate on the Dislocation Substructure in Deformed Niobium Single Crystals*

W. J. Gillich, U.S. Army Ballistic Research Laboratories, Aberdeen Proving Ground, Maryland: *Constitutive Relationships From Impact Studies*

Session IV

CHAIRMAN: C. D. Lundergan, Sandia Corporation, Albuquerque, N.M.

A. H. Jones, C. J. Maiden, S. J. Green, and H. Chin, General Motors Technical Center, Warren, Michigan: *Prediction of Elastic-Plastic Wave Profiles in Aluminum 1060-0 Under Uniaxial Strain Loading*

C. H. Karnes, Sandia Corporation, Albuquerque, N.M.: *The Plate Impact Configuration for Determining Mechanical Properties of Materials at High Strain Rates*

E. A. Ripperger, The University of Texas, Austin, Texas, and H. Watson, Jr., Southern Methodist University, Dallas, Texas: *The Relationship Between the Constitutive Equation and One-Dimensional Wave Propagation*

I. M. Fyfe, University of Washington, Seattle, Washington: *Plane-Strain Plastic Wave Propagation in a Dynamically Loaded Hollow Cylinder*

FRIDAY, SEPTEMBER 8

Session V 8:30 am

CHAIRMAN: D. M. Forney, Jr., AFML, Wright-Patterson Air Force Base, Ohio

N. Cristescu, Mathematical Institute, Bucarest, Romania: *Dynamic Plasticity Under Combined Stress*

K. C. Valanis, Iowa State University, Ames, Iowa: *Unified Theory of Thermomechanical Behavior of Viscoelastic Materials*

W. E. Jahsman, Lockheed Missiles and Space Company, Palo Alto, California: *Static and Dynamic Material Behavior of Syntactic Foam*

J. B. Cheatham, Jr., Rice University, Houston, Texas: *The Effect of Pressure, Temperature, and Loading Rate on the Mechanical Properties of Rocks*

The Symposium was held under the joint sponsorship of the Army Research Office—Durham and the Southwest Research Institute. The committee responsible for the planning and arrangements for the Symposium were:

Ulric S. Lindholm, Southwest Research Institute, San Antonio, Texas (CHAIRMAN)

Sudhir Kumar, Army Research Office—Durham, Durham, North Carolina (HONORARY CO-CHAIRMAN)

H. Norman Abramson, Southwest Research Institute, San Antonio, Texas (HONORARY CO-CHAIRMAN)

Philip H. Francis, Southwest Research Institute, San Antonio, Texas (SECRETARY)

David L. Black, Southwest Research Institute, San Antonio, Texas (SYMPOSIUM COORDINATOR)

U. S. LINDHOLM, Editor

INTRODUCTION
MECHANICS/MATERIALS, ASPECTS OF DYNAMIC LOADING

SUDHIR KUMAR
U.S. Army Research Office – Durham
Durham, N.C.

The behavior of materials under dynamic loads is obviously of considerable interest in most mechanical analyses of design problems where dynamic loads are present. Unfortunately, much of the engineering design today is still based on the static loading properties of the material rather than dynamic properties. Quite often this means over-design at best and incorrect design resulting in failure in the worst cases. The problem, however, has continued to exist due to insufficient basic knowledge and understanding of the behavior of different materials in spite of significant advances made in the recent past. This conference, and several other conferences [1–7] sponsored by ARO-D, AIME, ASTM, OSR, ASME and Battelle in the last eight years, is essentially an attempt towards such advancement.

The complex nature of the dynamic behavior problem can be seen from Fig. 1-1, which depicts the whole range of interaction of dynamic loads with various materials. Ideally, it will be desirable to know the mechanical response to the full range of dynamic loads for each material. However, certain load-material interactions have more relative importance for engineering design, and significant work on them exists already. Uniquely, the metals under both static and dynamic loads can be cited as the outstanding case. Both the continuum mechanics engineers and the metallurgical engineers found these materials to be most attractive to study. At the same time, it must be confessed that, relatively speaking, these materials were easier to handle for analysis and for scientifically planned and reproducible experiments. Even so, there is a great deal that we don't understand about them, in spite of voluminous scientific literature existing in this area. Each type of load response, e.g., creep, vibratory or hypervelocity impact, is a big field

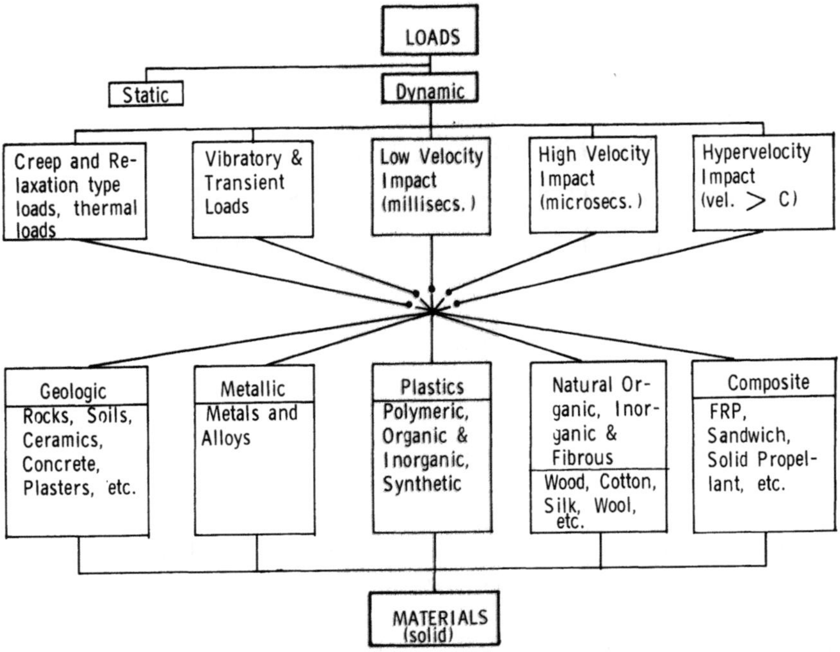

Fig. 1-1. Load Material Interaction.

in itself. Of course, the importance of plastics and composite materials has been steadily growing as they become more and more competitive with metals. Consequently, more work on them would be normally expected in the future. For certain engineering problems and applications, e.g., in foundations of buildings and other structures, the knowledge of behavior of soils and rock strata is of utmost importance. Much of the investigations for such materials have been in the creep type or slow strain rate studies. Engineering problems concerned with higher rates of loading have to be solved by empirical and trial-and-error methods while using very high factors of safety. As more tools for scientific investigation of dynamic behavior of materials become available, the activity and results in this field are also bound to grow. The subject of the symposium today is thus a most challenging one and it offers a lot of open fields for exploration.

The mechanical response of various materials may be classified in several types, as shown in Fig. 1-2. For each type of behavior there is a primary load environment. Considerable work has been done in creep, fatigue, fracture and hypervelocity impact in the past. During and after the World War II period, extensive activity in plastic wave propagation, notably by von Karman [8] and Taylor [9], also took place. It is perhaps noteworthy that about ten years back the knowledge of the true mechanical behavior of materials in the low and high energy

Types of Behavior of Materials Under Dynamic Loads	Primary Load-Environment
Ablation Creep, Relaxation	High and Moderate Temps.
Fatigue Viscoelastic Elastic	Elastic Range Vibration or Low Energy Impact
Elasto-Plastic	Vibration or High Energy Impact
Fluid-like Flow and State Transformation, e.g., Solid to Gas	Very High Vel. or Hypervel. Impact
Crack Propagation & Environmental Embrittlement Fracture - both brittle & ductile	Fracture

Fig. 1-2. Material Behavior – Load Environment Diagram.

impact range covering the elastic, elasto plastic and viscoelastic type behavior, the strain rate and thermodynamic effects, and the associated dislocation dynamics was quite elementary. Recognizing this, about seven years back ARO-D initiated a small basic research program to attempt to fill this gap. But before we get to that, it is proper to mention that load environment, which is primarily considered in this program, is only one of the four physical environments (Fig. 1-3) which influence the behavior of materials. Of course, the problem being quite complex, it was desirable to consider the environment parameters separately in order of their importance to the engineering problems. Moreover, limited funds required limitation of the scope of this program. At this point a brief mention of the ARO-D research programs is in order. The various basic research projects supported by this agency are in two major categories: one, Exploratory Basic Research, or basic research in any area of choice of the investigator; and two, Oriented Basic Research, or research in specific areas selected by ARO-D for existing gaps and relative importance. The area of "Mechanical Behavior of Materials under Dynamic Loads" was selected as one such area for Oriented Basic Research. Including this conference today, eight projects have been supported during the last six years. Five re-

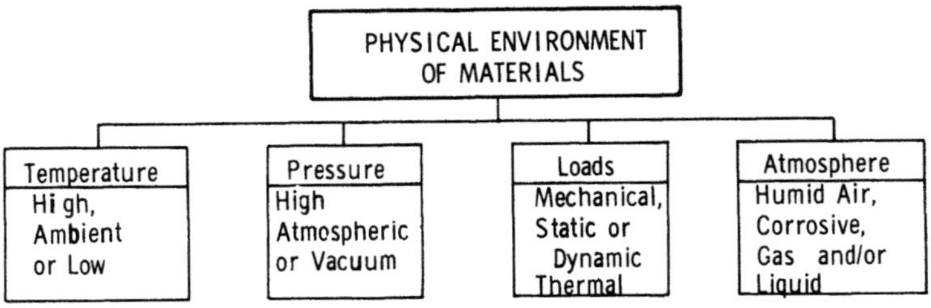

Fig. 1-3. Materials Environment Diagram.

search projects are presently active. A listing of all projects which have been supported under this program is as follows:

1. J. F. Bell,* "Finite Amplitude Wave Propagation in Solids," Johns Hopkins University, Baltimore, Md. (Active)
2. U. S. Lindholm and H. N. Abramson, "Experimental Investigation of Dynamic Material Properties," Southwest Research Institute, San Antonio, Texas (Active)
3. F. A. Donath, "Effect of Rate of Loading and Unloading on the Deformational Behavior of Rocks," Columbia University, New York (Terminated)
4. J. Marin, M. G. Sharma and V. H. Neubert, "Dynamic Stress-Strain Behavior of Engineering Materials," Pennsylvania State University, University Park, Pa. (Active)
5. E. A. Ripperger, "Stress-Strain Characteristics of Materials at High Rates of Strain," University of Texas, Austin, Texas (Recently Terminated)
6. H. Kolsky, "The Mechanical Behavior of Plastics and Metals at High Rates of Loading," Brown University, Providence, R.I. (Active)
7. I. M. Fyfe, "The Applicability of Stress-Strain Laws to Dynamic Problems," University of Washington, Seattle, Washington (Active)
8. U. Lindholm, "Symposium on the Mechanical Behavior of Materials under Dynamic Loads," Sep. 6, 7, 8, 1967, Southwest Research Institute, San Antonio, Texas.

There are two other projects which are related to this area and which are supported under a program on "Brittle Fracture," managed by the Metallurgy and Ceramics Division of ARO-D. These are:

1. J. J. Gilman, "Surface and Fracture Energies of Solids: Response to Impact," University of Illinois, Urbana, Ill.
2. G. T. Hahn, "Experiments to Test a System Analysis of Brittle Fracture," Battelle Memorial Institute, Columbus, Ohio.

You will hear presentations in this symposium on the work of Bell, Lindholm, Ripperger and Fyfe, and so it is not necessary here to go into details of their work. It will suffice to state that all four of these projects concerned themselves with the mechanical properties of metals under impact loads (low and high energy impacts leading to elasto-plastic behavior). All have studied strain rate effects. Bell and Ripperger have studied behavior in compression due to axial impact; Lindholm has studied both compression and tension due to axial im-

* This project is actually supported in the Exploratory Basic Research Program, but is included here because of its closeness with this OBR program.

pact as well as biaxial stress response; and Fyfe has studied the material behavior under conditions of cylindrical expansion. Different experimental technique, different types of specimens and different strain measurement methods have been used in all four projects. Since the apparent behavior of a material is quite often influenced by these conditions, it is important that the true behavior be studied by different methods to isolate any excessive influence of the technique and to verify the validity of the data. The work of Donath was concerned with behavior of rocks (Shale and others) under combined stress loads and very low (creep like) strain rates. The project work of Marin, Sharma and Neubert is in three distinct parts. The work of Marin, who regrettably died a couple of years back, was concerned with low strain rate studies of plastics in tension, bending and torsion. Sharma's work has been on the viscoelasticity of plastic materials with an elastic filler. He studied the dynamic behavior of these materials by determining relaxation modulus and creep compliance for a broad range of frequencies and then using the Boltzman superposition principle. Recently his work consisted of determing constitutive equations including inertia effects for a composite viscoelastic material. Effects of size and percentage of the volume of the filler are also being included in this study. Neubert's work was a study of mechanical properties of metals by axial impact of bars with slightly rounded impacting ends. Recently he has been working on dynamic bending of cold-worked mild steel bars to evaluate stress, strain and strain rate characteristics for both elastic and plastic deformations. Kolsky's work in the past was on the behavior of plastics under dynamic loads combined with static loading to produce a prestrained polymer. Vibration and rebound experiments, combined torsion and tension loading, and Hopkinson bar technique were used. Besides plastics, unicrystalline zinc (in Hopkinson bar tests) was also used. The projects of Gilman and Hahn are primarily concerned with brittle fracture. Understandably, however, dynamic loading being an inherent feature of fracture, their work is intimately related to the subject of this symposium. Since presentations related to this work will be made by both of these gentlemen, we will not go into any detail of their research.

Besides the work sponsored by ARO-D, some significant basic research work is being conducted by the Ballistic Research Laboratories (both in-house and externally by contract). Primary contacts for this work are Drs. R. Eichelberger and William Gillich. Presentation will be made to you by Dr. Gillich. Besides Gillich's work on "Plastic Wave Propagation Studies in Single Crystals and Polycrystalline Materials," two other research projects being conducted in-house are by A. S. Elder on "Determination of Mechanical Properties of Viscoelastic Materials," and by J. T. Frazier on "Elastic-Plastic Wave Propagation in Metallic Solids." These projects are closely related to the scope of the ARO-D program discussed above. Besides these, there are four more projects supported by BRL by contract with N. Davids at Pennsylvania State University, E. H. Lee at Stanford University, I. F. Weeks at General Atomics Division of General Dynamics Corporation, and C. J. Maiden, W. Isbell and A. H. Jones at General Motors Corporation. You will hear a presentation by Maiden at this meeting. The other three studies are all analytical and computational. The investigators of these studies approach the problem of determination of constitutive relations by a sort of inverse approach. The problem is to determine the constitutive relations by analytical and computation methods starting with the results of certain impacts experiments. The behavior of the materials used in these experiments is assumed to be elasto-plastic or thermo-elasto-visco-plastic. This is an interesting approach and some insight into material behavior that may not be otherwise possible may be achieved this way. Perhaps Dr. Gillich will tell us more about this approach in his presentation.

Some work on behavior of soils under dynamic loads is being conducted by and under the sponsorship of the Engineering Research and Development Laboratories, Ft. Belvoir, Virginia, and the Waterways Experiment Station, Vicksburg, Mississippi.

Work on metals and ceramic materials behavior under dynamic loads pertaining to yield, flow and rupture is being conducted at and under the sponsorship of Materials and Mechanics Research Center, Watertown, Mass. Primary contact for this is Dr. R. Beeuwkes. By contract, they are sponsoring the work of Volker Weiss at Syracuse University on constitutive relations and rupture locii of fcc metals, and the work of Kobayashi at University of Washington on notch stress analysis taking strain rate considerations into account.

In addition to the work in the above laboratories, there are several other research efforts conducted by the various Army laboratories which are either directly or indirectly related to dynamic behavior of materials. It will be hard to cite them all. Perhaps, the highlights may be mentioned as:

Picatinny Arsenal Dover, N.J.	Plastics materials
Cold Regions Research Laboratory Hanover, N.H.	Ice, ice composites and snow
Frankford Arsenal Philadelphia, Pa.	Dynamic behavior of birefringent materials
Natick Laboratories Natick, Mass.	Synthetic fabric and fibers
Research laboratories at Army Tank-Automotive Command, Warren, Michigan	Viscoelastic response of soils

In the growth of thought on the "Mechanical Behavior of Materials under Dynamic Loads," today's conference represents an important step. Starting from an elastic behavior study and later the elasto-plastic behavior, the strain rate dependence [10] became the important thing. It was recognized, however, that the thermodynamic considerations and dislocation dynamics were inherent to the problem and that strain rate sensitivity depended on whether the dislocation mechanisms were thermally or purely mechanically activated [11]. This conference today is supposed to highlight the constitutive relations of material behavior based on all the above considerations.

Gentlemen, let me end this talk with a greeting to all of you. I wish you Happy Constitutive Relations for the next three days.

References

1. "High Loading Rate Conference," March 25–26, 1959, OOR, Durham, N.C.
2. "Response of Metals to High Velocity Deformation," July 11, 12, 1960, Estes Park, Colorado. Sponsored by AIME.
3. "Dynamic Behavior of Materials and Structures," Sep. 26–28, 1962, Springfield, Mass. Sponsored by ARO-D.
4. "Dynamic Behavior of Materials," Sep. 27–28, 1962, Albuquerque, New Mexico. Sponsored by ASTM.
5. "Structural Dynamics under High Impulse Loading," Sep. 17–18, 1962, Dayton, Ohio. Sponsored by ASD, OAR.
6. "Behavior of Materials under Dynamic Loading," Nov. 9, 1965, Chicago, Ill. Sponsored by ASME.
7. "Dislocation Dynamics," May 1–6, 1967, Seattle, Washington, and Harrison, British Columbia. Sponsored by Battelle Memorial Institute.
8. T. von Karman and P. Duwez, "Propagation of Plastic Deformation in Solids," *Journal of Applied Physics*, *21*, 987 (1950).
9. G. I. Taylor, "Plastic Wave in a Wire Extended by an Impact Load," *Scientific Papers*, 1, Cambridge University Press (1958).
10. L. E. Malvern, "The Propagation of Longitudinal Waves of Plastic Deformation in a Bar of Material Exhibiting a Strain-Rate Effect," *J. App. Mechs.*, *18*, Trans. ASME, *73*, 203 (1951).
11. J. E. Dorn and F. Hauser, "Dislocation Concepts of Strain Rate Effects," *Proc. Symp. on Structural Dynamics under High Impulse Loading*, ASD-TDR-63-140 (1963).

CONTENTS

Preface v

Introduction: S. KUMAR: *Mechanics/Materials, Aspects of Dynamic Loading* ix

M. REINER: *Dynamical Strength of an Ideal Solid With Definite Constitutive Equation* 1

J. F. BELL: *An Experimental Study of Instability Phenomena in the Initiation of Plastic Waves in Long Rods* 10

O. W. DILLON, JR.: *Plastic Deformation Waves and Heat Generated Near the Yield Point of Annealed Aluminum* 21

P. PERZYNA: *On Thermodynamic Foundations of Viscoplasticity* . . . 61

U. S. LINDHOLM: *Some Experiments in Dynamic Plasticity Under Combined Stress* 77

G. T. HAHN, M. F. KANNINEN, A. K. MUKHERJEE, A. R. ROSENFIELD: *The Speed of Ductile-Crack Propagation and the Dynamics of Flow in Metals* 96

J. M. KRAFFT: *Dynamic Mechanical Behavior of Metal at the Tip of a Plane Strain Crack* 134

J. J. GILMAN: *Dynamical Behavior of Dislocations* 152

S. R. BODNER: *Constitutive Equations for Dynamic Material Behavior* . 176

J. W. EDINGTON: *Effect of Strain Rate on the Dislocation Substructure in Deformed Niobium Single Crystals* 191

W. J. GILLICH: *Constitutive Relationships From Impact Studies* . . . 241

C. J. MAIDEN, A. H. JONES, S. J. GREEN, H. CHIN: *Prediction of Elastic-Plastic Wave Profiles in Aluminum 1060-0 Under Uniaxial Strain Loading* 254

C. H. KARNES: *The Plate Impact Configuration for Determining Mechanical Properties of Materials at High Strain Rates* 270

E. A. RIPPERGER, H. WATSON, JR.: *The Relationship Between the Constitutive Equation and One-Dimensional Wave Propagation* 294

I. M. Fyfe: *Plane-Strain Plastic Wave Propagation in a Dynamically Loaded Hollow Cylinder* 314

N. Cristescu: *Dynamic Plasticity Under Combined Stress* 329

K. C. Valanis: *Unified Theory of Thermomechanical Behavior of Viscoelastic Materials* 343

W. E. Jahsman: *Static and Dynamic Material Behavior of Syntactic Foam* 365

J. B. Cheatham, Jr.: *The Effect of Pressure, Temperature, and Loading Rate on the Mechanical Properties of Rocks* 388

Appendix: Closing Comments by Session Chairmen

Session I W. Prager, University of California at San Diego . 403

Session II D. C. Drucker, Brown University, Providence, Rhode Island 405

Session III J. D. Campbell, University of Oxford, Oxford, England 410

Session IV C. D. Lundergan, Sandia Corporation, Albuquerque, New Mexico 415

Session V D. M. Forney, Jr., AFML, Wright-Patterson Air Force Base, Ohio 417

DYNAMICAL STRENGTH OF AN IDEAL SOLID WITH DEFINITE CONSTITUTIVE EQUATION

MARKUS REINER

Technion—Israel Institute of Technology
Haifa, Israel

ABSTRACT

The mechanical behavior of a solid is in general determined by the parameters of instantaneous elasticity, delayed elasticity with retardation time, flow with relaxation time and strength, the latter including resistance to plastic yield or fracture. A mechanical model which can represent these properties was proposed by Burgers [1]. Its constitutive equation was derived by Reiner [2]. In the present paper the thermodynamic theory of strength by Reiner and Weissenberg [3] is applied upon the strength behavior of a solid cylinder under the action of dynamic deformation by axial loads increasing in time at a given rate. According to this theory, failure will occur when the conserved part of the strainwork reaches a certain limit. It will be examined how the corresponding stress is affected by the rate of stress. It is known that, in general, with increased rate of stress the strength increases. This assertion will be examined under the conditions mentioned above.

1. INTRODUCTION

The mechanical behavior of a solid is, in general, determined by the parameters of instantaneous elasticity, delayed elasticity with a retardation time and recoverable deformation, flow with relaxation time and irrecoverable deformation and strength which we shall here understand as resistance to failure which is plastic yield or fracture. These properties refer to the deviators of stress and strain and it is assumed that in volumetric deformation d_v the material is perfectly elastic, described by the equation

$$-p = Kd_v \qquad (1\text{-}1)^*$$

not subjected to fracture when d_v is negative, i.e., under isotropic pressure. Fracture can take place under isotropic tension. In the present

* In a more general approach one would have to distinguish also solid volume viscosity and liquid volume viscosity, see Reiner [4], which we disregard here.

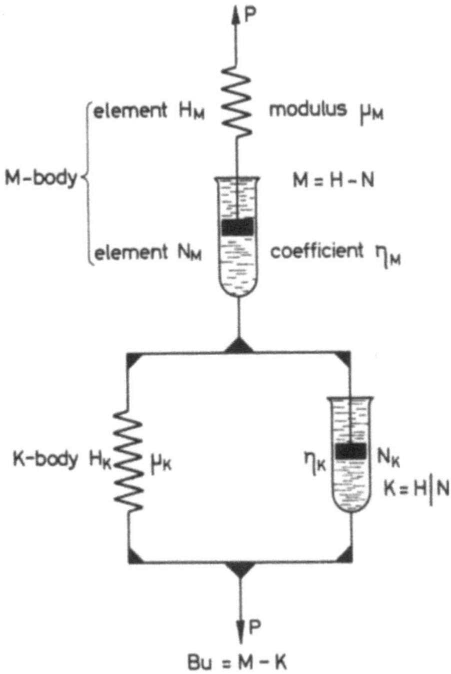

Fig. 1. Model for a Burgers-body
M-body = Maxwell body
K-body = Kelvin body
μ shear modulus
η coefficient of viscosity

paper we shall not consider volumetric deformations and restrict ourselves to the deviators of stress and strain and their time-derivatives.

As a model representing the rheological behavior of a solid, Zener [5] proposed one, postulated before by Poynting and Thomson [6]. However, this model cannot account for irrecoverable deformation. A model which permits of irrecoverable deformation was proposed by Burgers [1]. It is shown in Fig. 1. It is built through a combination in series of a Maxwell elastic liquid (M) and a Kelvin-viscous solid (K). These two result from combinations of a Hooke-solid (H) and a Newtonian liquid (N) in series and in parallel. The Burgers material accordingly has four rheological parameters in shear, namely two elastic μ_M and μ_K and two viscous η_M and η_K.

Reiner [2] has derived the constitutive equation of such a material with $\overset{\circ}{d}_{\ell m}$ for the deformation deviator and $\overset{\circ}{s}_{\ell m}$ for the stress deviator, where the superscript ° indicates the deviator and i and j are tensor indices, in two forms, namely

$$\overset{\circ}{\dot{d}}_{ij} = \frac{\overset{\circ}{s}_{ij}}{2}\left(\frac{1}{\eta_M}+\frac{1}{\eta_K}\right)+\frac{\overset{\circ}{\dot{s}}_{ij}}{2\mu_M}-\frac{\mu_K}{2(\eta_K)^2}\exp\left(-\frac{\mu_K}{\eta_K}t\right)\int_0^t \overset{\circ}{s}_{ij}\exp\left(\frac{\mu_K}{\eta_K}t\right)dt \tag{1-2}$$

$$\overset{\circ}{\dot{d}}_{ij}+\frac{\eta_K}{\mu_K}\overset{\circ}{\ddot{d}}_{ij}=\frac{\overset{\circ}{s}_{ij}}{2\eta_M}+\frac{\overset{\circ}{\dot{s}}_{ij}}{2}\left(\frac{1}{\mu_M}+\frac{1}{\mu_K}+\frac{\eta_K}{\eta_M\mu_K}\right)+\frac{\overset{\circ}{\ddot{s}}}{2}\frac{\eta_K}{\mu_M\mu_K}. \tag{1-3}$$

For simpler writing we shall omit the indices.

2. FAILURE CONDITION

Our problem is to find an expression for the dynamical strength of a material, for which we postulate the Burgers model, at different modes of loading, examining more especially the case when uniaxial loads tend to very large values.

We shall use the Reiner-Weissenberg [3] thermodynamic theory of strength.

In the deformation of a body, the external forces expend work which is converted into *stress work*. Part of the stress work is conserved as potential energy; this part we name *strainwork*. The other part is converted into heat energy and is dissipated. To produce rupture of the body, energy is required; this can come only from the potential energy or strainwork.

The Reiner-Weissenberg theory postulates that failure takes place when the deviatoric strainwork per unit volume reaches in the body a certain maximum value—the *resilience* R, which is a material property.

The stress work expended upon the Bu-body is the sum of the stress works expended upon its two components, or

$$W_{Bu} = W_M + W_K. \tag{2-1}$$

Both stress works are partly conserved and partly dissipated through viscous resistance. For the Maxwell-body we have

$$\dot{w}_M = s\,\dot{d}_M = \frac{s\,\dot{s}}{2\mu_M}+\frac{s^2}{2\eta_M} \tag{2-2 *}$$

where $\dot{w}$ is the stress power per unit volume.

The conserved part is

$$\dot{w}_{cM} = s_M\dot{s}_M/2\mu_M. \tag{2-3}$$

For the Kelvin-body

$$\dot{w}_K = s\dot{d}_K = 2\mu_K d_K\dot{d}_K + 2\eta_K(\dot{d}_K)^2 \tag{2-4}$$

* Reiner [7]. Note that s^2 is short for $s_{\alpha\beta}s_{\beta\alpha}$ with α and β summation indices in accordance with Einstein's summation convention.

with the conserved part

$$\dot{w}_{cK} = 2\mu_K d_K \dot{d}_K. \tag{2-5}$$

The conserved stress-power of the deviator is accordingly for the *Bu*-body

$$\dot{w}_{cBu} = \frac{s_M \dot{s}_M}{2\mu_M} + 2\mu_K d_K \dot{d}_K \tag{2-6}$$

and the conserved stress work starting from a stress-less and strain-less state

$$w_{cBu} = \frac{s_M^{\,2}}{4\mu_M} + \mu_K d_K^{\,2}. \tag{2-7}$$

For the Burgers body

$$\begin{aligned} s_M = s_K &= s \\ d = d_M &+ d_K \end{aligned} \tag{2-8}$$

or

$$d_K = d - d_M. \tag{2-9}$$

But

$$d_M = \frac{s}{2\mu_M} + \frac{1}{2\eta_M}\int s\,dt \tag{2-10}$$

and we finally have for the deviatoric part

$$\overset{\circ}{w}_c = \frac{\overset{\circ}{s}{}^2}{4\mu_M} + \mu_K\left[\overset{\circ}{d} - \frac{\overset{\circ}{s}}{2\mu_M} - \frac{1}{2\eta_M}\int \overset{\circ}{s}\,dt\right]^2. \tag{2-11}$$

The failure condition is found by introducing

$$\overset{\circ}{w}_c = R. \tag{2-12}$$

We shall denote by ϑ the failure stress and by θ the failure deformation. We assume that through appropriate experiments the parameters μ_M, μ_K, η_M, η_K have been determined.

3. STEP-TYPE LOADING

Let a cylindrical specimen of the considered material be subjected to a step-type loading, that is, an instantaneous increase in stress from $s = 0$ to $s = s_c =$ constant. There are three distinct cases of failure in such loading: 1) failure occurs instantaneously with the application of the load; 2) the applied load causes delayed failure; 3) there is no failure at all. These cases depend on the magnitude of the applied s_c.

During the instantaneous loading the response of the material can be only through the spring of the Maxwell element when $d = s_c/2\mu_M$. The time dependent deformation is obtained by integrating (1-2). We have for the total deformation

$$d = \frac{s_c t}{2\eta_M} - \frac{s_c}{2\mu_K} \exp\left(-\frac{\mu_K}{\eta_K} t\right) + \frac{s_c}{2}\left(\frac{1}{\mu_M} + \frac{1}{\mu_K}\right). \tag{3-1}$$

In this case we find from (2-11)

$$\frac{4w_c}{s_c^2} = \frac{1}{\mu_M} + \frac{1}{\mu_K} - \frac{2}{\mu_K} \exp\left(-\frac{\mu_K}{\eta_K} t\right) + \frac{1}{\mu_K} \exp\left(-2\frac{\mu_K}{\eta_K} t\right) \tag{3-2}$$

and the failure criterion gives

$$\vartheta_\ell^2 = 4R\mu_M \left[1 + \frac{\mu_M}{\mu_K} - \frac{2\mu_M}{\mu_K} \exp\left(-\frac{\mu_K}{\eta_K} t\right) + \frac{\mu_M}{\mu_K} \exp\left(-2\frac{\mu_K}{\eta_K} t\right)\right]^{-1} \tag{3-3}$$

When the stress ϑ_ℓ is applied, failure will take place after the lapse of time τ which can be calculated from

$$1 + \frac{\mu_M}{\mu_K} - \frac{2\mu_M}{\mu_K} \exp\left(-\frac{\mu_K}{\eta_K} \tau\right) + \frac{\mu_M}{\mu_K} \exp\left(-2\frac{\mu_K}{\eta_K} \tau\right) = 4R\mu_M/\vartheta_\ell^2. \tag{3-4}$$

We find

$$\tau = -\frac{\eta_K}{\mu_K} \, ln \, \left(1 \pm \sqrt{\mu_K\left(4R/\vartheta_\ell^2 - \frac{1}{\mu_M}\right)}\right). \tag{3-5}$$

We see that if the stress applied is

$$\overline{\vartheta}_\ell = 2\sqrt{\mu_M R} \tag{3-6}$$

$\tau = 0$ and failure will set in at *once*. On the other hand, when a smaller stress

$$\overline{\overline{\vartheta}}_\ell = 2\sqrt{\frac{\mu_M \mu_K}{\mu_M + \mu_K} R} \tag{3-7}$$

is applied, it will take infinite time ($\tau = \infty$) for failure to take place. If $s_c < \overline{\overline{\vartheta}}_\ell$ there will be no failure however long we may wait. If $\overline{\overline{\vartheta}}_\ell < s_c < \overline{\vartheta}_\ell$ failure will take place at time $0 < \tau < \infty$ to be found from (3-5). The corresponding failure deformations are $\overline{\theta} = \sqrt{R/\mu_M}$, $\overline{\overline{\theta}} = \infty$.

4. CONSTANT STRESS RATE

Let us now assume that a dynamical test is performed in which the stress applied is uniaxial with constant rate, or

$$s = a\,t. \tag{4-1}$$

Equation (1-2) with the introduction of (4-1) yields

$$d = \frac{a}{2}\left[\left(\frac{1}{\mu_M} + \frac{1}{\mu_K}\right) t + \frac{1}{2\eta_M} t^2 + \frac{\eta_K}{\mu_K^2}\left(\exp\left(-\frac{\mu_K}{\eta_K} t\right) - 1\right)\right]. \tag{4-2}$$

When (4-1) and (4-2) are introduced into (2-11), the deviatoric strain-work is obtained as:

$$\overset{\circ}{w}_c = \frac{a^2}{4}\left[\left(\frac{1}{\mu_M} + \frac{1}{\mu_K}\right) t^2 + 2\,\frac{\eta_K}{\mu_K^2}\, t\left(\exp\left(-\frac{\mu_K}{\eta_K} t\right) - 1\right)\right.$$
$$\left. + \frac{\eta_K^2}{\mu_K^3}\left(\exp\left(-2\,\frac{\mu_K}{\eta_K} t\right) - 2\,\exp\left(-\frac{\mu_K}{\eta_K} t\right) + 1\right)\right]. \tag{4-3}$$

Let τ be the time when failure takes place, then

$$\vartheta = a\,\tau. \tag{4-4}$$

The failure condition is

$$R = \frac{a^2\tau^2}{4}\left[\left(\frac{1}{\mu_M} + \frac{1}{\mu_K}\right) + \frac{2}{\tau}\,\frac{\eta_K}{\mu_K^2}\left(\exp\left(-\frac{\mu_K}{\eta_K}\tau\right) - 1\right)\right.$$
$$\left. + \frac{1}{\tau^2}\,\frac{\eta_K^2}{\mu_K^3}\left(\exp\left(-2\,\frac{\mu_K}{\eta_K}\tau\right) - 2\,\exp\left(-\frac{\mu_K}{\eta_K}\tau\right) + 1\right)\right] \tag{4-5}$$

from which

$$a = \frac{\frac{2}{\tau}\sqrt{R}}{\sqrt{\left(\frac{1}{\mu_M}+\frac{1}{\mu_K}\right)+\frac{2}{\tau}\frac{\eta_K}{\mu_K^2}\left(\exp\left(-\frac{\mu_K}{\eta_K}\tau\right)-1\right)+\frac{1}{\tau^2}\frac{\eta_K^2}{\mu_K^3}\left(\exp\left(-2\frac{\mu_K}{\eta_K}\tau\right)-2\exp\left(-\frac{\mu_K}{\eta_K}\tau\right)+1\right)}}. \tag{4-6}$$

The failure stress is:

$$\vartheta = \frac{2\sqrt{R}}{\sqrt{\left(\frac{1}{\mu_M}+\frac{1}{\mu_K}\right)+\frac{2}{\tau}\frac{\eta_K}{\mu_K^2}\left(\exp\left(-\frac{\mu_K}{\eta_K}\tau\right)-1\right)+\frac{1}{\tau^2}\frac{\eta_K^2}{\mu_K^3}\left(\exp\left(-2\frac{\mu_K}{\eta_K}\tau\right)-2\exp\left(-\frac{\mu_K}{\eta_K}\tau\right)+1\right)}} \tag{4-7}$$

and the failure strain is:

$$\theta = \frac{\sqrt{R}\left[\left(\frac{1}{\mu_M}+\frac{1}{\mu_K}\right)\tau+\frac{1}{2\eta_M}\tau^2+\frac{\eta_K}{\mu_K^2}\left(\exp\left(-\frac{\mu_K}{\eta_K}\tau\right)-1\right)\right]}{\sqrt{\left(\frac{1}{\mu_M}+\frac{1}{\mu_K}\right)\tau^2+\frac{2\eta_K}{\mu_K^2}\tau\left(\exp\left(-\frac{\mu_K}{\eta_K}\tau\right)-1\right)+\frac{\eta_K^2}{\mu_K^3}\left(\exp\left(-2\frac{\mu_K}{\eta_K}\tau\right)-2\exp\left(-\frac{\mu_K}{\eta_K}\tau\right)+1\right)}}. \tag{4-8}$$

We now consider two extreme cases.

(1) We assume $\tau \to 0$. Then from (4-6) $a \to \infty$ and

$$\vartheta_\infty = 2\sqrt{R\mu_M}; \quad \left.\frac{d\vartheta}{da}\right|_\infty = 0 \tag{4-9}$$

$$\theta_\infty = \sqrt{\frac{R}{\mu_M}}; \quad \left.\frac{d\theta}{da}\right|_\infty = 0. \tag{4-10}$$

(2) We assume $\tau \to \infty$. Then $a \to 0$ and

$$\vartheta_0 = 2\sqrt{\frac{R\mu_M\mu_K}{\mu_M+\mu_K}}; \quad \left.\frac{d\vartheta}{da}\right|_0 = \infty \tag{4-11}$$

$$\theta_0 = \infty; \quad \left.\frac{d\theta}{da}\right|_0 = \infty. \tag{4-12}$$

Conditions are shown in Fig. 2.

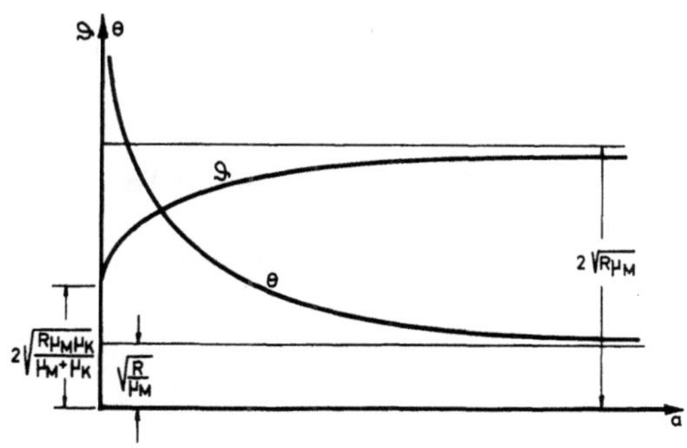

Fig. 2. Fracture stress and fracture strain as functions of the rate of stress

ϑ	fracture stress
θ	fracture strain
a	rate of stress

For any rate of stress $0 < a < \infty$ the fracture stress will be

$$2\sqrt{\frac{R\mu_M\mu_K}{\mu_M+\mu_K}} < \vartheta < 2\sqrt{R\mu_M} \tag{4-13}$$

and the fracture strain

$$\infty > \theta > \sqrt{\frac{R}{\mu_M}}. \tag{4-14}$$

5. CONCLUSIONS

We have assumed that the mechanical behavior of a solid material is adequately described by the constitutive equation of a Burgers body. It should be pointed out that there is no model and therefore no constitutive equation which is more general. This is also so with regard to a model often used in investigations consisting of two or more Maxwell bodies coupled in parallel. It is, of course, possible to use a number of *Bu* bodies coupled in parallel or in series and even an infinite number of such bodies. Conditions are then more general numerically, but not in principle.

Using the Reiner-Weissenberg theory of strength, we have shown that there is a delayed failure in a step-like stress application. The dependence of the failure stress on the stress rate was analyzed. It was found that in a dynamical test with a stress increasing at some rate a, the fracture stress depends not only upon the magnitude of the stress but also upon the rate of stress and increases with increasing rate of stress. This fact is well known, but has so far not been shown to be inherent in a theory of strength. The failure stress has a maximum value when the rate of stress is very high ($a \to \infty$) *, and a minimum when it is very low ($a \to 0$). The variation of the fracture strain is in the opposite direction. It is clear that it becomes infinitely great if we wait for an infinite time. At very high rate of stress it has a definite value. Conditions are shown in Fig. 2.

6. ACKNOWLEDGMENTS

I thank Dr. Amnon Foux for his help in the calculations.

NOTATION

a	rate of stress
Bu	Burgers body
d	deformation
d_0	initial deformation
d_v	volumetric deformation
e	base of natural logarithms
H	Hooke-solid
i,j	tensor indices
K	bulk modulus
K	Kelvin body

* It must be pointed out that in the theory as expounded here, the inertia of the material has been neglected.

ℓ, m	tensor indices
ℓ	longitudinal
M	Maxwell body
N	Newtonian liquid
$\circ$	as superscript indicates "deviator"
p	isotropic pressure
R	resilience
s	stress
s_0	initial stress
s_c	constant stress
W	stress work
$\dot{w}$	stress power per unit volume
$\dot{w}_c$	conserved stress power
η_M	coefficient of viscosity of M-body
η_K	coefficient of viscosity of K-body
ϑ	rupture stress
θ	rupture deformation
μ_M	shear modulus of M-body
μ_K	shear modulus of K-body
τ	time of rupture

References

1. J. M. Burgers, in *Committee for the Study of Viscosity*, Amsterdam (1935).
2. M. Reiner, in *Encyclopedia of Physics, VI*, Springer Verlag, 472 (1958).
3. M. Reiner and K. Weissenberg, "A thermodynamical theory of the strength of materials," Rheology Leaflet No. 10, 12 (1939).
4. M. Reiner, "On volume viscosity." Bull. Res. Counc. Israel, *3*, 67 (1953).
5. C. Zener, *Elasticity and Anelasticity of Metals*, Chicago (1948).
6. J. H. Poynting and J. J. Thomson, *Properties of Matter*, London (1902).
7. M. Reiner, *Lectures on Theoretical Rheology*, pp. 141–142, Amsterdam (1960).

AN EXPERIMENTAL STUDY OF INSTABILITY PHENOMENA IN THE INITIATION OF PLASTIC WAVES IN LONG RODS

JAMES F. BELL

The Johns Hopkins University
Baltimore, Md.

ABSTRACT

A series of experiments are considered for stress-time histories at the impact face for wave propagation in long rods. From these data it is shown that the finite deformation mode and transition stability structure of the writer's generalized parabolic stress-strain function is important in the problem of finite amplitude non-linear wave initiation and growth at the impact face and in the three-dimensional first diameter when considering the symmetrical free-flight impact of identical long cylindrical rods.

In 1949 the writer began a series of experiments in finite distortional deformation; the number of experiments now exceeds 1600. When these experiments are combined with the nearly 600 similar experiments from the 45-year literature, over 2200 experiments in 27 solids have been analyzed by the writer during the past two decades. Among the writer's experiments considered during this time have been one-microsecond long finite amplitude wave impact experiments at strain rates of 70,000.0 sec^{-1} using diffraction gratings for strain determination, and, at the other strain-rate extreme, a 357-hour continuously monitored quasi-static experiment at a strain rate of 0.000000006 sec^{-1}. Test temperatures have varied from 4.2° K to 1809° K, or from $T/T_m = .003$ to $T/T_m = .98$, where T is the test temperature and T_m is the melting point of the crystal of interest. Experiments have included 50% binary combinations of elements and a variety of purities to 99.999+%. Systematic experimental studies have been made of the effects of crystal structure, specimen geometry, prior deformation history, prior thermal history, grain size of aggregates, melting points, and type of experiment, i.e., whether uniaxial tension, compression, or torsion of hollow tubes.

In a monograph scheduled to be published in a few months, the writer [1] has shown from the detailed examination of 1200 individual experiments (both polycrystal and single crystal) in 27 crystalline solids that there exists a single generalized stress-strain function applicable to the uniaxial tension or compression finite distortion of all solids considered. These experiments have shown that without question the major finite distortion problem in crystals is that of the stability properties of a finite deformation mode and transition structure which characterizes all of these solids. Some aspects of this finite deformation behavior in single crystals have been described earlier [2]. The explicit form of the generalized uniaxial stress-strain function which these studies furnish is:

$$\sigma = (\tfrac{2}{3})^{r/2}\mu(0)B_0(1 - T/T_m)(\epsilon - \epsilon_b)^{1/2} \tag{1}$$

where σ is the uniaxial stress, ϵ is the uniaxial strain (both referred to the undeformed state of the material); B_0 is a dimensionless universal constant, $B_0 = 0.0280$; $\mu(0)$ is the zero point isotropic linear elastic shear modulus of any one of the 27 crystalline solids studied; the integer, $r(r = 1, 2, 3, 4 \cdots)$ designates the finite deformation mode; and ϵ_b designates the parabola intercept upon the strain abscissa of any particular deformation mode of interest. (For the initial finite deformation mode of dead annealed solids, $\epsilon_b = 0$.)

Extensive, as yet unpublished experimental studies contained in the monograph referred to above [1], have been made by the writer to determine the factors which control the finite deformation mode index, r, for initial parabolas, and whether or not a transition from one value to another will occur at any one of a series of specified critical transition strains. Two illustrations of this behavior are shown in Fig. 1 and Fig. 2.

A comparison of quasi-static experimental data with (1) is best accomplished in a σ^2 vs ϵ plot in which deformation modes appear as a series of straight lines. Such plots of a series of constant stress-rate uniaxial tension experiments (circles) are compared with prediction from (1) (solid lines) for completely annealed commercial purity aluminum polycrystals at room temperature. The variable strain rate is in the range $\dot{\epsilon} = 10^{-5}$ sec^{-1}. The initial finite deformation mode for all of these experiments is $r = 3$, with transitions to $r = 5$, and then to $r = 7$ at the known critical strains (known from hundreds of dynamic and quasi-static experiments) of $\epsilon_N = 4.2\%$ and $\epsilon_N = 7.5\%$. Initial finite deformation modes of $r = 1$, $r = 2$, and $r = 4$ have been produced in this same completely annealed solid under controlled conditions.

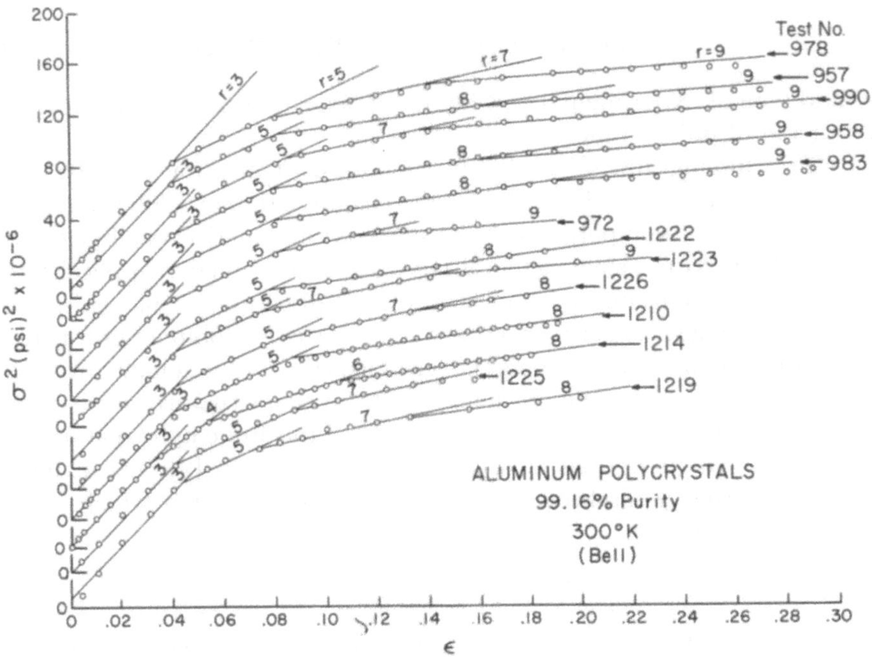

Fig. 1. Constant stress rate uniaxial tension quasi-static experiments (circles) for strain rates of 10^{-5} sec^{-1} compared with parabolic deformation mode and transition structure, Eq. (1) (solid lines) where transitions occur at specified critical strains.

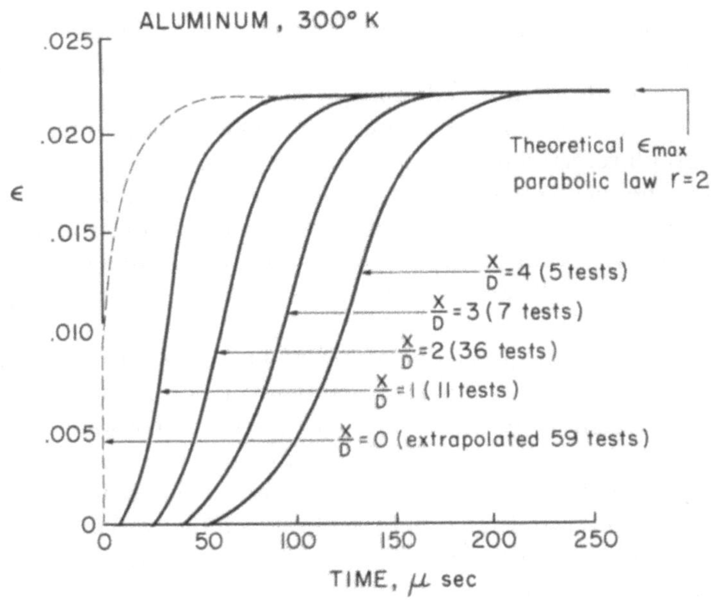

Fig. 2. Averaged experimental diffraction grating strain-time measurements for 59 symmetrical free-flight impact experiments at 2030 cm/sec impact velocity in annealed polycrystalline aluminum. The mode index of the governing stress-strain function is $r = 2$.

In general the finite deformation mode governing dynamic finite amplitude wave propagation in this commercial purity completely annealed aluminum polycrystal is $r = 2$, as may be seen from the averaged diffraction grating dynamic plasticity experiments in Fig. 2. These 59 symmetrical free-flight impact experiments had 0.005 in. long, 30,720 lines per inch diffraction gratings from which the strain-time data were determined at four positions from the impact face. All of these experiments were performed at the same impact velocity of 2030 cm/sec. It was the observed constant wave speeds and the observed invariable relationship between particle velocity and strain which established the applicability of the finite amplitude wave theory of Taylor [3], von Karman [4], and Rakhmatulin [5] in aluminum of various purities and in copper, zinc, lead, magnesium, nickel, and 70–30 α-brass. The dynamic data in aluminum shown in Fig. 2 provided a finite deformation mode index of $r = 2$, as the comparison of predicted and measured maximum strain reveals.

In the present paper the influence of the stability properties of this finite deformation mode and transition structure upon the initiation and growth of finite amplitude waves at the impact face is considered for the axial collision of long cylinders. Of all the many dynamic plasticity problems the writer has studied during the past two decades, the phenomenon of plastic wave three-dimensional initiation and growth in the first diameter has been one of the most interesting and, at the same time, perplexing. In numerous earlier papers [6–14] experimental studies were aimed at providing at least a description of this phenomenon.

These experimental studies have revealed that at impact a high-peaked stress uniaxial strain front parallel to the impact face is inaugurated. This high-stressed wave front collapses in the first diameter due to the growth of finite distortional deformation at the free-surfaced side walls of the cylinders. By one diameter from the impact face (independent of the actual dimensions) in all of the crystalline solids studied, there is a stable one-dimensional uniaxial stress front satisfying the constant wave speed and invariable particle velocity finite strain conditions of the finite amplitude wave theory and governed by (1) for some measurable finite deformation mode index, r.

As an example of this behavior a piezo crystal measurement of the stress-time history at the impact face for a symmetrical free-flight impact of completely annealed commercial purity polycrystalline aluminum is shown in Fig. 3. Following the initial peak stress in this experiment there is an initial collapse to an intermediate dynamic overstress before a later reduction to the theoretical maximum stress for the finite amplitude wave theory, with a mode index $r = 2$ in the governing stress-strain function from (1).

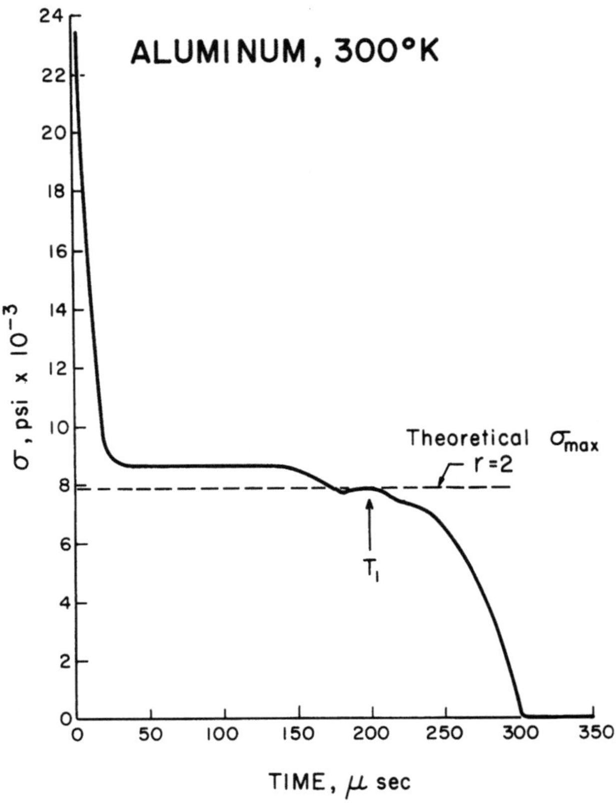

Fig. 3. A piezo crystal measurement of stress-time history at impact face for symmetrical free-flight impact in annealed polycrystalline aluminum compared with predicted maximum stress for parabolic stress-strain function of mode index $r = 2$. The impact velocity is 1900 cm/sec.

Much of the writer's earlier experimental work related to this first diameter wave initiation and growth was concerned with the quantitative determination of the magnitude of the peak stress, the dynamic overstress, and the study of the fact that above critical velocities, now known to coincide with quasi-static transition strains (see Fig. 1), the intermediate dynamic overstress disappears and the peak stress falls to the predicted parabolic maximum stress for the finite deformation mode index given from the constant wave speed data. In aluminum this intermediate dynamic overstress disappears above an impact velocity of 5080 cm/sec [10, 14] which for $r = 2$, in terms of the finite amplitude wave theory, corresponds to the critical strain of 7.5% (see Fig. 1). A symmetrical free-flight impact piezo crystal measurement of Filbey [15, 16, 17] at an impact velocity of 7100 cm/sec, Fig. 4, shows this collapse of the very high peak stress to the theoretically predicted maximum parabolic stress from (1) for the finite amplitude wave theory. One notes that the mode index is $r = 2$, and there is no intermediate dynamic overstress.

The invariable relation between finite strain and longitudinal particle velocity is given by:

$$\dot{u} = \int_0^{\epsilon} c_p \, d\epsilon \tag{2}$$

where $c_p(\epsilon) = \sqrt{\dfrac{d\sigma/d\epsilon}{\rho}}$ are the wave speeds which depend upon strain but are constant for any particular strain amplitude. Introducing (1) into the relation for c_p, (2) becomes (3) or (4) for a parabolic law:

$$\dot{u} = \frac{8}{9\rho} \times (\tfrac{2}{3})^{r/2} \mu(0) B_0 (1 - T/T_m) \epsilon^{3/2} \tag{3}$$

or

$$\dot{u} = \frac{8\sigma^3}{9\rho(\tfrac{2}{3})^r \mu^2(0) B_0^2 (1 - T/T_m)^2}. \tag{4}$$

For a given maximum particle velocity $\dot{u} = v_0$. The maximum stress depends upon the mode index, r. For example, for $v_0 = 2030$ in./sec, as in Fig. 2, $\sigma_{max}(r = 2) = 8300$ psi, while $\sigma_{max}(r = 1) = 9520$ psi. This is close to the measured dynamic overstress at this impact velocity and suggests an initial finite deformation mode instability in the finite amplitude wave initiation.

Recent experiments of the writer for axial impacts of annealed solids in which hitter and struck specimens are different elements, and recent finite amplitude wave propagation studies by Hartman [18] in symmetrical impacts in 70–30 α-brass have shown that it is possible to produce finite amplitude waves satisfying both the constant wave speed

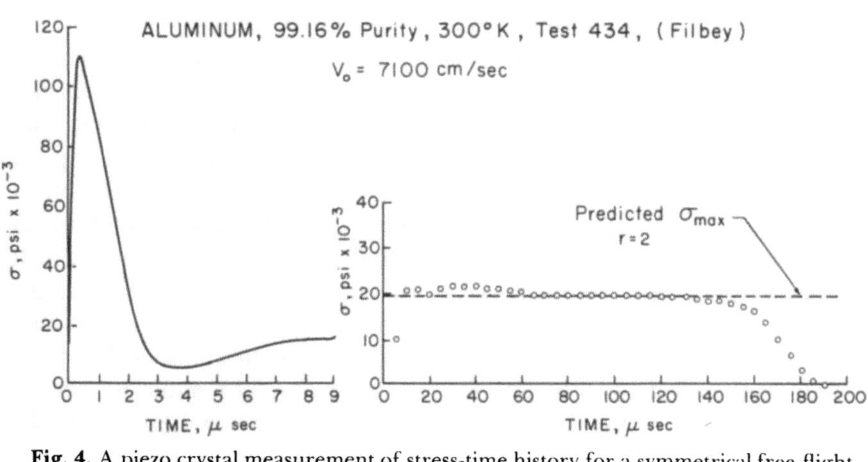

Fig. 4. A piezo crystal measurement of stress-time history for a symmetrical free-flight impact experiment above the critical velocity of 5080 cm/sec. For clarity the initial peak stress portion of the experiment is shown on a different time-scale. The mode index is $r = 2$.

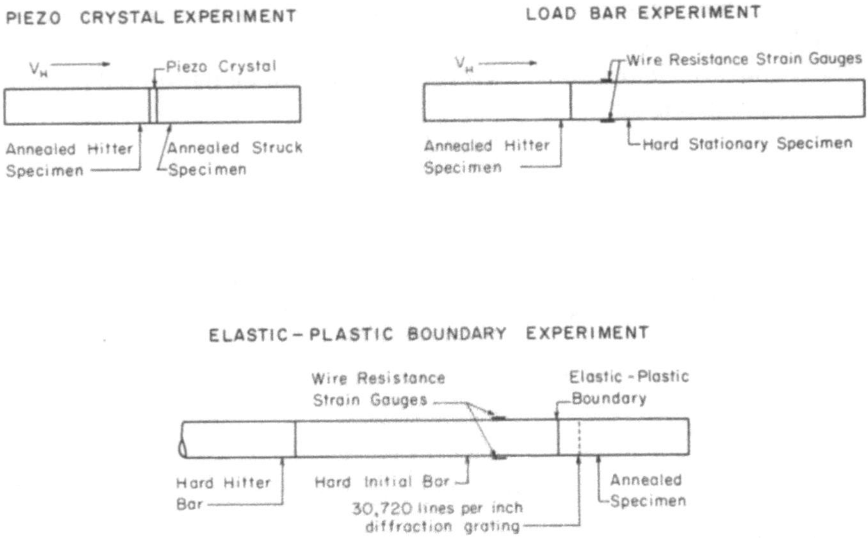

Fig. 5. Experiment for determining boundary stress-time histories.

and invariable finite strain vs particle velocity conditions, but with different stable finite deformation mode indices, r, from one experiment to another in the same solid.

In Fig. 5 one sees three types of experiments which give stress-time histories at the impact face for differing impact conditions. Each of these experiments, as may be shown from diffraction grating measurements of strain, generates non-linear wave fronts beyond the first diameter, satisfying the finite amplitude wave theory. In the symmetrical free-flight impact experiment, $v_0 = v_H/2$.

Figures 3 and 4 have shown examples of the piezo crystal experiment in completely annealed polycrystalline aluminum. Figures 6 and 7 show the data from a load bar experiment and an elastic-plastic boundary experiment for nearly the same impact velocity as the piezo crystal experiment of Fig. 3. In the elastic-plastic boundary experiment the theoretical dynamic overstress for $r = 1$ very rapidly falls to the theoretical maximum stress for $r = 2$. Extensive diffraction grating measurements in the annealed specimen of this experiment, which are described elsewhere, have shown that the non-linear wave fronts have constant wave speeds and maximum strains for the mode index $r = 2$. The predicted stresses of Fig. 6 were determined by equating stresses and particle velocities at the elastic-plastic boundary, where $\sigma = E\epsilon$ for the hard bar, and the annealed bar has the parabolic stress-strain function of (1). In completely annealed polycrystalline aluminum the dynamic overstress is always present above the first critical velocity of 582 in./sec [7] and below the third critical velocity of 2000 in./sec [10].

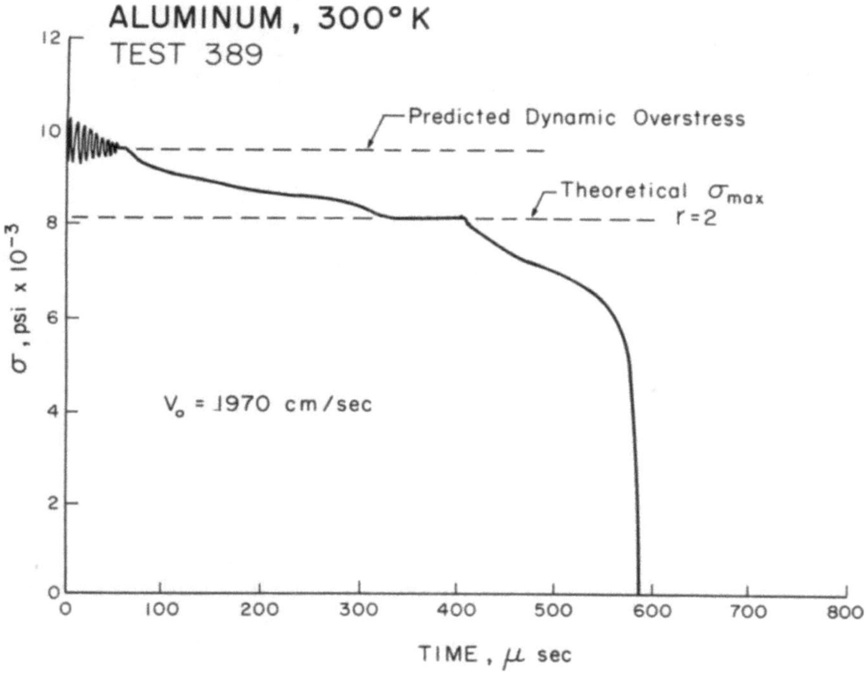

Fig. 6. A load bar experiment in annealed polycrystalline aluminum showing both final predicted stress level for mode index $r = 2$ and intermediate dynamic overstress.

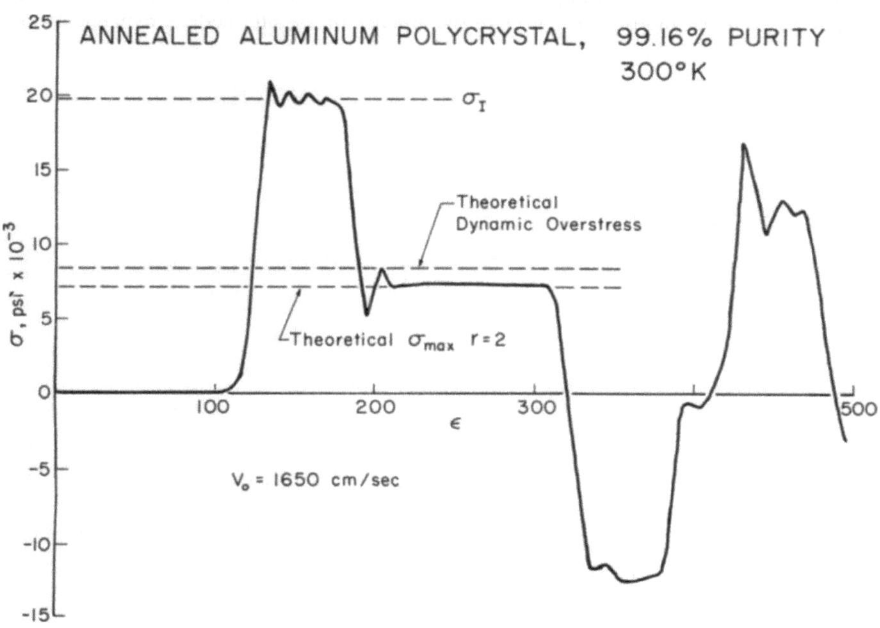

Fig. 7. The stress-time history in the hard bar for an elastic-plastic boundary experiment showing the rapid reduction from the intermediate dynamic overstress. The theoretical stress maximum is for a mode index of $r = 2$.

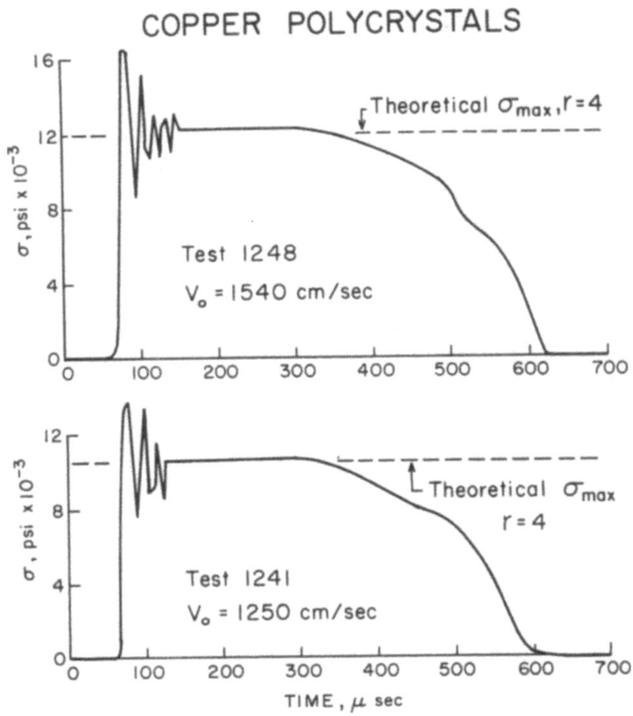

Fig. 8. Two load bar experiments in annealed polycrystalline copper at the designated impact velocities showing the agreement of the maximum stress with the predicted value for a mode index of $r = 4$. Note the absence of an intermediate dynamic overstress.

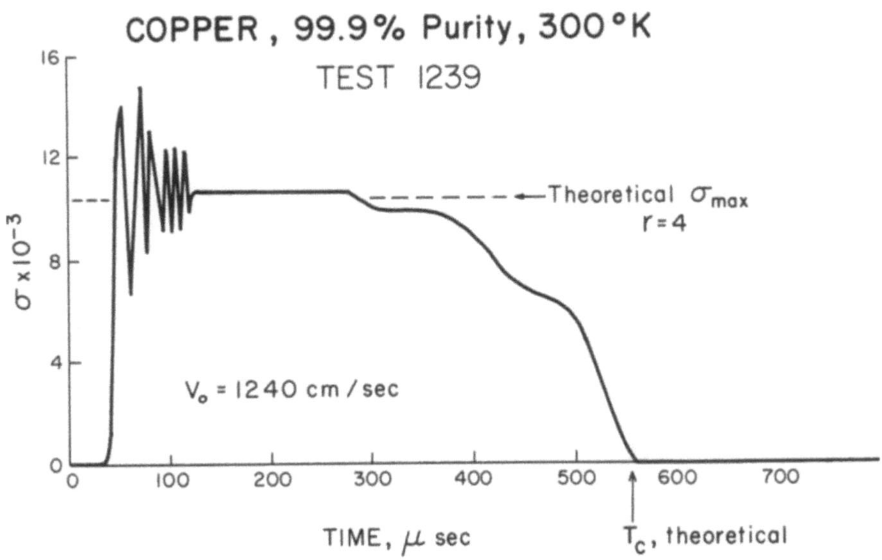

Fig. 9. A load bar experiment in annealed polycrystalline copper for which an intermediate dynamic overstress occurs. The theoretical maximum stress shown is for a mode index of $r = 4$. T_c designates the predicted time of contact [7].

In Figs. 8 and 9 are shown load bar experiments in annealed copper for which the most usual finite deformation mode index is $r = 4$ but for which in dynamic tests, unlike aluminum but similar to α-brass, different stable dynamic mode indices of $r = 3$ and $r = 5$ are occasionally seen. These load bar data in copper show two experiments in which no dynamic overstress is present. The maximum experimental stress has the mode index, $r = 4$, determined from wave speed studies in this solid [19]. One experiment (Fig. 9) in copper has a higher stress, approximating an $r = 3$ mode index.

SUMMARY AND CONCLUSIONS

It was the purpose of this paper to show that finite deformation mode stability is of importance in the finite amplitude wave *first diameter* wave initiation and growth, even though the finite amplitude wave propagation *beyond* the first diameter is stable and completely specified by the one-dimensional uniaxial stress finite amplitude wave theory.

The initial collapse of the uniaxial strain front may be directly to the stable finite deformation mode outside the first diameter, as was shown in Fig. 4 and Fig. 7 for aluminum and copper; or, the first collapse may be to an intermediate unstable deformation mode which undergoes an immediate transition to the final stable mode of the finite amplitude wave propagation outside the first diameter, as shown in Figs. 3, 6, and 7 for aluminum.

Each of the three experiments considered has different impact-face conditions. An inspection of the data reveals that the duration of the intervening deformation mode and the manner of its transition to the stable deformation mode is affected by these differences in experimental impact conditions. The aluminum load bar experiment of Fig. 6, for example, required a specimen twice as long as that for the piezo crystal measurement of Fig. 3 or the elastic-plastic boundary experiment of Fig. 7 to establish a stress plateau at the stress maximum of the stable parabolic mode index of the finite amplitude wave theory.

Previous empirical calculations of the dynamic overstress [6, 10, 14] may be shown to be related to the present stability calculation.

The fact that in axial cylindrical impact the intermediate dynamic overstress sometimes is present and sometimes is not, introduces a probability aspect which eliminates geometric radial effects and viscosity as plausible explanations of the phenomenon. This type of behavior is entirely consistent with a finite deformation mode and transition structure which is now known to characterize quasi-static deformation in the same solids. Further support for this proposed explanation of the dynamic overstress arises when it is observed that changes in the behavior always occur above and below the specified critical strains known from quasi-static experiment. These same critical strains through the finite amplitude wave theory have been shown earlier to have corresponding critical velocities at which the marked changes in finite wave initiation were noted several years ago [6, 10, 13, 14].

That the stability properties are important to the study of the initiation and growth of non-linear finite amplitude waves offers an interesting new set of theoretical problems in dynamic plasticity where the major issue is finite elastic stability rather than viscosity.

ACKNOWLEDGMENTS

The author would like to thank Research Assistants John Gottschalk, Mary Cory, and Faith Pacquet for their careful work.

This research was sponsored by the United States Army, Army Research Office (Durham).

References

1. J. F. Bell, *The Physics of Large Deformation of Crystalline Solids*, Springer Tracts in Natural Philosophy, Springer-Verlag New York Inc. (1968).
2. J. F. Bell, Phil. Mag., *11*, 1135 (1965).
3. G. I. Taylor, British Official Report, RC 329 (1945).
4. T. von Karman, NDRC Report A29, OSRD 365 (1942).
5. K. A. Rakhmatulin, Soviet J. Appl. Math. & Mech. (Prikl. Mat. Mekh.), *9*, 91 (1945).
6. J. F. Bell, J. Appl. Phys., *31*, 2188 (1960).
7. J. F. Bell, J. Appl. Phys., *32*, 1982 (1961).
8. J. F. Bell, J. Mech. and Phys. of Solids, *9*, 261 (1961).
9. J. F. Bell, Proc. I.U.T.A.M. Symposium on 2nd Order Effects in Elasticity, Plasticity, and Fluid Dynamics, 173, Haifa (1962).
10. J. F. Bell, Proc. I.U.T.A.M. Symposium on Stress Waves in Anelastic Solids, 166, Brown University (1963).
11. J. F. Bell, Colloquium on Behavior of Materials Under Dynamic Loading, ASME, New York, 19 (1965).
12. J. F. Bell, J. Mech. & Phys. of Solids, *14*, 309 (1966).
13. J. F. Bell, Experimental Mechanics, *7*, 8 (1967).
14. J. F. Bell and J. H. Suckling, Proc. 4th U.S. Natl. Cong. of Appl. Mech., 877 (1962).
15. G. L. Filbey, Jr., Ph.D. Dissertation, The Johns Hopkins University (1961).
16. G. L. Filbey, Jr., TR No. 8, U.S. Army BRL Contract No. DA-36-034-21X4992, 509-ORD-3104RD, The Johns Hopkins University (1961).
17. G. L. Filbey, Jr., Proc. Symposium on Structural Dynamics Under High Impulsive Loading, 147 (1963).
18. W. F. Hartman, Ph.D. Dissertation, The Johns Hopkins University (1967).
19. J. F. Bell and W. M. Werner, J. Appl. Phys., *33*, 2416 (1962).

PLASTIC DEFORMATION WAVES AND HEAT GENERATED NEAR THE YIELD POINT OF ANNEALED ALUMINUM

Oscar W. Dillon, Jr.

University of Kentucky
Lexington, Kentucky

ABSTRACT

Experimental data on the propagation of deformation waves and the heat generated during the plastic deformation of annealed aluminum are presented. Particular emphasis is given to the transition from the elastic to the plastic state. Wave propagation data that are consistent with annealed aluminum being mechanically unstable under impact loading are given. Some wave propagation results for specimens made by gluing soft and hard sections together are included.

Experimental data on the heat generated in annealed aluminum tubes undergoing torsional oscillations at about one cycle per second are also presented.

INTRODUCTION

In this paper we are concerned with the dynamic behavior of a single material-annealed aluminum. This is the same material used by the author in several papers in recent years and, furthermore, is as close as possible to that used by Bell for many years. However, the range of strains involved and the experimental method are entirely different than his.

In order to put the present work into focus certain aspects of the author's research are reviewed in more or less chronological order. This program began [1] with a completely theoretical interest in coupled thermoelasticity. The lack of relevant experimental data in the literature suggested an attempt be made to obtain such information. These data have never been obtained but it soon became interesting to measure the heat generated as metals deform in the plastic range, and therefore in coupled thermoplasticity [2, 3]. This work had been done by G. I. Taylor [4] in the early days of plasticity but modern instruments permitted one to obtain different data.

It is quite common to discuss the difference between thermodynamics and thermostatics as applied to continuous media. Therefore an experiment in which thermal and mechanical equilibrium were more nearly approached was a rather natural thing to attempt. The end result is a test where it takes three weeks to do the experiment needed to obtain a stress-strain curve for a single specimen. We were very lucky that this work began with a material in which creep was minimal in slowly loaded tests. The results of three experiments in which the load is applied very slowly are shown in Fig. 1. Two of these specimens were thought to be identical and in view of some later results it is emphasized that variations between specimens do exist in the static tests.

The primary point of interest in Fig. 1 are the long plateaus which one observes as one very slowly loads the specimen. A test which lasts an hour does not exhibit as many sharp steps as are shown in Fig. 1. The closer one comes to having equilibrium conditions, the sharper are these changes in the tangent to the stress-strain curve. It is relevant to point out that the phenomena of a serrated stress-strain curve is not original with us and that there are many metallurgical mechanisms which are consistent with such a response. It is very natural to consider what a serrated constitutive relation means to problems of wave propagation. Clearly incremental waves can travel at the elastic velocity. However, it is also easy to observe very slow waves if one is making measurements when the plateaus actually develop. One "wave" was observed to take twenty minutes to propagate four inches. One simply has to instrument to obtain data for the times needed in order to observe these slow waves.

Ordinarily I wouldn't dwell much on this next point but it seems to be a point of confusion in discussions with many people. If one assumes that a serrated constitutive relation applies to a given material, then it is an easy theoretical problem to show that one *must* use a slow loading experiment in order to observe the details of the response sufficiently well that one can use the data as a constitutive relation. In other words serrated constitutive relations *exist for all rates of loading* but one *can properly observe* them only in slowly loaded tests. However, by itself the observation of slow waves does not conclusively prove that the serrated relation is a constitutive one—it is conceivable that the material doesn't have a simple constitutive relation or that the approach to it is not stable with respect to the rate of loading. The Instron test simply is a "dynamic" experiment for unstable materials and causes the stress to be associated with the wrong strain.

A few people have now become convinced that serrated relations are real enough but frequently say to me "surely this does not apply to real dynamic problems, such as those of impact." That is saying in different

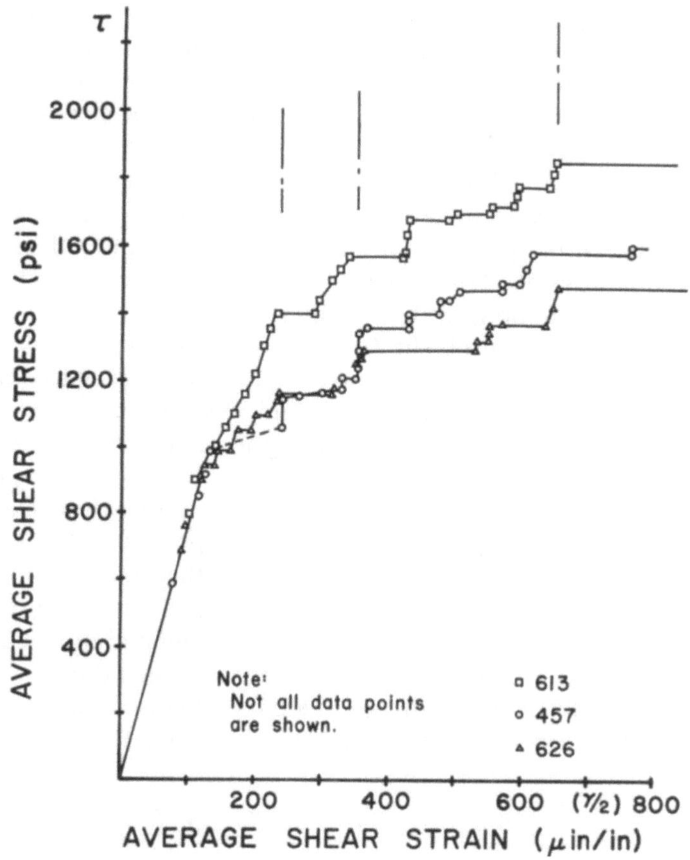

Fig. 1. Shear stress-shear strain data obtained in slowly loaded tests. Specimen Nos. 457 and 626 were thought to be identical while No. 613 was made by drilling a rod and therefore slightly different. The strain measure is the tensor component and hence one-half the strength of materials value.

words that Fig. 1 is *not* a constitutive relation. We shall be concerned in the first part of the paper with providing data which we believe demonstrate that Fig. 1 is applicable to impact problems. In fact it is very applicable, because the dynamic problems have a wide variation in their response which is thought to be nothing more nor less than the variation in the tangents of data like that shown in Fig. 1. Because of the variation in the response of "identical" specimens, we present data averaged for several tests together with a discussion of the statistical aspects of the observations. There is a very wide variation in responses in certain types of impact. In addition to these wave propagation results, data on the dynamic elastic-plastic interface are presented.

The second section of the paper reports additional information on the heat generated as a metal (annealed aluminum) deforms and describes our attempts to measure the heat generated in the plastic range. Said differently we are attempting to obtain data on the energy equation for annealed aluminum. In this respect we are sometimes reprimanded because we are looking at something (the change in temperature) which everyone knows is very *small*—and this is taken to be synonymous with *trivial.* We heartily agree on the smallness but not at all with the triviality of small quantities. In many respects there is an analogy between temperature in a continuous medium and damping in a mechanical system. For some phenomena in the mechanical system one can ignore damping while for others (long time) it is a very important feature of the problem. Our key point however is that we believe it is important to the very basis of plasticity to establish explicit forms for the conservation of energy and this means measuring the heat generated. Certainly the history of the temperature is approximately equal to the plastic work and therefore if one considers plastic work to be important, *so is the heat.* For static problems these may very well be interchangeable but from a basic viewpoint the distinction is believed to be important. Our second point is that in continuum mechanical theories of materials in which coupling is possible it makes a considerable difference whether one puts the temperature equal to zero before or after certain differentiations have been done.

We close the paper by relating our recent attempts at measuring the heat generated during impact—that is by combining the two previous sections.

PLASTIC DEFORMATION WAVES

Experimental Method

Commercially pure aluminum (1100 alloy) tubes which have an 0.50 in. outer diameter and a 0.05625 in. wall thickness are used for all tests reported in this section. The specimens are annealed at 1100° F for two hours and furnace cooled which produces a fine grained structure in the material. The specimen lengths change and some tests use short soft (annealed) sections glued to long hard (as received) ones.

Strains are measured at several axial stations by means of SR4-A-8 wire resistance gauges attached to the specimen with Post Yield cement. The changing resistance of the gauge is converted into a suitable oscilloscope signal by means of an Ellis Associates Model BAM-1 Bridge Amplifier. The oscilloscope uses a Tektronix Type A-74 (four channels) plug-in unit on the vertical plates. We do not attempt any measurement of boundary data since we prefer to compare strain

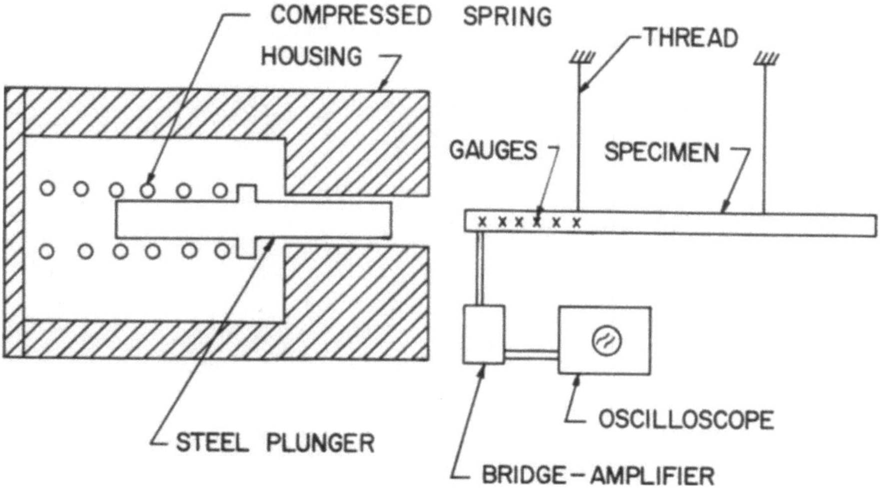

Fig. 2. Schematic diagram of the system used.

histories at interior points. Such measurements indicate that there is an alteration in the wave form near the ends of the specimens that drastically alters the internal response so that one can be seriously misled by using boundary data together with approximate theories to predict the response under other conditions. This alteration in wave form is due in part to three-dimensional effects and in our experimental arrangement also to an impedance mismatch between the steel plunger and the soft specimen.

A predominantly axial deformation wave is produced by impacting the end of the aluminum specimen with a steel plunger that is driven by releasing a compressed spring. The amplitudes of the strains are controlled by adjusting the magnitude of the spring compression. The specimen is hung from an A frame by long strings and is aligned with the plunger prior to each test. The system is simple and direct, if one accepts strain gauges, and can easily be reproduced by anyone interested in doing so. A schematic diagram of the system used is shown in Fig. 2.

Experimental Data

Two reports [5, 6] have recently been prepared which contain certain experimental data which complements that which is presented here. One of these reports [5] contains a study of a plastic wave impinging on an elastic-plastic interface while the other [6] emphasizes the wide variation in the response that one observes in specimens

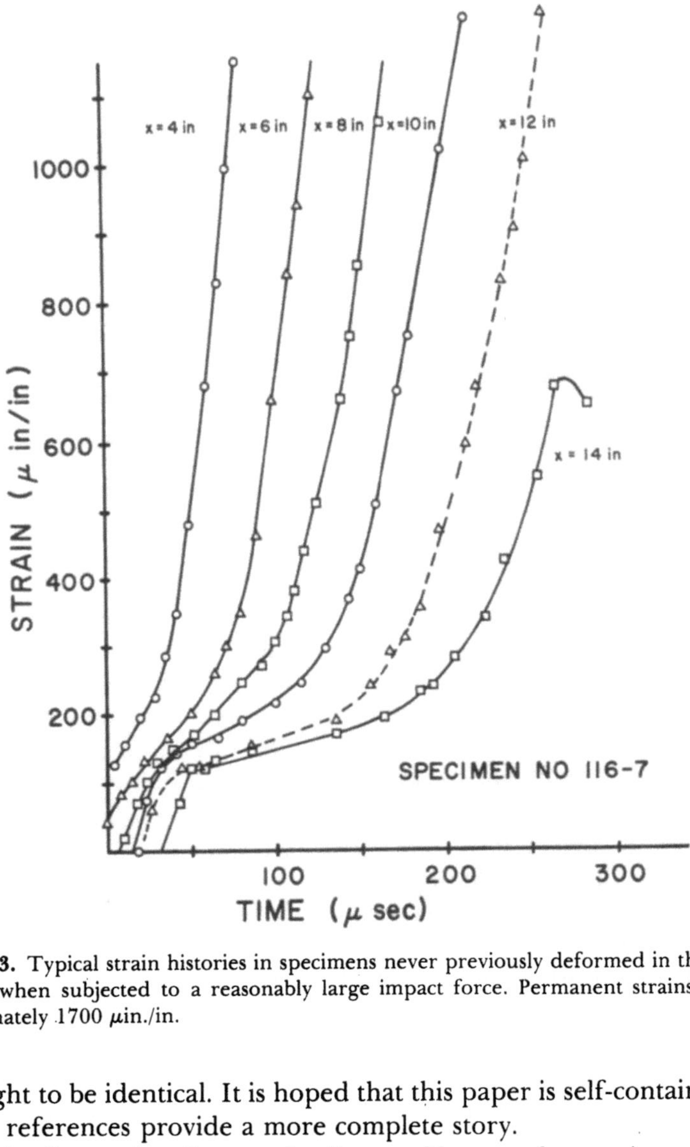

Fig. 3. Typical strain histories in specimens never previously deformed in the plastic range when subjected to a reasonably large impact force. Permanent strains are approximately 1700 μin./in.

thought to be identical. It is hoped that this paper is self-contained but these references provide a more complete story.

In the first series of tests to be discussed here twelve specimens which were thought to be identical were subjected to impacts with the spring compressed the same amount for every test. These tests involved relatively large final plastic deformations (2000 μin./in.), but since we are primarily interested in the response near the yield point; oscilloscope settings are made for this region rather than to permit one to observe the final strains. A typical result is shown in Fig. 3 where the strain histories of six axial stations are presented. Reflections from the free end arrive at $x = 14$ in. at 270 μsec. Interest is confined to times less than this and the time origin is arbitrary. At the larger values of strain shown in this figure, the speed of propagation is not too far from being uniform throughout the region of measurement. For example the time required for a strain equal to 1020 μin./in. to propagate the two inches between gauge positions are: 48, 42, 40 and 49 microseconds. However if one takes another specimen one finds that these numbers are: 47, 56, 47 μsec and infinity since the reflected wave arrived at $x = 12$ in. prior to 1020 μin./in. Hence one must average the response in Fig. 3 with several others in order to achieve data that can be meaningfully compared with theoretical results. The next relevant question has to do with the scatter in the data; how much of it is really in the material and how much is in the experimental method? Figure 3 also indicates a rapid slowing down of the wave near 125 μin./in. which is an apparent yield point. However, closer examination also reveals that, for strains just above and below this value, the time interval required for a given value of strain to propagate two inches is very far from being uniform.

Table 1 is a summary of the data obtained from these twelve identical specimens subjected to identical impacts. These data are obtained from histories similar to those shown in Fig. 3. Table 1 contains the average value of the time required to propagate the indicated two inches, the standard deviation of the data and the maximum positive and negative deviations from the average. Clearly the data in Table 1 indicate that strains above 210 μin./in. propagate at sensibly constant speed for the first ten inches but then something drastic happens. Furthermore, the standard deviation for the propagation times is almost the same for the elastic region and for strains of 765 μin./in. This is considered adequate to establish that the present system provides data with less than 5 μsec error and probably much less than this for averaged data. Therefore the rest of the variation in the data near the yield point (say at 170 μin./in.) *is due to variations in the material response.* This is believed to be the dynamic version of the variation indicated in Fig. 1 in which specimens 457 and 626 have different responses. The nature of the distribution of the time increments required for a given strain to propagate two inches are shown in Fig. 4 for a small value of the strain and in Fig. 4a for a larger value. The width of these distributions is about the same for both strains despite the larger mean time for the higher value of strain. The data in these large initial impacts will be compared with other data in a later section. The data in Table 1, especially in the first ten inches, are uniform and therefore could be consistent with a strain-rate independent constitutive relation and a smooth stress-strain curve.

Table 1

THE TIME INTERVAL, Δt, FOR A GIVEN VALUE OF STRAIN, ϵ, TO PROPAGATE BETWEEN THE DESIGNATED AXIAL LOCATIONS. THE NUMBER OF SPECIMENS, n, THE STANDARD DEVIATION, S.D., THE MAXIMUM POSITIVE DEVIATION, MAX. POS. DEV. AND THE MAXIMUM NEGATIVE DEVIATION, MAX. NEG. DEV. THESE SPECIMENS WERE ALL SUBJECTED TO THE PLUNGER BEING PROPELLED BY THE SAME LARGE SPRING COMPRESSION AND WERE THOUGHT TO BE IDENTICALLY PREPARED.

	$x=4-x=6$	$x=6-x=8$	$x=8-x=10$	$x=10-x=12$	$x=12-x=14$
			$\epsilon=85$ μin./in.		
Δt (μsec)	10.0	9.6	6.8	9.7	–
n	2	5	6	6	–
S.D. (μsec)	0	4.5	3.5	2.1	–
max. pos. dev. (μsec)	0	7.4	4.2	3.3	–
max. neg. dev. (μsec)	0	6.6	6.8	1.7	–
			$\epsilon=127$ μin./in.		
Δt (μsec)	14.8	8.0	6.6	12.1	11.5
n	5	6	7	7	2
S.D. (μsec)	4.1	7.0	2.8	9.4	3.5
max. pos. dev. (μsec)	6.2	24.0	4.4	20.9	3.5
max. neg. dev. (μsec)	4.8	8.0	5.6	12.1	3.5
			$\epsilon=150$ μin./in.		
Δt (μsec)	16.8	11.4	6.5	20.0	21.2
n	5	7	6	6	4
S.D. (μsec)	6.6	8.8	3.9	14.1	14.0
max. pos. dev. (μsec)	8.2	14.6	5.5	23.0	22.8
max. neg. dev. (μsec)	8.8	11.4	6.5	16.0	16.8
			$\epsilon=170$ μin./in.		
Δt (μsec)	17.6	11.0	9.6	25.5	28.5
n	11	10	10	10	2
S.D.(μsec)	10.7	6.7	4.5	21.5	–
max. pos. dev. (μsec)	19.4	14.0	5.4	30.5	1.5
max. neg. dev. (μsec)	7.6	13.0	9.6	25.5	1.5
			$\epsilon=213$ μin./in.		
Δt (μsec)	22.7	19.6	22.0	44.2	39.0
n	11	10	10	10	2
S.D. (μsec)	8.9	5.4	7.0	14.8	–
max. pos. dev. (μsec)	14.3	10.4	8.0	21.8	7.0
max. neg. dev. (μsec)	11.0	5.6	15.0	30.2	7.0

Table 1 (*Cont.*)

	$x=4-x=6$	$x=6-x=8$	$x=8-x=10$	$x=10-x=12$	$x=12-x=14$
		$\epsilon=255$ μin./in.			
Δt (μsec)	27.0	23.1	28.1	46.0	43.0
n	11	10	10	10	2
S.D. (μsec)	4.6	6.8	7.4	10.6	–
max. pos. dev. (μsec)	8.0	11.9	13.9	16.0	7.0
max. neg. dev. (μsec)	10.0	15.1	13.1	19.0	7.0
		$\epsilon=340$ μin./in.			
Δt (μsec)	30.8	31.2	30.7	50.0	40.0
n	11	11	11	11	1
S.D. (μsec)	5.2	9.6	5.9	11.2	
max. pos. dev. (μsec)	7.2	6.8	13.3	32.0	
max. neg. dev. (μsec)	9.7	15.2	10.7	10.0	
		$\epsilon=425$ μin./in.			
Δt (μsec)	33.2	32.0	34.3	48.2	
n	11	11	11	10	
S.D. (μsec)	5.0	6.6	4.4	8.8	
max. pos. dev. (μsec)	5.8	13.0	10.7	23.8	
max. neg. dev. (μsec)	10.2	12.0	7.3	8.2	
		$\epsilon=510$ μin./in.			
Δt (μsec)	34.8	32.8	36.7	45.6	
n	11	11	11	10	
S.D. (μsec)	4.9	6.1	5.7	6.7	
max. pos. dev. (μsec)	5.2	12.2	9.3	5.4	
max. neg. dev. (μsec)	9.2	10.8	7.7	12.6	
		$\epsilon=595$ μin./in.			
Δt (μsec)	37.0	34.2	37.2	48.2	
n	12	11	10	10	
S.D. (μsec)	4.8	6.3	5.7	9.8	
max. pos. dev. (μsec)	5.0	12.8	9.8	26.8	
max. neg. dev. (μsec)	11.0	10.2		9.2	
		$\epsilon=680$ μin./in.			
Δt (μsec)	38.2	35.6	37.7	45.9	
n	12	11	10	10	
S.D. (μsec)	5.0	6.4	7.4	10.9	
max. pos. dev. (μsec)	6.8	13.4	13.3	26.1	
max. neg. dev. (μsec)	10.2	9.4	8.7	16.9	
		$\epsilon=765$ μin./in.			
Δt (μsec)	40.1	37.2	39.3	49.5	
n	12	11	10	9	
S.D. (μsec)	5.2	6.5	7.1	4.2	
max. pos. dev. (μsec)	6.9	13.8	14.7	5.5	
max. neg. dev. (μsec)	10.1	11.2	10.3	7.5	

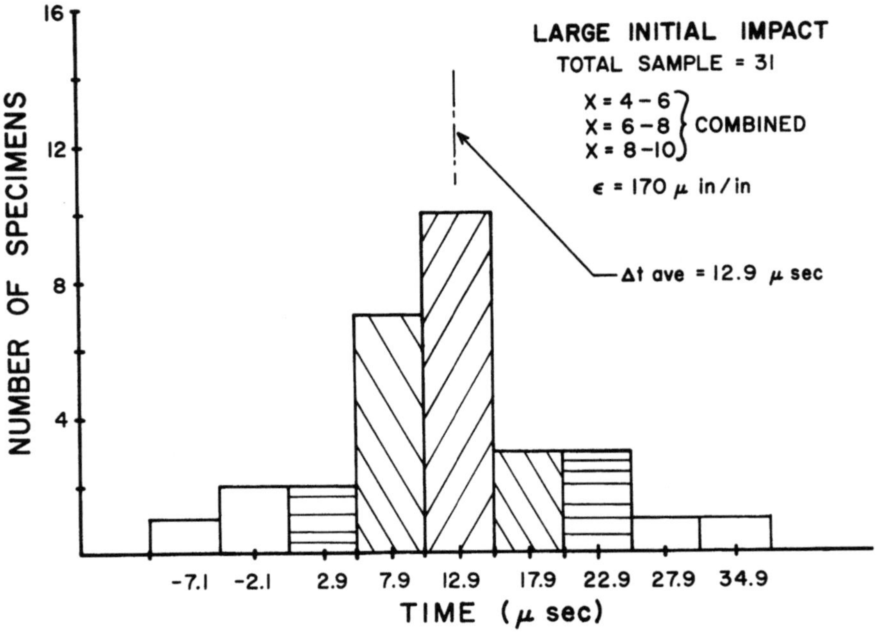

Fig. 4. The distribution function for the time increments required for a strain of 170 μin./in. to propagate two inches in impacts like the one used to produce Fig. 3.

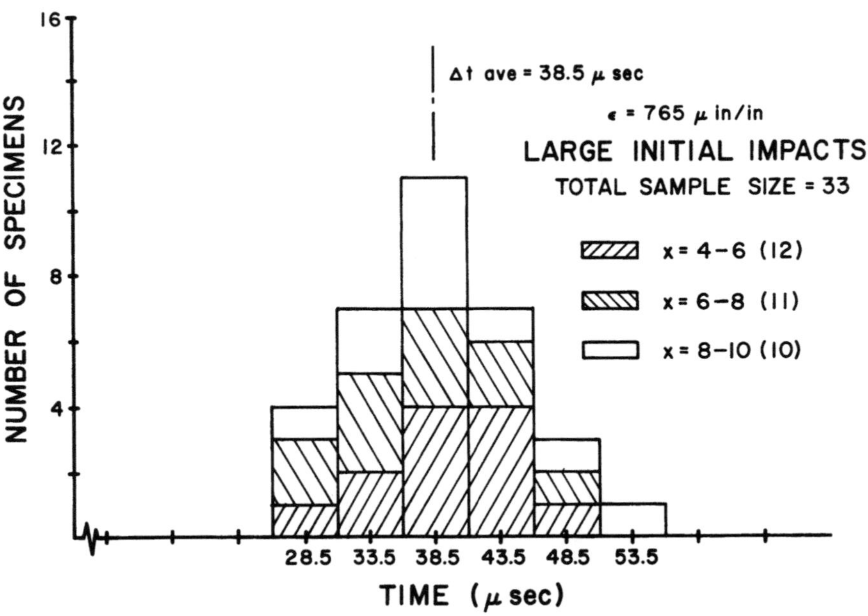

Fig. 4a. The distribution function for the time increment required for a strain of 765 μin./in. to propagate two inches in impacts like the ones used in Figs. 3 and 4.

Table 1a

THE SAME INFORMATION AS IN TABLE 1 EXCEPT THESE SPECIMENS WERE SLIGHTLY PLASTICALLY DEFORMED IN A PREVIOUS IMPACT. THE PURPOSE IS TO VALIDATE THE DATA IN TABLE 1 FOR THIS SPATIAL INTERVAL.

	$\epsilon = 255$ μin./in.	$\epsilon = 340$ μin./in.	$\epsilon = 510$ μin./in.
	$x = 12 - x = 14$		
Δt (μsec)	33.0	41.5	44.2
n	4	4	4
S.D. (μsec)	16.8	15.5	9.9
max. pos. dev. (μsec)	17.0	13.5	9.8
max. neg. dev. (μsec)	22.0	26.5	16.2

In order to better establish the validity of the data at distances from $x = 10$ in. to $x = 14$ in. data from other specimens is given in Table 1a. These other specimens were previously impacted slightly into the plastic range (i.e., Fig. 5) and *then* subjected to an impact with the spring the same compression as the data in Table 1.

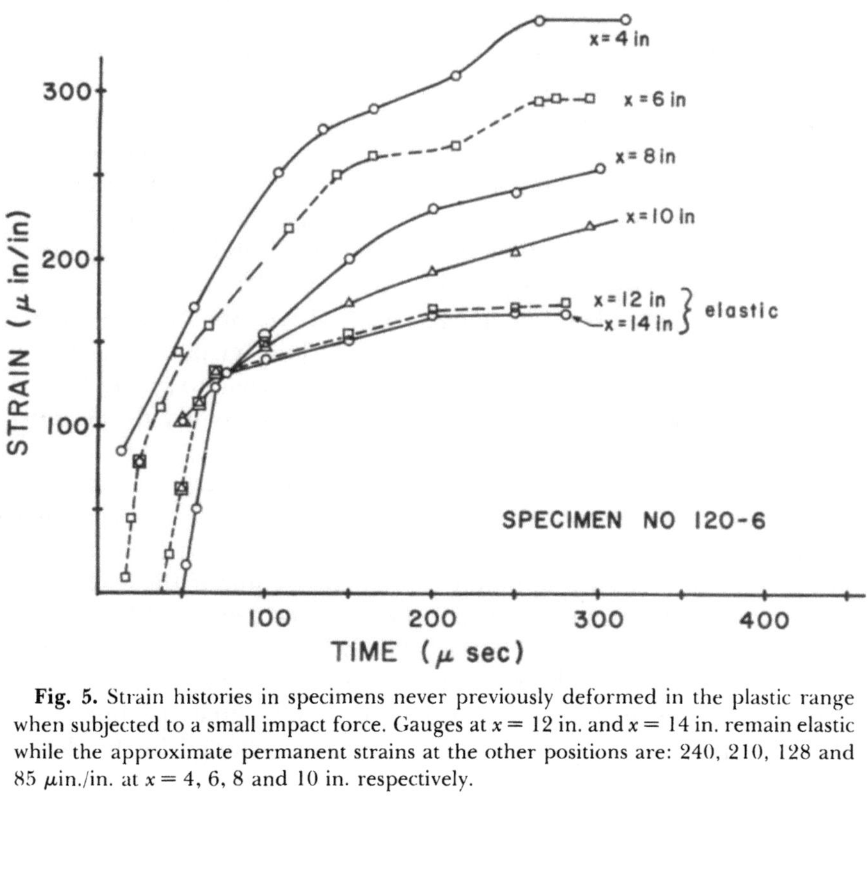

Fig. 5. Strain histories in specimens never previously deformed in the plastic range when subjected to a small impact force. Gauges at $x = 12$ in. and $x = 14$ in. remain elastic while the approximate permanent strains at the other positions are: 240, 210, 128 and 85 μin./in. at $x = 4$, 6, 8 and 10 in. respectively.

Very Small Impacts

The writer has shown [7] that data obtained in large impacts are not sensitive to the difference between smooth stress-strain relations and a relation such as that represented by Fig. 1. The wave speeds for materials with the same averaged curves are approximately the same for large amplitude waves, but are not for smaller ones. Hence if one doesn't know whether the appropriate constitutive relation has continuous derivatives or not, and one is interested in such questions; it becomes necessary to do experiments in which the stress is just barely above one of the plateaus. Clearly this is easy enough to do in static tests if one has patience and creep can be ignored. In an impact situation this is more difficult, because of the variation in the location of the plateaus from specimen to specimen. Hence one does many tests and *expects* to observe significant deviations from an average in only a few of them. Then one is faced with the problem that the exceptional cases may possibly be regarded as "flukes." Being interested in this problem, we initially compress the spring to values just above and just below where we expect to produce a plastic strain. There really isn't any "typical" result for such experiments and moreover one is interested in the extremes not the "typical" data. However, Fig. 5 is as "typical" as one can find for these small impact tests. In this particular experiment the gauges located at $x = 4$, 6, 8 and 10 in. showed permanent strains after the test while those at $x = 12$ and 14 in. indicated none. The time required for a strain of 170 μin./in. to propagate the inches between gauge locations are: 15, 42, 26, 60 and something in excess of 80 μsec. The time required for a strain of 213 μin./in. to propagate the two inches between gauge positions are: 30, 56 and 98 μsec. Clearly these are not uniform along the bar. The key point is that *there is ample time* for a strain of 180 μin./in. to propagate to $x = 12.0$ in. if the stress-strain curve is smooth and the material is strain-rate independent.

To illustrate the variety of responses obtained in these small impacts, Figs. 6 and 6a are presented for two other specimens. In the specimen whose response is shown in Fig. 6a gauges located at $x = 10$ in. and beyond remained elastic. The time increments required for a strain of 170 μin./in. to propagate the two inches for specimen No. 136-5 shown in Fig. 6 are: 30, 7, 28 and 75 μsec while in Fig. 6a the time increments are: 10 and 42 μsec. The wide spread in the data for a strain of 170 μin./in. in these small hits is shown in Fig. 7. Of course specimens like those represented by Figs. 6 and 6a are not included in the down-

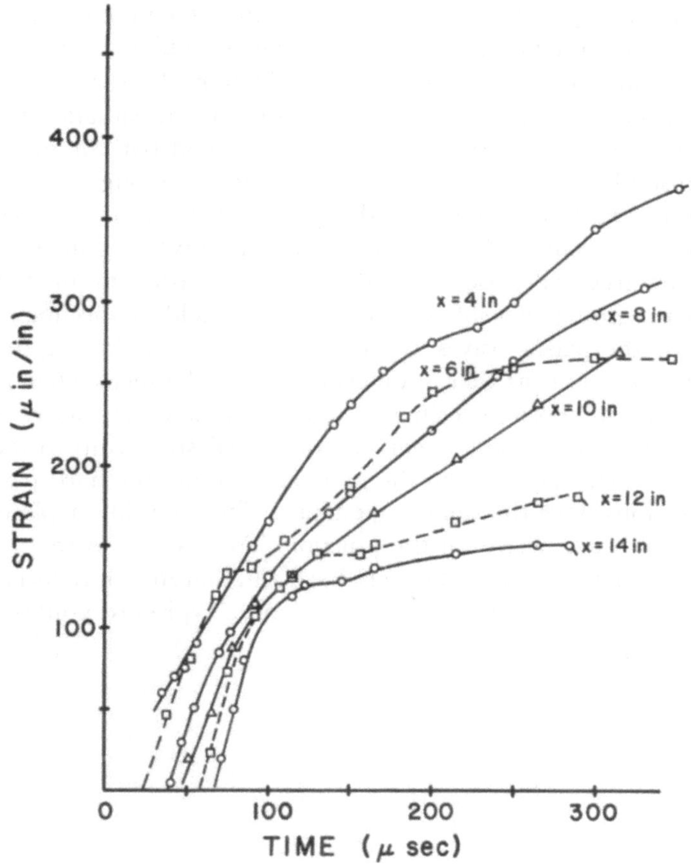

Fig. 6. Strain histories in specimens never previously deformed in the plastic range when subjected to a small impact force. The approximate permanent strains are: 280, 150, 200, 160, 75 and 15 μin./in. at $x = 4, 6, 8, 10, 12$ and 14 in. respectively.

stream positions. The mean time for a strain of 170 μin./in. to propagate two inches in small impacts is 20.1 μsec as compared to 12.9 μsec for the large initial impacts. The data for these small initial hits are summarized in Table 2. It is emphasized that what is perhaps the most significant factor—that of the certain level of strain never reaching 8, 10 or 12 inches—is not adequately reflected in the data in Table 2.

Composite Specimens

In plasticity the solution of boundary value problems involving the transition from the elastic to the plastic response is always interesting.

Consider the case of a rod which is soft on one end and hard on the other. Such a situation is created by welding or gluing an annealed section to another made of the same material but in a harder state. Besides being an interesting problem in its own right, this situation is closely related to the validity of data taken in a "load bar" or in the split Hopkinson bar experiment. We have developed enough technique of gluing the two sections together that the elastic wave is the same as in a continuous specimen. Under impact loadings involving small stresses, all specimens that are used act as though there wasn't any junction. Furthermore, gluing two soft sections together yields a specimen just as good for the plastic waves as a completely soft one having the same total length, so far as we can determine. What happens when the soft end of this composite specimen is given a blow which causes it to go plastic? With the arrangement just described strains up to the yield point pass right on through the junction and into the hard part since both sections have the same impedance. Then the slower plastic portion of the wave arrives at the junction. To this part of the wave the hard section acts as though it were hard and therefore it reflects, carrying an *increased* stress back into the soft end. Typical responses for the

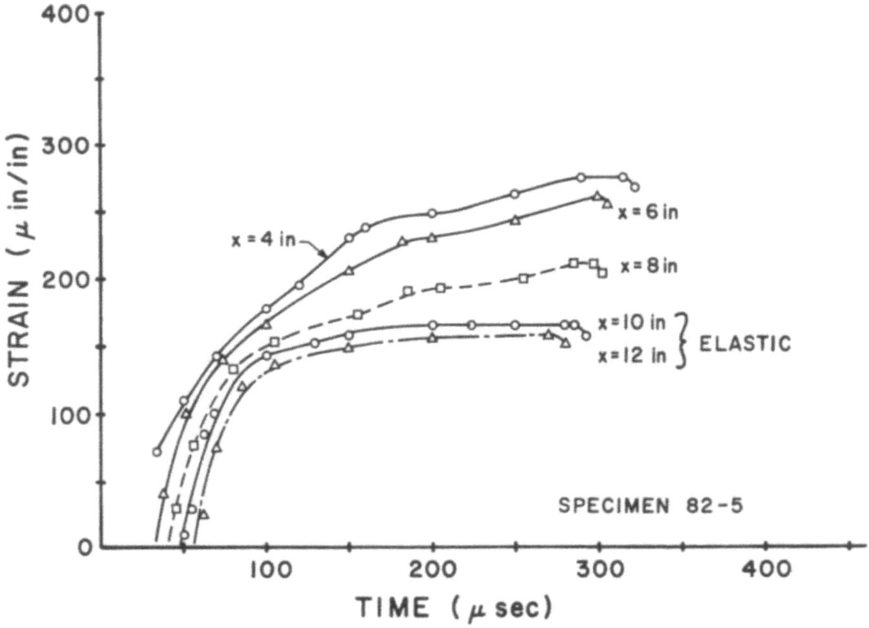

Fig. 6a. Strain histories in a specimen never previously deformed in the plastic range when subjected to a small force. The approximate permanent strains are: 170, 85 and 50 μin./in. at $x = 4$, 6 and 8 in. respectively.

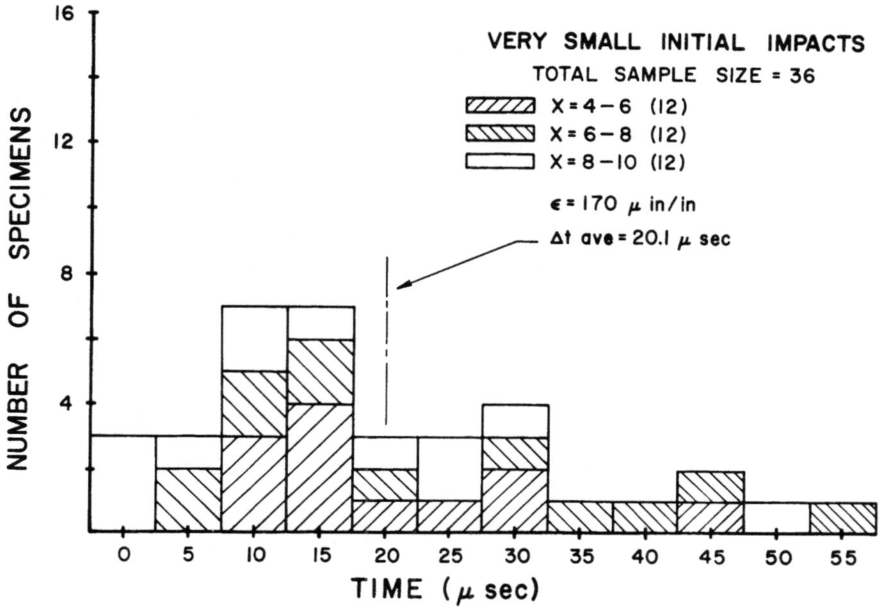

Fig. 7. The distribution function for the time intervals required for a strain of 170 μin./in. to propagate two inches in very small impacts such as those represented by Figs. 5, 6 and 6a.

gauges located in the soft section are shown in Fig. 8 for the impact where the spring is compressed the same amount * as used in obtaining Fig. 3. There is an apparent slight speeding up of the wave between the last two gauges. However, these data are within the variation of the raw results used in developing Table 1 and therefore not especially significant. The significant feature of Fig. 8 is the second wave that one observes in which the gauge responses *have reversed their order.* The first plateau in Fig. 8 is at 1900 μin./in. which is very close to the maximum obtained in the continuous soft specimen of the same total length. Clearly the second wave is a reflection of the plastic part from the junction. At other time sensitivities one observes still a third wave where the gauges return to their original order. This is a reflection from the plunger face. The strain measured in the hard section is sometimes used as a measure of the "stress" in the soft section. Typical data for our experiments are shown in Fig. 9 for the same impact as illustrated in Fig. 8. One can extrapolate the strain histories observed in the soft section forward to the junction position and the data in the

* However, this specimen had been subjected to a smaller force which caused some plastic deformation prior to this test (see Fig. 13).

Table 2

THE TIME INTERVAL, Δt, FOR A GIVEN LEVEL OF STRAIN, ϵ, TO PROPAGATE BETWEEN THE DESIGNATED AXIAL LOCATIONS. THE NUMBER OF SPECIMENS, n, THE STANDARD DEVIATION, S.D., THE MAXIMUM POSITIVE DEVIATION, MAX. POS. DEV., AND THE MAXIMUM NEGATIVE DEVIATION, MAX. NEG. DEV. OF THE DATA OBTAINED IN HITS WITH VERY SMALL PERMANENT STRAINS. AN IMPORTANT FEATURE OF THE DATA, SOME POSITIONS REMAINING ELASTIC SUCH AS IN FIG. 6A, IS NOT REFLECTED IN THE TABLE.

	$x=4-x=6$	$x=6-x=8$	$x=8-x=10$	$x=10-x=12$	$x=12-x=14$
			$\epsilon=85$ μin./in.		
Δt (μsec)	11.0	12.9	8.6	8.6	9.0
n	10	8	10	9	3
S.D. (μsec)	6.0	7.1	3.9	5.7	1.5
max. pos. dev. (μsec)	11.0	16.1	9.4	11.4	–
max. neg. dev. (μsec)	11.0	7.9	4.6	8.6	–
			$\epsilon=127$ μin./in.		
Δt (μsec)	11.5	15.6	6.7	10.8	8.5
n	11	9	10	9	2
S.D. (μsec)	9.4	10.7	6.2	10.7	–
max. pos. dev. (μsec)	24.5	20.4	13.3	20.2	–
max. neg. dev. (μsec)	16.5	15.6	11.7	10.8	–
			$\epsilon=170$ μin./in.		
Δt (μsec)	20.1	24.0	16.3	37.3	–
n	12	12	12	10	–
S.D. (μsec)	9.9	15.7	14.3	24.4	–
max. pos. dev. (μsec)					–
max. neg. dev. (μsec)					–
			$\epsilon=255$ μin./in.		
Δt (μsec)	43.5	63.1	37.6	46.0	–
n	11	10	6	3	–
S.D. (μsec)	11.8	41.5	19.6	11.3	–
max. pos. dev. (μsec)	17.5	91.0	24.4	8.0	–
max. neg. dev. (μsec)	18.5	48.1	30.6	16.0	–
			$\epsilon=240$ μin./in.		
Δt (μsec)	59.4	57.4	44.0	–	–
n	7	5	4	–	–
S.D. (μsec)	25.8	19.8	23.8	–	–
max. pos. dev. (μsec)	52.6	38.4	38.0	–	–
max. neg. dev. (μsec)	27.4	15.4	22.0	–	–

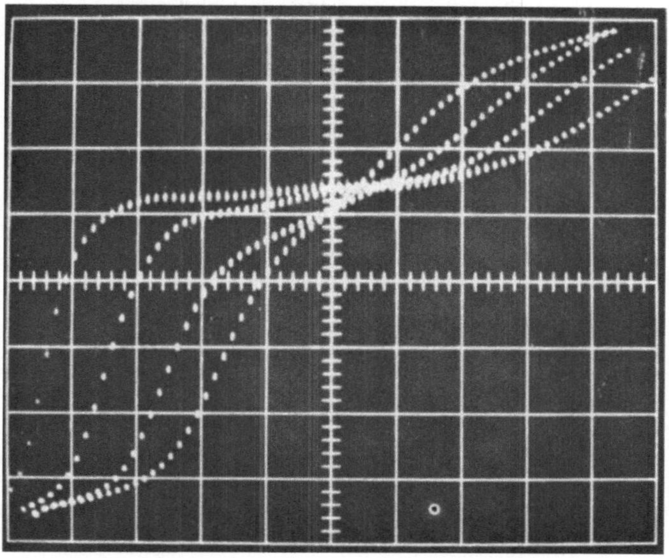

Fig. 8. Actual experimental record for the strain histories in a composite specimen (see Fig. 9). Each major division is 340 μin./in. strain (vertical) and 50 μsec in time (horizontal). The plunger setting is the same as used in producing Fig. 3, but this specimen had been very slightly deformed in the plastic range on a previous impact.

hard section back to this location; and thereby obtain a "stress-strain" curve for the material which isn't too bad. The results of such an operation are shown in Fig. 9a for the specimen whose response is shown in Figs. 8 and 9. Also shown in Fig. 9a is a reference curve * which will be developed below from the data in Table 1 and a particular theory. The strain shown in Fig. 9 for $x = 13.6$ in. was translated to the left by 15 μsec and then multiplied by Young's modulus (10.5×10^6 psi) to give the "stress" at the junction location. Without some prior knowledge of the material it is a difficult thing to know how to extrapolate the strains in the soft section since there are known to be some reflections between $x = 8.0$ in. and the junction. However, the material response is frequently what is being sought in the test. In constructing Fig. 9a the strain used was obtained by using the average wave speed between $x = 2.0$ and $x = 8.0$ in. and therefore these reflections were ignored. However, the logic of how one extrapolates the strains in the soft section escapes me and I do not recommend this procedure despite its modest

* The difference between these two curves can be viewed as the increase due to reflections. It agrees with the predictions [5] of the strain rate independent theory in a general way. The reference curve is developed later and shown in Fig. 17.

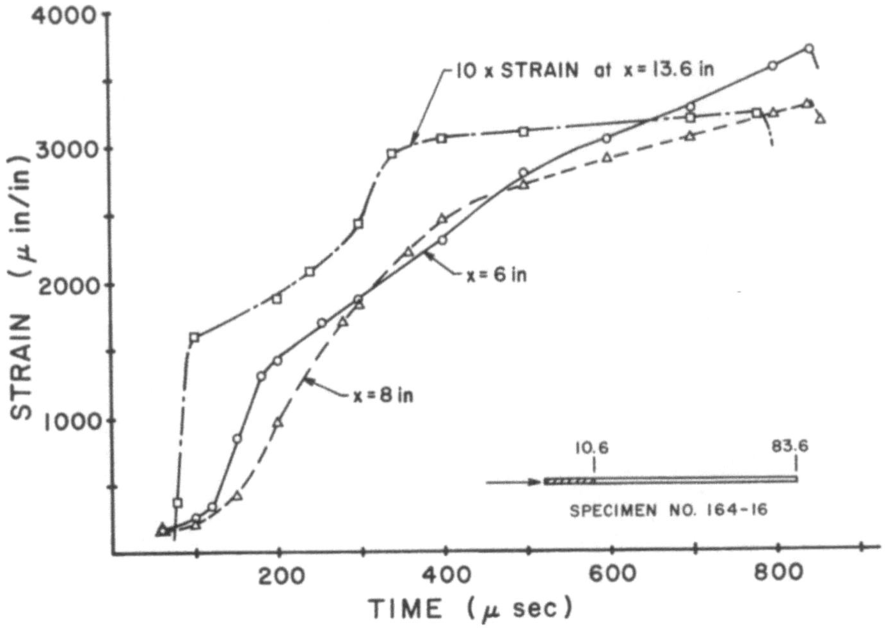

Fig. 9. The strain histories of points near the junction and in the hard section for the same impact as shown in Fig. 8.

success. Another point of interest to us is the homogeneity of the final strains in the soft section. It is very good; typically being within five percent when the hard section is long. Finally, a systematic study [5] of the strains below 1800 μin./in. shows that they propagate in these short ten-inch soft sections just as though the specimen were a continuous soft one, provided one stays two inches away from either end.

A rather novel aspect of composite specimens is shown in Fig. 10 where the results of a soft-hard-soft-hard type specimen are shown. This multi-section specimen was subjected to an initial impact where the spring was compressed the same amount as used for the test in Fig. 3. Consider first the time required for a given strain to propagate between $x = 4.0$ in. and $x = 8.0$ in. For this specimen the time required for a strain of 765 μin./in. to propagate this distance is 98 μsec and therefore greater than the 77.1 μsec indicated in Table 1. Probably this * is greater than the scatter in the data. Thus we have the situation where a single junction does not *affect* the propagation of strains which are below 1800 μin./in. but *more than one does*. One can show [5] that this happens in strain-rate independent materials because the length

* This is a better established [5] point than can be obtained from this single specimen.

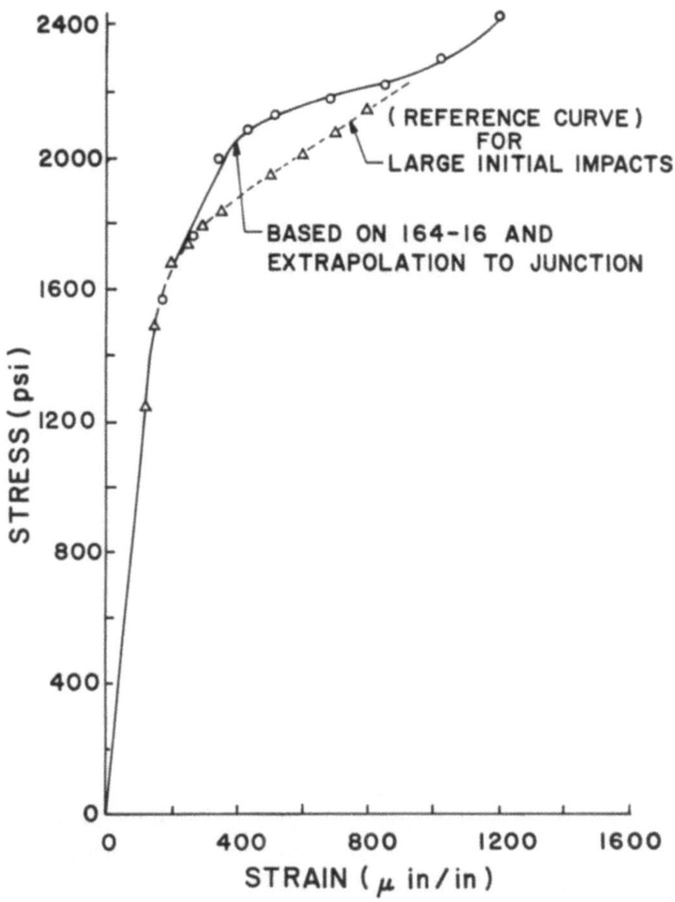

Fig. 9a. The stress-strain curve obtained from extrapolations of the data contained in Fig. 9 compared with one based on wave data (see Fig. 17).

of the hard section in Fig. 10 is less than in Fig. 8. When the wave propagates through the hard section and arrives at the junction with the second soft one, the stress decreases in value and this reflects back into the first soft section along a negative characteristic at the elastic bar velocity and thus penetrates far into the *first* soft section. Another interesting situation is shown in Fig. 10 where it takes 130 μsec for a strain of 150 μin./in. to propagate the four inches from 16.6 in. to 20.6 in. This is a much longer time than one would expect from the data in Table 1. In Fig. 10 it can also be seen that there is a plateau in the strain histories of both hard sections at 192 μin./in. (960 ÷ 5) and that this strain is constant at this value for significant time intervals. Therefore

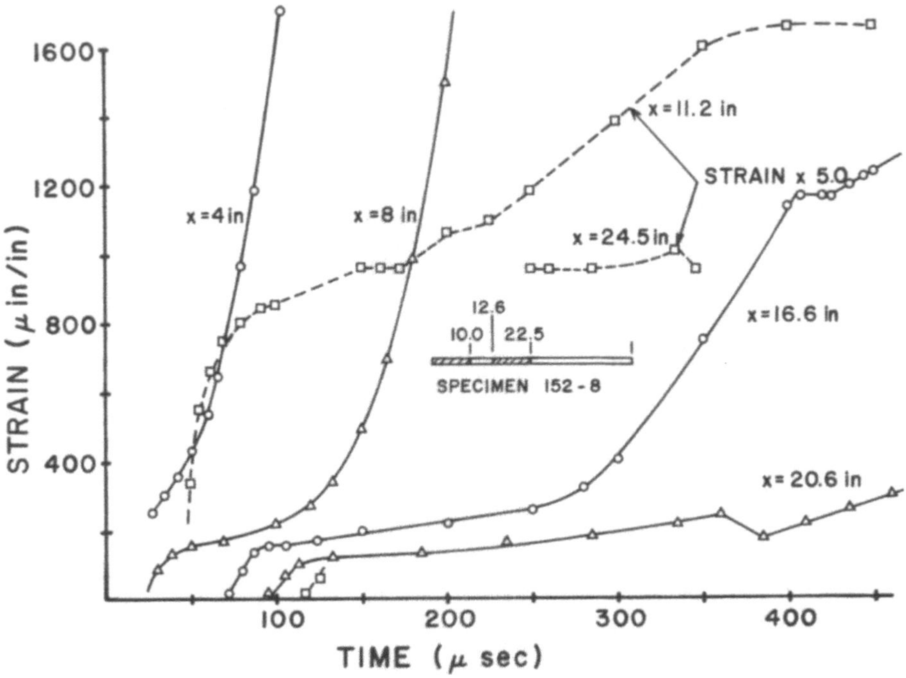

Fig. 10. Strain histories in a multi-section composite specimen when the spring is compressed the same amount as used for Figs. 3 and 8.

we conclude composite specimens are interesting boundary value problems but not very good for obtaining constitutive data for the material.

We consider now data obtained in smaller initial impacts. One specimen consisting of four soft (i.e., annealed) sections glued together has been tested. The results are shown in Fig. 11. The time required for a strain of 170 μin./in. to propagate between $x = 4.3$ in. and $x = 11.1$ in. is 80 μsec which compares very favorably with the averaged data in Table 2 for the same positions. Furthermore, subsequent larger impacts also gave data having very good agreement with averaged results. Thus we conclude that the plastic parts of the wave are not appreciably affected by the junctions. Observe however in Fig. 11 that it takes 68 μsec for a strain of 170 μin./in. to propagate two inches beyond the 11.1 in. station and this is within a single section. Station $x = 19.0$ in. remains at 155 μin./in. for 100 μsec before an unloading wave arrives from the free end and reduces the stress there. There appeared to be no permanent strain at $x = 19.0$ in. (or at positions $x = 23.0$ and $x = 30.0$ in.) in this impact. Observe also how sharply the data approach and leave the plateaus. It is also likely that all three downstream sec-

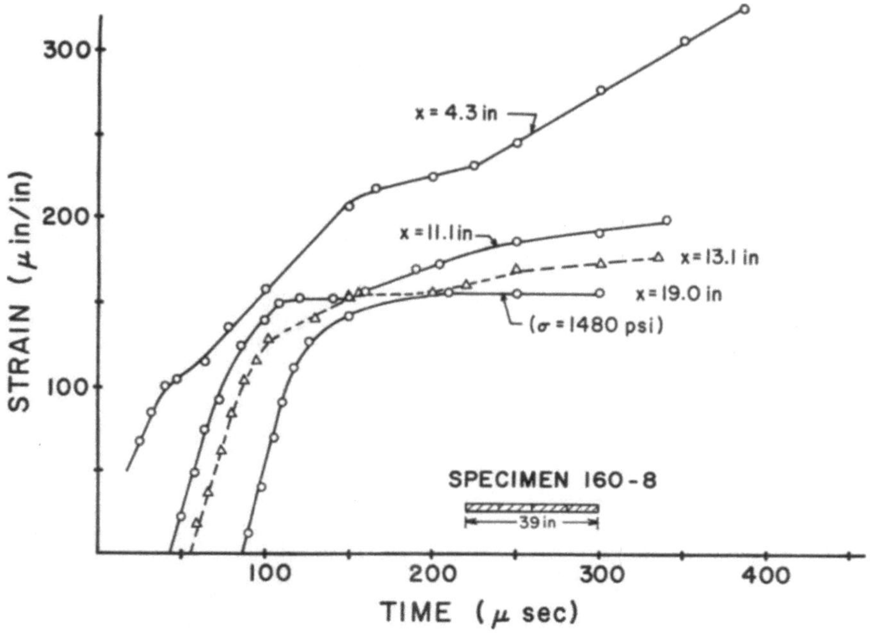

Fig. 11. Strain histories in a specimen consisting of four soft sections glued together subjected to a small impact force.

tions yielded at very close to the same value but this is possibly not a typical result.

Because the specimen whose response is shown in Fig. 11 did not yield at $x = 19.0$ in. we have done more extensive testing with a specimen consisting of two soft sections glued to a longer hard one. A reasonably typical result for small initial plastic impacts of these specimens is shown in Fig. 12. Observe in this figure that there is a slight increase in the strain from 155 μin./in. at 140 μsec to 170 μin./in. at 275 μsec. It is possible that this is due to the plastic wave impinging on the hard section since this is a slightly harder hit than used in Fig. 11. However, it is also possibly due to the material, and the increase in stress is just the value associated with a change from one plateau to another in Fig. 1. At any rate there is a considerable increase in strain at $x = 16.2$ in. with only a very modest change in the hard section. Furthermore, it is not impossible that the speeding up of the plastic wave near a strain of 300 μin./in. is real.

We consider now data obtained in small initial plastic impacts for the simplest of the composite specimens. A very typical result is shown in Fig. 13 for specimen No. 164-14 which is the same one used for the

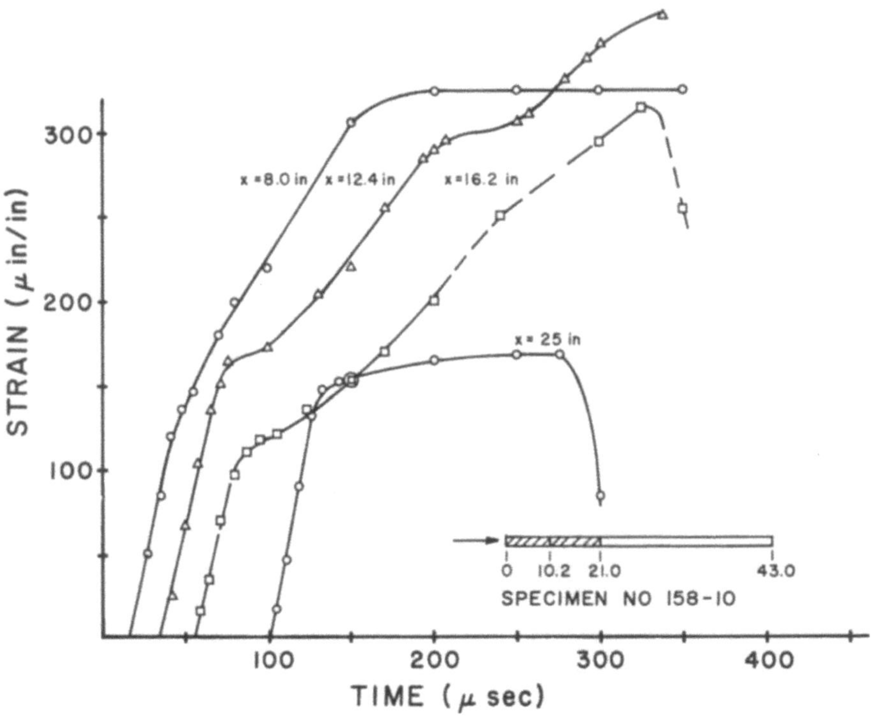

Fig. 12. Strain histories in a specimen consisting of two soft sections glued to a single (long) hard one subjected to a small impact force.

data in Fig. 8. Observe in Fig. 13 the very large (relative) changes in strain which occur in the soft side of the junction while the strain (stress) in the hard section remains virtually constant at 155 μin./in. (1450 psi). The key point is that the 155 μin./in. is established in the hard section at the time when the strain in the soft section is only 235 μin./in. While the strain in the hard section remains constant at a value of 155 μin./in., the soft section *continues to deform* to 380 μin./in. Without being firmly committed, an examination of Fig. 1 reveals several plateaus with approximately this value for the change in strain at constant stress. It is desirable to emphasize that the exact value of the strain in the plateau in tests like that represented by Fig. 13 varies somewhat, but in all other respects it is very typical. We have many results which are virtually identical with Fig. 13.

The response in more complicated multi-section specimens is varied. One interesting one is shown in Fig. 14 which involves an impact that is somewhat harder than used for Figs. 11–13 but is less than used to obtain Fig. 10. In Fig. 14, the response for the gauge at $x = 25.0$ in. is greater than for the one located at $x = 12.5$ in. However, the time incre-

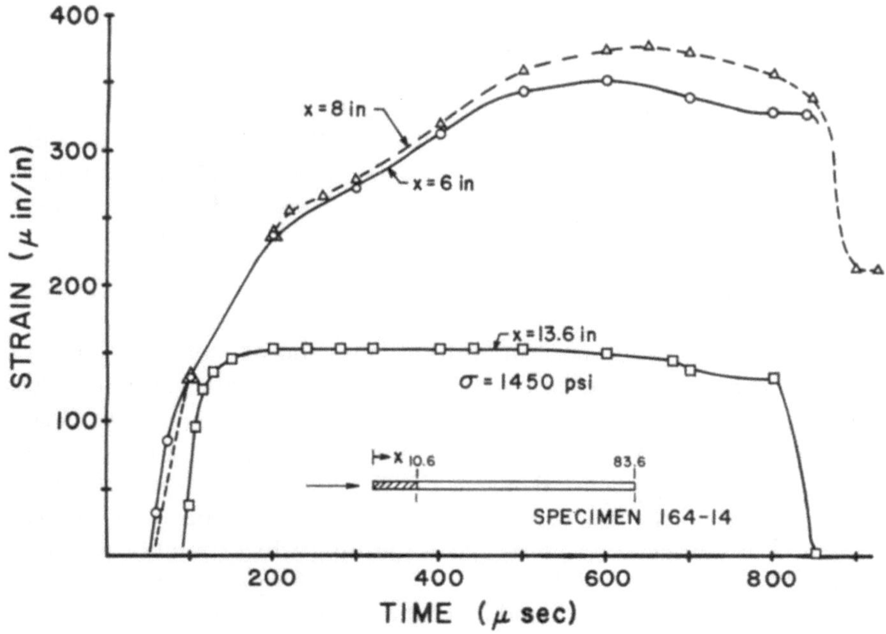

Fig. 13. Strain histories in a composite specimen when subjected to a stress slightly above the yield point. The most important feature is the large additional strain which develops in the soft specimen without an increase in the hard one.

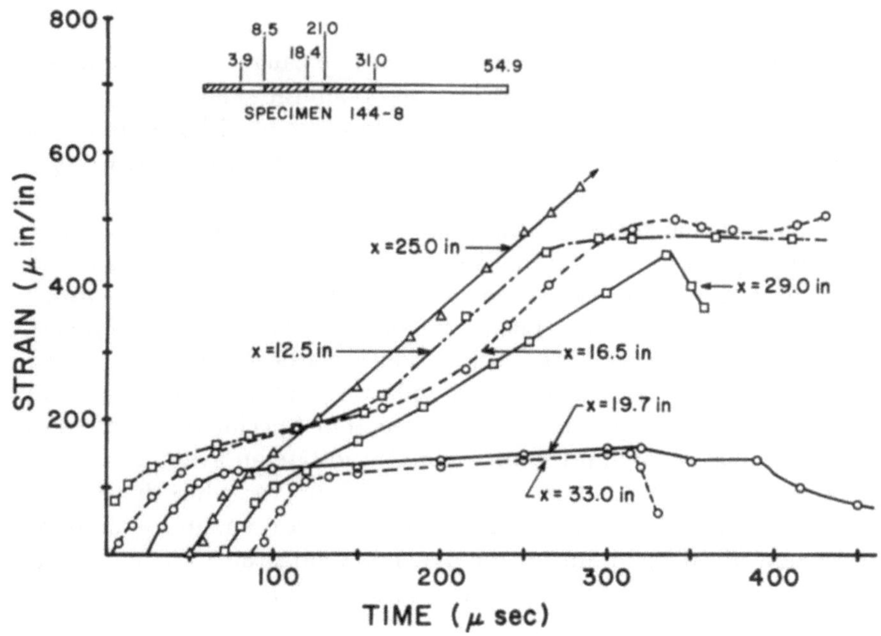

Fig. 14. Strain histories in a multi-section specimen when subjected to an intermediate size impact. The approximate permanent strains are: 340, 310, 650 and 280 μin./in. at $x = 12.5$, 16.5, 25 and 29 in. respectively.

ment for a given level of strain to propagate four inches is much less between $x = 12.5$ in. and $x = 16.5$ in. than it is between $x = 25.0$ in. and $x = 29.0$ in. All of which illustrates that one must be careful in drawing conclusions from composite specimens of the soft-hard-soft-hard type.

Dynamic Yield Strain

The writer recently presented [6] data on the dynamic yield strain for the same material used in the present series of tests and indicated that there was a very wide spread in the data for this parameter. The definition of the dynamic yield strain used in interpreting the experimental wave propagation data was the slowing down of the wave to a speed less than the bar velocity. We indicated in our previous report some of the difficulties we found in deciding which numerical value should be called the yield strain. Part of the interpretation difficulties are illustrated by Figs. 3, 5 and 6a of the present paper. In Fig. 3 there is a very sharp slowing down at a strain 127 μin./in. and so this value is used. In Figs. 5 and 6a there is an appreciable slowing down at strains of 140 and 130 μin./in., but part of these specimens remain elastic in the sense that no permanent strains were observed after the test at $x = 12$ and 10 in.

In the earlier report every specimen where gauges had been placed at $x = 4$ and $x = 10$ inches was used to obtain the statistical variation. This resulted in an average dynamic yield strain of 173 μin./in. with a standard deviation of 33 μin./in. In this paper we use the opposite extreme and select only "good data." That is, we use only data where we are satisfied that there is a sharp decrease in the wave speed, where bending appears to be very small, where the sensitivity of strain and time permit accurate measurements and where the overall picture * just looks good to us. The results are shown in Fig. 15 for eleven very small and six large impacts which produce plastic deformations in the initial plastic hit. We use the response histories of the front gauge and the last one on the specimen in this evaluation. Specimens subjected to a large initial plastic impact (like Fig. 3) indicate an average value of 134 μin./in. with a standard deviation of 8.3 μin./in., while those given a small blow indicate an average value of 139 μin./in. and a standard deviation of 16.1 μin./in. for the dynamic yield strain. These data do not involve any composite specimens and are shown in Fig. 15. These

* Our technique has also improved so we use smaller size hits in the elastic range in order to reduce the possibility of causing plastic deformation with a mistake. Furthermore, it depends on where the measurements are taken. For example in Table 1, 127 μin./in. have a very uneven propagation speed even when averaged results are used.

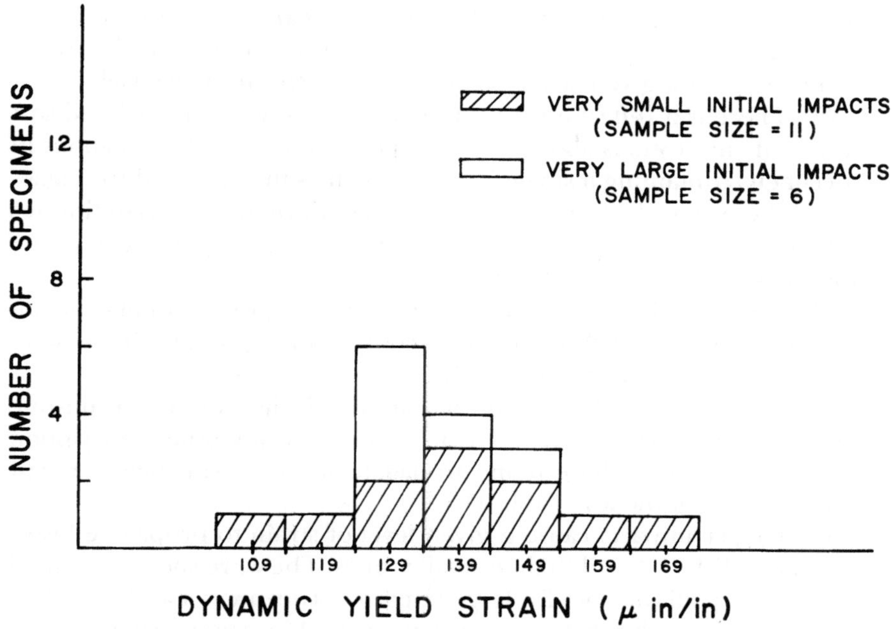

Fig. 15. Distribution function for the dynamic yield strain for "good" tests.

values are in better agreement with data reported by other investigators for annealed aluminum than our previous results [6]. While the difference between the averages of 173 and 139 μin./in. may appear to some as being large, let me emphasize that the experiment for ascertaining the yield point is a very difficult one in dynamic tests unless one has very long specimens. Even in this case it is also difficult because one cannot easily obtain long specimens which are homogeneous and straight. Observe also that in Fig. 5 the level of the plateau is about 170 μin./in. indicating that this might even be a very good alternative definition for the yield point and lead to a different numerical average.

Discussion

The data presented above are new experimental results which are intended to be reliable evidence but as free of interpretation as possible. This section will contain the interpretation and quite a lot of personal bias on the nature of inelastic behavior and in particular as applied to annealed aluminum. The major points which it is believed that the data given above establish are:

(1) There is a real, relatively large variation in the response (both

static and dynamic) of annealed aluminum near the yield point (Figs. 1, 4 and 7). By response we especially mean the tangent of the stress-strain curve at a given stress, as contrasted with the curve itself.

(2) There is a reflection from the junction between a hard and soft section if the stress is above the yield point. The final deformations in specimens containing a single junction, such as the one used for Fig. 8, are of the order of $1\frac{3}{4}$ times those which are developed in a continuous soft one. The lengths of the soft and hard sections in the composite specimen are relevant to this increase.

(3) There is a large amount of plastic straining which occurs in composite specimens which is *not* observed as a corresponding increase in stress in the hard section (Fig. 13).

(4) It is as reasonable to say that annealed aluminum is strain-rate independent in large impacts as it is to give it any other description since strains above 200 μin./in. propagate at a uniform speed, at least for the first ten inches (Table 1).

(5) Larger impacts cause a given level of strain to propagate faster (compare Tables 1 and 2) than smaller ones. The wave speeds obtained by averaging all the data for large impacts between $x = 4$ and $x = 12$ in. and for the small hits between $x = 4$ and $x = 14$ in. are shown in Fig. 16.

(6) There is a reasonably good chance of being able to initiate a small plastic deformation wave which propagates more or less normally for some distance, and then suddenly stops or slows down very drastically (Figs. 5 and 11 as well as a previous paper [6]).

If one now assumes the strain-rate independent bar theory of Karman, Taylor and Rakhmatulin and *that there exists some smooth* stress-strain relation which governs the propagation of plastic deformation waves, then one *deduces* that this relation is given by

$$\sigma = \int_0^\epsilon \rho c^2 \, d\epsilon$$

where σ, ϵ, ρ and $c(\epsilon)$ are the stress, strain, mass density and wave speed expressed as a function of strain, respectively. One should use this relation to obtain the governing stress-strain relation only after it *has been demonstrated* that the wave speed is indeed uniform down the bar. We have used the totality of our averaged experimental wave speed data for each of the very small impacts and the large ones. The results are shown in Fig. 17 where these derived curves are compared with two static compressive stress-strain relations taken from the literature [8, 9]. Recalling the scatter in the static data (Fig. 1) and in the dynamic results (Tables 1 and 2); the writer concludes from Fig. 17 that one might just as well *use the strain-rate independent theory* with a smooth stress-strain curve as any other he knows about, if one has *to predict* the

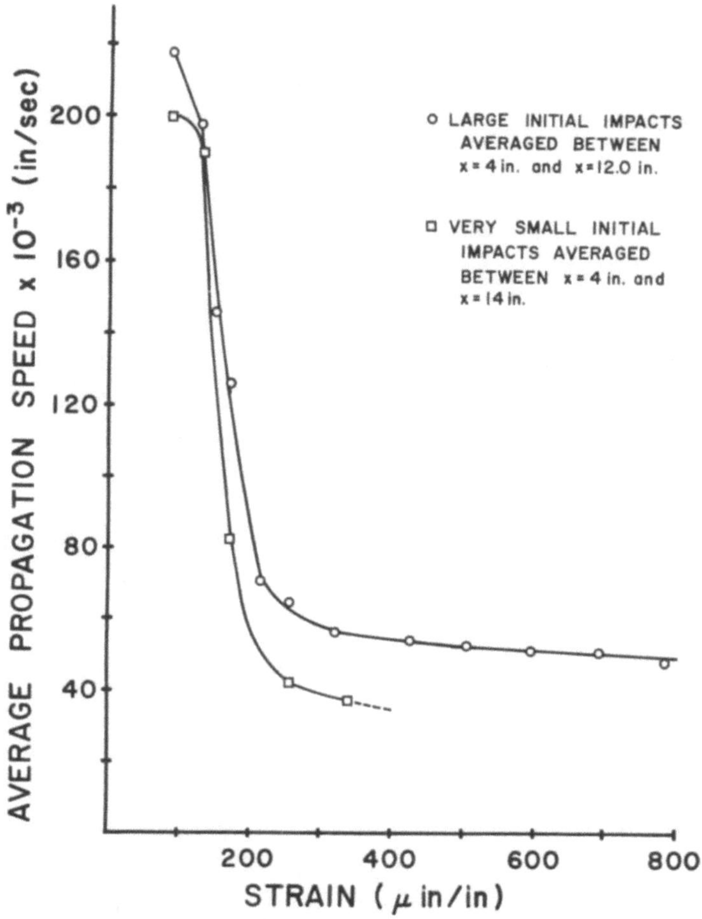

Fig. 16. Average speed of propagation between $x = 4$ in. and $x = 12$ (or $x = 14$ in.) for large (and small) initial impacts.

deformations in an untested structure of annealed aluminum. However, there is considerable evidence which points out that one must also recognize that such a prediction can be very much in error. More importantly one must be extremely cautious about looking too closely at detailed results of a single experiment and drawing far reaching conclusions. If one subjects an untested structure to a large impact loading, the experimental results will agree "reasonably" well with the predictions. This includes the reflections from the junction of a composite specimen as can be seen by comparing Fig. 9a with the theoretical result in a previous paper [5].

The strain-rate independent theory of plastic wave propagation as

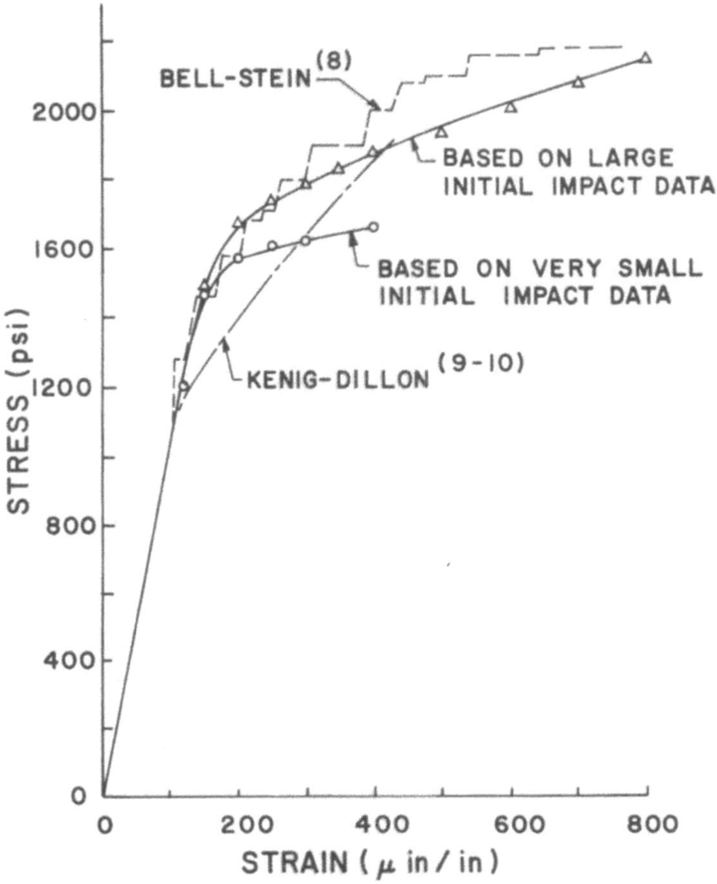

Fig. 17. A comparison of stress-strain relations in compression as obtained by two investigators [8, 9] in a static test and in the large and small impacts. The dynamic data is based on the strain-rate independent theory of wave propagation.

commonly used is known to be inadequate for problems involving *incremental loading*. Among the results which must be made consistent with its applicability to the large impacts are:

(1) The speed of propagation of incremental waves in a prestressed region being of the order of magnitude of the bar velocity [8].

(2) The stopping or sudden slowing down of waves as illustrated by Figs. 11 and 13.

(3) The existence [7] of slow waves such as illustrated in Fig. 18a where the specimen was prestressed to 4300 psi in torsion and subjected to an increment of 3.6 psi in stress. The wave speed is 0.45 in./sec. These slow waves are associated with the actual development of the plateaus such as exist in Fig. 1.

(4) A pulse of 20 milli-second duration can produce 800 μin./in. of plastic strain at one point in a specimen while a gauge located 1.5 in. away remains elastic. Figure 18b is an illustration [10] of this situation in which a prestressed specimen is subjected to an additional loading pulse of 20 milli-second duration. The prestressing is gradually increased by small amounts interlaced with numerous applications of the pulse type loading. Under these conditions one obtains elastic response data for most pulse applications, but occasionally results like Fig. 18b are observed.

(5) Static bending, torsion of a solid rod, tension, compression and

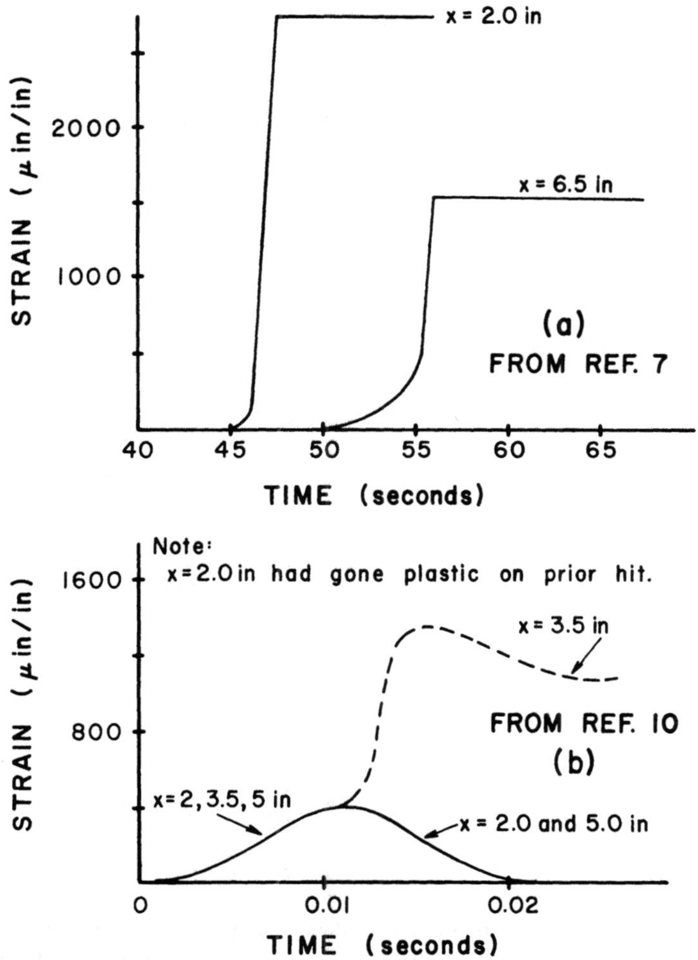

Fig. 18. A slow wave (a) which is simply one of the plateaus in Fig. 1 developing and (b) the response due to impacting in torsion with a pulse of 20 multi-sec duration.

biaxial states [9] of stress all produce load deflection curves such as Fig. 1 in the case of slow loading.

Thus there is a wide variety of data both static *and* dynamic which can be made consistent by the *concept that annealed aluminum is mechanically unstable.* That is, large increments of strain are produced by very small changes in stress—at certain discrete stresses. Furthermore, the

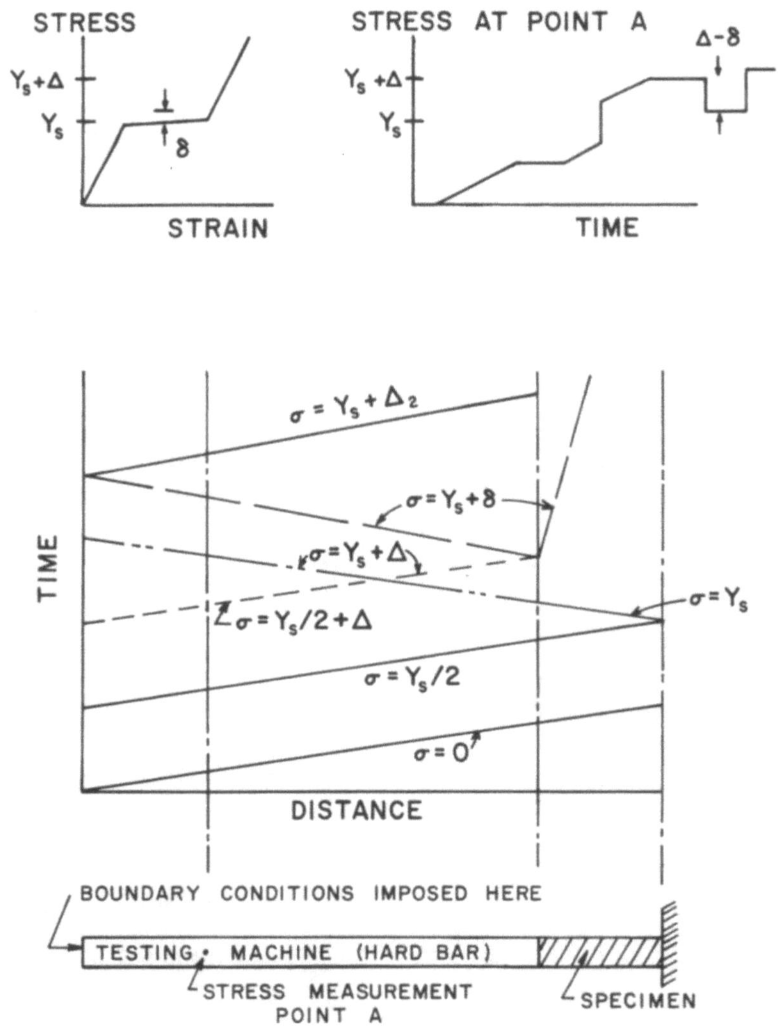

Fig. 19. A schematic solution of a boundary value problem of the Instron machine with an unstable specimen. The key point is that there is an "unloading" when the increment Δ arrives at the junction with the soft material.

results found in slow Instron tests can also be made consistent with these other results if one considers the distributed elasticity of the machine and views the problem as one in dynamics. I will rephrase one of the statements made above; if one assumes that a serrated constitutive relation applies to a given material, *a static test is almost impossible,* all tests are dynamic and involve wave propagation. An Instron type test is illustrated in Fig. 19 where the testing machine is regarded as a hard rod in contact with the soft specimen (clearly the specimen is soft when it is unstable). The load is applied until the stress is one half the yield stress of the specimen; upon reflection this doubles and the specimen reaches the yield point. A small time later an increment of load Δ is added which propagates through the machine to the interface. At the interface, *the specimen is virtually a free end* with respect to increments and there is an *unloading* in the hard bar and a slowly propagating wave in the specimen. The unloading reflects back to the crosshead position and something else happens. The unloading can be significant, even if the material isn't unstable, but in that case the whole thing happens so fast that one establishes homogeneous straining on the scale used to make the observations. However, when the material is unstable it takes a very long time from the instant when the load is applied until the specimen finally becomes homogeneously strained. It once took more than twenty minutes. One can see that one cannot approximate the results in Fig. 19 with a quasi-static analysis.

The stopping of the waves in the small hits (Figs. 11 and 13) and the slowing down in the large ones (Tables 1 and 1a) between $x = 10$ and $x = 12$ in. is believed to require additional consideration of the state of stress or the use of a more sophisticated material description.

There are *many metallurgical explanations* of this phenomenon, the problem is to sort out which mechanism actually is operative. The main difficulty in applying the concept of unstable materials is that it is *wonderful hindsight* but lacks predictive value until one can establish precisely when the instabilities will develop. Attempts in this direction have been made [11, 12], but the writer is not completely happy with them because equilibrium conditions are far from existing in the related experiments, and yet it is necessary to assume that they do to obtain useful data. On the other hand he doesn't have anything better to offer except the philosophy that the response is in fact a statistical one in the plastic range and the expected value of the wave speed ranges from practically zero to virtually infinity. Furthermore, despite the fact that I tend to prefer the strain-rate independent theory for annealed aluminum structures, I suspect that strain-rates *are* important parameters in the details of the plateaus and how *they* propagate. That is to say that some dislocation mechanisms are relevant to the plateaus and to the "structure" of the wave and these may very well be strain-rate sensitive. However, standard strain-rate models do not apply to annealed aluminum because certain discrete stresses have responses which are very different than their neighbors.

HEAT GENERATION

Experimental Method

In order to study the heat generated and its relation to the mechanical work done on the specimen as metals deform in the plastic range; it is necessary to produce a homogeneous deformation in which the stress, strain and temperature can be simultaneously measured. In previous papers [2, 3] we have given data on the heat generated as annealed aluminum and annealed copper are deformed. The experimental technique has also been discussed [13] for the case of coupled thermoelasticity. Basically we twist a tube in torsion, slow enough that we hope mechanical equilibrium conditions are approximately realized but fast enough that heat conduction is negligible. Of course neither situation is fully obtained but one hopes that the error committed is not important. In our system the torque that is being applied to the specimen is measured by a strain gauge on the load bar, strain is measured at two interior points by means of SR4-A-8 wire gauges and temperature is measured by a fine wire thermocouple. It is necessary to measure strain at two points because some specimens deform nonhomogeneously and these are not suitable for constitutive data. The thermocouple wires are held in contact with the specimen by standard electrical tape. The changing resistance of the strain gauges is made into a suitable oscilloscope signal by means of the Ellis Associates Model BAM-1 Bridge Amplifier. The torque is converted into average applied stress by means of strength of materials formula for thin walled tubes. The copper-constantan thermocouple output is changed into a measurable voltage* by means of an Astrodata Model 120 Nanovolt Amplifier with a gain of 50,000. The electrical outputs are observed on an oscilloscope with a four channel plug-in and recorded with an oscilloscope camera. The only modification of the basic method since our papers [3, 13] is the use of an additional oscilloscope in which one of the strains drives the horizontal plates and the stress and temperature are connected to the vertical ones. Thus we obtain the hysteresis

* One of our favorite tricks for visitors in the laboratory is to adjust electrical gains so that the temperature exceeds the strain on the oscilloscope. We then ask "how can this be so trivial?"

loop directly without the cumbersome procedure of eliminating the time parameter.

All specimens used in this heat generation section have a 0.25 in. outer diameter, a wall thickness of 0.031 in. and are 8.5 in. long. They are annealed the same way as the larger specimens used above. They are obtained from a different supplier and are made by a drawing process which yields a structure having larger grains than the 0.50 in. diameter tubes.

The specimen is initially centered in the torsion apparatus and then twisted between equal values of the angular displacement of one end. The other end is fixed against rotation but free to move axially. A typical experimental record is shown in Fig. 20. This particular record is for a specimen previously deformed for a few cycles. The top part of Fig. 20 contains the hysteresis loop and the temperature–strain relation. The initial condition in the fully annealed specimen is at the center of the top photograph.

In the bottom photo, the top trace is "stress" (actually torque), the middle two traces are the strains at two points while the lower trace is the temperature. The two strain traces differ because the gains of their electrical systems are unequal.

Experimental Data

In Fig. 20 one can observe that there are plateaus in the temperature history which occur when the specimen "unloads" indicating that the material is indeed elastic at these times. If cooling is important the temperature decreases during the times corresponding to the plateaus in Fig. 20. The most significant point is a near perfect balance between the work represented by the area of the hysteresis loops and the heat generated. That is to say that the plastic work per cycle is almost equal to heat generated per cycle. This is especially true after the third cycle from the fully annealed state. Typical results for a case where the total strain is 0.0100 in./in. (±0.0050 in./in.) are

$$\text{Plastic work} = 32.5 \text{ in.-lb/cu in.}$$

$$\text{Heat generated} = 30.6 \text{ in.-lb/cu in.}$$

Thus the heat generated is 94% of the work required to deform the specimen.

We have tested several specimens at amplitudes of ±0.0075 in./in. and others at ±0.0050 in./in. In group I we oscillate three cycles at ±0.0075 in./in., then three more cycles at 0.0050 in./in., back to ±0.0075 in./in. and then oscillate for about fifty cycles at ±0.0075

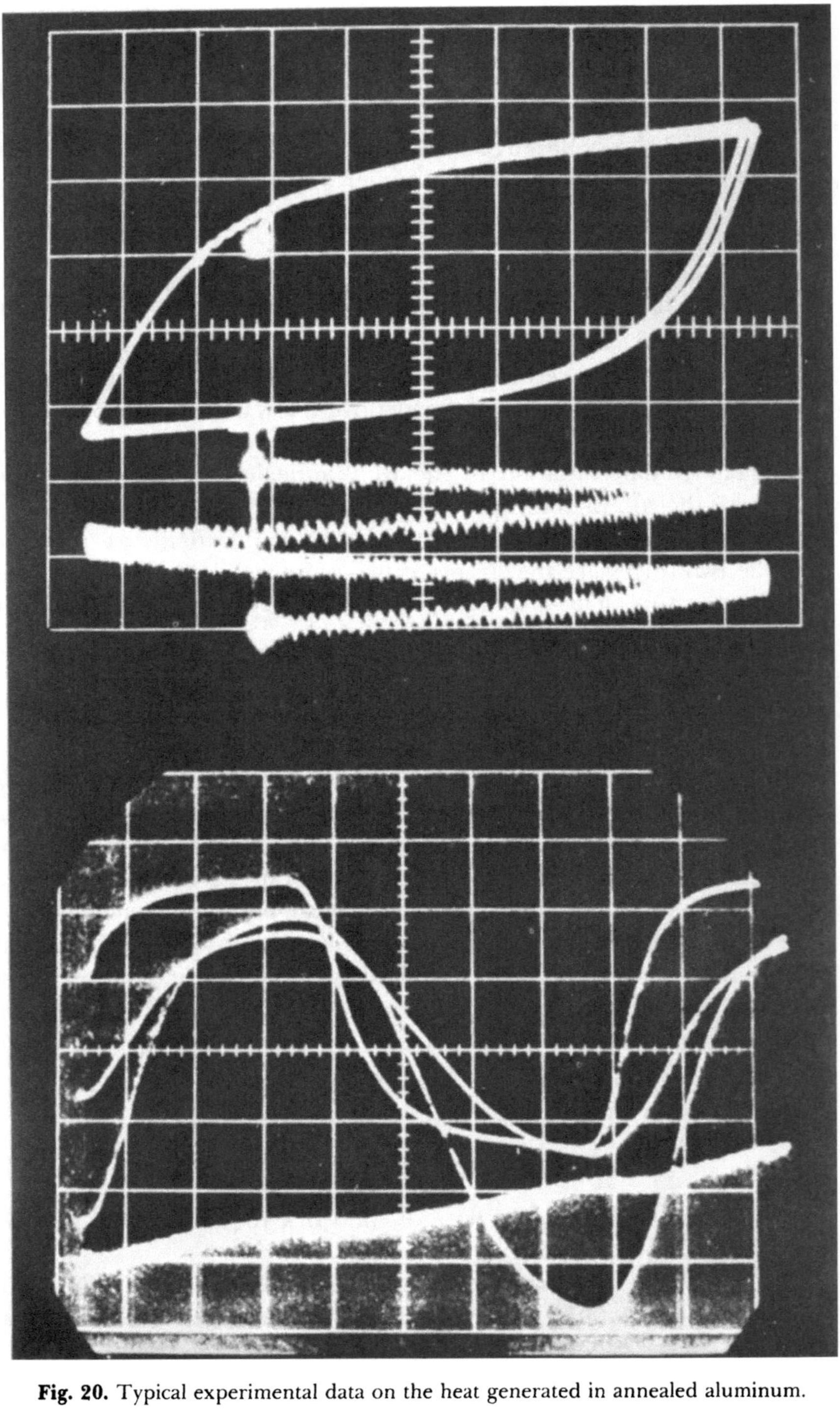

Fig. 20. Typical experimental data on the heat generated in annealed aluminum.

in./in. Group II is tested in the reverse order; i.e., three cycles at ±0.0050 in./in., then three more at 0.0075 in./in., back to 0.0050 in./in. and then finally fifty cycles at ±0.0075 in./in. The specimens are allowed to reach thermal equilibrium between each of the series listed above. The results are listed in Table 3. The program for Group III is the same as Group II except the fifty cycles are run at an amplitude of ±0.0050 in./in. and *then* the amplitude changed to ±0.0075 in./in.

Table 3

THE AVERAGE TEMPERATURE INCREASE PER CYCLE (°F) AFTER THE STRAINING PROGRAM DESCRIBED IN THE TEXT.

GROUP I	
Average temperature rise in the first cycle	0.102° F
Average temperature rise in the third cycle	0.129° F
Average temperature rise in the first cycle at ±0.0050 in./in.	0.072° F
Average temperature rise/cycle upon returning to ±0.0075 in./in.	0.133° F
Average temperature rise/cycle after 50 cycles at ±0.0075 in./in.	0.137° F
Average temperature rise/cycle at ±0.0050 in./in. after 50 cycles at ±0.0075 in./in.	0.083° F
GROUP II	
Average temperature rise in the first cycle	0.055° F
Average temperature rise in the third cycle	0.062° F
Average temperature rise in the first cycle at ±0.0075 in./in.	0.127° F
Average temperature rise/cycle upon returning to ±0.0050 in./in.	0.074° F
Average temperature rise/cycle after 50 cycles at ±0.0075 in./in.	0.132° F
GROUP III	
Average temperature rise/cycle after 50 cycles at ±0.0050 in./in. and *then* oscillating at ±0.0050	0.082° F
Average temperature rise/cycle after 50 cycles at ±0.0050 in./in. and *then* oscillating at ±0.0075	0.150° F

A study of the data in Table 3 results in the conclusion that the temperature rise (after the first two cycles) is related to the present amplitude rather than the previous one. That is there is a sort of independence of path for the heat generated per cycle for the range of strains used in this study. Except for the first two cycles from the fully annealed state, we could find no significant change in the percent of the mechanical work that appeared as heat. It is always between 95% and 100%. The magnitude of the experimental error is estimated to be five percent.

Coupled Thermoplasticity

The Coleman-Noll approach to the thermodynamics of deformable bodies has recently permitted [14] an *analytical* description of materials which are possible prototypes for plastically deforming solids. That is, we attempt to provide an answer to the question: "In a completely analytical world how does one describe plastically deforming solids?" We illustrate how their approach works for the case of a body B in which acts only a single shearing stress τ and where all quantities are functions of a single spatial coordinate, X, and time, t. A thermodynamic process is described by the deformation function (strain), the free energy per unit mass ϕ, the heat flux vector q, the hydrostatic stress σ, the temperature θ, and the heat supply per unit mass per unit time r. For simplicity we shall limit to deformations where the gradient of the deformation function is equivalent to the small strain tensor. Under the present restrictions, the energy equation and the second law are

$$\rho_0\dot{\phi} = \sigma\dot{v} + \tau\dot{e} - \rho_0\dot{\theta}s - \rho_0\dot{\theta}s - \partial q/\partial X + \rho_0 r$$

and

$$\rho_0\dot{s} + \partial(q/\theta)/\partial X + \rho_0 r/\theta \geqslant 0$$

where v is the dilatation, e is the shearing strain conjugate to τ, ρ_0 is the undeformed material density, and superposed dots indicate partial differentiation with respect to time.

Because plastic materials do not have a unique constitutive relation we do not impose that condition in their analytical definition. Rather the parameters which are functions of space and time are regarded as many-valued functions of the dilatation, temperature, shearing stress and shearing strain acting as intermediate variables. Thus the free energy is

$$\phi(X, t) = \phi(v, \theta, \tau, e)$$

and similarly for all the other parameters except the heat flux vector which contains the temperature gradient as well. The inclusion of τ in this relation distinguishes plastic substances from nonlinearly elastic ones. We define a *prototype plastic material* as a plastic material where the shearing stress is given by

$$\tau = \rho_0(\partial\phi/\partial e) + 2\mu\rho_0(\partial\phi/\partial\tau)$$

which is used largely for convenience. By considering certain special processes, the balance laws are trivially satisfied but the second law is not. For example when the temperature rates can change but the stresses and strains are held constant, the second law becomes

$$\theta^{-1}\rho_0(s + \partial\phi/\partial\theta)\dot{\theta} \geqslant 0.$$

Since $\dot{\theta}$ can be of either sign the term in parentheses must vanish and we have the general result that

$$s + \partial\phi/\partial\theta = 0.$$

Similarly one finds that

$$\sigma = \rho_0\partial\phi/\partial v.$$

These two results are introduced into the second law and combined with the constitutive relation for the prototype plastic material with the result that

$$(\partial\phi/\partial\tau)(\dot{e} - \dot{\tau}/2\mu) \geqslant 0.$$

Therefore, if $\partial\phi/\partial\tau$ is not zero on loading—and since it does not change value with the rates—the material must unload as a linearly elastic material. Thus we deduce the result that a specimen made of a material that does not involve history effects can load according to one stress-strain relation and *must* unload via another; this is the distinguishing feature of plasticity. Of course the constitutive relation for the prototype plastic material is obviously not the only one which will lead to the unique aspects of plasticity nor is it obviously applicable to any real material.

If one incorporates the results given above into the energy equation, one obtains

$$-\partial q/\partial x = \rho_0 C_D\dot{\theta} - \rho_0 r - \rho_0\theta(\partial^2\phi/\partial\theta\partial v)\dot{v} - \rho_0\theta(\partial\tau/\partial\theta)\dot{e}$$
$$- 2\mu(\partial^2\phi/\partial\tau\partial\theta)\dot{e} - (\partial^2\phi/\partial\theta\partial\tau)\dot{\tau} - 2\mu\rho_0\partial\phi/\partial\tau(\dot{e} - \dot{\tau}/2\mu)$$

where C_D is the specific heat at constant strain. In our experiments $\dot{v}$ is approximately zero and so are $\partial q/\partial x$ and r. Thus the energy equation reduces to

$$0 = \rho_0 C_D\dot{\theta} - 2\mu\rho_0(\partial\phi/\partial\tau)(\dot{e} - \dot{\tau}/2\mu) - \rho_0\theta(\partial\tau/\partial\theta)\dot{e}$$
$$- 2\mu(\partial^2\phi/\partial\tau\partial\theta)\dot{e} - (\partial^2\phi/\partial\theta\partial\tau)\dot{\tau}.$$

The experimental results for annealed aluminum given above for the heat generated can be made approximately consistent with this result by equating

$$2\mu\rho_0(\partial\phi/\partial\tau) = \tau$$

and

$$\partial\tau/\partial\theta = \partial^2\phi/\partial\tau\partial\theta = 0.$$

Finally we illustrate the general results with a specific example. Consider a material whose free energy is

$$\phi = \tau^2/2\mu\rho_0 - 4\beta e^{\frac{3}{2}}/3\rho_0 - C_D\theta^2/2.$$

Then the prototype constitutive relation reduces to

$$\tau = \beta e^{1/2}$$

and of course

$$\partial\phi/\partial\tau = \tau/2\mu\rho_0.$$

Hence in this example the second law becomes

$$\tau(\dot{e} - \dot{\tau}/2\mu) > 0$$

which is just the condition of positive plastic work. Hence the main features of plasticity theory are completely consistent with the prototype constitutive relation and annealed aluminum is a real material which is very closely approximated by the results. An essential feature of the unified picture thus obtained is the experimental data on the amount of heat generated in the deformation. It could not have been obtained if the experiment was approximated as an isothermal one even though the temperature changes only by 0.1° F. I find it also comforting to explicitly examine the conservation of energy principle and to be consistent with it. For static work it is unlikely there is any advantage to the coupling with the thermal field but for dynamic tests this may not be the case.

Rapid Deformation Data

We have recently been attempting to combine the two previous phases of our work by measuring the heat generated under the impact conditions. One soon finds that the heat generated in a uniform long specimen is appreciably smaller than expected. Presumably the heat is rapidly conducted to those sections of the specimen which are not strained as much as the impact end. For this reason we have changed to the composite specimens where the strain is more uniform and where the glue acts as a thermal insulator between the hard and soft sections. So far our technique does not lead to reproducible results of the type suitable for publication. However, we do measure reasonable temperature histories and hopefully can soon provide data on how fast plastic deformations become permanent. In the impact studies one must also include the change in volume due to the elastic part of the deformation rather than regarding the plastic work as the only source term.

In conclusion we have here implied the use of a smooth stress-strain relation in the heat generation part of the paper while in the first section we insisted that it was discontinuous. We also used load bar type data to obtain stress, a practice we criticized others for doing. Obviously our point is to use the best data that one has available for a given purpose. If adequate instruments and apparatus were available we would like to examine the hysteresis loops and temperature generated into more detail to see if the latter comes in bursts (at a point) as one might expect from Fig. 1. In our own view we have here done the best we could and we would reiterate comments in a recent paper by Pipkin [15] in which "second order approximations lead to inconsistencies more easily than first order approximations do." Clearly this view is at least as applicable to plasticity as it is to visco-elasticity which Pipkin was considering.

ACKNOWLEDGMENT

The author acknowledges the generous support of this research by the Air Force Office of Scientific Research by means of contract No. AF49(638)-1646 and the National Aeronautics and Space Administration Grant No. NGR-18-001-020. He also appreciates the help of D. Weaver, B. Taylor and M. Miller in conducting the experiments and Mrs. J. Murphy in preparing the manuscript.

References

1. O. W. Dillon, Jr., A Nonlinear Thermoelasticity Theory, J. Mech. Phys. Solids, *10,* 123 (1962).
2. O. W. Dillon, Jr., Coupled Thermoplasticity, J. Mech. Phys. Solids, *11*, 21 (1963).
3. O. W. Dillon, Jr., The Heat Generated During Torsional Oscillations of Copper Tubes, Intl. J. Solids Structures, *2,* 181 (1966).
4. G. I. Taylor and W. S. Farren, The Heat Developed During Plastic Extension of Metals, Proc. Roy. Soc., *A107,* 422 (1925).
5. O. W. Dillon, Jr., The Dynamic Elastic-Plastic Interface and Related Topics, submitted for publication (1967).

6. O. W. Dillon, Jr., Experimental Data on Elastic Plastic Deformation Waves in Annealed Aluminum, submitted for publication (1967).
7. O. W. Dillon, Jr., Waves in Bars of Mechanically Unstable Materials, J. Appl. Mech., *33*, 267 (1966).
8. J. F. Bell and A. Stein, The Incremental Loading Wave in the Pre-Stressed Field, J. de Mecanique, *1*, 395 (1962).
9. M. J. Kenig, Experiments on Annealed Aluminum, Ph.D. dissertation, Princeton University (1965).
10. M. J. Kenig and O. W. Dillon, Jr., Shock Waves Produced by Small Stress Increments in Annealed Aluminum, J. Appl. Mech., *33*, 907 (1966).
11. W. N. Sharpe, Jr., The Portevin-Le Chatelier Effect in Aluminum Single Crystals and Polycrystals, J. Mech. Phys. Solids, *14*, 187 (1966).
12. S. R. Bodner and A. Rosen, Discontinuous Yielding of Commercially-Pure Aluminum, J. Mech. Phys. Solids, *15*, 63 (1967).
13. O. W. Dillon, Jr., and T. R. Tauchert, The Experimental Technique for Observing the Temperatures Due to the Coupled Thermoelastic Effect, Int. J. Solids Structures, *2*, 385 (1966).
14. O. W. Dillon, Jr., A Thermodynamic Basis of Plasticity, Acta Mech. *3*, 2, 182 (1967).
15. A. C. Pipkin, Approximate Constitutive Equations, *Modern Developments in Mechanics of Continua,* Academic Press, 89 (1966).

ON THERMODYNAMIC FOUNDATIONS OF VISCOPLASTICITY

PIOTR PERZYNA

Institute of Basic Technical Research
Polish Academy of Sciences
Warsaw, Poland

ABSTRACT

The aim of the present paper is to discuss the thermodynamic approach to combined treatment of rheologic and plastic phenomena and to construct a thermodynamic theory of non-linear viscoplastic materials which may be used to describe the behavior of metals under dynamic loads.

In the first part of this paper the discussion is given of three different thermodynamic approaches to continuous media. It is shown that the thermodynamic foundations of viscoplasticity may be considered within the framework of the continuum mechanics of materials with memory. A non-linear material with memory is defined by a system of constitutive equations in which some state functions such as the stress tensor, the internal energy, the heat flux, etc., are determined as functionals of a function which represents the time history of the local configuration of a material particle.

As a result of simultaneous introduction of elastic, viscous and plastic properties of a material, a description of the actual state functions involves the history of the local configuration expressed as a function of the time and of the path.

The restrictions which impose the second law of thermodynamics and the principle of material objectivity have been analyzed.

In the second part of this paper some particular cases of constitutive equations are discussed. Among others, a viscoplastic material of the rate type and a strain-rate sensitive plastic material are examined.

INTRODUCTION

In agreement with J. Meixner's well founded opinion [1] three different approaches to a thermodynamic theory of continuum can be distinguished. These approaches differ from each other by the fundamental postulates, on which the theory is based. All of them are characterized by the same fundamental requirement that the results should be obtained without having recourse to statistical or kinetic theories. None of these approaches is concerned with the atomic structure of the material. Therefore, they represent a pure phenomenological approach.

The principal postulates of the first approach, initiated by Onsager's works and usually called the classical thermodynamics of irreversible processes, are as follows (cf. S. R. De Groot and P. Mazur [2]). 1) The principle of local state is assumed to be valid. 2) The Gibbs' relation is satisfied. 3) The equation of entropy balance is assumed to involve a term expressing the entropy production which can be represented as a sum of products of fluxes and forces. This term is zero for a state of equilibrium and positive for an irreversible process. 4) The fluxes are function of forces, not necessarily linear. However, the Onsager-Casmir reciprocity relations concern only coefficients of the linear terms of the series expansions. Using methods of this approach, a thermodynamic description of elastic, rheologic and plastic materials was obtained. Let us mention the works by M. A. Biot [3], D. C. Drucker [4], H. Ziegler [5–14], A. A. Vakulenko [15, 16], A. E. Green and J. E. Adkins [17], O. W. Dillon [18], G. A. Kluitenberg [19–25], S. L. Koh and A. C. Eringen [26], J. Kestin [27] and J. F. Besseling [28].

The second approach, called the thermodynamic theory of materials with memory, was initiated by the work of B. D. Coleman and W. Noll [29]. The fundamental postulates of this approach are as follows: 1) The temperature and entropy functions are assumed to exist for non-equilibrium states. 2) The principal restriction imposed on the constitutive equations is the Clausius-Duhem inequality. 3) The notion of the thermodynamic state is modified by assuming that the state of a given particle at time t is characterized, in general, by the time history of the local configuration of that particle. It should be emphasized, however, that in particular cases the history of the local configuration of a particle can be determined by giving the actual values of this configuration and its time derivatives [30]. 4) No limitations are introduced for the processes considered. The constitutive equations are in general nonlinear. Within the framework of this approach, thermodynamic foundations of rheologic materials were established [31–34]. The same was done for plastic materials also [35–37].

The third approach has been developed by J. Meixner [38–40] and is called the thermodynamic theory of passive systems. It is based on the following postulates: 1) The introduction of the notion of entropy is avoided for non-equilibrium states and the principle of local state is not assumed. 2) The Clausius-Duhem inequality is replaced by an inequality expressing the fundamental property of passivity. This inequality follows from the second law of thermodynamics and the condition of thermodynamic stability. Further the inequality is known to have sense only for states of equilibrium. 3) The temperature is assumed to exist for non-equilibrium states. 4) As a consequence of the fundamental inequality the class of processes under consideration is limited to processes in which deviations from the equilibrium conditions are small. This enables full linearization of the constitutive equations. An important feature of this approach is the clear physical interpretation of all the quantities introduced.

Each of the three approaches above has its weaknesses and none is commonly accepted.* The first is subject to excessive limitations in the form of the assumptions of the Onsager-Casmir relations. Its present development does not appear to be promising for the overcoming of the difficulties that are encountered in nonlinear mechanics.† The second approach is criticized principally from the point of view of physical foundations [1]. Indeed, we must agree with the opinion, that the problem of physical interpretation of quantities such as the temperature or entropy has not found a detailed treatment within the framework of this approach. The advantages of the first approach are the mathematical foundations which are very well developed and offer a possibility of analysis of many interesting processes. They can also be used for the description of nonlinear materials. It is also worth mentioning that the theories of elastic and viscoelastic materials can be obtained as particular cases of the theory of materials with memory [31, 32]. This theory enables the description of many important mechanical phenomena, such as elastic instability and phenomena accompanying wave propagation [41]. The applicability of the methods of the third approach is, on the other hand, limited to linear problems. It does not seem likely that further generalization to nonlinear problems is possible within the framework of the assumptions of this approach. The results obtained concern problems of linear viscoelasticity only [38–40, 43].

It is worth mentioning that recent works concerned with axiomatic foundations of continuum thermodynamics have shown in a clear manner the correctness of the conception of the second approach.‡ Although they have not removed the objections against the physical foundations of the theory, they have formulated in a mathematically accurate manner the applicability conditions of the methods of rational thermodynamics.

* This fact was pointed out by a detailed discussion at the IUTAM Symposium on Irreversible Aspects of Continuum Mechanics in Vienna, June 1966 [1, 41].

† A detailed critical analysis of this approach can be found in the monograph of C. Truesdell and R. Toupin [42].

‡ The axiomatic foundations of continuum thermodynamics have been presented in the papers by M. E. Gurtin and W. O. Williams [44, 45]. They are a generalization to thermodynamics problems of the earlier conceptions of W. Noll [46, 47] concerning the purely mechanical theory of a continuum.

The aim of the present paper is to discuss the thermodynamic foundations of the theory of viscoplasticity, the essential feature of which is the simultaneous description of rheologic and plastic effects of a material. The necessity of simultaneous consideration of viscoelastic and plastic properties of a material is indicated by the results of experimental investigations of dynamic loads. These investigations show clearly that during dynamic loading of a test piece the plastic and viscoelastic effects are coupled and play roles of equal importance.* The viscous properties of the material introduce a time dependence of the states of stress and strain. The plastic properties, on the other hand, make these states depend on the deformation path. Different results will be obtained for different deformation paths and for different time durations of the process.

It appears that by investigating thermodynamic processes in viscoplastic materials, characterized by a nonlinearity resulting from dependence on the time and the path, their description can be obtained within the assumptions of the second approach. We shall try to show that the methods of thermodynamics of materials with memory can be used for the establishment of the thermodynamic foundations of viscoplasticity.

It will be proposed to describe a viscoplastic material as a material with memory, for which the history of the local configuration depends on the time as well as the path. Viscoplastic materials of the rate type, for which the path-dependency is characterized by different sets of constitutive equations, for the loading and unloading process, will be discussed in more detail. As a particular case of an elastic-viscoplastic material of the rate type, a rate sensitive plastic material will be discussed.

Finite deformations of a body during the general thermodynamic process will be considered. All the constitutive equations obtained will be invariant under a change of the reference frame.

PRELIMINARY CONSIDERATIONS

In general we shall use a similar notation as in monograph of C. Truesdell and W. Noll [33]. The motion of the body B with points $\underset{\sim}{X}$ is described by the equation

$$\underset{\sim}{x} = \underset{\sim}{\chi}(\underset{\sim}{X}, t) \tag{1}$$

where x denotes the spatial position occupied by the material point X at time t.† The gradient F of χ with respect to X, i.e.,

* A discussion of the results of experimental investigations in the domain of dynamic loads acting on metals can be found in the paper [48].

† We identify the material point X with its position $\underset{\sim}{X}$ in the reference configuration.

$$\underset{\sim}{F} = \partial\underset{\sim}{\chi}(\underset{\sim}{X}, t)/\partial\underset{\sim}{X} \tag{2}$$

is the deformation gradient at the points $\underset{\sim}{X}$ relative to the reference configuration. It is assumed that $\det\underset{\sim}{F} \neq 0$. By the polar decomposition of $\underset{\sim}{F}$, viz.,

$$\underset{\sim}{F} = \underset{\sim}{R}\underset{\sim}{U} \tag{3}$$

we define the orthogonal rotational tensor $\underset{\sim}{R}$ and positive definite and symmetric right stretch tensor $\underset{\sim}{U}$. Similarly, the relation

$$\underset{\sim}{C} = \underset{\sim}{F}^T\underset{\sim}{F} = \underset{\sim}{U}^2 \tag{4}$$

where $\underset{\sim}{F}^T$ denotes the transpose of $\underset{\sim}{F}$, defines the right Cauchy-Green tensor.

If $\underset{\sim}{F}$ is replaced by the relative deformation gradient $\underset{\sim}{F}_{(t)}$, then the notations $\underset{\sim}{R}_{(t)}$, $\underset{\sim}{U}_{(t)}$ and $\underset{\sim}{C}_{(t)}$ are used respectively for the corresponding relative rotation tensor, relative stretch tensor, and relative Cauchy-Green tensor. The expressions $\underset{\sim}{W}(t) = \underset{\sim}{R}_{(t)}(t)$ and $\underset{\sim}{D}(t) = \dot{\underset{\sim}{U}}_{(t)}(t)$ define the spin and the stretching tensor, respectively.

Let us denote the Cauchy stress tensor by $\underset{\sim}{T}(t)$. We shall introduce, after W. Noll [46], the following co-rotational stress rate:

$$\overset{\circ}{\underset{\sim}{T}} = \dot{\underset{\sim}{T}} - \underset{\sim}{W}\underset{\sim}{T} + \underset{\sim}{T}\underset{\sim}{W} \tag{5}$$

where $\underset{\sim}{W}$ denotes spin.

A thermodynamic process in B is described by eight functions $\{\underset{\sim}{\chi}, \underset{\sim}{T}, b, \epsilon, \eta, \vartheta, \underset{\sim}{q}, r\}$ of $\underset{\sim}{X}$ and time t. The values of these functions have the following physical interpretation [29]. The function $\underset{\sim}{\chi}(\underset{\sim}{X}, t)$ describes the motion of the body B and is called the deformation function: $\underset{\sim}{T}(\underset{\sim}{X}, t)$ is the symmetric stress tensor; $\underset{\sim}{b}(\underset{\sim}{X}, t)$ is the body force per unit mass; $\epsilon(\underset{\sim}{X}, t)$ denotes specific internal energy per unit mass; $\eta(\underset{\sim}{X}, t)$ the specific entropy and $\vartheta(\underset{\sim}{X}, t)$ is the local absolute temperature; $\underset{\sim}{q}(\underset{\sim}{X}, t)$ is the heat flux vector and $r(\underset{\sim}{X}, t)$ the heat supply per unit mass and unit time.

The set of eight functions $\{\underset{\sim}{\chi}, \underset{\sim}{T}, \underset{\sim}{b}, \epsilon, \eta, \vartheta, \underset{\sim}{q}, r\}$ is called a thermodynamic process if, and only if, it is compatible with the condition for the balance of linear momentum (Cauchy's first law of motion) *

$$\operatorname{div}\underset{\sim}{T} - \rho\ddot{\underset{\sim}{x}} = -\rho\underset{\sim}{b} \tag{6}$$

and the law of balance of energy (the first law of thermodynamics)

$$tr\{\underset{\sim}{T}\underset{\sim}{L}\} - \operatorname{div}\underset{\sim}{q} - \rho\dot{\epsilon} = -\rho r \tag{7}$$

where ρ denotes the mass density, $\underset{\sim}{L} = \partial\dot{\underset{\sim}{x}}/\partial\underset{\sim}{x}$ and the trace operator is denoted by tr.

* Cauchy's second law of motion requires $\underset{\sim}{T} = \underset{\sim}{T}^T$, which we assumed in advance. Couple stresses and body couples are assumed to be absent.

In order to define a thermodynamic process, it suffices to prescribe the six functions $\{\chi, \underset{\sim}{T}, \epsilon, \eta, \vartheta, \underset{\sim}{q}\}$. The two remaining functions $\underset{\sim}{b}$ and r are then uniquely determined by (6) and (7).

Two thermodynamic processes $\{\chi, \underset{\sim}{T}, \epsilon, \eta, \vartheta, \underset{\sim}{q}\}$ and $\{\chi^*, \underset{\sim}{T}^*, \epsilon^*, \eta^*, \vartheta^*, \underset{\sim}{q}^*\}$ are equivalent if they are related only by a change of a reference frame.

We assume that all constitutive equations describing the physical properties of the material satisfy the following principle of material frame-indifference [46, 33]: If process $\{\chi, \underset{\sim}{T}, \epsilon, \eta, \vartheta, \underset{\sim}{q}\}$ is compatible with a system of constitutive equations, then all processes $\{\chi^*, \underset{\sim}{T}^*, \epsilon^*, \eta^*, \vartheta^*, \underset{\sim}{q}^*\}$ equivalent to it must be compatible with the same system of constitutive equations.

A thermodynamic process in B, compatible with the constitutive equations at each point $\underset{\sim}{X}$ of B and for all time t, will be called an admissible process in B.

Thus, the principle of material frame-indifference states, that if a thermodynamic process $\{\chi, \underset{\sim}{T}, \epsilon, \eta, \vartheta, \underset{\sim}{q}\}$ is admissible in B, then also all thermodynamic processes $\{\chi^*, \underset{\sim}{T}^*, \epsilon^*, \eta^*, \vartheta^*, \underset{\sim}{q}^*\}$ must be admissible in B. The physical meaning of this principle is simply that the material properties of a body should not depend on the observer, irrespective of how he moves.

We assume the following postulate: For every admissible thermodynamic process in a body B the production of entropy must be non-negative.

Under suitable smoothness assumptions we can write

$$-\dot{\psi} + \frac{1}{\rho}\, tr\{\underset{\sim}{F}^{-1}\underset{\sim}{T}\dot{\underset{\sim}{F}}\} - \eta\dot{\vartheta} - \frac{1}{\rho\vartheta}\, \underset{\sim}{q} \cdot \operatorname{grad} \vartheta \geqq 0 \tag{8}$$

where $\psi = \epsilon - \vartheta\eta$ is the specific free energy function. This is a local form of the Clausius-Duhem inequality which is a mathematical statement of the second law of thermodynamics.*

TIME AND PATH DEPENDENT MATERIALS

The non-linear material with memory is defined by a system of constitutive equations [31–33, 49]

$$\Pi(t) = \mathop{\mathscr{F}}_{\tau=-\infty}^{t} (\Psi(\tau)) \tag{9}$$

in which $\Pi(t)$ represents the actual values of some state functions such as the stress tensor $\underset{\sim}{T}$, the specific free energy ψ, the heat flux $\underset{\sim}{q}$ and the

* For the discussion of the conditions under which a local form of the Clausius-Duhem inequality [8] is valid see M. E. Gurtin and W. O. Williams [45].

specific entropy $\hat{\eta}$, is a function which represents the history of the local configuration of a material. It is assumed that the history of the local configuration $\Psi(\tau)$ can be characterized by the history of the local deformation gradient $\underset{\sim}{F}(\tau)$ with $\tau \in (-\infty, t]$; the history of the local temperature $\vartheta(\tau)$ with $\tau \in (-\infty, t]$, and the local temperature gradient, grad ϑ.* Thus we have

$$\Pi(t) = \{\underset{\sim}{T}(t), \psi(t), \underset{\sim}{q}(t), \eta(t)\} \tag{10}$$

$$\Psi(t) = \{\underset{\sim}{F}(\tau), \vartheta(\tau), \text{grad}\,\vartheta(t)\}, \tau \in (-\infty, t]. \tag{11}$$

The response of the material is characterized by the functional $\mathscr{F}$, called the constitutive functional. The functional $\mathscr{F}$ must satisfy an invariance requirement relative to a change of the observer, some requirements of the symmetry and special smoothness requirements [31–32].

We shall now concern ourselves with simultaneous description of the elastic, rheologic and plastic properties of a material. To this end, let us accurately define the meaning of the plastic behavior of a material. Unloading is an important feature which distinguishes the behavior of a plastic material from that of a nonlinear material with memory. We shall introduce the distinction between unloading, neutral state, and loading phenomena.

In inviscid plasticity, it is assumed that the material deforms elastically until the state of what is called an initial yield surface or the loading surface is reached. Unloading, neutral state and loading in this theory are defined as follows [35]. An unloading process has taken place if the deformation from an existing elastic-plastic state takes place elastically so that the stress point lies inside the yield surface. Neutral state has taken place if no additional plastic strain is produced when the stress point lies on the yield surface. Similarly a loading process has taken place, if due to additional deformation the stress point reaches a subsequent yield surface.

The determination of the yield condition for a nonlinear material with memory at finite deformations is very difficult and has not yet been achieved. Thus, in the formulation of the general constitutive equations of an elastic-viscoplastic material, we do not use the yield condition. We shall assume that the material of a body B will show plastic effects from the initial configuration $\underset{\sim}{\chi}_{t_0}$. This initial configuration $\underset{\sim}{\chi}_{t_0}$ will be called the configuration of yielding.

Thus, as a result of simultaneous consideration of rheologic and plastic properties of a material a description of the actual state functions involves the history of the local configuration expressed as a func-

* Different assumptions have been discussed by M. E. Gurtin [50].

tion of time and of path. To make this idea clear, let us introduce the definition of path in the ten-dimensional space of deformation and temperature as follows *

$$s(\tau) = \int_{-\infty}^{\tau} [tr(\dot{\underset{\sim}{F}}\dot{\underset{\sim}{F}}^T) + \dot{\vartheta}^2]^{1/2}\, d\tau'. \tag{12}$$

We shall introduce the history of the local configuration $\Xi(\tau, s)$ in the following form

$$\Xi(\tau, s) = \{\Psi(\tau), s\},\ \tau \in (-\infty, t],\ s \in [0, s(t)]. \tag{13}$$

The constitutive equations for time and path dependent material can be defined by the system of the form

$$\Pi(t) = g(\Xi(\tau, s)),\ \tau \in (-\infty, t],\ s \in [0, s(t)]. \tag{14}$$

ELASTIC-VISCOPLASTIC MATERIAL OF THE RATE TYPE

Unloading from an elastic-viscoplastic state follows a path in the deformation and temperature space different from that of loading. We assume the following definition.

A thermodynamic process represents unloading if the condition

$$\frac{1}{\rho} tr\{\underset{\sim}{T}\underset{\sim}{L}\} - \eta\dot{\vartheta} < 0 \tag{15}$$

is satisfied, i.e., if the rate of work of the generalized stress is negative. The case

$$\frac{1}{\rho} tr\{\underset{\sim}{T}\underset{\sim}{L}\} - \eta\dot{\vartheta} = 0 \tag{16}$$

is called neutral state, whereas the positive rate of work of the generalized stress

$$\frac{1}{\rho} tr\{\underset{\sim}{T}\underset{\sim}{L}\} - \eta\dot{\vartheta} > 0 \tag{17}$$

determines the loading process.†

Let us introduce the following definition.

A non-linear material has plastic properties if its behavior is described by different constitutive equations for loading and unloading.

This formal definition shows that the loading and unloading phe-

* This idea is similar to that of A. C. Pipkin and R. S. Rivlin [51].

† During the isothermal process, the criteria of unloading, neutral state and loading take the respective forms $tr\{\underset{\sim}{T}\underset{\sim}{L}\} \lesseqgtr 0$, i.e., they coincide with those first introduced by A. E. Green [52].

nomena for the same material are characterized by different features.

We shall not bother to introduce the distinction between elastic, viscous and plastic deformations. Total deformation will be treated as a combined result of elastic, rheologic and plastic effects.

We now intend to obtain, on the basis of the thermodynamic theory of the rate type material, the general constitutive equations of an elastic-viscoplastic material.* We shall use only the first order differential equations. This, of course, implies some restrictions concerning the memory of a material. We assume the following.

An elastic-viscoplastic material is a simple material of the rate type of the first order which is characterized by different properties during the loading and unloading processes.

Thus, in the thermodynamic process in a body B, which represents loading, i.e., the condition (17) is satisfied, we postulate for an elastic-viscoplastic material the following system of the constitutive equations

$$\dot{\Pi}(t) = \bar{f}(\Pi(t), \Psi(t), \dot{\Psi}(t)) \tag{18}$$

with initial values

$$\Pi(t_0) = \{\psi(t_0), \eta(t_0), \underset{\sim}{T}(t_0), \underset{\sim}{q}(t_0)\}. \tag{19}$$

For the thermodynamic process in a body B, which is unloading, i.e., satisfying the condition (15), we assume the following system of the constitutive equations

$$\dot{\Pi}(t) = \hat{f}(\Pi(t), \Psi(t), \dot{\Psi}(t)). \tag{20}$$

The response functions $\bar{f}$ and $\hat{f}$ must satisfy the following condition for a neutral state

$$\bar{f} = \hat{f} \text{ if } \frac{1}{\rho}\, tr\{\underset{\sim}{T}\underset{\sim}{L}\} - \eta\dot{\vartheta} = 0. \tag{21}$$

This condition is the continuity condition for the first derivatives $\dot{\psi}$, $\dot{\eta}$, $\dot{\underset{\sim}{T}}$ and $\dot{\underset{\sim}{q}}$. Additionally, we have to assume the continuity condition for the functions ψ, η, $\underset{\sim}{T}$ and $\underset{\sim}{q}$. This assumption determines the initial values for the system (20).

It is worth noting, that the path dependence in the constitutive equations of an elastic-viscoplastic material of the rate type is expressed by the following fact. The properties of this material are described by a different system of equations for the loading and for the unloading path.

After satisfying the principle of material frame-indifference, we can write the systems of constitutive equations (18) and (20) respectively in the following reduced and explicit form

* Cf. the previous papers of the author [53, 54].

$$
\begin{aligned}
\dot{\psi} &= \bar{p}(\psi, \underset{\sim}{D}^*, \underset{\sim}{U}, \vartheta, \dot{\vartheta}, \underset{\sim}{R}^T \operatorname{grad} \vartheta) \\
\dot{\eta} &= \bar{h}(\eta, \underset{\sim}{D}^*, \underset{\sim}{U}, \vartheta, \dot{\vartheta}, \underset{\sim}{R}^T \operatorname{grad} \vartheta) \\
\overset{\circ}{\underset{\sim}{T}}{}^* &= \bar{T}(\underset{\sim}{T}^*, \underset{\sim}{D}^*, \underset{\sim}{U}, \vartheta, \dot{\vartheta}, \underset{\sim}{R}^T \operatorname{grad} \vartheta) \\
\underset{\sim}{R}^T \overset{\circ}{\underset{\sim}{q}} &= \bar{q}(\underset{\sim}{R}^T \underset{\sim}{q}, \underset{\sim}{D}^*, \underset{\sim}{U}, \vartheta, \dot{\vartheta}, \underset{\sim}{R}^T \operatorname{grad} \vartheta)
\end{aligned}
\tag{22}
$$

and

$$
\begin{aligned}
\dot{\psi} &= \hat{p}(\psi, \underset{\sim}{D}^*, \underset{\sim}{U}, \vartheta, \dot{\vartheta}, \underset{\sim}{R}^T \operatorname{grad} \vartheta) \\
\dot{\eta} &= \hat{h}(\eta, \underset{\sim}{D}^*, \underset{\sim}{U}, \vartheta, \dot{\vartheta}, \underset{\sim}{R}^T \operatorname{grad} \vartheta) \\
\overset{\circ}{\underset{\sim}{T}}{}^* &= \hat{T}(\underset{\sim}{T}^*, \underset{\sim}{D}^*, \underset{\sim}{U}, \vartheta, \dot{\vartheta}, \underset{\sim}{R}^T \operatorname{grad} \vartheta) \\
\underset{\sim}{R}^T \overset{\circ}{\underset{\sim}{q}} &= \hat{q}(\underset{\sim}{R}^T \underset{\sim}{q}, \underset{\sim}{D}^*, \underset{\sim}{U}, \vartheta, \dot{\vartheta}, \underset{\sim}{R}^T \operatorname{grad} \vartheta)
\end{aligned}
\tag{23}
$$

where $\underset{\sim}{T}^* = \underset{\sim}{R}^T \underset{\sim}{T} \underset{\sim}{R}$, $\overset{\circ}{\underset{\sim}{T}}{}^* = \underset{\sim}{R}^T \overset{\circ}{\underset{\sim}{T}} \underset{\sim}{R}$ and $\underset{\sim}{D}^* = \underset{\sim}{R}^T \underset{\sim}{D} \underset{\sim}{R}$.

We assume that the constitutive equations of an elastic-viscoplastic material in both the loading and the unloading ranges satisfy the thermodynamic postulate. By (8) we have the following inequalities:

$$
-\bar{p}(\psi, \underset{\sim}{D}^*, \underset{\sim}{U}, \vartheta, \dot{\vartheta}, \underset{\sim}{R}^T \operatorname{grad} \vartheta) + \frac{1}{\rho} tr\{\underset{\sim}{T}\underset{\sim}{L}\} - \eta\dot{\vartheta} - \frac{1}{\rho\vartheta} \underset{\sim}{q} \cdot \operatorname{grad} \vartheta \geqq 0 \tag{24}
$$

for

$$
\frac{1}{\rho} tr\{\underset{\sim}{T}\underset{\sim}{L}\} - \eta\dot{\vartheta} > 0 \tag{25}
$$

and

$$
-\hat{p}(\psi, \underset{\sim}{D}^*, \underset{\sim}{U}, \vartheta, \dot{\vartheta}, \underset{\sim}{R}^T \operatorname{grad} \vartheta) + \frac{1}{\rho} tr\{\underset{\sim}{T}\underset{\sim}{L}\} - \eta\dot{\vartheta} - \frac{1}{\rho\vartheta} \underset{\sim}{q} \cdot \operatorname{grad} \vartheta \geqq 0 \tag{26}
$$

for

$$
\frac{1}{\rho} tr\{\underset{\sim}{T}\underset{\sim}{L}\} - \eta\dot{\vartheta} < 0. \tag{27}
$$

The inequalities (24–27) represent the basic restrictions imposed on the constitutive equations for an elastic-viscoplastic material.

RATE SENSITIVE PLASTIC MATERIAL

Recent theoretical and experimental research in the domain of the dynamical properties of materials has shown the significant sensitivity of some materials to the rate of deformation. This effect is disregarded in the inviscid theory of plasticity. The influence of strain rate may, however, be taken into account, within the framework of assumptions of an elastic-viscoplastic material of rate type.

Every material displays more or less definite viscous properties. For many materials, however, these properties are more pronounced after the plastic state has been reached. In these cases it may be assumed that material displays viscous properties in the plastic range only.

General foundations for the study of problems connected with rate sensitive plastic material were given by K. Hohenemser and W. Prager [55]. Further development of this idea is contained in the papers [48, 56–61].

The basic assumption in the theory of rate sensitive plastic materials is the additivity of the elastic and inelastic parts of the rate of deformation tensor

$$\underset{\sim}{D} = {}^{e}\underset{\sim}{D} + {}^{i}\underset{\sim}{D} \tag{28}$$

where ${}^{e}\underset{\sim}{D}$ and ${}^{i}\underset{\sim}{D}$ are the elastic and inelastic parts of the stretching tensor, respectively.

To obtain the elastic response from rate type material we should assume that the constitutive equation $(22)_3$ is invariant under a change of time-scale and is independent on the stretch tensor $\underset{\sim}{U}$. Thus, the constitutive equation of a rate type for elastic response has form similar to that of hypoelastic material

$$\overset{\circ}{\underset{\sim}{T}} = \underset{\sim}{H}_1(\underset{\sim}{T}^*, \vartheta)[{}^{e}\underset{\sim}{D}] + \underset{\sim}{H}_2(\underset{\sim}{T}^*, \vartheta)\dot{\vartheta}. \tag{29}$$

After substituting (28) into this equation we have the following result

$$\overset{\circ}{\underset{\sim}{T}} = \underset{\sim}{H}_1(\underset{\sim}{T}^*, \vartheta)[\underset{\sim}{D} - {}^{i}\underset{\sim}{D}] + \underset{\sim}{H}_2(\underset{\sim}{T}^*, \vartheta)\dot{\vartheta}. \tag{30}$$

Since the material has no viscous properties in the elastic region, the choice of an adequate yield criterion will be much simpler than in the case of an elastic-viscoplastic material. The initial yield condition, which will be called the static yield criterion, will not differ from the known condition of the inviscid theory of plasticity at finite strains.

In order to keep our considerations sufficiently general, we now introduce a static yield function in the form

$$\mathscr{F}(\underset{\sim}{T}^*, {}^{i}\underset{\sim}{E}, \vartheta) = \frac{f(\underset{\sim}{T}^*, {}^{i}\underset{\sim}{E}, \vartheta)}{\kappa(\underset{\sim}{T}^*, {}^{i}\underset{\sim}{E}, \vartheta)} - 1 \tag{31}$$

where the function $f(\underset{\sim}{T}^*, {}^{i}\underset{\sim}{E}, \vartheta)$ depends on the state of stress $\underset{\sim}{T}$ and the state of inelastic strain ${}^{i}\underset{\sim}{E}$ and temperature ϑ. The strain tensor $\underset{\sim}{E}$ is defined by the relation $2\underset{\sim}{E} = \underset{\sim}{C} - 1$, and it is assumed that $\underset{\sim}{E} = {}^{e}\underset{\sim}{E} + {}^{i}\underset{\sim}{E}$. The work-hardening parameter κ is defined by the expression [35]

$$\dot{\kappa} = tr\{\underset{\sim}{N}(\underset{\sim}{T}^*, {}^{i}\underset{\sim}{E}, \vartheta){}^{i}\underset{\sim}{D}^*\} \tag{32}$$

where $\underset{\sim}{N}$ is a tensor function. The flow surface, $\mathscr{F} = 0$, in the ten-dimensional stress and temperature space is assumed regular and convex.

We propose for the inelastic part of the rate of deformation tensor the equation

$$ {}^{i}\underset{\sim}{D} = \gamma(\vartheta)\langle\Phi(\mathscr{F})\rangle \underset{\sim}{M}(\underset{\sim}{T}^*, \vartheta, {}^{i}\underset{\sim}{E}) \tag{33} $$

where the tensor function $\underset{\sim}{M}$ satisfies the relation

$$ \underset{\sim}{M} = \underset{\sim}{M}^T \tag{34} $$

$\gamma(\vartheta)$ denotes a viscosity coefficient and the symbol $\langle\Phi(\mathscr{F})\rangle$ is defined as follows:

$$ \langle\Phi(\mathscr{F})\rangle = \begin{cases} 0 \text{ for } \mathscr{F} \leqq 0 \\ \Phi(F) \text{ for } \mathscr{F} > 0. \end{cases} \tag{35} $$

The function $\Phi(\mathscr{F})$ may be chosen to represent the results of tests on the behavior of metals under dynamic loading. The proper choice of $\Phi(\mathscr{F})$ at the same time enables a description of the influence of the rate of deformation and the temperature on the yield limit of the material.

By (30) and (33) we have

$$ \overset{\circ}{\underset{\sim}{T}} = \underset{\sim}{H}_1(\underset{\sim}{T}^*, \vartheta)[\underset{\sim}{D} - \gamma(\vartheta)\langle\Phi(\mathscr{F})\rangle \underset{\sim}{M}(\underset{\sim}{T}^*, \vartheta, {}^{i}\underset{\sim}{E})] + \underset{\sim}{H}_2(\underset{\sim}{T}^*, \vartheta)\dot{\vartheta}. \tag{36} $$

This constitutive equation involves the assumption that the inelastic part of the rate of deformation tensor is a function of excess stresses above the static yield criterion. This function of stress above the static yield criterion generates the inelastic rate of deformation tensor according to a viscosity law of the Maxwell type.

It can easily be seen that the constitutive equation leads to the following dynamic yield condition

$$ f(\underset{\sim}{T}^*, {}^{i}\underset{\sim}{E}, \vartheta) = \kappa(\underset{\sim}{T}^*, {}^{i}\underset{\sim}{E}, \vartheta)\left\{1 + \Phi^{-1}\left[\frac{(tr\,{}^{i}\underset{\sim}{D}^2)^{1/2}}{\gamma(\vartheta)}(tr\underset{\sim}{M}^2)^{1/2}\right]\right\}. \tag{37} $$

This relation determines the change in the actual yield surface during the inelastic deformation process. This change is caused by isotropic and anisotropic work-hardening effects and by influence of the rate of deformation tensor and temperature on the yield point of the material.

The full system of constitutive equations for a rate sensitive plastic material has been postulated in the following form *

$$ \begin{aligned} \dot{\psi} &= p_1(\psi, \vartheta)tr(\underset{\sim}{D} - {}^{i}\underset{\sim}{D}) + p_2(\psi, \vartheta)\dot{\vartheta} \\ \dot{\eta} &= h_1(\eta, \vartheta)tr(\underset{\sim}{D} - {}^{i}\underset{\sim}{D}) + h_2(\eta, \vartheta)\dot{\vartheta} \\ \overset{\circ}{\underset{\sim}{T}} &= \underset{\sim}{H}_1(\underset{\sim}{T}^*, \vartheta)[\underset{\sim}{D} - {}^{i}\underset{\sim}{D}] + \underset{\sim}{H}_2(\underset{\sim}{T}^*, \vartheta)\dot{\vartheta} \\ \underset{\sim}{R}^T\overset{\circ}{\underset{\sim}{q}} &= \underset{\sim}{q}_1(\underset{\sim}{R}^T\underset{\sim}{q}, \vartheta, R^T \operatorname{grad}\vartheta)(\underset{\sim}{D} - {}^{i}\underset{\sim}{D}) + \underset{\sim}{q}_2(\underset{\sim}{R}^T\underset{\sim}{q}, \vartheta, \underset{\sim}{R}^T \operatorname{grad}\vartheta)\dot{\vartheta} \\ {}^{i}\underset{\sim}{D} &= \gamma(\vartheta)\langle\Phi(F)\rangle \underset{\sim}{M}(\underset{\sim}{T}^*, \vartheta, {}^{i}\underset{\sim}{E}) \end{aligned} \tag{38} $$

* A different approach for establishing the constitutive equations for a rate sensitive plastic material has been recently presented in the paper [61].

with initial values

$$\psi(t_0),\ \eta(t_0),\ \underset{\sim}{T}(t_0),\ \underset{\sim}{q}(t_0). \tag{39}$$

The constitutive equations must satisfy the thermodynamic postulate

$$-p_1(\psi,\vartheta)tr\{\underset{\sim}{D}-\gamma(\vartheta)\langle\Phi(F)\rangle\underset{\sim}{M}(\underset{\sim}{T}^*,\vartheta,{}^i\underset{\sim}{E})\}$$

$$-p_2(\psi,\vartheta)\dot{\vartheta}+\frac{1}{\rho}tr\{\underset{\sim}{T}\underset{\sim}{L}\}-\eta\dot{\vartheta}-\frac{1}{\rho\vartheta}q\cdot\operatorname{grad}\vartheta\geqq 0. \tag{40}$$

After assuming the infinitesimal deformations [42] and perfectly plastic material with Huber-Mises initial yield condition we obtain from $(38)_3$ and $(38)_5$ the constitutive equations for a temperature and rate sensitive plastic material which were discussed in the paper [59]. In that paper a detailed analysis of some particular cases of the constitutive equations and a comparison of theoretical and experimental results for metals can be found. Reference [59] also presents a complete discussion of the problem of an appropriate selection of the temperature dependent coefficients (cf. also the review paper [48]).

References

1. J. Meixner, TIP has many faces, IUTAM Symposium on Irreversible Aspects of Continuum Mechanics, Vienna, Austria, June 22–25 (1966).
2. S. R. De Groot and P. Mazur, *Non-Equilibrium Thermodynamics,* North-Holland, Amsterdam (1962).
3. M. A. Biot, Linear thermodynamics and the mechanics of solids, Proc. Third U.S. National Congress of Applied Mechanics, June 11–14, 1 (1958).
4. D. C. Drucker, Stress-strain-time relations and irreversible thermodynamics, Proc. Inter. Symposium on Second Order Effects in Elasticity, Plasticity and Fluid Dynamics, Haifa, April 21–29, 1962, Pergamon Press, Oxford, 331 (1964).
5. H. Ziegler, Thermodynamik and rheologische Probleme, Ing. Arch., *25,* 58 (1957).
6. H. Ziegler, An attempt to generalize Onsager's principle, and its significance for rheological problems, ZAMP, *9b,* 748 (1958).
7. H. Ziegler, Zwei Extremalprinzipien der irreversiblen Thermodynamik, Ing. Arch., *30,* 410 (1961).
8. C. Wehrli und H. Ziegler, Einige mit dem Prinzip von der grössten Dissipationsleistung verträgliche Stoffgleichungen, ZAMP, *13,* 372 (1962).
9. H. Ziegler, Über ein Prinzip der grössten specifischen Entropie produktion und seine Bedentung für die Rheologie, Rheologica Acta, *2,* 230 (1962).
10. H. Ziegler, Die statistischen Grundlagen der irreversiblen Thermodynamik, Ing. Arch., *31,* 317 (1962).
11. H. Ziegler, Some extremum principles in irreversible thermodynamics,

with application to continuum mechanics, *Progress in Solid Mechanics,* vol. 4, North Holland, Amsterdam, 91 (1963).

12. H. Ziegler, Thermodynamics aspects of continuum mechanics, CIME Colloquium, Bressanone (1963).
13. H. Ziegler, Thermodynamic considerations in continuum mechanics, Minta Martin Lecture, Dep. of Aeronautics and Astronautics, MIT, Cambridge, Mass. (1964).
14. H. Ziegler, Thermodynamik der Deformationen, Applied Mechanics, Proc. of the Eleventh International Congress of Applied Mechanics, Munich 1964, Springer, Berlin, 99 (1966).
15. A. A. Vakulenko, On stress-strain relations for inelastic bodies, Doklady Akad. Nauk SSSR, *118,* 665 (1958).
16. A. A. Vakulenko, Thermodynamic investigation of stress-strain relations in isotropic elasto-plastic bodies, Doklady Akad. Nauk SSSR, *126,* 736 (1959).
17. A. E. Green and J. E. Adkins, Large Elastic Deformations and Non-Linear Continuum Mechanics, Oxford (1960).
18. O. W. Dillon, Coupled thermoplasticity, J. Mech. Phys. Solids, *11,* 21 (1963).
19. G. A. Kluitenberg, Thermodynamical theory of elasticity and plasticity, Physica, *28,* 217 (1962).
20. G. A. Kluitenberg, A note on the thermodynamics of Maxwell bodies, Kelvin bodies /Voigt bodies/, and fluids, Physica, *28,* 561 (1962).
21. G. A. Kluitenberg, On rheology and thermodynamics of irreversible processes, Physica, *28,* 1173 (1962).
22. G. A. Kluitenberg, On the thermodynamics of viscosity and plasticity, Physica, *29,* 633 (1963).
23. G. A. Kluitenberg, A unified thermodynamic theory for large deformations in elastic media and in Kelvin /Voigt/ media, and for viscous fluid flow, Physica, *30,* 1945 (1964).
24. G. A. Kluitenberg, Application of the thermodynamics of irreversible processes to continuum mechanics, Non-Equilibrium Thermodynamics, Variational Techniques and Stability, Proc. Symposium held at the University of Chicago, May 17–19, 1965, The University of Chicago Press, 91 (1966).
25. G. A. Kluitenberg, On heat dissipation due to irreversible mechanical phenomena in continuous media, to be published in Physica.
26. S. L. Koh and A. C. Eringen, On the foundations of non-linear thermoviscoelasticity, Int. J. Engng. Sci., *1,* 199 (1963).
27. J. Kestin, On the application of the principles of thermodynamics to strained solid materials, Brown University, Report (1966); IUTAM Symposium on Irreversible Aspects of Continuum Mechanics, Vienna, Austria, June 22–25, 1966.
28. J. F. Besseling, A thermodynamic approach to rheology, IUTAM Symposium on Irreversible Aspects of Continuum Mechanics, Vienna, Austria, June 22–25, 1966.
29. B. D. Coleman and W. Noll, The thermodynamics of elastic materials with heat conduction and viscosity, Arch. Rat. Mech. Anal., *13,* 167 (1963).

30. B. D. Coleman and V. J. Mizel, Existence of caloric equations of state in thermodynamics, J. Chemical Physics, *40,* 1116 (1964).
31. B. D. Coleman, Thermodynamics of materials with memory, Arch. Rat. Mech. Anal., *17,* 1 (1964).
32. B. D. Coleman, On thermodynamics, strain impulses, and viscoelasticity, Arch. Rat. Mech. Anal., *17,* 230 (1964).
33. C. Truesdell and W. Noll, The Non-Linear Field Theories of Mechanics, *Encyclopedia of Physics,* vol. III/3, Springer, Berlin (1965).
34. R. M. Christensen and P. M. Naghdi, Linear non-isothermal viscoelastic solids, Acta Mech., *3,* 1 (1967).
35. A. E. Green and P. M. Naghdi, A general theory of an elastic-plastic continuum, Arch. Rat. Mech. Anal., *18,* 251 (1965).
36. A. E. Green and P. M. Naghdi, Plasticity theory and multipolar continuum mechanics, Mathematica, *12,* 21 (1965).
37. O. W. Dillon, A thermodynamic basis of plasticity, Acta Mech., *3,* 182 (1967).
38. J. Meixner, On the theory of linear viscoelastic behavior, Rheologica Acta, *4,* 77 (1965).
39. J. Meixner, Bemerkungen zur Theorie der Wärmeleitung, Zeitschrift für Physik, *193,* 366 (1966).
40. J. Meixner, Consequences of an inequality in nonequilibrium thermodynamics, J. Appl. Mech., *33,* 481 (1966).
41. C. Truesdell, Letter to the Participants in the IUTAM Symposium on Thermodynamics at Vienna in June 1966 and Comments on Professor Meixner's Lecture; October 26, 1966.
42. C. Truesdell and R. Toupin, The Classical Field Theories, *Encyclopedia of Physics,* vol. III/1, Springer, Berlin, 226 (1960).
43. J. Meixner, On the theory of linear passive systems, Arch. Rat. Mech. Anal., *17,* 278 (1964).
44. M. E. Gurtin and W. O. Williams, On the Clausius-Duhem Inequality, ZAMP, *17,* 626 (1966).
45. M. E. Gurtin and W. O. Williams, An axiomatic foundation for continuum thermodynamics, Carnegie Institute of Technology, Report (1967).
46. W. Noll, A mathematical theory of the mechanical behavior of continuous media, Arch. Rat. Mech. Anal., *2,* 117 (1958).
47. W. Noll, The foundations of classical mechanics in the light of recent advances in continuum mechanics, The Axiomatic Method with Special Reference to Geometry and Physics, Proc. Int. Symposium, University of California, Berkeley, December 26, 1957–January 4, 1958, North-Holland, Amsterdam, 266 (1959).
48. P. Perzyna, Fundamental problems in viscoplasticity, *Advances in Applied Mechanics, 9* (1966).
49. C.-C. Wang and R. M. Bowen, On the thermodynamics of nonlinear materials with quasi-elastic response, Arch. Rat. Mech. Anal., *22,* 79 (1966).
50. M. E. Gurtin, Thermodynamics and the possibility of spatial interaction in elastic materials, Arch. Rat. Mech. Anal., *19,* 339 (1965).
51. A. C. Pipkin and R. S. Rivlin, Mechanics of rate-independent materials, ZAMP, *16,* 313 (1965).

52. A. E. Green, Hypo-elasticity and plasticity, J. Rat. Mech. Anal., *5*, 725 (1956); Proc. Royal Soc. London, *234A*, 46 (1956).
53. P. Perzyna, On thermodynamics of the rate type material, Bulletin de L'Académie Polon. Scien., Sér. scien. tech., *14*, 397 (1966).
54. P. Perzyna, On thermodynamics of elastic-viscoplastic material, Bulletin de L'Académie Polon. Scien., Sér. scien. tech., *14*, 409 (1966).
55. K. Hohenemser and W. Prager, Über die Ansätze der Mechanik isotroper Kontinua, ZAMM, *12*, 216 (1932).
56. P. Perzyna, The constitutive equations for rate sensitive plastic materials, Quart. Appl. Math., *20*, 321 (1963).
57. P. Perzyna, The constitutive equations for work-hardening and rate sensitive plastic materials, Proc. Vibr. Probl., *4*, 281 (1963).
58. P. Perzyna, On the constitutive equations for work-hardening and rate sensitive plastic materials, Bull. Acad. Polon. Sciences, Sér. scien. tech., *12*, 199 (1964).
59. P. Perzyna and T. Wierzbicki, On temperature dependent and strain sensitive plastic materials, Bull. Acad. Polon. Sciences, Sér. scien. tech., *12*, 225 (1964).
60. P. Perzyna and W. Wojno, On the constitutive equations of elastic/viscoplastic materials and finite strain, Arch. Mech. Stos., *18*, 85 (1966).
61. W. Wojno, On thermodynamics of elastic/viscoplastic materials, Thesis, Institute of Basic Technical Research, Polish Academy of Sciences (1967).

SOME EXPERIMENTS IN DYNAMIC PLASTICITY UNDER COMBINED STRESS

ULRIC S. LINDHOLM

Southwest Research Institute
San Antonio, Texas

ABSTRACT

The deformation of aluminum at strain rates from 10^{-3} sec^{-1} to 10^{3} sec^{-1} and temperatures from 300° K to 700° K is studied experimentally under a range of stress states including tension, compression, torsion, and combined tension and torsion. The results from these tests are compared with a generalized constitutive equation developed from the thermally-activated dislocation model of deformation. Predicted functional relationships between the stress, strain and strain-rate invariants and temperature are supported by the experimental data.

INTRODUCTION

This paper will present some experimental results on the plastic flow of aluminum subject to a rather wide spectrum of loading conditions. This spectrum includes stress states of pure compression, tension and torsion as well as combined stress states incorporating tension and torsion. In order to demonstrate the rate sensitive or viscoplastic behavior of the metal, the rate of loading is varied to produce strain-rates within the range of 10^{-3} sec^{-1} to 10^{3} sec^{-1}. Finally, for compression and tension, elevated temperature data from approximately 300° K to 700° K is obtained over the complete range in strain rates. The purpose for performing these comprehensive stress/strain/strain rate/temperature tests on aluminum is to determine to what extent the material behavior can be described with a single generalized constitutive equation.

It has previously been shown [1, 2] for 1100-0 aluminum in compression, at room temperature, and over the same range in strain-rate that the response can be described by a constitutive relation based upon the assumption that a single, thermally-activated mechanism governs the dislocation motion. The same assumption is made herein, with the one-dimensional relations being generalized to include combined states of stress. The generalized constitutive equations are constructed so as to be analogous in form to those given by Perzyna [3, 4] for viscoplastic materials. Basing the generalized constitutive equations on a particular mechanism of deformation serves the purpose of defining explicitly the functional relations and gives physical meaning to the parameters involved.

The experimental data presented is in general agreement with the constitutive relations derived. Aluminum 1100-0 was chosen as the specimen material because of the considerable amount of background information already available on this metal. Also, it has a moderate rate-sensitivity and it yields and strain-hardens in a very uniform manner at stress levels convenient for the experimental procedures used. Most of the experimental data is obtained with either uniaxial or proportional loading paths and at approximately constant plastic strain-rate. Some experiments are presented, however, to illustrate the behavior under sudden changes in the direction of the applied stress or the rate of deformation.

CONSTRUCTION OF GENERALIZED CONSTITUTIVE EQUATIONS

With the assumptions that thermal activation as well as the applied stress is effective in producing dislocation motion and that the activation energy is linearly stress dependent, the strain rate can be expressed in the following form [2, 5] for a simple state of shear;

$$\dot{\gamma}^{P} = \eta \exp\left[-\frac{H_0 - v^*(\tau - \tau^*)}{kT}\right]$$

$$= \eta \exp\left(-\frac{H_0}{kT}\right) \exp\left[\frac{v^*\tau^*}{kT}\left(\frac{\tau}{\tau^*} - 1\right)\right] \tag{1}$$

where $\dot{\gamma}^{P}$ = plastic shear strain-rate
τ = applied shear stress
T = absolute temperature
η = frequency parameter
H_0 = total activation energy
k = Boltzmann constant (1.22×10^{-22} in. lb/°K)
v^* = activation volume
τ^* = internal athermal stress component opposing dislocation motion

In the most general case, all four internal material parameters H_0, η, v^* and τ^* may be considered functions of the four independent variables, γ^P, $\dot{\gamma}^P$, τ and T. In the present development, however, we will restrict H_0 and η to be constants and v^* and τ^* to be functions only of the plastic strain, γ^P. More precisely, v^* and τ^* will be dependent upon the internal dislocation structure in the metal, but we may assume that the externally measurable parameter γ^P represents or is proportional to this internal structure and its change with gross deformation.

In order to extend (1) to generalized states of stress, certain common assumptions will be made. It is assumed that components of the total strain-rate tensor may be resolved into independent elastic and inelastic components such that

$$\dot{\epsilon}_{ij} = \dot{\epsilon}_{ij}{}^E + \dot{\epsilon}_{ij}{}^P \tag{2}$$

where the superscripts E and P denote the elastic and inelastic parts, respectively. The elastic component of deformation is taken to be independent of the rate of strain; i.e., elastic deformation involves no net resultant motion of dislocations. The inelastic components, $\dot{\epsilon}_{ij}{}^P$, contain both plastic and viscous effects. Similarly, the stress tensor and the elastic and plastic components of the strain and strain-rate tensor may be resolved into their respective spherical (volumetric) and deviatoric (distortional) components. In the following, a single prime will denote deviatoric components and a double prime will denote spherical components. For stresses of the order of magnitude of the yield stress in metals, a spherical or hydrostatic state of stress is assumed to be of no influence on the plastic component of deformation. On the basis of recent experiments [6], this assumption will be extended to inelastic viscous effects also.

For temperature-dependent and strain-rate sensitive materials, Perzyna [2] has suggested a yield function of the form

$$F = \frac{f(\sigma'_{ij}, \epsilon'_{k\ell})}{\kappa(W^P, T)} - 1$$

where the surface defined by $F = 0$ is assumed regular and convex. The strain-hardening parameter, κ, is taken as a function of both the plastic work W^P and the temperature. For the present purposes, we will employ a somewhat simplified, explicit yield function of the form

$$F = \frac{f(J'_2)}{\kappa(I'^P_2)} - 1 = \frac{|J'_2|^{1/2}}{\tau^*(I'^P_2)} - 1 \tag{3}$$

where $J'_2 = \frac{1}{2}\sigma'_{ij}\sigma'_{ij}$ and $I'^P_2 = \frac{1}{2}\epsilon'^P_{ij}\epsilon'^P_{ij}$ are the second invariants of the stress and plastic strain deviatoric tensors, respectively. The yield function based upon J'_2 only is analogous to a Mises' yield criteria. The strain-hardening parameter, $\kappa = \tau^*(I'^P_2)$, is the same as the athermal component of the stress in (1) and will similarly be assumed a function of the plastic strain invariant only. The assumed dependence of τ^* on $I_2'^P$ rather than on W^P is one of convenience in the later determination of material parameters from the reduced experimental data. The two assumptions are not equivalent in viscoplasticity, since the plastic work will depend upon the rate of deformation as well as the total deformation. The question remains open as to whether the plastic work or a measure of the plastic deformation such as $I_2'^P$ correlates best with the internal structure of the metal. We have arbitrarily chosen the more convenient of the two from the standpoint of computation. Temperature effects will be introduced through the functional dependence of the plastic flow rate on F.

Again following Perzyna, the plastic component of strain-rate is given by

$$\dot{\epsilon}_{ij}'^{P} = \bar{\eta}\Phi(F)\,\frac{\partial f}{\partial \sigma_{ij}'} \tag{4}$$

where $\bar{\eta}$, F and the function Φ may all be temperature dependent. The choice of $\bar{\eta}$ and $\Phi(F)$ will be selected based upon the relationship of (1) for simple shear. Thus,

$$\begin{aligned}\dot{\epsilon}_{ij}'^{P} &= 2\eta \exp\left(-\frac{H_0}{kT}\right)\exp\left(\frac{v^*\tau^*}{kT}F\right)\frac{\partial f}{\partial \sigma_{ij}'} \\ &= \eta \exp\left(-\frac{H_0}{kT}\right)\exp\left[\frac{v^*\tau^*}{kT}\left(\frac{|J_2'|^{1/2}}{\tau^*}-1\right)\right]\frac{\sigma_{ij}}{|J_2'|^{1/2}}.\end{aligned} \tag{5}$$

The constants η and H_0 remain the same as in (1) while the parameters v^* and τ^* are now functions of $I_2'^P$ rather than of γ. Upon adding the elastic components, the total constitutive equations become

$$\begin{aligned}\dot{\epsilon}_{ij}' &= \frac{1}{2\mu}\dot{\sigma}_{ij}' + \eta \exp\left(-\frac{H_0}{kT}\right)\exp\left[\frac{v^*\tau^*}{kT}\left(\frac{|J_2'|^{1/2}}{\tau^*}-1\right)\right]\frac{\sigma_{ij}}{|J_2'|^{1/2}} \\ \dot{\epsilon}_{ii}'' &= \frac{1}{3K}\dot{\sigma}_{ii}'' + \alpha\dot{T}\end{aligned} \tag{6}$$

where μ and K are the elastic shear and bulk moduli and α is the coefficient of thermal expansion.

The dependence of the yield function on strain rate and temperature is determined by squaring both sides of (5). This yields the relation

$$|\mathscr{I}_2'^P|^{1/2} = \eta \exp\left(-\frac{H_0}{kT}\right)\exp\left[\frac{v^*\tau^*}{kT}\left(\frac{|J_2'|^{1/2}}{\tau^*}-1\right)\right] \tag{7}$$

where $\mathscr{I}_2'^P = \frac{1}{2}\dot{\epsilon}_{ij}'^P\dot{\epsilon}_{ij}'^P$, is the second invariant of the inelastic strain-rate deviator.* Equation (7) may be solved explicitly for the stress invariant to give

$$|J_2'|^{1/2} = \tau^* + \frac{H_0}{v^*} - \frac{kT}{v^*}\log_e\left(\frac{\eta}{|\mathscr{I}_2'^P|^{1/2}}\right). \tag{8}$$

Viscoplastic flow will occur for $F > 0$ and $0 < T < T_c$, where $T_c = H_0/k \log_e (\eta/|\mathscr{I}_2'^P|^{1/2})$. At the temperature T_c the thermal energy available is itself enough to overcome the potential barriers without the assistance of additional applied stress, i.e., the activation barriers become essentially transparent to the motion of the dislocation. For $T \geqslant T_c$, the applied stress required to produce plastic flow becomes independent of the rate of deformation and equal to τ^*. Thus, $F = 0$ for $T \geqslant T_c$.

In the present approach, the transition from elastic to inelastic behavior will be governed by the functions $\tau^*(I_2'^P)$ and $v^*(I_2'^P)$. For small values of $I_2'^P$, the plastic strain-rates can be made vanishingly small. Actually, as τ approaches the value of τ^* in (1), it becomes necessary to include the probability that a dislocation may move by means of thermal activation in a direction opposed to the applied stress. For this case, (1) becomes

$$\begin{aligned}\dot{\gamma}^P &= \eta\left\{\exp\left[-\frac{H_0 - v^*(\tau-\tau^*)}{kT}\right] - \exp\left[-\frac{H_0 + v^*(\tau-\tau^*)}{kT}\right]\right\}\\ &= 2\eta\exp\left(-\frac{H_0}{kT}\right)\sinh\left[\frac{v^*(\tau-\tau^*)}{kT}\right].\end{aligned} \tag{1a}$$

Correspondingly, for the general case

$$\dot{\epsilon}_{ij}'^P = 4\eta\exp\left(-\frac{H_0}{kT}\right)\sinh\left(\frac{v^*\tau^*}{kT}F\right)\frac{\partial f}{\partial\sigma_{ij}'}. \tag{5a}$$

Equations (5a) and (5) differ significantly in magnitude only for values of $v^*\tau^*F/kT < 3$. This may occur for small values of F or at high temperatures. Equations (1a) or (5a) often are used to describe creep phenomena.

The material parameters H_0, η, v^* and τ^* may be determined from mechanical testing at constant strain-rate and temperature. For this purpose, certain derivatives of (8) are useful. The activation volume $v^*(I_2'^P)$ is determined from the slope of the stress-log strain-rate curve at constant temperature and strain by the relation

$$\left(\frac{d|J_2'|^{1/2}}{d\log_e|\mathscr{I}_2'^P|^{1/2}}\right)_{T,\,I_2'^P} = \frac{kT}{v^*}. \tag{9}$$

* When taking square roots of the invariants, absolute value signs are used since the invariants may have negative values.

Once v^* is determined the frequency parameter η can be obtained from the slope of the stress-temperature curves at constant strain and strain-rate by the relation

$$\left(\frac{d|J_2'|^{1/2}}{dT}\right)_{\mathscr{I}_2'^P,\, I_2'^P} = \frac{k}{v^*} \log_e \frac{|\mathscr{I}_2'^P|^{1/2}}{\eta}. \tag{10}$$

The other two parameters τ^* and H_0 can also be determined from the plot of $|J_2'|^{1/2}$ vs T. At $T = 0$, $|J_2'|^{1/2} = \tau^* + H_0/v^*$ and is independent of the rate of deformation. The additional relationship needed to determine τ^* and H_0 independently is that for T_c. T_c is not always easy to determine and may, in fact, be greater than the melting temperature. In the case of annealed aluminum, τ^* is evidently very small as will be seen. Determination of these material parameters will be more apparent when the experimental results are presented.

EXPERIMENTAL METHODS

Experimental results were obtained with three different types of machine. In uniaxial tension and compression, an Instron machine was used for low strain-rates and a split Hopkinson pressure bar system was used for high strain rates. The Hopkinson pressure bar system for tension and compression has recently been described in detail by the author [7]. Both the Instron and pressure bar tests were performed at room and at elevated temperatures to 700° K. Specimen geometries and dimensions for the uniaxial tension and compression tests are shown in Fig. 1.

Biaxial tests at low to intermediate strain rates were performed on a pneumatic piston-type machine also previously described in the literature [8]. This machine loads a thin-walled, tubular specimen (see Fig. 1) in combined tension and torsion. The pressure systems controlling the tension and torsion modes are independently controlled so that any ratio of combined stress between pure tension and pure torsion is possible. Rate control is afforded by means of replaceable, fixed orifice plates in the discharge lines of the actuating pressure reservoirs. Maximum strain rates obtained on this machine with aluminum were of the order of 50 $\sec^{-1}$, approximately one order of magnitude less than the minimum rate of the split pressure bar test. The lower rates of the biaxial machine overlap with the range of the Instron.

Additional capabilities of the biaxial machine include the capacity for abrupt changes in the rate of loading or in the direction of loading. The rate of loading is changed by control of the pressure differential across the driving piston. Change in the direction of loading, such as from pure torsion to tension, is achieved by the opening or closing of

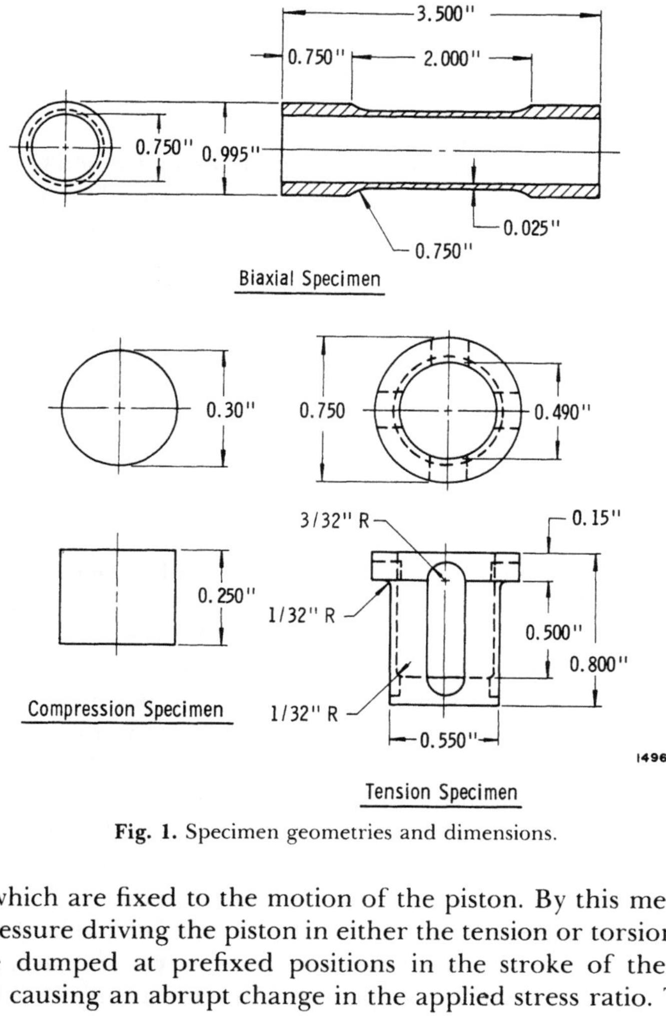

Fig. 1. Specimen geometries and dimensions.

valves which are fixed to the motion of the piston. By this means, the high pressure driving the piston in either the tension or torsion modes may be dumped at prefixed positions in the stroke of the piston, thereby causing an abrupt change in the applied stress ratio. This results in a rapid rotation of the principal axes of the stress and strain rate.

All results reported are for 1100-0 aluminum specimens. Dimensions of the specimens are given in Fig. 1. After machining, all specimens were heat-treated at 650° F for two hours in air.

AMBIENT TEMPERATURE, CONSTANT STRAIN-RATE TESTS

These tests include all the modes of loading previously described. For states of stress incorporating only axial tension or compression, $\pm\sigma$, and shear, τ, the stress, strain, and strain-rate invariants are given by,

$$J_2' = -(\tfrac{1}{3}\sigma^2 + \tau^2)$$
$$I_2'^P = -\tfrac{1}{4}(3\epsilon^{p2} + \gamma^{p2}) \tag{11}$$
$$\mathcal{I}_2'^P = -\tfrac{1}{4}(3\dot{\epsilon}^{p2} + \dot{\gamma}^{p2})$$

where ϵ^P and γ^P are the plastic components of the axial and shear strains, respectively. Since we are considering finite deformations, the axial components of the stress and strain are computed as true stress and logarithmic strain in all the subsequent data. Also, we will want to consider the orientation of the principal axes of stress and strain rate with respect to the longitudinal axis of the specimen. These angles are given by the relations

$$\Theta = \frac{1}{2}\tan^{-1}\frac{2\tau}{\sigma}$$

and (12)

$$\psi = \frac{1}{2}\tan^{-1}\frac{2\dot{\gamma}}{3\dot{\epsilon}}.$$

We may now compare the functional relationships between the three invariants and temperature given by (8) with the experimental results. Figure 2 presents typical stress-strain curves for the aluminum plotted in terms of the invariants. Each curve is obtained at constant temperature (294° K) and strain-rate. For all the tests reported in this section the stress ratio, τ/σ, was maintained nearly constant during the deformation. The flags on the points plotted in this and subsequent figures indicate the direction of the principal stress axes. The horizontal axis is tension to the right and compression to the left, with the vertical axis being pure shear, similar to a Mohr's circle representation. Combined stress states lie between the tension and shear axes.

In Fig. 3, the data of Fig. 2 and a large number of additional tests are replotted to show the relation between the stress and strain-rate invariants at constant strain amplitude and temperature. According to (8), this should be a linear relationship on the semi-logarithmic plot. The straight line through the experimental points at each strain amplitude is a best least-squares fit. The standard estimate of error of the data about the mean is 5.4% for the lowest strain and less than 3% for the two higher strain amplitudes. This can be considered good agreement when it is recalled that the data were obtained on three different types of machine, with three different specimen geometries, and in a variety of modes of loading. The scatter that does occur in the data would not appear to be reducible by a more complex choice of the yield function, such as inclusion of a term proportional to J_3'.

The slopes of the straight lines in Fig. 3 are used to determine v^* as a function of $I_2'^P$ by the relation (9). These values of v^* are plotted in Fig. 4. The magnitude of v^* is of the order of $100b^3$, where b is the

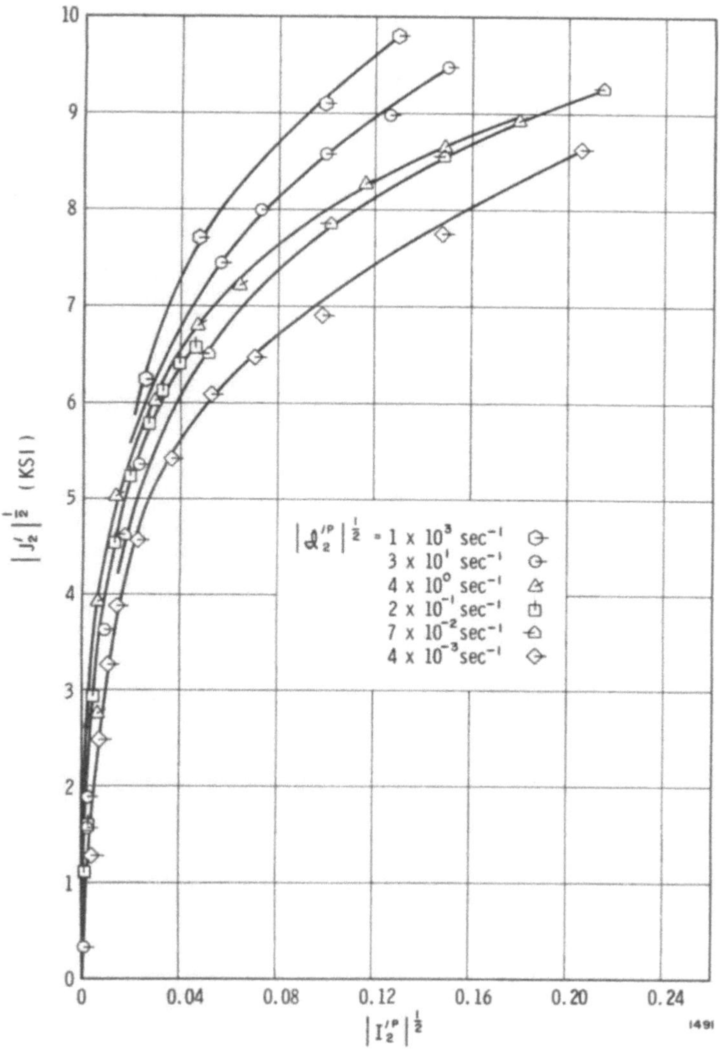

Fig. 2. Relation between the stress and strain invariants at $T = 294°$ K.

Burger's vector for slip in aluminum. The decrease in activation volume with strain is due to an increase in the number of activation sites resulting from dislocation multiplication.

ELEVATED TEMPERATURE TESTS

In order to assess the thermal dependence of the constitutive relations, elevated temperature tests were performed in tension and compression on the Instron machine and the split pressure bar. These results are presented in plots of stress vs. strain rate in Fig. 5 and stress

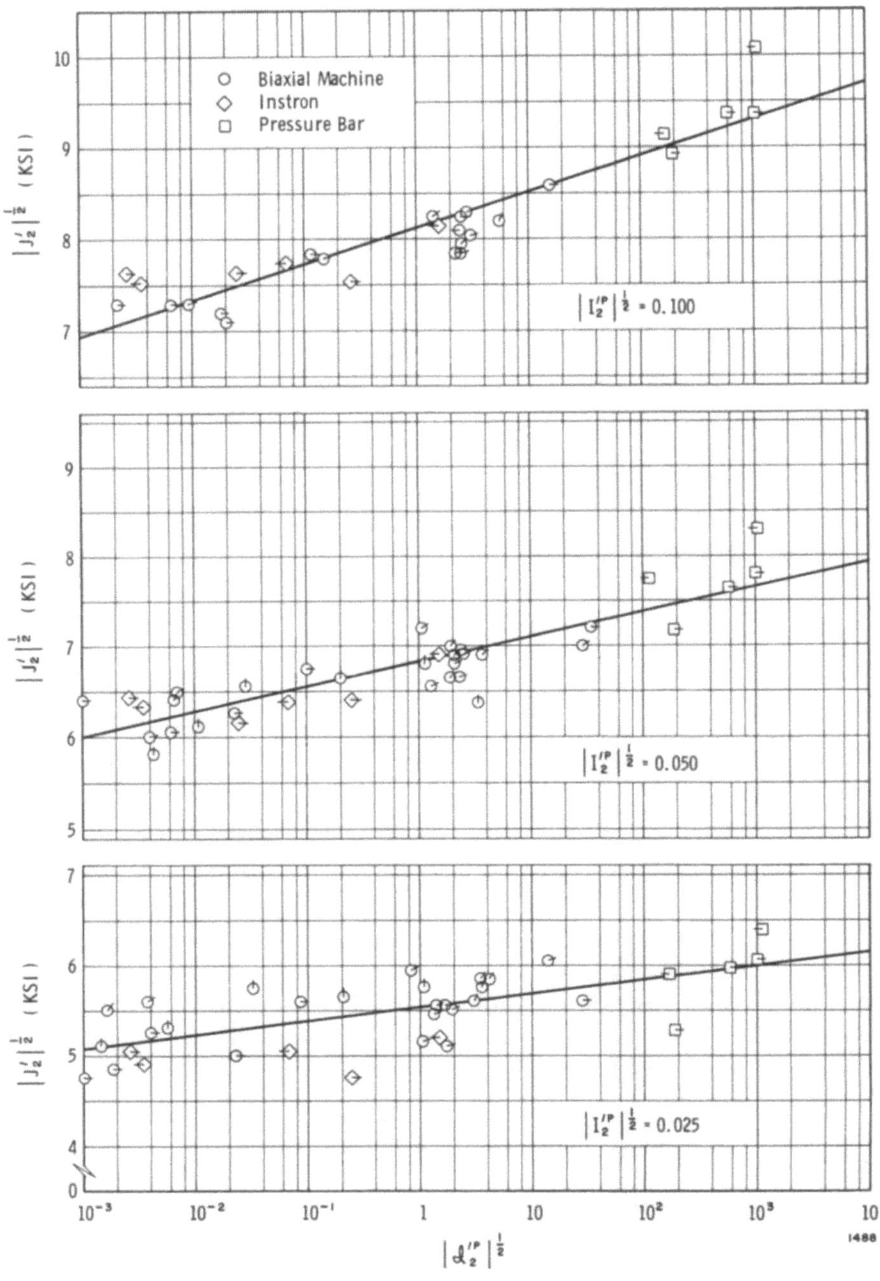

Fig. 3. Relation between the stress and strain-rate invariants at $T = 294°$ K.

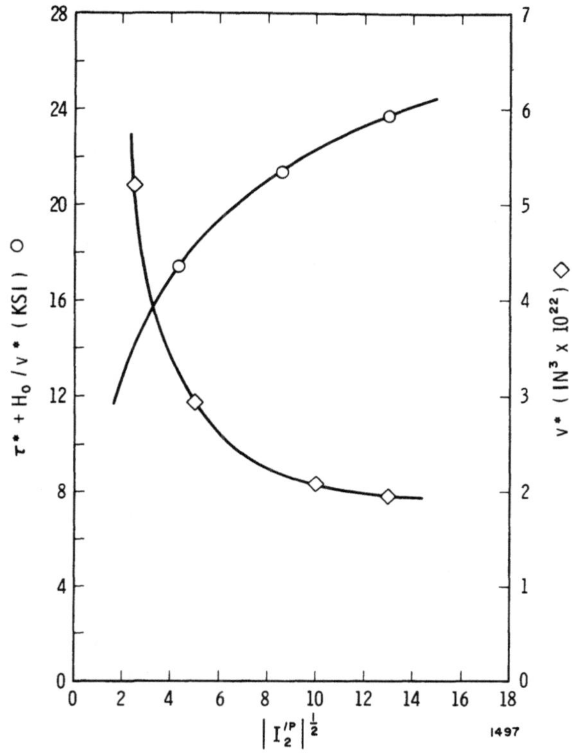

Fig. 4. Activation volume and stress, extrapolated to zero temperature, vs. strain invariant.

vs. temperature in Fig. 6. Figure 6 includes additional data taken from a paper by Chiddister and Malvern [9], who also used the split pressure bar method at elevated temperatures. The slopes of the lines in Fig. 5 are not determined by a best fit through the data points but are those determined by the values of v^* computed from the room temperature tests and given in Fig. 4. At constant strain, the change in slope is directly proportional to the change in temperature since v^* has been assumed independent of temperature. The rate sensitivity at elevated temperatures is thus seen to be predicted relatively well from the room temperature results.

Again, the values of $d\sigma/dT$ in Fig. 6 are based upon the room temperature activation volume as well as a constant value of the frequency parameter η and the consideration that at absolute zero temperature the stress is independent of the rate of deformation. For the straight lines in Fig. 6, $\eta = 4.5 \times 10^{11}$ sec^{-1}. This value for η is determined from the above considerations and the use of (10). The fact that $\eta > \dot{\epsilon}$ results in a positive magnitude for the logarithmic term in (8). The stress-strain relation at $T = 0$ is given by $\sigma/\sqrt{3} = \tau^* + H_0/v^*$. These values,

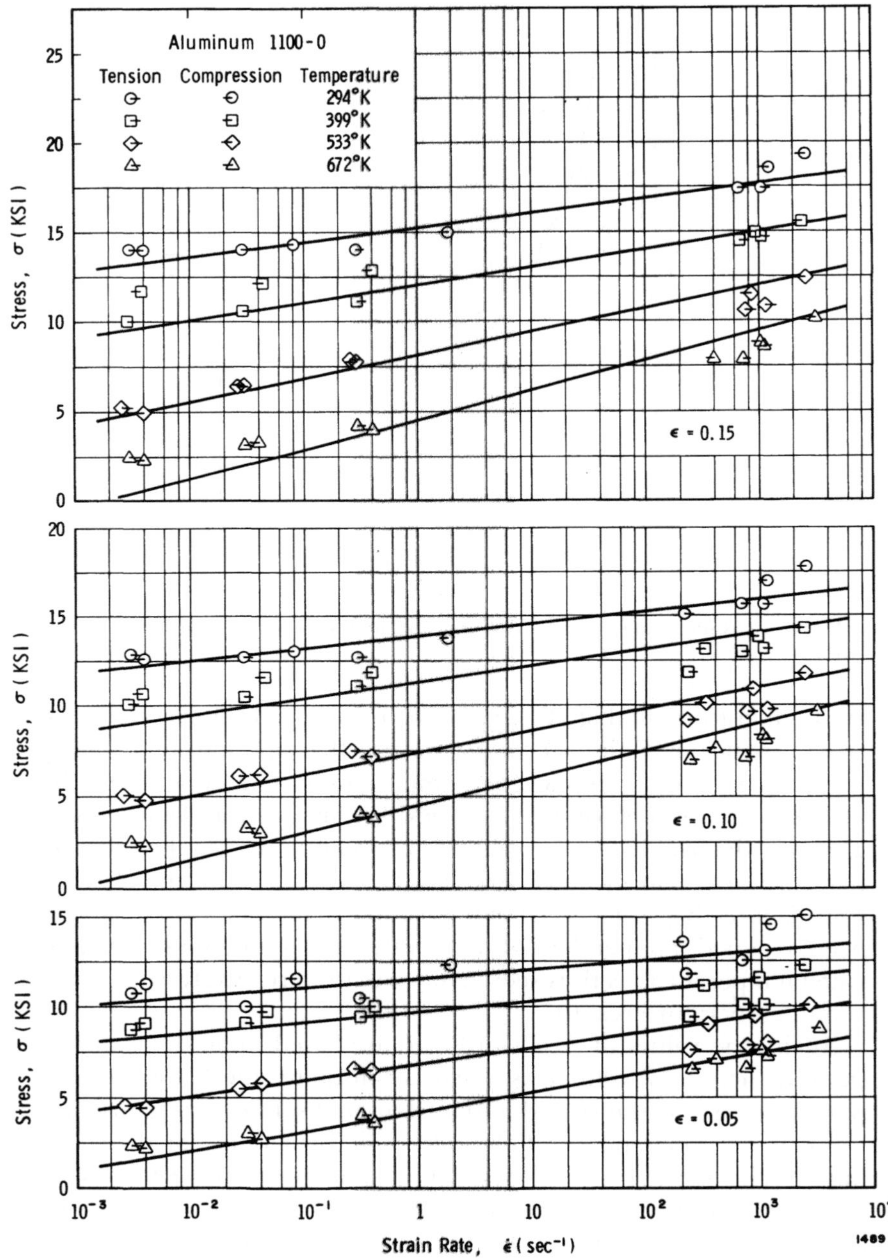

Fig. 5. Stress vs. strain-rate at constant temperature and strain.

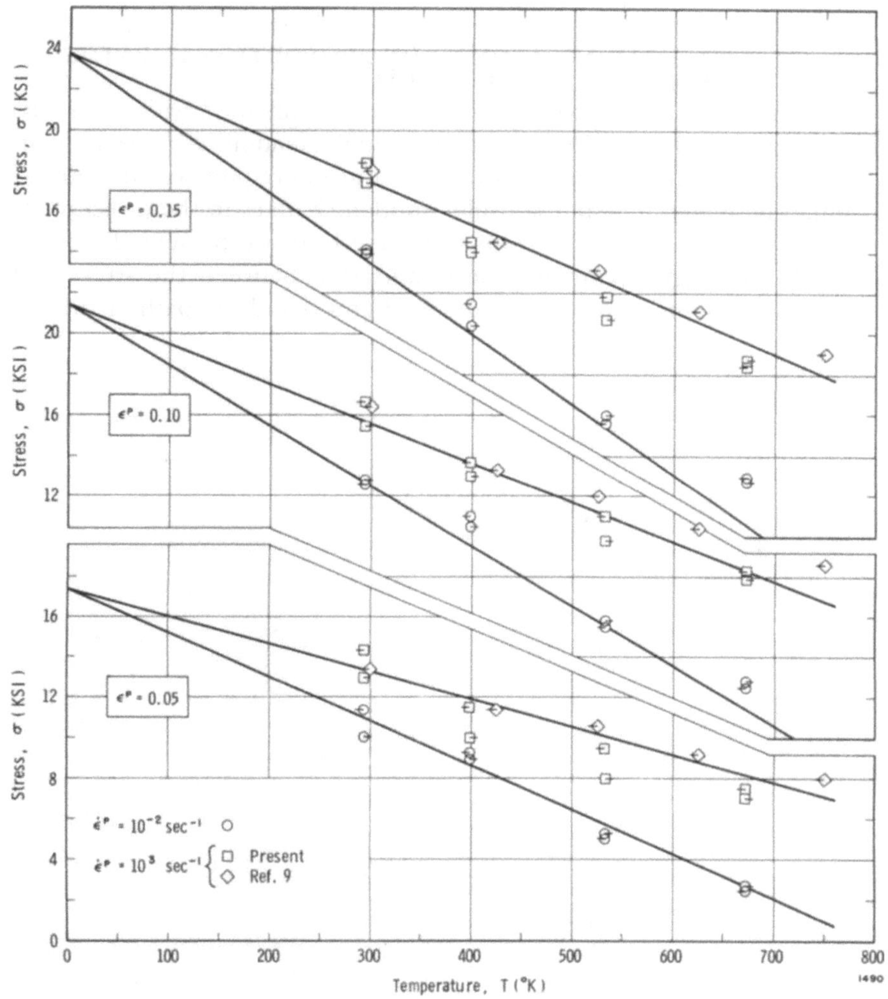

Fig. 6. Stress vs. temperature at constant strain-rate and strain.

determined from the zero intercepts of Fig. 6, are plotted in Fig. 4.

For annealed 1100 aluminum, the magnitude of τ^* is very small and it is not possible to determine the temperature T_c from the data presented. Thus, we can only estimate that τ^* has a magnitude of 2000 psi or less and that the total activation energy of the barriers is then of the order of 5×10^{-18} in. lb.

Based upon the elevated temperature data presented, the temperature dependence expressed in the constitutive Equation (5) appears to be qualitatively correct and the material parameters may be chosen to provide reasonable quantitative predictions. Extrapolation beyond the temperature or strain-rate range considered may be dangerous, however, as other mechanisms of deformation may become effective.

BIAXIAL LOADING ALONG NON-PROPORTIONAL PATHS

All the preceding experimental data was obtained with the stress ratio, τ/σ or Θ, and the rate of plastic deformation, $\mathscr{I}_2'^P$, nearly constant. Several tests of an exploratory nature were performed to determine the material response under sudden changes in the direction or the rate of loading. The results of these tests are presented in Figs. 7–10. Each figure contains the relationship between the stress and strain invariants, value of the angles Θ and ψ and the path of the de-

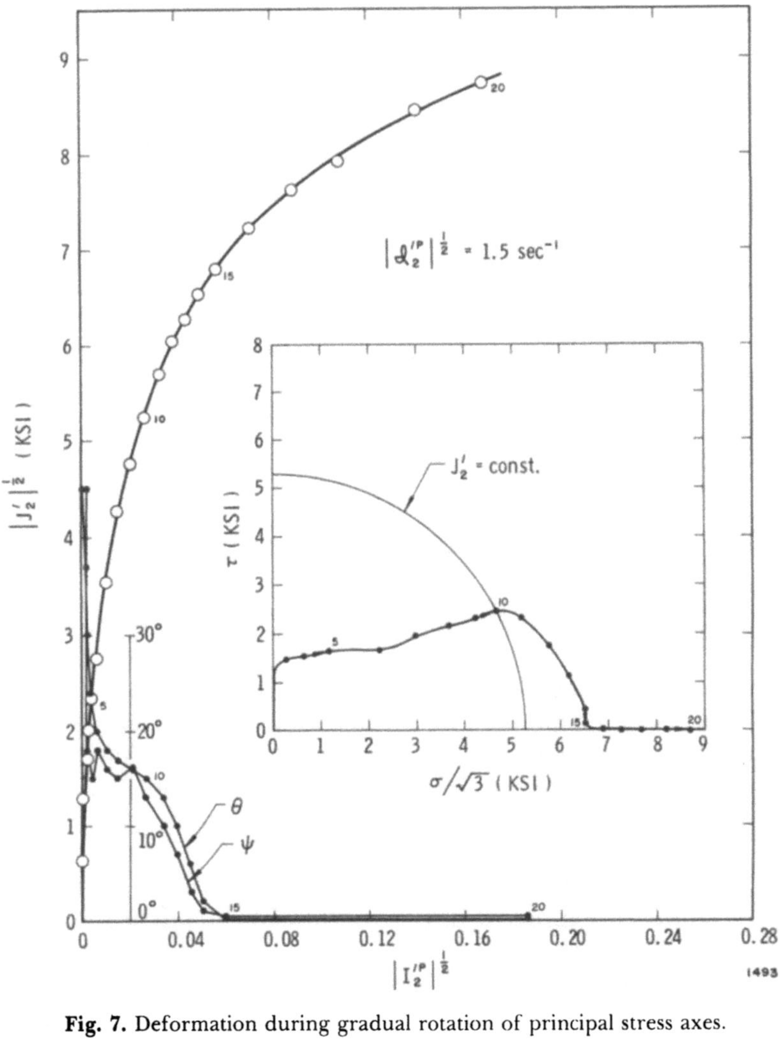

Fig. 7. Deformation during gradual rotation of principal stress axes.

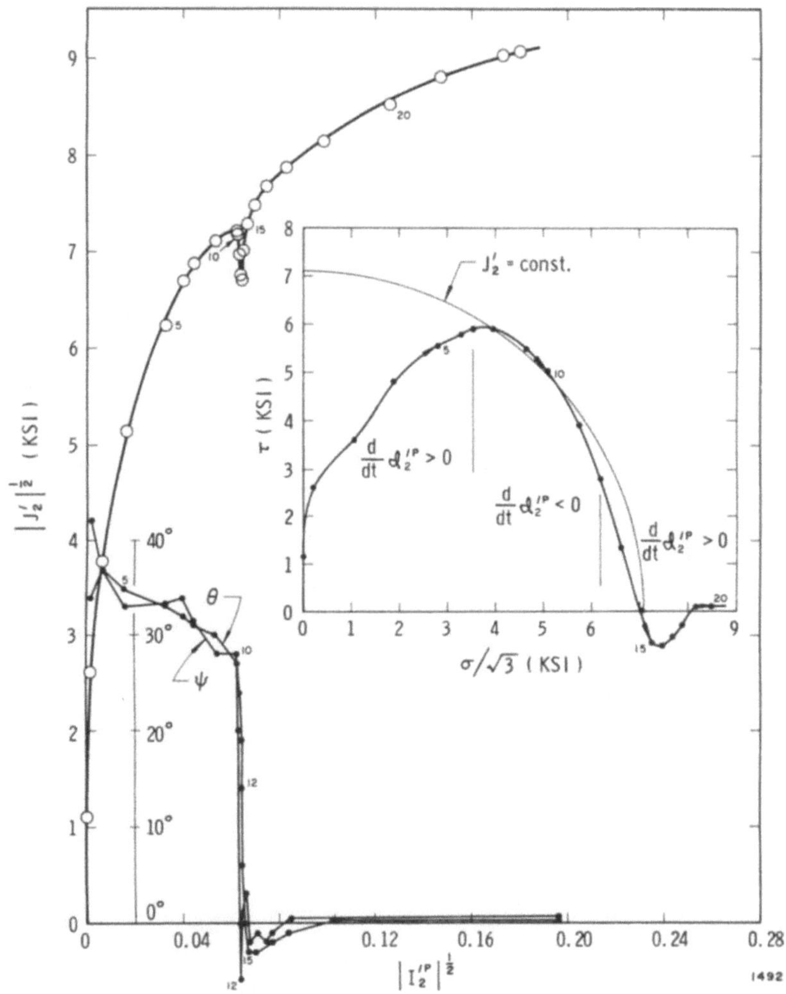

Fig. 8. Deformation during rapid rotation of principal stress axes.

formation in the $\tau - \sigma$ plane. Points on each curve are numbered sequentially so that corresponding points may be identified.

Figure 7 illustrates a test where the direction of the principal stresses is rotated a full 45° from the shear to the tensile stress axis. This occurs, however, under continuously increasing values of J_2' and nearly constant rate of deformation, $|\mathscr{I}_2'^P|^{1/2} = 1.5$ sec^{-1}. Magnitudes of J_2' are consistent with those obtained from proportional loading paths at equivalent stages of the deformation. In Fig. 8 a more abrupt change in the direction of loading was achieved by means of a more rapid decrease in the applied torsional loading. In this case the rota-

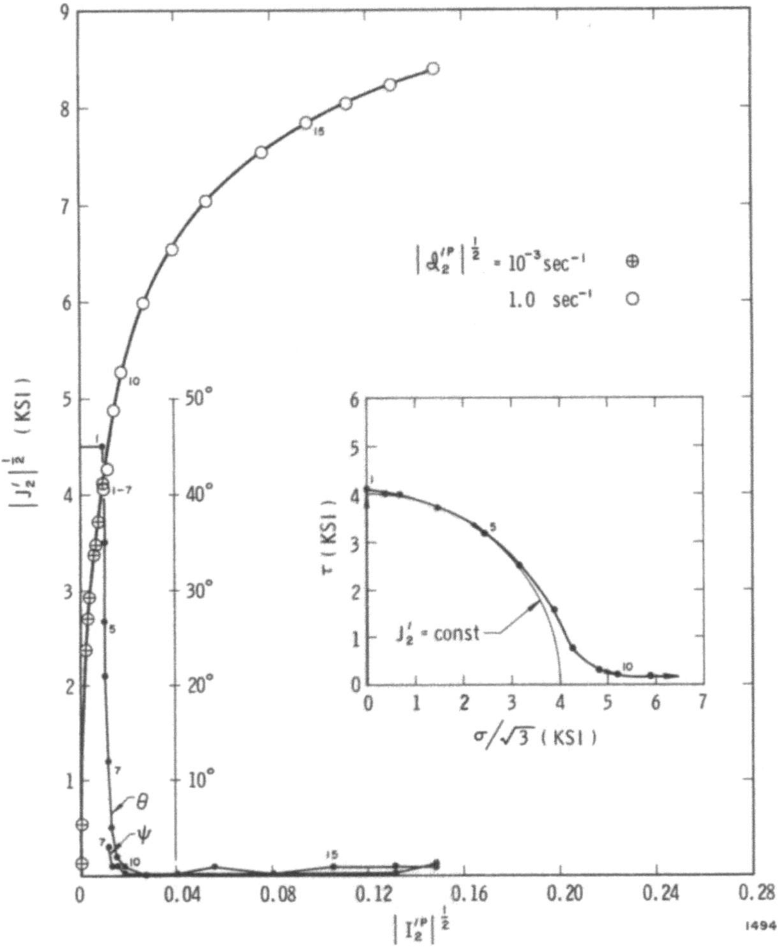

Fig. 9. Deformation during change in rate and direction of loading.

tion of the principal axes is accompanied by unloading and a decrease in the rate of deformation. During the initial portion of the curves the strain is accelerating slightly up to point 7 where $|\mathscr{I}_2'^P|^{1/2} = 4.9$ sec^{-1}. For points 7 through 12, the rate of deformation decreases approximately one order of magnitude to $|\mathscr{I}_2'|^{1/2} = 0.43$ sec^{-1} at point 12. This is accompanied by an unloading or decrease in J_2'. While unloading occurs in the sense that $dJ_2'/dt < 0$, the response is still viscoplastic since $F > 0$. In this unloading region the time interval between successive points is less than 2 milliseconds. Positive loading commences again at point 13 and the rate of deformation accelerates to the value of $|\mathscr{I}_2'|^{1/2} = 4.2$ at point 17 and remains relatively constant thereafter.

In both Figs. 7 and 8, the angles Θ and ψ are approximately equal throughout most of the deformation as is required by (5). Small differences of the order of 3° are not considered significant because of limitations in the experimental accuracy. A wide divergence of 20° does occur in Fig. 8 at point 12 where there is a sudden change in the direction and rate of loading. However, at this time dynamic effects in the extensometer or dynamometer cannot definitely be ruled out as a cause for this divergence. More extensive experiments on the possible role of $d\Theta/dt$ in the constitutive relation are perhaps warranted.

The final two experiments to be reported include a sudden change in the rate of loading simultaneously with a change in direction. In

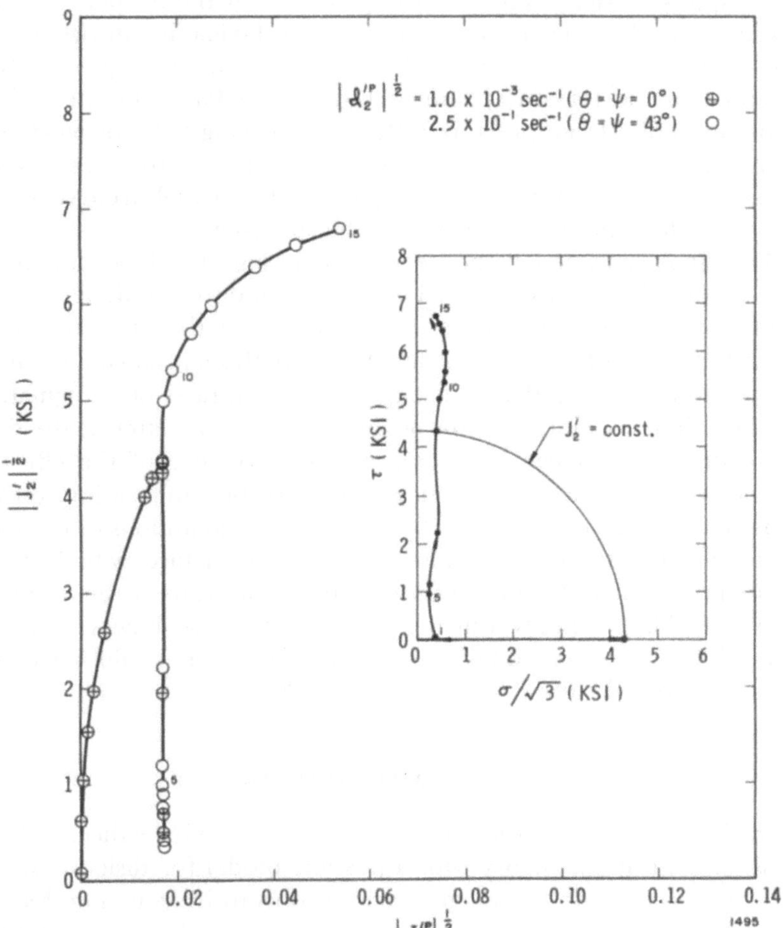

Fig. 10. Deformation during change in rate and direction of loading after unloading.

Fig. 9 the specimen was loaded at a very slow rate in torsion to a shear stress of 4000 psi. The loading was then transferred dynamically to the tensile axis, reaching a deformation rate of $|\mathscr{I}_2'^P|^{1/2} = 1.0\ \text{sec}^{-1}$ at about point 10. During this period of acceleration and rotation of the principal axes, J_2' remains essentially constant. The plastic strain developed during this period of rotation is very small because the rotation occurs in a short time interval (less than 50 milliseconds). After the sudden change in deformation rate and direction of loading, the magnitude of J_2' rapidly approaches those values to be expected if the test had been performed continuously at the higher rate. One final comment on this test is that the constancy of J_2' during the transfer of loading from the torsion to the tension axis is not a programmed loading path. Rather, the response is determined to a large extent by the properties of the specimen itself as well as by the dynamics of the loading device. In this case, the loading was developed by a slow buildup of the pressure differential in the torsion system followed immediately by the sudden application of a large pressure differential driving the tension system. Only the resulting load transfer and deformation of the specimen are measured. The fact that the loading path follows closely a curve of constant J_2' reflects on the properties of the specimen.

The final specimen response shown in Fig. 10 shows an initial, slow loading in tension, a partial unloading, and then a higher-rate reloading in torsion. The interesting point here is that the stress levels in each portion of the stress-strain curve are those predicted from the proportional loading data at the corresponding rates of deformation.

These few tests indicate that within the normal scatter of the data, there does not appear to be any marked evidence of the effect of previous loading history, either rate or direction of loading, on the subsequent material response. Thus, under the conditions of these experiments in aluminum, expansion of the yield surface by both strain-hardening and increasing deformation rate appears to be relatively isotropic. This certainly will not be generally true, however, as evidenced by the many quasi-static experiments on strain-hardening under combined stress.

CONCLUDING REMARKS

In this paper the author has attempted to combine the results of classical continuum theory with a possible model for dislocation motion, in order to construct a generalized constitutive equation. Experimental results on the deformation of aluminum under a wide range of the deformation rate, temperature and load condition are then shown to be in fair agreement with the constitutive relation derived. The agreement between experiment and theory is based upon relationships between the pertinent invariants of the stress, strain and strain-rate tensors only. Prediction of detailed response of a structure to complex loading paths requires the actual integration of the constitutive equations. The form of the proposed constitutive relation cannot be expected to fit all metals and possibly not even all possible mechanisms of deformation in aluminum. Other models for dislocation motion have been proposed which may be generalized in a similar manner. The thermal activation model chosen here is particularly useful because it describes the effects of both temperature and stress on the dislocation motion in a relatively simple manner.

ACKNOWLEDGMENT

The work reported in this paper was supported by the Army Research Office, Durham.

References

1. U. S. Lindholm, "Some Experiments with the Split Hopkinson Pressure Bar," J. Mech. and Phys. Solids, *12,* 317 (1964).
2. U. S. Lindholm and L. M. Yeakley, "Dynamic Deformation of Single and Polycrystalline Aluminum," J. Mech. and Phys. Solids, *13,* 41 (1965).
3. P. Perzyna, "The Constitutive Equations for Rate Sensitive Plastic Materials," Quart. Appl. Math., *20,* 321 (1963).
4. P. Perzyna and T. Wierzbicki, "Temperature Dependent and Strain Rate Sensitive Plastic Materials," Arch. Mech. Stos., *16,* 135 (1964).
5. U. S. Lindholm, "Dynamic Deformation of Metals," BEHAVIOR OF MATERIALS UNDER DYNAMIC LOADING, ASME, New York, 42 (1965).
6. J. D. Chalupnik and E. A. Ripperger, "Dynamic Deformation of Metals Under High Hydrostatic Pressure," Experimental Mechanics, *6* (1966).
7. U. S. Lindholm and L. M. Yeakley, "High Strain-Rate Testing: Tension and Compression," Experimental Mechanics, 7, 1 (1967).
8. U. S. Lindholm and L. M. Yeakley, "A Dynamic Biaxial Testing Machine," Experimental Mechanics, *7,* 1 (1967).
9. J. L. Chiddister and L. E. Malvern, "Compression-Impact Testing of Aluminum at Elevated Temperatures," Experimental Mechanics, *3,* 81 (1963).

THE SPEED OF DUCTILE-CRACK PROPAGATION AND THE DYNAMICS OF FLOW IN METALS

M. F. KANNINEN, A. K. MUKHERJEE *,
A. R. ROSENFIELD, AND G. T. HAHN

Battelle Memorial Institute
Columbus, Ohio

ABSTRACT

In this paper the connection between the speed of ductile-crack propagation and the dynamic-flow properties of metals is examined. A theoretical analysis based on a dynamic solution for the Dugdale crack model and employing descriptions of 1) the strains within the plastic zone, 2) the rate dependence of the flow stress, and 3) a simple criterion for ductile fracture is developed. The calculations are found to compare favorably with observed crack speeds of 1.6 to 410 ft/sec in 0.00175-in.-thick steel foil. It is concluded that ductile-crack speed is limited by the increased resistance to plastic flow at high strain rates. The key factors determined in the analysis are used to show that flow stress data for strain rates exceeding 10^4 sec^{-1} can be extracted from ductile-crack-propagation experiments.

INTRODUCTION

Ductile-crack propagation in metals—whether by the shear or by the fibrous mode—is preceded by intense local plastic deformation. The crack propagates rapidly and imposes a steep strain gradient on the material immediately ahead of it. Consequently, the metal just in front of a propagating ductile crack deforms at high rates: 10^3 to 10^5 sec^{-1}, rates for which the resistance to plastic flow is markedly increased. Taken together, these observations suggest the following possibilities:

1) The speed at which a ductile crack propagates is limited by the increased resistance to plastic flow encountered at high strain rates, i.e., plastic deformation is the rate-controlling process.
2) Ductile-crack speeds can be calculated by a method of analysis

* Presently at the University of California, Davis, California.

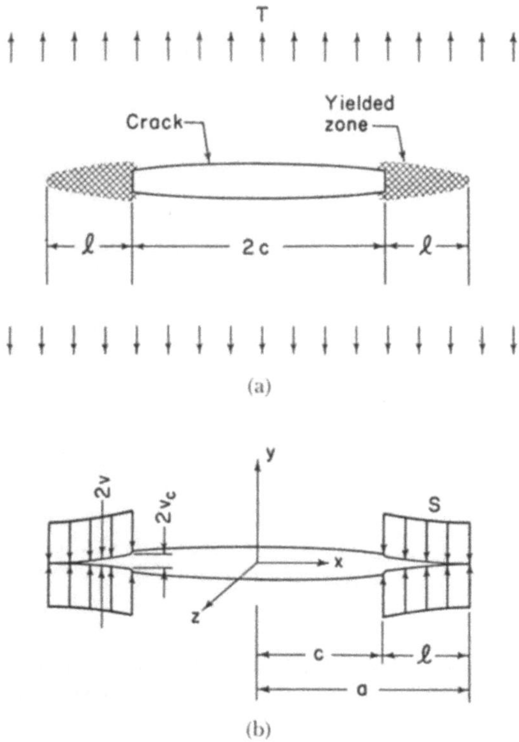

Fig. 1. The Dugdale Crack: (a) Schematic drawing of a crack with narrow wedge-shaped plastic zones, and (b) The Dugdale model of a crack [1].

that takes into account the attendant local deformation and the rate dependence of the flow stress.

3) Information about the flow properties of metals at high strain rates can be extracted from crack-speed measurements with the aid of a suitable analysis.

To test these ideas, experimental measurements of the crack speed are compared with calculated values that are based on the dynamic solution for a Dugdale crack. This model [1] represents the plastically deformed regions as thin extensions of the crack (Fig. 1). A uniform internal stress equal to the yield stress of the material is imposed in the yielded region. In Dugdale's analysis [1], which employs a solution given by Muskhelishvili [2], the length of the zone is adjusted such that the stress singularities at the crack tips are abolished. With no further refinements, the model predicts values of plastic-zone size and crack-tip displacement. These predictions are found to be in agreement with experimental results under conditions where narrow and wedge-shaped plastic zones are observed [1, 3].

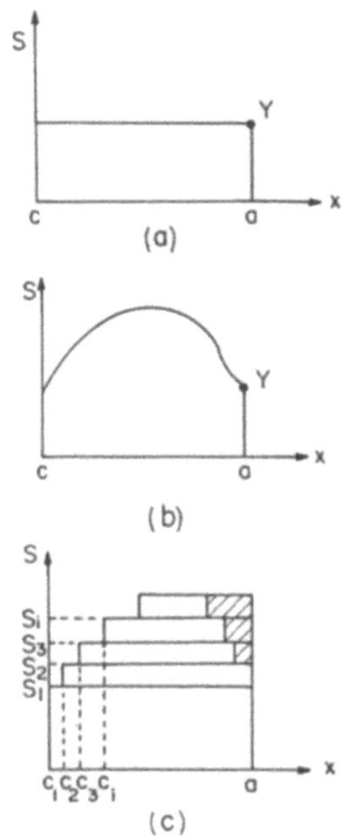

Fig. 2. Distributions of internal tension. (a) Uniform distribution treated by Dugdale [1], (b) Varying distribution, and (c) Stepwise approximation of (b).

The possibility of using the Dugdale model to treat the deformation within the plastic region, produced by both stationary and running cracks, was first examined by Goodier and Field [4]. Further progress was made by Rosenfield, Dai, and Hahn [5], who extended the Dugdale model to handle a flow stress that varies along the length of the plastic zone as a result of strain-hardening or rate effects. Their model was formed by superposing simple Dugdale solutions to obtain a stepwise distribution, as shown in Fig. 2. A simple model for straining within the zone was also introduced to relate the displacements to the strains and the internal stress to the flow stress. The final link was provided by the stress/strain/strain-rate characteristics of the material. Using this model, Rosenfield, Dai, and Hahn [5] actually calculated ductile-crack speeds in a hypothetical material, but no attempt was made to check the calculations with experiments.

In the present paper, a theoretical analysis of the dynamic solution of the Dugdale crack is derived. These calculations are compared with crack-speed measurements on 0.00175-in.-thick, cold-worked steel foil, a material selected because it displays narrow wedge-shaped zones like the model. The calculations and the experiments lend support to the idea that ductile-crack speed is controlled by the rate of plastic deformation. Finally, a simplified theory based on the key factors influencing crack speed is constructed. This theory, together with Kobayashi and Engstrom's [6] crack-speed data on 7075-T6 aluminum, is used to demonstrate the possibility of deriving flow-stress data from crack-propagation experiments.

A DYNAMIC-CRACK-PROPAGATION MODEL

The starting point for the development reported in this paper is a solution for a Dugdale crack propagating at a constant speed, U, in an infinite sheet subjected to a uniform tensile loading, T. This solution, which includes the static solution as a special case, was developed by following the work of Radok [7] and Sneddon [8]. In other respects, however, the following development parallels that given by Rosenfield, Dai, and Hahn [5].

An outline of the derivation of the dynamic solution appears as Appendix A. The key results given there are, first, the finiteness condition

$$\frac{c}{a} = \cos \frac{\pi}{2} \frac{T}{Y} \tag{1}$$

where Y is the yield stress of the material and c and a are as shown in Fig. 1. The normal stress on $y = 0$ at time $t = 0$ is

$$\sigma_y = \begin{cases} O & |x| < c \\ Y & c \leqslant |x| \leqslant a \\ \frac{2}{\pi} Y \tan^{-1} \frac{x}{c} \sqrt{\frac{a^2 - c^2}{x^2 - a^2}} & a \leqslant x| \end{cases} \tag{2}$$

while the normal displacement, v, is

$$v = \frac{2}{\pi} \frac{Y}{E} \mathscr{L}(U) \begin{cases} x \log \left| \frac{x\sqrt{a^2 - c^2} - c\sqrt{a^2 - x^2}}{x\sqrt{a^2 - c^2} + c\sqrt{a^2 - x^2}} \right| + c \log \left| \frac{\sqrt{a^2 - c^2} + \sqrt{\bar{a}^2 - x^2}}{\sqrt{a^2 - c^2}\ \sqrt{a^2 - x^2}} \right| & |x| \leqslant a \\ O & |x| \geqslant a \end{cases} \tag{3}$$

where $\mathscr{L}(U)$ is a function of the crack speed and the wave velocities in the infinite medium U_1 and U_2. It is readily seen that not only are (1) and (2) independent of U, but that they are exactly the same as in the static problem. Furthermore, with the exception of the factor $\mathscr{L}(U)$, the expression obtained for the displacement is the same as that obtained for the Dugdale model by Goodier and Field.

The function $\mathscr{L}(U)$, which multiplies the static displacements, is shown in Fig. 3. It can be seen that as the crack speed increases from zero $\mathscr{L}(U)$, the values of displacement steadily increase from their

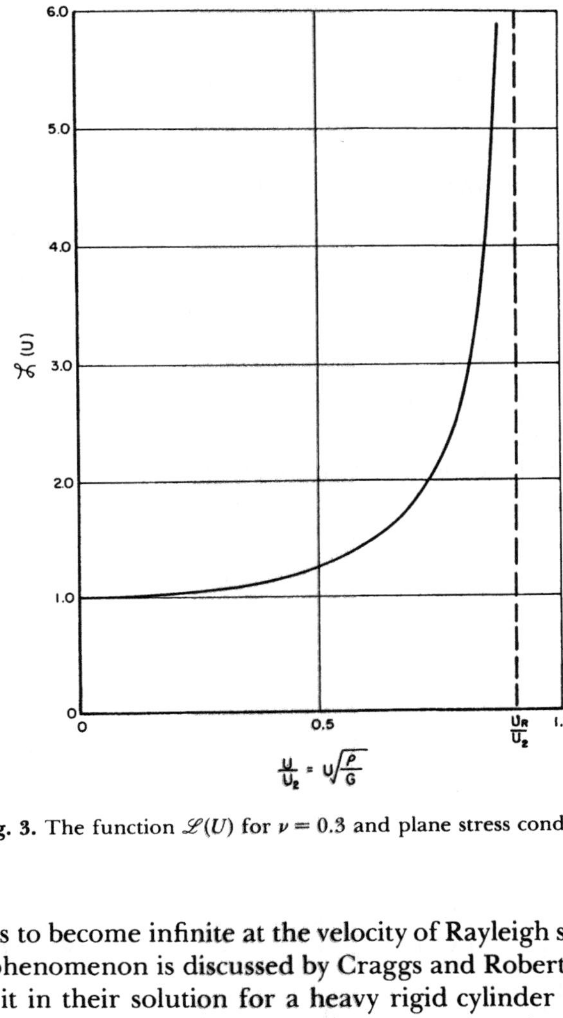

Fig. 3. The function $\mathscr{L}(U)$ for $\nu = 0.3$ and plane stress conditions.

static values to become infinite at the velocity of Rayleigh surface waves U_R. (This phenomenon is discussed by Craggs and Roberts [9] who encountered it in their solution for a heavy rigid cylinder moving over the surface of an elastic half space.) However, when the crack speeds are small in comparison with U_R($U_R = 9800$ ft/sec in steel, 9600 ft/sec in aluminum [10]), the dynamic effects are not significant, as Fig. 3 shows. In the present work, the crack speeds are always small enough so that the function $\mathscr{L}(U)$ can be set equal to its static value (unity for plane stress).

Following Rosenfield, Dai, and Hahn, a stepwise varying distribution of internal stress $S(x)$, as shown in Fig. 2, can be envisaged. Expressions for the plastic-zone length and the displacements are then obtained by linear superposition of (1) and (3). Despite the fact that the equations can only be solved when a finite number of increments are specified, it is also useful to provide the more exact expressions obtained for the limiting case of a continuously varying distribution. These are:

$$T = \frac{2}{\pi} \int_c^a \frac{S(\zeta)}{\sqrt{a^2 - \zeta^2}} \, d\zeta \tag{4}$$

which is the generalization of (1), and

$$v = -\frac{2}{\pi E} \int_c^a S(\zeta) \frac{\partial}{\partial \zeta} \psi(x, \zeta) \, d\zeta \tag{5}$$

where

$$\psi(x, \zeta) = x \log \left| \frac{x\sqrt{a^2 - \zeta^2} - \zeta\sqrt{a^2 - x^2}}{x\sqrt{a^2 - \zeta^2} + \zeta\sqrt{a^2 - x^2}} \right| + \zeta \log \left| \frac{\sqrt{a^2 - \zeta^2} + \sqrt{a^2 - x^2}}{\sqrt{a^2 - \zeta^2} - \sqrt{a^2 - x^2}} \right| \tag{6}$$

which is the generalization of (3).

In order to properly determine $S(x)$, which must depend on the strain and the strain-rate distribution in the plastic zone, a relation between displacement, $v(x)$, and the maximum plastic strain, $\epsilon(x)$, at each point must be developed. Then, by relating ϵ and $\dot{\epsilon}$ to S at each point, a unique determination of all the quantities of interest can be made. It should be emphasized, however, that these essential parts of the solution are not included in the solution to the boundary value problem represented by the propagating Dugdale crack, but are obtained from a supplemental model.

An approximate relation between v and ϵ at each point along the zone can be derived by treating the zone as a collection of tensile bars. Each bar is considered to be elongated an amount $2v$ by straining in the y-z-plane only, i.e., through the thickness.* Because the material of interest displays so little strain hardening, a neck will tend to form as soon as yielding begins. Initially, the necking will be confined within a distance, d, comparable to the sheet thickness. As can be seen by referring to Figs. 4a and 4b, the deformation will occur at constant volume, so that $A_0 = A_1$, and it will tend to be confined to the minimum section of the neck. For these circumstances the following relation can be derived:

$$\frac{b}{b_1} = \left(\frac{1}{1 - Bv}\right).$$

Here, b is the original sheet thickness, b_1 is the minimum thickness of a particular section, and B is a constant related to the shape of the neck (the derivation can be found in a companion paper [11]). By definition, the maximum true strain is

$$\epsilon = \log \left(\frac{b}{b_1}\right) \tag{7}$$

* Large strains in the plane of the sheet are precluded by a zone whose shape is narrow and elongated in the x-direction.

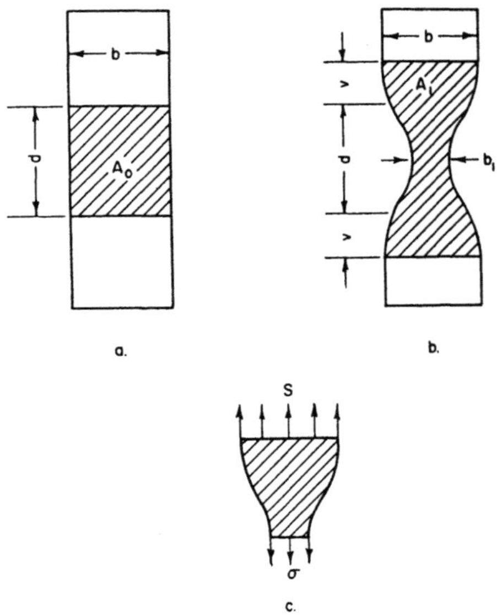

Fig. 4. Schematic drawing of the sheet cross section in front of the crack (a) Before yielding, (b) After some yielding, and (c) Showing the relation between the internal stress S, and the true flow stress σ. Constancy of volume requires that $A_0 = A_1$; the symbol b represents the original thickness of the sheet.

so that

$$\epsilon = \log\left(\frac{1}{1 - Bv}\right) \tag{8}$$

which is the strain-displacement relation used in this paper.

To complete the mathematical formulation of the problem, the strain rate, $\dot{\epsilon}$, is, using (8),

$$\dot{\epsilon} = U\frac{d\epsilon}{dx} = \frac{BU}{1 - Bv}\frac{dv}{dx}. \tag{9}$$

The true stress at the minimum section is then given by the function describing the flow properties of the material: $\sigma = \sigma(\epsilon, \dot{\epsilon})$.* Finally, the true stress is related to the internal stress by noting that from Fig. 4c, $S = \sigma\dfrac{b_1}{b}$, or, that from (7),

$$S = \sigma(\epsilon, \dot{\epsilon}) \exp(-\epsilon). \tag{10}$$

* Throughout this paper we assume that stress is a single-valued function of strain and strain rate, independent of history; i.e., a mechanical equation of state exists. To the extent that the experimental material fails to obey the mechanical equation of state, the numerical results will be in error. However, it is not expected that this error will be large.

CONSTITUTIVE RELATIONS

The velocity measurements described in this paper were performed or cold-worked, 0.08% carbon steel foil, 0.00175 in. thick.* The material was especially selected because cracks in the foil produce plastic zones shaped like those of the Dugdale model (as shown in Fig. 5). The foil is not only experimentally easy to handle, but also enables the existing data on flow properties of steel to be used.

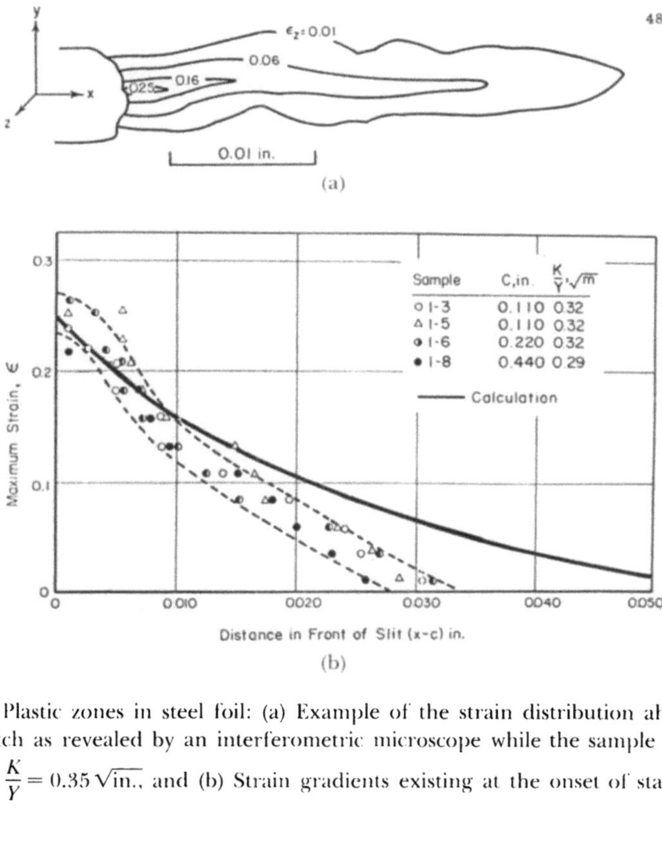

Fig. 5. Plastic zones in steel foil: (a) Example of the strain distribution ahead of a sharp notch as revealed by an interferometric microscope while the sample was fully loaded to $\frac{K}{Y} = 0.35\sqrt{\text{in.}}$, and (b) Strain gradients existing at the onset of stable crack growth.

Plastic Flow

It was assumed that the following relation, based on results for a number of steels [12–29], is also applicable to the foil:

* The foil, which was made available by Dr. E. T. Stephenson of the Bethlehem Steel Corporation, is also the subject of a study of local yielding and crack extension reported in a companion paper [11]. The as-received thickness of the foil was 0.00185 ± 0.0001 in., which was reduced to 0.00175 ± 0.0001 in. as a result of electropolishing.

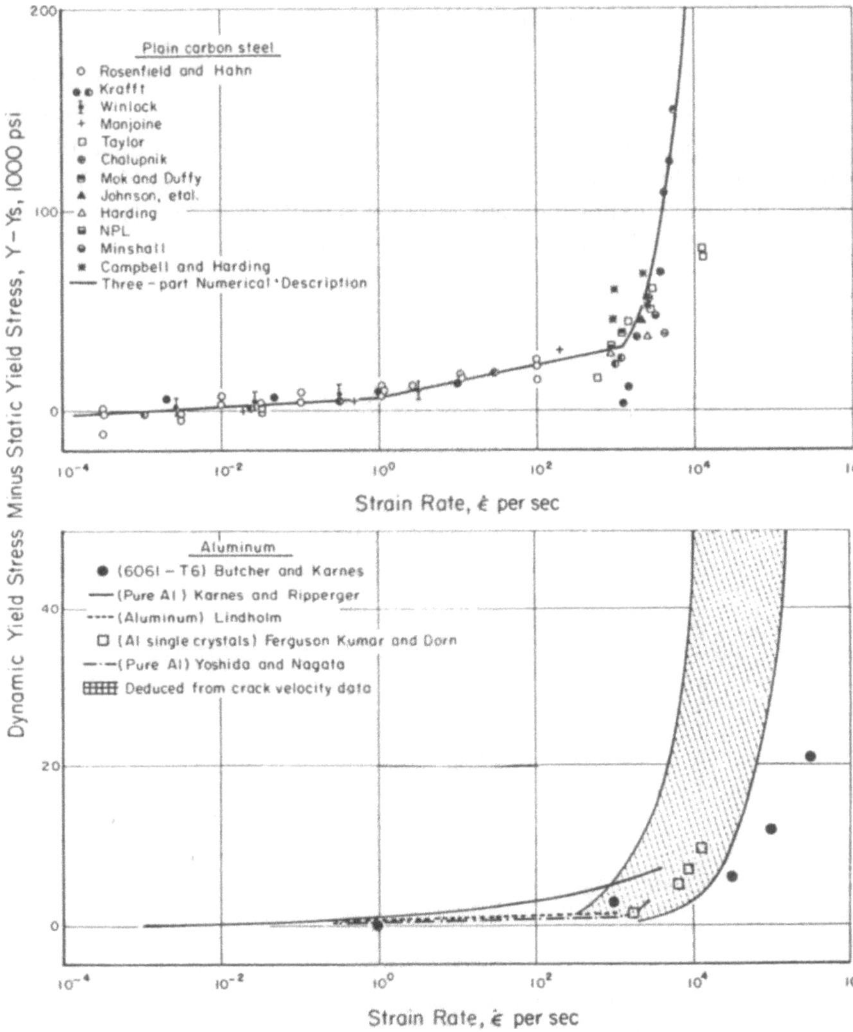

Fig. 6. The strain rate dependence of the yield stress of steel[13-24] and aluminum[25-29] at room temperature.

$$Y - Y_s = \begin{cases} 6000 + 2000 \log_{10} \dot{\epsilon} & 1 \geq \dot{\epsilon} \\ 6000 + 8000 \log_{10} \dot{\epsilon} & 1110 \geq \dot{\epsilon} \geq 1 \\ 25\dot{\epsilon} & \dot{\epsilon} \geq 1110. \end{cases} \tag{11}$$

Here, Y(psi) is the yield stress, Y_s(psi) is the yield stress for a strain rate of 10^{-3} sec^{-1} (the "static" yield stress), and $\epsilon(\text{sec}^{-1})$, as previously defined, is the plastic strain rate. This relation, together with the data from which it was derived, is shown in Fig. 6.

The value $Y_s = 105{,}000$ psi was obtained from tensile tests. The re-

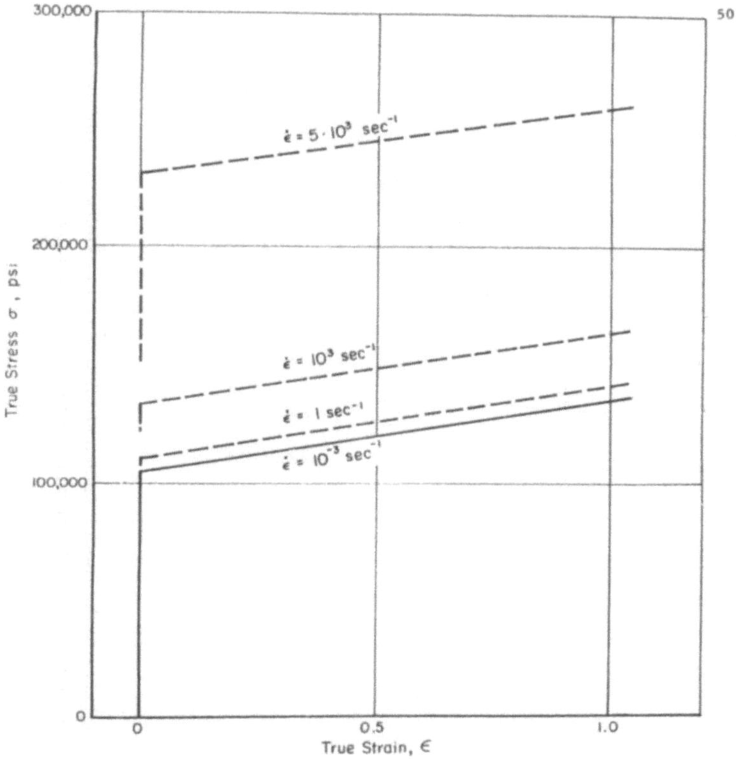

Fig. 7. Examples of flow curves of steel foil as defined by equations (11) and (12).

lation between flow stress and plastic strain after the onset of yielding could not be obtained from tensile tests,* but was deduced from data reported by Embury, Keh, and Fisher [30]. Their results indicate that a mild steel, cold reduced to the 105,000-psi strength level, continues to deform with a modest, approximately linear rate of strain hardening:

$$\sigma = Y + \phi\epsilon \tag{12}$$

where σ(psi) is the true stress, ϵ the true plastic strain, and ϕ (= 30,000 psi) is a strain-hardening coefficient, which was assumed to be independent of the strain rate. The function $\sigma(\epsilon, \dot{\epsilon})$ is obtained by combining (11) and (12). Examples of flow curves calculated in this way are shown in Fig. 7.

* Foil samples begin to neck soon after the onset of yielding. The energy stored in the testing machine immediately extends the neck to fracture in an uncontrolled manner without further crosshead motion because the overall extension accompanying necking of the foil is extremely small.

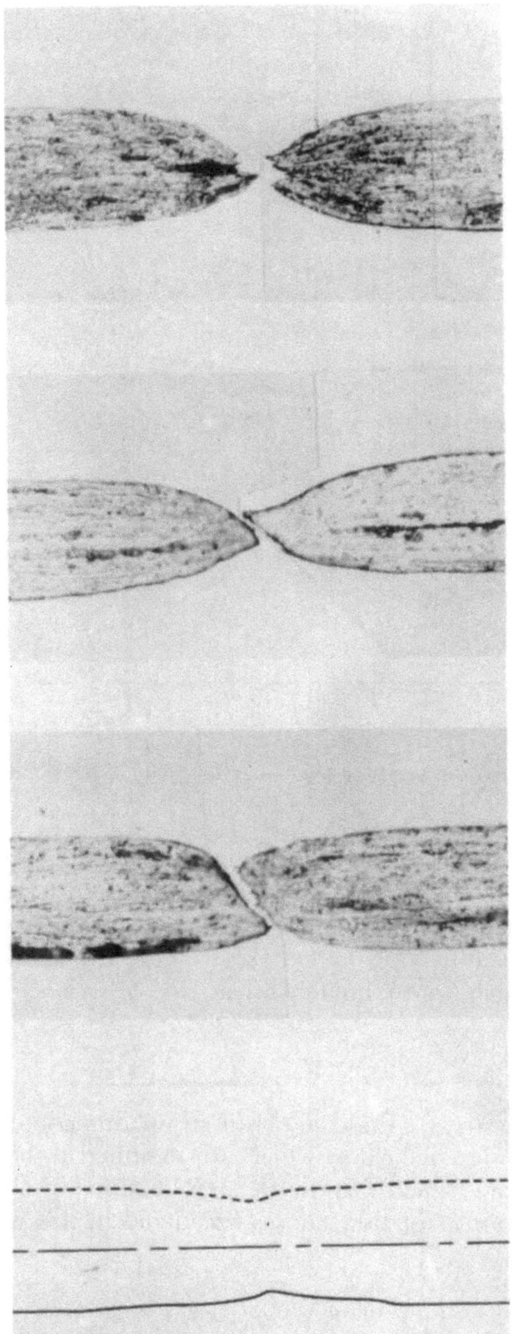

a b c d

Fig. 8. Cross sections through actual plastic zones in the steel foil: (a) close to the root of a notch just before crack extension (from interferometric pattern, (b) same section as (a) after penetration by a slow moving crack, (c) Section fractured by a fast-moving crack, and (d) Cross section of an unnotched tensile coupon.

Strain-Displacement Relation

The value of the strain-displacement coefficient, B, appropriate for steel foil was determined by fitting (8) to actual strain and displacement measurements. As described in the companion paper [11], the value $B = 950$ in.$^{-1}$ gives a reasonably good description of the deformation all along the zone before crack extension, and also of the deformation observed after fast fracture. Examples of the foil cross section just before and after slow and fast fracture are shown in Fig. 8.

Ductile Fracture

The broken sections shown in Fig. 8 reveal that the foil fractures mainly by the ductile shear mode. For a given stress state, the development of a ductile shear crack is probably associated with a critical strain that is related to the microstructure and the flow properties of the foil – at least to a first approximation [31]. Results of crack-extension experiments performed on the same foil and reported separately [11] support this idea. As shown in Table 1, crack extension begins when the strain at the crack tip reaches a value $\epsilon_c^* = 0.24$. This value is independent of the average applied stress on the foil and of the crack length. Furthermore, this value is not exceeded during the period of stable crack growth.

Table 1

CRACK-EXTENSION PROPERTIES OF STEEL FOIL [11]

c, in.[a]	0.110	0.220	0.440
ϵ_c^* (measured)[b]	0.21–0.25	0.22–0.24	0.23–0.265
T^*, ksi (measured)[b]	58–63	41–44	28.4–30.4
T^*, ksi [calculated for $\epsilon_c^* = 0.24$[c]]	57	42	30
K_c, ksi $\sqrt{\text{in.}}$ (measured)[b]	34–37.0	36–36.5	33.4–42.3
K_c, ksi $\sqrt{\text{in.}}$ [calculated for $\epsilon_c^* = 0.24$[d]]	34	35	36
K_c, ksi $\sqrt{\text{in.}}$ [calculated for $\epsilon_c^* = 1.10$[c,e]]	43	49	53

(a) Centrally located, spark-machined slit 0.006 in. wide in a sheet.
(b) The range quoted here reflects experimental scatter only.
(c) Obtained from calculations described in Fig. 11.
(d) The two values quoted reflect the increase in crack length and stress during growth stage.
(e) Hypothetical K_c value associated with $\epsilon_c^* = 1.10$, a value displayed only by fast-moving cracks.

Sections of the foil penetrated by fast-moving cracks and also unnotched tensile coupons experienced a much larger strain before frac-

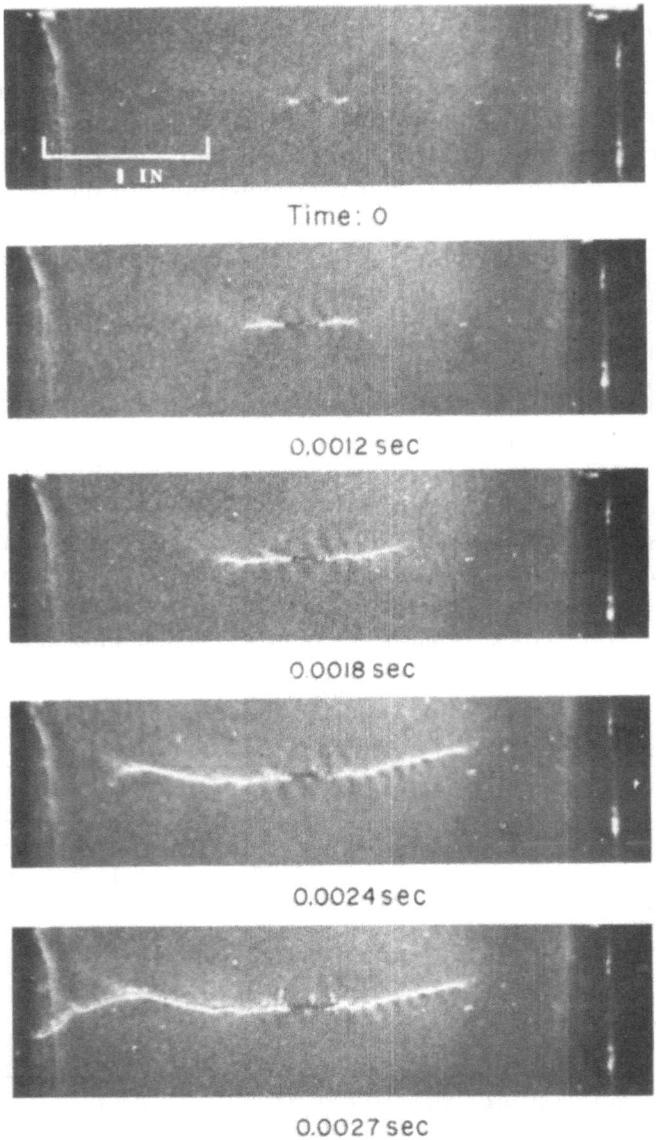

(a)

Fig. 9. Crack speed data for a sample loaded to $T/Y = 0.87$: (a) Successive frames from 3300 fps motion picture film, and (b) Crack propagation record derived from the film showing the initial constant speed stage AB.

ture—$\epsilon_c^* \simeq 1.10$ (see Fig. 8)—and this is again independent of stress level or crack length. Since both the precracked foils and the unnotched coupons break very quickly, the value of $\epsilon_c^* = 1.10$ may be a consequence of the speed of deformation. One possible explanation is that it is related to the greater rate sensitivity displayed by steel at rates

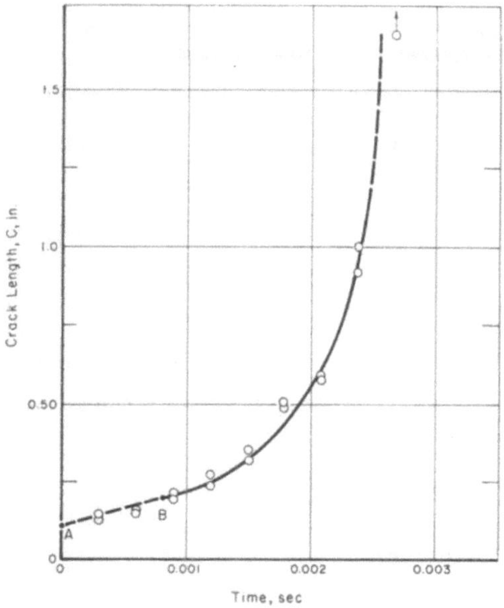

Fig. 9 (b)

above 1110 sec^{-1}.* On the basis of these observations, the following ductile-fracture criterion was adopted for the steel foil:

$$\epsilon_f^* = \begin{cases} 0.24, & U \sim 0\dagger \\ 1.10, & U > 1 \text{ ft/sec.} \end{cases} \tag{13}$$

EXPERIMENTAL MEASUREMENTS AND RESULTS

Crack-speed measurements were performed on 3.25 × 3.5-in.-foil coupons initially containing a centrally located slit, 0.220 in. long by 0.006 in. wide ($c = 0.110$ in.), introduced by spark machining. The samples were cemented into grips, mounted in a creep machine, and dead loaded to a point just below the level for crack extension. Then, a high-speed motion-picture camera (3300 frames per sec) was activated and, simultaneously, an additional load increment was applied. In this way, crack propagation was recorded at a number of different load levels. Sample frames from one of the films are shown in Fig. 9a. Figure 9b is a record of crack propagation obtained by measuring the crack length on successive frames. Velocities were calculated from the slope of the curve.

In all cases the cracks accelerated after an initial period of constant speed $U = 1.6$ to 9.0 ft/sec (Region A-B in Fig. 9b), which terminated

* Crack-velocity measurements described in the next section indicate that the larger fracture strain is encountered when the crack speed exceeds 1.6 to 9.0 ft/sec. According to the calculations, it is in this velocity range that the crack-tip strain rate first approaches 10^3 sec^{-1}.

† The limit $U \sim 0$ (rather than $U = 0$) is used to draw attention to the fact that, during stable crack growth, the velocity is small but finite. It is not clear whether the change in ϵ_f^* from 0.24 to 1.10 is continuous at intermediate speeds $0 \lesssim U < 1$ ft/sec.

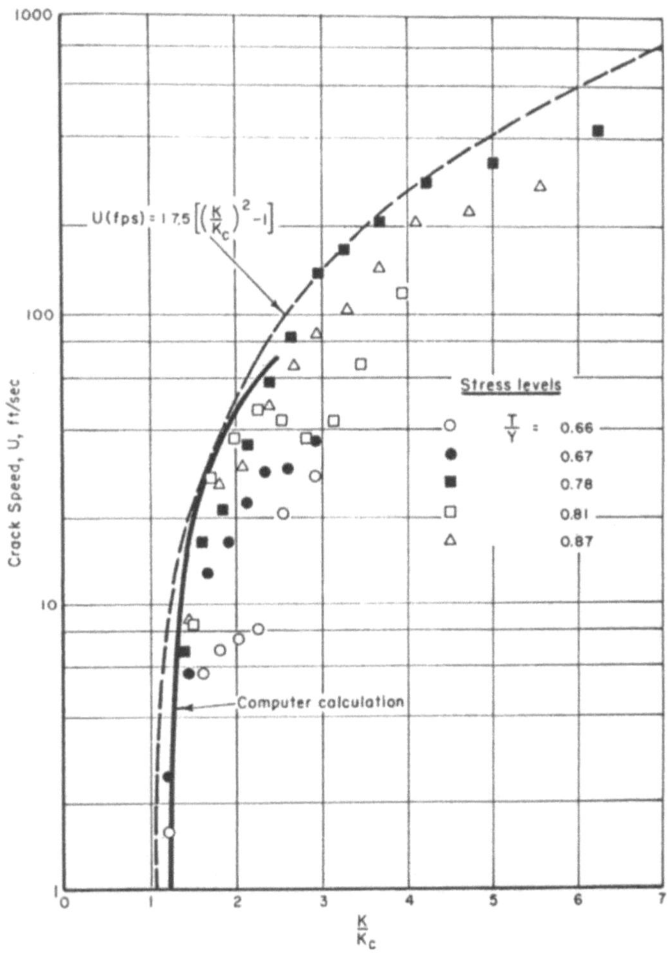

Fig. 10. Comparison of measured and calculated ductile crack speed on steel foil at different load levels. Computer calculations are for a plastic zone divided into 8 intervals.

when the crack attained a half-length $c = 0.22$–0.27 in. The stress-intensity values corresponding to these lengths are $K = 67$ to 78 ksi $\sqrt{\text{in.}}$, ($K = T\sqrt{\pi c}$), and this is consistent with $K_c \approx 50$ ksi $\sqrt{\text{in.}}$ at the onset of crack extension and $\epsilon_c^* = 1.10$ if the subsequent enhancement attending stable crack growth * and the initial acceleration are taken into account (see the last row of Table 1, $K_c = T^*\sqrt{\pi c}$ refers to the conditions at the onset of propagation). In other words, the crack does not

* Note that the steel foil displays some stable crack growth amounting to about 10 percent of the initial crack length before unstable propagation begins [11]. This region of stable growth is not recognized by the model, which becomes unstable as soon as the crack begins to extend. However, the stress-level difference observed between the onset of stable growth and the onset of unstable propagation is relatively small, about 10 percent, and largely within the experimental scatter. Consequently, slow growth is not a serious complication.

begin to accelerate (Region A-B in Fig. 9b) until the stress requirements for the larger strain are satisfied. This result, together with metallographic sections taken through the fractures (Fig. 8), suggests that the initial period of constant speed coincides with a change in fracture strain from $\epsilon_c^* = 0.24$ to $\epsilon_c^* = 1.10$. Beyond this point, the crack propagates with a constant value of $\epsilon_c^* = 1.10$. It is this phase that is amenable to analysis.

The motion-picture films provide speed values and the corresponding instantaneous crack lengths. To simplify the presentation, each crack-length–stress combination was reduced to a stress-intensity value $K = \beta T\sqrt{\pi c}$, where β is the static finite-plate correction factor [32]. The results are presented in Fig. 10 as a function of $\frac{K}{K_c}$. Finally, by taking $K_c = 53$ ksi $\sqrt{\text{in.}}$ (an estimate of K_c for $\epsilon_c^* = 1.10$, and $U = 0$; see the last row in Table 1)* and by ignoring the constant-velocity region, complications attending the change of ϵ_c^* are avoided.

Figure 10 shows that cracks in the foil accelerate continuously at constant load. Although the data points display considerable scatter, it is evident that results for diffierent stress levels overlap. This illustrates that the basic equivalence of T and $\sqrt{c}$ established for stationary cracks also extends to running ductile cracks. The origin of the scatter has not been explored, but it seems likely that it is related to misalignment of the foil in the grips, the problem of cementing them evenly, and, possibly, slight differences in the thickness of the foil.

Finally, the question of temperature deserves comment. Although the experiments were performed at room temperature, other work [33, 34] has shown that nearly 100 percent of the work expended to produce plastic deformation within the zone is converted to heat. This is enough to raise the temperature of the plastic zone in the steel foil to about 400° C under adiabatic conditions. A crude estimate of the heat conducted to the cold material on either side of the plastic zone as the crack propagates suggests that the peak temperature rise for the highest speeds is substantially less than 400° C.† Furthermore, the rise will occur near the tip of the crack, where rates in excess of 1110 $\sec^{-1}$ are experienced and where the linear rate dependence is obeyed. Recent results of both Ferguson, Kumar, and Dorn [28] and Mukherjee, Ferguson, Barmore, and Dorn [35] suggest that even a 400° C rise would not influence the flow stress significantly in this strain-rate regime.

* The value $K_c = 53$ ksi $\sqrt{\text{in.}}$ is an estimate of the value of K, intermediate between the values for the onset and end of stable crack growth, and corresponding to $c = 0.220$ in.

† Approximate calculations by N. Levy and J. R. Rice (Brown University, private communication, 1967) based on a moving Dugdale zone 1 cm in length lead to temperature rises from 25 to 100° C in the 100 to 1000 fps range of speeds for mild steel and 2024 Aluminum alloys. These calculations are based on room temperature values of yield strength and thermal properties, with rate dependence neglected.

CRACK-SPEED COMPUTATIONS

There are three distinct phases of the general problem that may be considered: subcritical loading, critical loading, and unstable propagation. These are indicated in Fig. 11. The simplest situation, subcritical loading, occurs when the crack has produced local yielding but no rupturing, and this uses only a portion of the available information. The critical loading case (in reality, the transition point between stable and propagating cracks) is obtained by supplementing these equations with a criterion for fracture. Finally, adding the strain-rate dependence to the stress-strain behavior of material enables the moving crack solution to be obtained.

From the computational point of view, in the most general case, there are three more variables that must be specified arbitrarily. In the static-crack case, $U = 0$. This leaves two free choices in the subcritical case, i.e., T and c. Only one free choice can be made in the critical case, either T^* or c. Finally, in the dynamic case where U must be determined, two free choices, T and c, are again available. This results from the fact that the rate dependence introduces just as many equations as unknowns.

The relevant equations (see Fig. 11) were transformed into a set of nonlinear simultaneous equations by subdividing the plastic zone into N intervals. These equations were programmed for Battelle's CDC 6400 computer. The technique used to obtain a solution for the dynamic problem is quite similar to that used for the subcritical and critical loading problems, which are described in the companion paper [11]. The portions of the computational procedure pertaining to the dynamic problem are presented in Appendix B.

A comparison of a subcritical calculation with actual measurements on the foil taken from [11] is shown in Fig. 5. These results are reproduced to show that the calculations offer a good picture of both the strain and the strain gradient near the crack tip. Since the fracture criterion is based on the strain at the crack tip, the calculations also reproduce measured values of T^* and K_c, the fracture stress and fracture toughness; this is illustrated in Table 1.

In Fig. 12 are shown the individual values of ϵ, $\dot{\epsilon}$, σ, and S in the plastic region for each of five different crack speeds (each corresponding to a particular crack length). The most important conclusions that can be drawn from these results are that 1) the strain gradients at the crack tip are relatively insensitive to crack speed and 2) because the strain gradient is much steeper near the crack tip, the strain rates and hence the flow stresses are much greater at the crack tip than elsewhere. Thus, the model shown in Fig. 5 to be most accurate at the

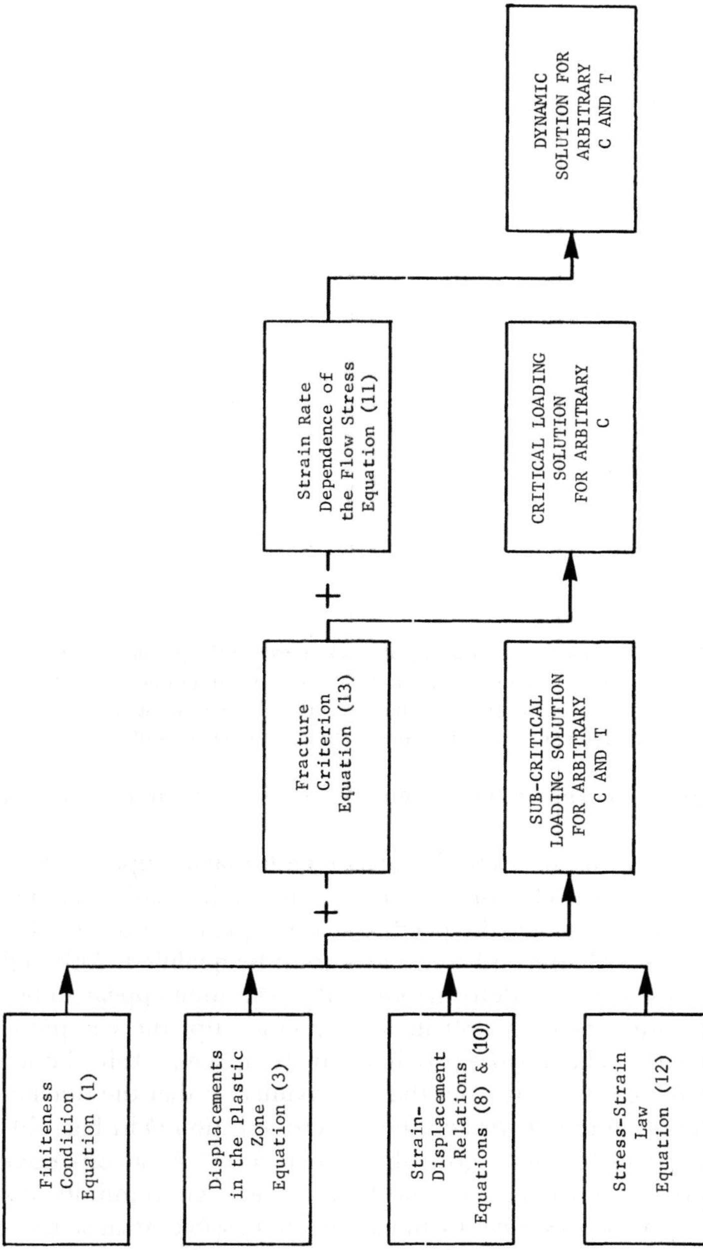

Fig. 11. Schematic representation of the computational procedure.

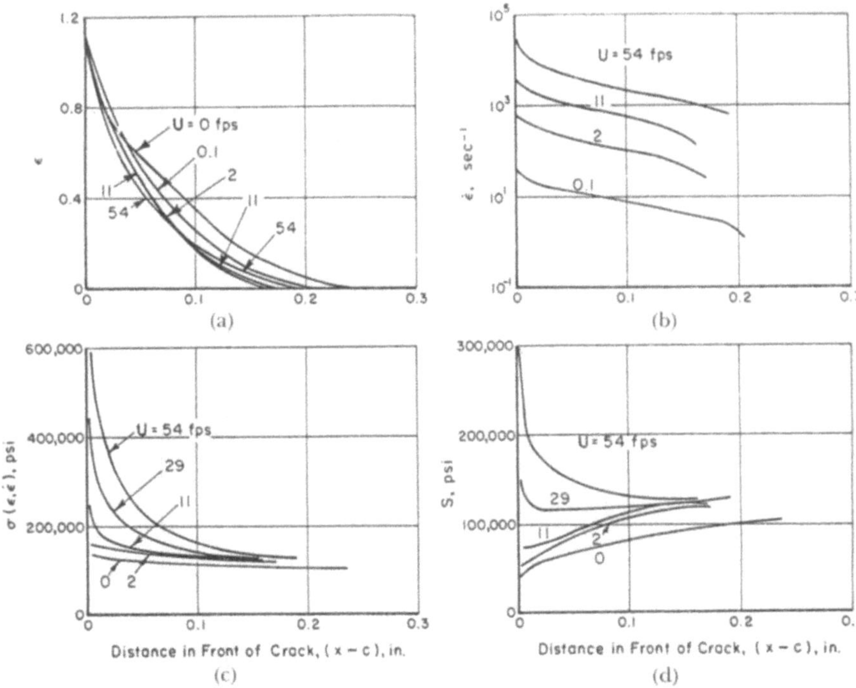

Fig. 12. Calculations of the influence of crack speed on the plastic zone generated by a propagating ductile crack in the steel foil: (a) Plastic strain gradient, (b) Plastic strain-rate gradient, (c) Flow stress gradient, and (d) Internal stress gradient. The calculations employed $N = 32$ intervals within the zone, each about 0.007 in. wide.

crack tip is therefore most reliable in the region of greatest importance.

The first of the above conclusions suggests that a simple theory could be developed that relies on the fact that the strain gradient at the crack tip is insensitive to crack speed. Such a representation will be considered later. The second conclusion must be qualified. Although the calculations of plastic deformation within the zone appear to be most reliable within about 0.050 in. of the crack tip, the computational scheme is not valid at close proximity to the tip, i.e., within about 0.010 in. The difficulty stems from the approximation that the ductile-fracture process does not weaken the foil cross section (b in Fig. 4b) until $\epsilon_c = \epsilon_c^*$, at which point b goes abruptly to zero.* As a consequence S goes discontinuously to zero and both σ and $\dot{\epsilon}$ go to infinity at $x = c$. However, crack speed calculations are still possible because the stress singularity is integrable, and this is discussed in Appendix B.

A comparison between the computed crack speeds and the experi-

* In real cracks the deterioration of the cross section is likely to occur over a small, but finite range of strain [36]. The S-gradient will be steep, but finite and ϵ will take on finite values at $x = c$.

mentally observed values for the steel foil is shown in Fig. 10. It can be seen that although the computed values are consistently high, the deviations are not large. This discrepancy could be caused by a number of factors: the complications responsible for stable crack growth, and the fact that the real zone shape, strain gradient, and constraints within the zone are not faithfully reproduced. Greater precision cannot really be expected. The model gives reasonable agreement with the measurements and this can be regarded as indirect evidence that the ductile crack speed is controlled by increased resistance to plastic flow at high strain rates. The fact that the experimental material was carefully tailored to match the model strengthens the conclusion, but does not detract from its generality. The same mechanism can govern crack speed in materials when the shape of the plastic zone does not conform to the Dugdale model. Some of the equations derived in this paper will then be invalid, but the general influence of the dynamic flow properties will be the same. Evidence along these lines is presented in the next section.

CALCULATING FLOW-STRESS DATA FROM CRACK-SPEED MEASUREMENTS

The dynamic problem outlined in Fig. 11 also offers the possibility of calculating an unknown parameter of the strain-rate law when U, T, and c are known. This can be demonstrated with the help of a simple analysis, developed in Appendix C, which is based on the following key assumptions:

(1) The plastic-strain gradient is relatively insensitive to the crack speed, and
(2) A linear rate dependence of the form $Y = Y_s + F\dot{\epsilon}$ predominates.

The resulting expression is:

$$U \approx \frac{Y_s}{F\left(\dfrac{\Delta\epsilon}{\Delta x}\right)}\left[\frac{K^2}{K_c^2} - 1\right] \tag{14}$$

where U = crack speed
Y_s = static yield stress
F = linear rate coefficient
$\left(\dfrac{\Delta\epsilon}{\Delta x}\right)$ = effective plastic strain gradient
$K_c = T_1\sqrt{\pi c_1}$ and
$K = T_2\sqrt{\pi c_2}$.

The subscripts 1 and 2 refer to the values of T and c at the onset of cracking and at the velocity U, respectively.

For example, the appropriate values for the steel foil are:

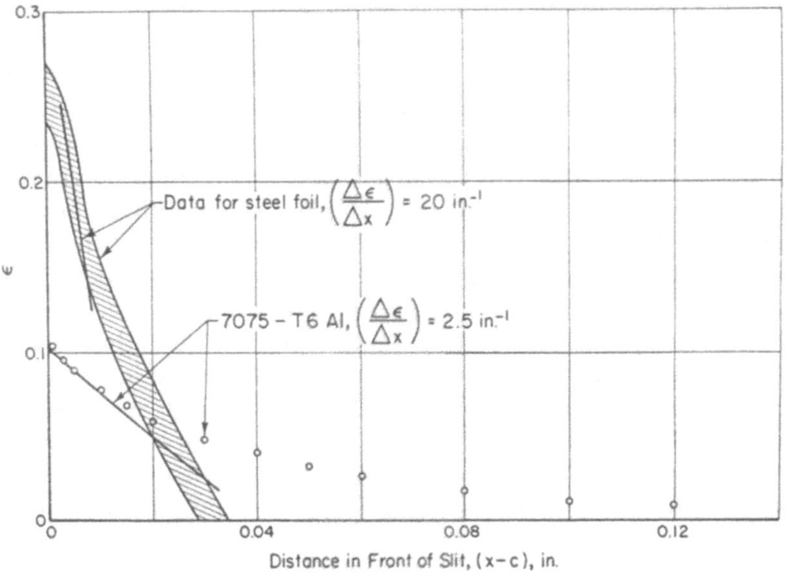

Fig. 13. Strain gradients observed in 0.00175 in.-thick steel foil and 1/16 in.-thick 7075-T6 Al under plane stress conditions. Results for the foil are taken from Figure 5 [11]. Results for the aluminum alloy are from unpublished work at Battelle on center-notched coupons. The gradient shown corresponds to a stress level close to fracture, i.e., $\frac{K}{K_c} = 0.95$, and was obtained from measurements of a plastic replica of the coupon surface taken while the coupon was fully loaded.

$$Y_s = 105{,}000 \text{ psi}, \; F = 25 \text{ psi-sec, and } \left(\frac{\Delta\epsilon}{\Delta x}\right) \sim 20 \text{ in.}^{-1}$$

(see Figs. 12a and 13), or

$$U(\text{ft/sec}) \approx 17.5\left(\frac{K^2}{K_c^{\,2}} - 1\right). \tag{15}$$

As shown in Fig. 10, this simple expression offers a very good description of the experimental results. With this in mind the simple analysis was applied to crack speed measurements in $\frac{1}{16}$-in.-thick 7075-T6 aluminum reported by Kobayashi and Engstrom [6]. As shown in Fig. 14, (14) can be fitted to their measurements for velocities up to about 2000 ft/sec,* provided

$$F \approx \frac{Y_s}{480\left(\frac{\Delta\epsilon}{\Delta x}\right)}. \tag{16}$$

* Figure 13 illustrates that cracks in the 7075-T6 sheet attain a velocity of about 2000 ft/sec when $\frac{K}{K_c} = 2.5$, and do not accelerate with further increases in the stress intensity. This limiting speed may be associated with the experimental problem of maintaining the stress on the plate as it fractures. It does not seem to be connected with the Rayleigh wave velocity, which is ~9600 ft/sec for aluminum.

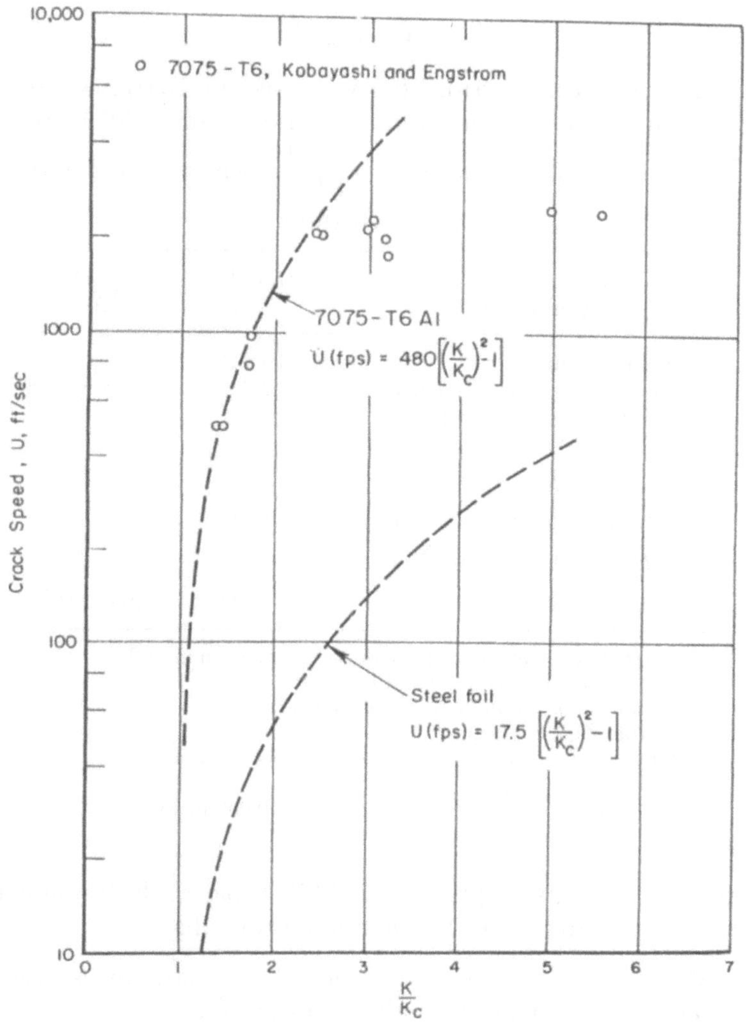

Fig. 14. Comparison of crack speed data and calculations. Data for 7075-T6 Al were reported by Kobayashi and Engstrom [6].

The value of Y_s for 7075-T6 aluminum is ~70,000 psi. Consequently, the coefficient F can be determined from the crack speed measurements once the gradient is known.

The evaluation of the gradient for 7075-T6 aluminum illustrates another interesting complication. As shown in Appendix C, the quantity $\left(\frac{K}{Y}\right)^2$ decreases as the crack accelerates according to the relation

$$\left(\frac{K}{Y}\right)^2 \approx \left(\frac{K_c}{K}\right)^2 \cdot \left(\frac{K_c}{Y_s}\right)^2. \tag{17}$$

This means that the character of the zone tends to change in the direc-

Table 2

RELATIONS AMONG THE CHARACTER OF THE DEFORMATION WITHIN THE PLASTIC ZONE IN $\frac{1}{16}$-INCH-THICK 7075-T6 ALUMINUM, THE PARAMETER $\left(\frac{K}{Y}\right)^2 \frac{1}{b}$, AND THE EFFECTIVE STRAIN GRADIENT

Condition	$\left(\frac{K}{Y}\right)^2 \frac{1}{b}$	$\left(\frac{\Delta\epsilon}{\Delta x}\right)$
Plane stress zone	>10.4 (a)	
Slowly growing crack in 7075-T6 Al at the onset of fracture, $\frac{K}{K_c} = 1$	8.2 (b)	2.5 in.$^{-1}$ (d)
Rapidly propagating in 7075-T6 Al, $U = 2000$ ft/sec and $\frac{K}{K_c} = 2.3$	1.5 (c)	
Plane strain zone	<1.3 (a)	~ 40 in.$^{-1}$ (e)

(a) Based on a treatment of stationary cracks given in [37].
(b) Measured by Kobayashi and Engstrom.
(c) Estimate based on (b) and Equation (17).
(d) From Fig. 13.
(e) Based on a treatment of stationary cracks given in [37]:

$$\left(\frac{\Delta\epsilon_y}{\Delta x}\right) = \frac{2\epsilon_c^*}{\ell^*}, \quad \epsilon_c^* \sim 0.12 \text{ [37]}, \ \ell^* \sim 0.006 \text{ in. [37].}$$

tion from plane stress to plane strain [37]. Table 2, which summarizes values estimated for 7075-T6 aluminum, shows that

(a) At the onset of crack extension plastic relaxation is predominantly plane stress (through the thickness).
(b) By the time the crack attains a speed of 2000 ft/sec, the relaxation in front of the crack is approaching plane strain.

Evidence that such a change actually does occur is presented in Table 3, which shows that the through-the-thickness strain diminishes as the crack propagates.* The values quoted were obtained at Battelle on a $\frac{1}{16}$ in.-thick center-notched 7075-T6 coupon similar to the one tested by Kobayashi and Engstrom.

* The fracture displayed by this coupon was predominantly shear with a narrow (~20 percent of the thickness) ribbon of flat, fibrous fracture in the center. It is possible that the propagating crack front is not straight, the front in the center of the plate (associated with the flat fracture) leading the front near the plate surface. In that case, the unbroken portion of the plate behind the leading edge of the crack is less than half as thick

Table 3

INFLUENCE OF CRACK PROPAGATION ON PLANE STRESS (THROUGH-THE-THICKNESS) DEFORMATION IN A $\frac{1}{16}$-INCH-THICK, 7075-T6, CENTER-NOTCHED COUPON

Distance Propagated, in.	0	0.005	0.050	0.100	0.250	0.500	0.750
Relative Stress Intensity, $\left(\frac{K}{K_c}\right)$	1	1.03	1.07	1.13	1.29	1.53	1.8
Estimated Crack Speed, ft/sec [(a)]	0	20	70	140	270	650	1080
Maximum Through-the-Thickness Strain, percent [(b)]	11.0	6.9	6.0	4.8	5.4	5.9	5.7

(a) Calculated from $U = 480\left[\left(\frac{K}{K_c}\right)^2 - 1\right]$.

(b) Measured.

Results quoted in Table 2 also indicate that the change from plane stress to plane strain relaxation within the zone could produce a larger strain gradient. The results in Table 2 suggest that the gradient at $U = 2000$ ft/sec and $\frac{K}{K_c} = 2.3$ should fall somewhere between the values for plane stress and plane strain. $2.5\text{ in.}^{-1} < \left(\frac{\Delta\epsilon}{\Delta x}\right) < 40\text{ in.}^{-1}$. Combining this with (16) gives the result:

$$5 \text{ psi sec} > F > 0.3 \text{ psi sec.}$$

(Because $\left(\frac{\Delta\epsilon}{\Delta x}\right)$ is probably closer to 40 in.$^{-1}$, the value of F is likely nearer to 0.3 than 5 psi sec.) These two limits define the shaded band in Fig. 6, and are consistent with other measurements for aluminum [34]. Finally, it should be noted that the two limits imply an effective strain rate between $6 \cdot 10^4$ to $9.6 \cdot 10^5$ sec^{-1}. These values, while speculative, serve to delineate the upper bands of the strain rate regime involved in ductile crack propagation for aluminum.

ACKNOWLEDGMENTS

This study was sponsored by the Army Research Office, Durham, and the authors are grateful for this agency's support. The authors

as the plate, a situation making through-the-thickness relaxation more favorable. In that case, the leading edge of the crack front can be associated with a plane-strain plastic zone consistent with the data in Table 2. At the same time, the trailing edge can be associated with plane stress, and this could account for the residual 5 to 6 percent through-the-thickness strain displayed by the broken plate in Table 3.

also wish to thank C. R. Barnes for his expert handling of the laboratory work, P. F. Forgrave for the high-speed motion-picture photography, and Miss C. J. Pepper for her work on the manuscript.

References

1. D. S. Dugdale, J. Mech. Phys. Solids, *8*, 100 (1960).
2. N. I. Muskhelishvili, *Some Basic Problems of the Mathematical Theory of Elasticity*, p. 340, Noordhoff, Gronigen (1953).
3. G. T. Hahn and A. R. Rosenfield, Acta Met., *13*, 293 (1965).
4. J. N. Goodier and F. A. Field, *Fracture of Solids* (Edited by Drucker and Gilman), Interscience, New York, p. 103 (1963).
5. A. R. Rosenfield, P. K. Dai and G. T. Hahn, Proc. First Int. Conf. Fracture, *1*, 223 (1966).
6. A. S. Kobayashi and W. L. Engstrom, "Transient Analysis in Fracturing Aluminum Plate," Technical Report No. 3, Contract Nonr-477(39), (November 1966).
7. J. R. M. Radok, Quart. Appl. Math., *14*, 289 (1956).
8. I. N. Sneddon, *Crack Problems in the Mathematical Theory of Elasticity*, North Carolina State College, Raleigh, North Carolina, p. 121 (1961).
9. J. W. Craggs, and A. M. Roberts, Jour. Appl. Mech., *34*, 207 (1967).
10. H. Kolsky, *Stress Waves in Solids*, Dover (1963).
11. G. T. Hahn, M. F. Kanninen, A. K. Mukherjee and A. R. Rosenfield, "Plane Stress Crack Extension in Steel Foil," to be published.
12. A. R. Rosenfield and G. T. Hahn, Trans. ASM, *59*, 962 (1966).
13. J. Krafft, Paper presented to Subcommittee III of ASTM, E24 (February 1966).
14. J. Krafft, Appl. Mat'ls Res., *3*, 88 (1964).
15. M. J. Manjoine, Trans. ASME, *66*, A-211 (1944).
16. J. Winlock, Trans. AIME, *197*, 797 (1953).
17. G. I. Taylor, J. Inst. Civ. Eng., *26*, 486 (1946).
18. J. D. Chalupnik, Ph.D. Thesis, University of Texas (1964).
19. C. H. Mok, and J. Duffy, Technical Report Nonr 562(20), (1964).
20. P. C. Johnson, B. A. Stein, and R. S. Davis, Technical Report No. WAL TR 111.2/20-6, to U.S. Army Materials Research Agency (1962).
21. J. Harding, J. Mech. Eng. Sci., *7*, 163 (1965).
22. National Physical Laboratory, Metallurgy Division, "Report for the Year 1963," p. 167.
23. F. S. Minshall, J. Appl. Phys., *26*, 463 (1955).
24. J. D. Campbell and J. Harding, *Response of Metals to High Velocity Deformation*, Shewmon and Zackay (editors), Interscience, New York, p. 51 (1961).
25. B. M. Butcher and C. H. Karnes, J. Appl. Phys., *37*, 402 (1966).
26. C. H. Karnes and E. A. Ripperger, J. Mech. Phys. Solids, *14*, 75 (1966).
27. U. S. Lindholm, J. Mech. Phys. Solids, *12*, 317 (1964).
28. W. G. Ferguson, A. Kumar and J. E. Dorn, J. Appl. Phys., *38*, 1863 (1967).

29. S. Yoshida and N. Nagata, Trans. (Japan) Nat. Res. Inst. Met., *9*, 20 (1967).
30. J. D. Embury, A. S. Keh and R. M. Fisher, Trans. AIME, *236*, 1252 (1966).
31. F. A. McClintock, S. M. Kaplan and C. A. Berg, Int. J. of Fracture Mech., *2*, 614 (1966).
32. M. Isida, Proceedings Fourth U.S. Congress on Appl. Mech. (1962), see also P. C. Paris and G. Sih, ASTM STP 381, p. 41 (1965).
33. A. L. Tichener and M. B. Bever, Progress in Met. Phys., *7*, 247 (1954).
34. A. L. Tichener and M. B. Bever, Trans. AIME, *215*, 326 (1959).
35. A. K. Mukherjee, W. G. Ferguson, W. L. Barmore and J. E. Dorn, J. Appl. Phys., *37*, 3707 (1966).
36. J. I. Bluhm and R. J. Morrissey, Proceedings First Int. Conf. Fracture, *3*, 1739 (1966).
37. G. T. Hahn and A. R. Rosenfield, "Sources of Fracture Toughness: The Relation Between K_{I_c} and the Ordinary Tensile Properties of Metals." Presented at the ASTM Symposium on Applications Related Phenomena in Titanium and Its Alloys, April 1967, and to be published in an ASTM STP.
38. M. F. Kanninen, to be published.
39. J. W. Craggs, J. Mech. Phys. Solids, *8*, 99 (1960).

APPENDIX A

A DYNAMIC SOLUTION FOR AN ELASTIC-PLASTIC CRACK PROPAGATING AT A CONSTANT SPEED IN AN INFINITE SHEET SUBJECTED TO A UNIFORM TENSILE LOAD

Consider a crack of length $2c$ having thin plastic zones of length ℓ (pictured as extensions of the crack itself) at each of its ends in an infinitely large sheet. The overall length of the crack and the plastic zones is $2a$. The sheet is subjected to a uniform tensile loading at infinity in a direction normal to the crack line. In the general case the crack is supposed to be extending uniformly at a speed U.

The solution to this problem can be obtained by superposing the solutions to two subproblems which will be called problems A and B. In problem A a crack of length $2a$ is propagating at a speed U in a sheet under a uniform load T at infinity. In problem B, a uniform tension Y acts upon the portion of the crack faces in the intervals of length ℓ directly behind the crack tips. The length of the free crack is $2c$ and it is propagating at a speed U, but here the tractions at infinity are supposed to vanish.

The solutions to problems A and B will be combined such that the singularities at the crack tips in each problem will cancel each other. A condition which insures finiteness of stresses will result just as in Dugdale's [1] stationary crack solution. In addition to the finiteness condition, the boundary conditions for the combined problem are, at $y = 0$

$$\sigma_y = \begin{cases} O & |x \pm Ut| < c \\ Y & c < |x \pm Ut| < a \end{cases} \qquad \text{(A-1)}$$

$$\tau_{xy} = O \qquad |x \pm Ut| < a$$

and, at infinity

$$\sigma_y = T$$

$$\sigma_x = \tau_{xy} = O. \qquad \text{(A-2)}$$

Equivalently, the region under consideration can be confined to the upper half plane in which case Equations (A-1) must be supplemented by the equations

$$v = \tau_{xy} = O \qquad a < |x \pm Ut|.$$

The equations of motion in plane elasticity are:

$$\frac{\partial \sigma_x}{\partial x} + \frac{\partial \tau_{xy}}{\partial y} = \rho \frac{\partial^2 u}{\partial t^2}$$

$$\frac{\partial \tau_{xy}}{\partial x} + \frac{\partial \sigma_y}{\partial y} = \rho \frac{\partial^2 v}{\partial t^2} \tag{A-3}$$

while the constitutive equations for plane strain are

$$\sigma_x = (\lambda + 2\mu) \frac{\partial u}{\partial x} + \frac{\partial v}{\partial y}$$

$$\sigma_y = (\lambda + 2\mu) \frac{\partial v}{\partial y} + \frac{\partial u}{\partial x} \tag{A-4}$$

and

$$\tau_{xy} = \mu \left(\frac{\partial u}{\partial y} + \frac{\partial v}{\partial x}\right).$$

Radok [7] has shown that when disturbances move at a constant speed U parallel to the x axis a general solution to Equations (A-3) and (A-4) can be obtained by introducing the transformation

$$\xi = x \pm Ut.$$

The solution so obtained is in terms of two analytic functions of the two complex variables

$$z_1 = \xi + i\beta_{1y}$$

and

$$z_2 = \xi + i\beta_{2y}$$

where

$$\beta_1^2 = 1 - U^2/U_1^2 \tag{A-5}$$

and

$$\beta_2^2 = 1 - U^2/U_2^2.$$

Here, U_1 and U_2 are, respectively, the velocity of dilatation waves and of distortion waves in an infinite medium. Radok's solution in terms of the derivatives of the arbitrary and analytic functions $F_1(z_1)$ and $F_2(z_2)$ is then

$$\sigma_y = (1+\beta_2{}^2)\ \mathrm{Re}\ \{F_1''(z_1) + F_2''(z_2)\}$$

$$\sigma_x + \sigma_y = -2(\beta_1{}^2 - \beta_2{}^2)\ \mathrm{Re}\ \{F_1''(z_1)\} \tag{A-6}$$

$$\tau_{xy} = 2\ \mathrm{Im}\left\{\beta_1 F_1''(z_1) + \frac{(1+\beta_2{}^2)^2}{4\beta_2} F_2''(z_2)\right\}$$

and

$$\mu u = -\mathrm{Re}\ \{F_1'(z_1) + \tfrac{1}{2}(1+\beta_2{}^2)F_2'(z_2)\}$$

$$\mu v = \mathrm{Im}\left\{\beta_1 F_1'(z_1) + \frac{(1+\beta_2{}^2)}{2\beta_2} F_2'(z_2)\right\}$$

where the prime notation indicates differentiation of the function with respect to its argument.

The solution satisfying the Equations (A-3) and (A-4) and taking on the boundary values given by (A-1) and (A-2) will be obtained by superposing the solutions to the two subproblems. The boundary conditions for problem A are at $y = 0$

$$\sigma_y = \tau_{xy} = O \qquad |\xi| < a$$

and, at infinity

$$\sigma_y = T,\ \sigma_x = \tau_{xy} = O.$$

The solution satisfying these conditions is given by Radok as an example of the application of Equations (A-6). His result can be written as

$$F_1''(z_1) = \left[A_1^* + B_1^* - \frac{A_1^* z_1}{\sqrt{z_1{}^2 - a^2}}\right]$$

and (A-7)

$$F_2''(z_2) = \left[A_2^* + B_2^* - \frac{A_2^* z_2}{\sqrt{z_2{}^2 - a^2}}\right]$$

where

$$A_1^* = \frac{T}{D}(1+\beta_2{}^2)$$

$$A_2^* = -\frac{T}{D}\frac{4\beta_1\beta_2}{1+\beta_2{}^2} \tag{A-8}$$

$$B_1^* = -\frac{T}{2(\beta_1{}^2 - \beta_2{}^2)}$$

$$B_2^* = \frac{T}{1+\beta_2{}^2} + \frac{T}{2(\beta_1{}^2 - \beta_2{}^2)}$$

and

$$D = 4\beta_1\beta_2 - (1 + \beta_2{}^2)^2.$$

The boundary conditions for problem B are at $y = 0$.

$$\sigma_y = \begin{cases} O & |\xi| < c \\ Y & c < |\xi| < a \end{cases}$$

$$\tau_{xy} = O \qquad |\xi| < a$$

and, at infinity

$$R\sigma_x, R\sigma_y, R\tau_{xy} \to O$$

as

$$R = \sqrt{\xi^2 + y^2} \to \infty.$$

The solution satisfying these conditions has been obtained from Equations (A-6) by Kanninen using a technique given by Sneddon. His result is

$$F_1''(z_1) = B_1 \left[\frac{\alpha z_1}{\sqrt{z_1{}^2 - a^2}} - \pi + i \log \frac{z_1\sqrt{a^2 - c^2} - ic\sqrt{z_1{}^2 - a^2}}{z_1\sqrt{a^2 - c^2} + ic\sqrt{z_1{}^2 - a^2}} \right]$$

and (A-9)

$$F_2''(z_2) = B_2 \left[\frac{\alpha z_2}{\sqrt{z_2{}^2 - a^2}} - \pi + i \log \frac{z_2\sqrt{a^2 - c^2} - ic\sqrt{z_2{}^2 - a^2}}{z_2\sqrt{a^2 - c^2} + ic\sqrt{z_2{}^2 - a^2}} \right]$$

where

$$B_1 = \frac{Y}{\pi D}(1 + \beta_2{}^2)$$

$$B_2 = -\frac{Y}{\pi D}\frac{4\beta_1\beta_2}{(1 + \beta_2{}^2)} \qquad \text{(A-10)}$$

and

$$\alpha = 2\cos^{-1}\frac{c}{a}.$$

Combining the solutions to problems A and B gives, from Equations (A-7) and (A-9)

$$F_1''(z_1) = (A_1^* + B_1^* - \pi B_1) + iB_1 \log \frac{z_1\sqrt{a^2 - c^2} - ic\sqrt{z_1{}^2 - a^2}}{z_1\sqrt{a^2 - c^2} + ic\sqrt{z_1{}^2 - a^2}}$$

$$+ (\alpha B_1 - A_1^*)\frac{z_1}{\sqrt{z_1{}^2 - a^2}} \qquad \text{(A-11)}$$

and

$$F_2''(z_2) = (A_2^* + B_2^* - \pi B_2) + iB_2 \log \frac{z_2\sqrt{a^2-c^2} - ic\sqrt{z_2{}^2-a^2}}{z_2\sqrt{a^2-c^2} + ic\sqrt{z_2{}^2-a^2}} + (\alpha B_2 - A_2^*)\frac{z_2}{\sqrt{z_2{}^2-a^2}}.$$

The first two terms in each equation are nonsingular while the remaining term is singular at $z_1 = z_2 = \pm a$. The singularities can be completely removed,* therefore, by simply setting

$$\alpha B_1 = A_1^*$$

$$\alpha \beta_2 = A_2^*.$$

Upon substituting from Equations (A-8) and (A-10) these two singularity cancelling equations are found to be identical.

$$\frac{c}{a} = \cos\frac{\pi}{2}\frac{T}{Y}. \qquad \text{(A-12)}$$

This is exactly the same result obtained by Dugdale in the static problem. It may be noted that Goodier and Field similarly obtain a finiteness condition in a dynamic problem which is independent of the crack speed, but their solution, which is based on Craggs' semi-infinite crack solution, otherwise differs from the above.

The final form of the equations for the combined solution can now be written out.

$$F_1''(z_1) = C_1 + iB_1 \log \frac{z_1\sqrt{a^2-c^2} - ic\sqrt{z_1{}^2-a^2}}{z_1\sqrt{a^2-c^2} + ic\sqrt{z_1{}^2-a^2}}$$

and $\qquad$ (A-13)

$$F_2''(z_2) = C_2 + iB_2 \log \frac{z_2\sqrt{a^2-c^2} - ic\sqrt{z_2{}^2-a^2}}{z_2\sqrt{a^2-c^2} + ic\sqrt{z_2{}^2-a^2}}$$

where

$$C_1 = A_1^* + B_1^* - \pi B_1$$

$$C_2 = A_2^* + B_2^* - \pi B_2$$

or, using Equations (A-8) and (A-10)

$$C_1 = (T - Y)\frac{1+\beta_2{}^2}{D} - \frac{T}{2(\beta_1{}^2 - \beta_2{}^2)}$$

and $\qquad$ (A-14)

$$C_2 = (Y - T)\frac{4\beta_1\beta_2}{D(1+\beta_2{}^2)} + \left(\frac{1}{1+\beta_2{}^2} + \frac{1}{2(\beta_1{}^2-\beta_2{}^2)}\right)T.$$

* It is still possible that the stresses may become infinite at some particular crack speed, e.g., the Rayleigh velocity.

The displacements are related to $F_1'(z_1)$ and $F_2'(z_2)$, which can be obtained by integrating Equation (A-12).

$$F_1'(z_1) = C_1 z_1 + iB_1 \left\{ z_1 \log \frac{z_1\sqrt{a^2-c^2} - ic\sqrt{z_1{}^2 - a^2}}{z_1\sqrt{a^2-c^2} + ic\sqrt{z_1{}^2 - a^2}} + c \log \frac{\sqrt{a^2-c^2} + i\sqrt{z_1{}^2-a^2}}{\sqrt{a^2-c^2} - i\sqrt{z_1{}^2-a^2}} \right\} \qquad \text{(A-15)}$$

and

$$F_2'(z_2) = C_2 z_2 + iB_2 \left\{ z_2 \log \frac{z_2\sqrt{a^2-c^2} - ic\sqrt{z_2{}^2 - a^2}}{z_2\sqrt{a^2-c^2} + ic\sqrt{z_2{}^2 - a^2}} + c \log \frac{\sqrt{a^2-c^2} + i\sqrt{z_2{}^2-a^2}}{\sqrt{a^2-c^2} - i\sqrt{z_2{}^2-a^2}} \right\}.$$

By substituting Equations (A-13) and (A-15) into Equations (A-6), the stresses and displacements can be determined everywhere in the sheet. In particular, on the crack line where $z_1 = z_2 = \zeta$,

$$\sigma_y = \begin{cases} O & |\xi| < c \\ Y & c \leqslant |\xi| \leqslant a \\ \dfrac{2}{\pi} Y \tan^{-1} \dfrac{\xi}{c} \sqrt{\dfrac{a^2-c^2}{\xi^2-a^2}} & a \leqslant |\xi| \end{cases} \qquad \text{(A-16)}$$

$$\tau_{xy} = O$$

and

$$v = \frac{2}{\pi}\frac{Y}{E}\mathscr{L}(U) \begin{cases} \xi \log \left| \dfrac{\xi\sqrt{a^2-c^2} - c\sqrt{a^2-\xi^2}}{\xi\sqrt{a^2-c^2} + c\sqrt{a^2-\xi^2}} \right| + c \log \left| \dfrac{\sqrt{a^2-c^2} + \sqrt{a^2-\xi^2}}{\sqrt{a^2-c^2} - \sqrt{a^2-\xi^2}} \right| & |\xi| \leqslant a \\ O & |\xi| \geqslant a \end{cases}$$

where

$$\mathscr{L}(U) = \frac{\beta_1(1-\beta_2{}^2)(1+\nu)}{4\beta_1\beta_2 - (1+\beta_2{}^2)^2}. \qquad \text{(A-17)}$$

It is readily seen that the surface tractions on the crack line are independent of the crack speed and are just equal to their static values. Apart from the presence of the factor $\mathscr{L}(U)$, the same is true for the displacement normal to the crack line. It can be shown that the static value is achieved in the limit as follows. Substituting Equation (A-5) in (A-17) gives

$$\mathscr{L}(U) = \frac{U^2/U_2{}^2\sqrt{1-U^2/U_1{}^2}\,(1+\nu)}{4\sqrt{1-U^2/U_1{}^2}\,\sqrt{1-U^2/U_2{}^2} - (2-U^2/U_2{}^2)^2}$$

and, for $U \ll U_2 < U_1$

$$\mathscr{L}(U) = \frac{[1 - \frac{1}{2}\kappa^2(U/U_2)^2](1+\nu)}{2(1-\kappa^2) - (\frac{3}{2} - \kappa^2 + \frac{1}{2}\kappa^4)(U/U_2)^2} \qquad \text{(A-18)}$$

where $\kappa = U_2/U_1$. Noting that

$$\kappa^2 = \begin{cases} \dfrac{1}{2}\dfrac{1-2\nu}{1-\nu} & \text{plane strain} \\ \dfrac{1}{2}(1-\nu) & \text{plane stress.} \end{cases}$$

Equation (A-18) becomes upon setting $U = 0$

$$\mathscr{L}(0) = \begin{cases} 1 - \nu^2 & \text{plane strain} \\ 1 & \text{plane stress.} \end{cases}$$

Thus, the normal displacement reduces to the static solution as $U \to 0$ (cf. Goodier and Field). Note that just as in the static case, plane stress can be obtained from plane strain by appropriate changes in the elastic constants.

APPENDIX B

DESCRIPTION OF THE CRACK-SPEED COMPUTATIONS

The Basic Equations

Consider that the plastic zone is divided in N equal intervals and that S takes on the constant value S_i in the ith interval, $i = 1, 2, \cdots, N$. For convenience, set $S_0 = 0$. Then, by simple superposition, the condition that guarantees finite stresses [Equation (A-12)] in the simple theory is

$$T = \frac{2}{\pi}\sum_{i=1}^{N} (S_i - S_{i-1}) \cos^{-1}\frac{c_i}{a} \qquad \text{(B-1)}$$

while the normal displacement [from Equation (A-16)] is

$$v(x) = \frac{2}{\pi}\frac{\mathscr{L}(U)}{E}\sum_{i=1}^{N} (S_i - S_{i-1})\psi(c_i, x) \qquad \text{(B-2)}$$

where

$$\psi(c, x) = x \log \left| \frac{x\sqrt{a^2 - c^2} - c\sqrt{a^2 - x^2}}{x\sqrt{a^2 - c^2} + c\sqrt{a^2 - x^2}} \right| + c \log \left| \frac{\sqrt{a^2 - c^2} + \sqrt{a^2 - x^2}}{\sqrt{a^2 - c^2} - \sqrt{a^2 - x^2}} \right|. \tag{B-3}$$

If x_i denotes the midpoint of the ith interval and if $v_i = v(x_i)$ then

$$v_j = \frac{2}{\pi} \frac{\mathscr{L}(U)}{E} \sum_{i=1}^{N} (S_i - S_{i-1})\psi_{ij} \qquad j = 1, 2, \cdots, N \tag{B-4}$$

where

$$\psi_{ij} = \psi(c_i, x_j).$$

In order to relate the S_i's to the v_i's we also have

$$\epsilon_i = -\log(1 - Bv_i) \qquad i = 1, 2, \cdots, N \tag{B-5}$$

and

$$S_i = \sigma_i \exp(-\epsilon_i) \qquad i = 1, 2, \cdots, N \tag{B-6}$$

where $\sigma_i = \sigma(\epsilon_i, \dot{\epsilon}_i)$ is given by the stress-strain rate behavior of the material.

The strain rates are determined by the relation

$$\dot{\epsilon} = U \frac{d\epsilon}{dx} = U \frac{d\epsilon}{dv} \frac{dv}{dx}$$

or, using Equations (B-3) and (B-5)

$$\dot{\epsilon}_j = \frac{2}{\pi} \frac{BU}{1 - Bv_j} \frac{\mathscr{L}(U)}{E} \sum_{i=1}^{N} (S_i - S_{i-1})\psi'(c_i, x_j) \qquad j = 1, 2, \cdots, N \tag{B-7}$$

where

$$\psi'(c, x) = \frac{\partial\psi}{\partial x} = \log \left| \frac{x\sqrt{a^2 - c^2} - c\sqrt{a^2 - x^2}}{x\sqrt{a^2 - c^2} + c\sqrt{a^2 - x^2}} \right|. \tag{B-8}$$

The Computer Program

The basic equations programmed for Battelle's CDC 6400 computer are Equations (B-1), (B-4), (B-5), and (B-7). These constitute a system of $4N + 1$ nonlinear simultaneous equations in the $4N + 4$ variables: $S_1, S_2, \ldots, S_N, v_1, v_2, \ldots, v_N, \epsilon_1, \epsilon_2, \ldots, \epsilon_N, \dot{\epsilon}_1, \dot{\epsilon}_2, \ldots, \dot{\epsilon}_N, c, a, T$ and U. To obtain a solution, therefore, three quantities must be specified. For the static crack $U = 0$. Then, either c and T or c and ϵ_1 are prescribed. In the dynamic problem c, $T = T^*$, and $\epsilon_1 = \epsilon_c^*$ are the three given quantities where T^* is the load that gives the critical-crack tip strain at the initial crack length.

The key idea in the dynamic crack computation is that by fixing U

the procedure becomes identical to that of determining T given c and ϵ_1 in the static crack computation. Thus, using the method of false position, the value of U corresponding to T^* can be determined arbitrarily closely. The details of the computational procedure designed for static cracks are given in the author's companion paper.

As in most nonlinear problems it is essential here to have a close initial estimate if the procedure is to converge. To provide this the computations for a number of crack lengths were performed sequentially beginning with the static crack. In this way the initial estimate of the solution for each individual crack length was provided by the previously determined values of the displacements together with extrapolated values of a and U. With these values, the ϵ_i's, the $\dot{\epsilon}_i$'s, the σ_i's, and the S_i's are, in turn, computed. New values of a, T, and the v_i's are computed and the cycle repeated until the change in T from one iteration to the next is suitably small. Then, T is compared with T^*, U adjusted appropriately, and the computation repeated.

Results of Computations

In the computational procedure outlined above N, the number of increments into which the plastic zone is divided, is selected arbitrarily and not changed during a particular run. Because there is no physical basis for choosing N, the only way to determine whether it is sufficiently large is to compare the solutions obtained using different values. Such a comparison is shown in the following table.

Table B-1

COMPARISON OF CRACK SPEEDS COMPUTED USING DIFFERENT NUMBERS OF POINTS IN THE PLASTIC ZONE

	U, ft/sec			
c	$N = 8$	$N = 16$	$N = 32$	$N = 64$
0.110	0	0	0	0
0.165	1.2	1.8	2.1	2.4
0.220	16.5	17.9	18.7	19.2
0.275	26.1	28.0	29.2	29.9
0.330	34.2	36.6	38.2	39.3

It is apparent from these results that the computed speeds are converging, although not rapidly. It should be noted that the CDC 6400 time required (approximately 3 sec for each crack length when $N = 8$)

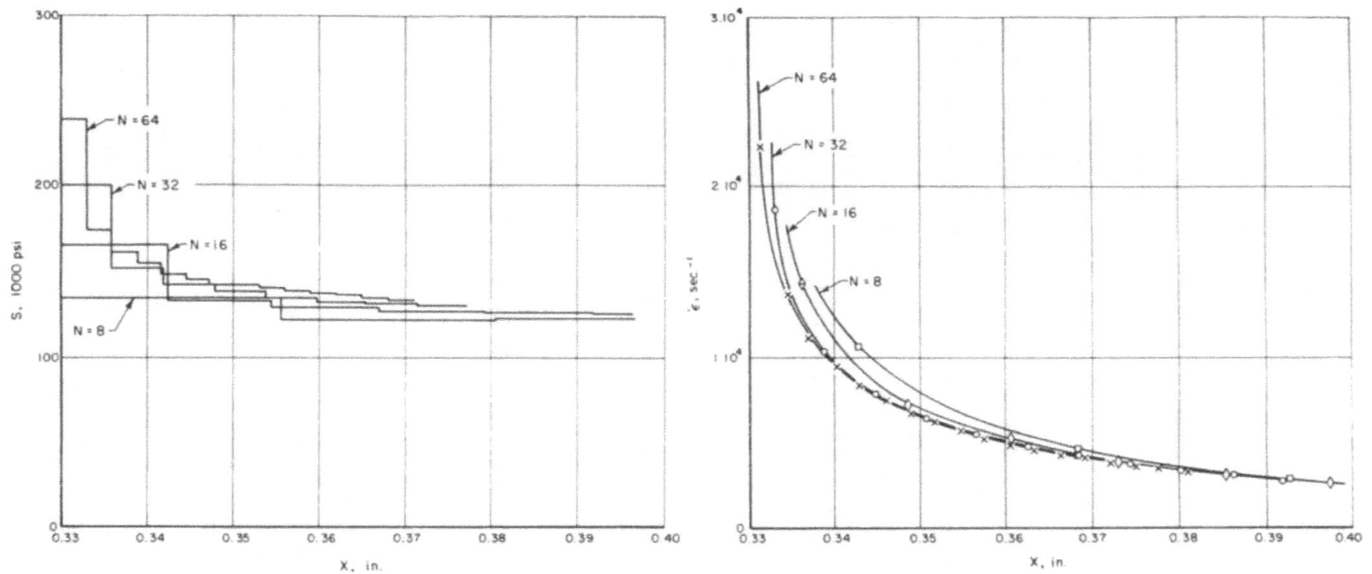

(a) (b)

Fig. B-1. Examples showing the effect of interval size on the computed results near the different crack tips: (a) Internal stress, and (b) Strain rate. The calculations are for $c = 0.330$ in., $T = 73{,}000$ psi, $\epsilon_c^* = 1.10$, and $U = 34$–39 ft/sec (see Table B-1).

roughly trebles as N is doubled so that the precise determination of these values is unwarranted at present. In retrospect one reason for the slow rate of convergence is probably a result of setting $\epsilon_1 = \epsilon_c^*$ because the location of x_1, the point at which ϵ_1 is measured, depends on N (i.e., $x_1 = c + \ell/2N$). Thus, ϵ at $x = c$ varies with N and only in the limit is $\epsilon(c) = \epsilon_c^*$.

A second noteworthy point is that without further refinement this analysis is not able to predict the strain rate at the crack tip. This is a consequence of imposing a discontinuous S distribution since at the edges of each interval the displacement curve has a vertical tangent. (It is for this reason that the strain rates are measured at the midpoints of the intervals.) The values of the strain rate near the crack tip computed for different values of N are shown for a typical case in Fig. B-1.

Because S increases with $\dot{\epsilon}$, the results shown in Fig. B-1 indicate that the values of S at the crack tip similarly increase without bound as N is increased. That this is actually the case is shown in Fig. B-1. Here it can be seen, however, that in spite of this the average value of S over the entire plastic zone is not changing significantly. If this were not the case the crack velocities would not converge.

APPENDIX C

A SIMPLIFIED ANALYSIS OF CRACK SPEED

A simplified analysis of ductile crack speed can be derived with the aid of the following assumptions:

(1) The strain gradient near the crack tip, $\frac{\Delta\epsilon}{\Delta x}$, is relatively insensitive to crack speed, U. This assumption receives support from the present computer calculations (see Fig. 12a). Accordingly, the plastic strain rate, $\dot{\epsilon}$, is

$$\dot{\epsilon} = \left(\frac{\Delta\epsilon}{\Delta x}\right) U. \tag{C-1}$$

(2) At high strain rates a linear rate dependence dominates; other rate laws can be neglected:

$$Y = Y_s + F\dot{\epsilon} \tag{C-2}$$

where F is the linear strain rate coefficient.

(3) The crack tip strain, ϵ_c^*, is a single-valued, rate-insensitive function of displacement and is constant during crack propagation, i.e., $v_c^* =$ constant. Further, the crack tip displacement can be approximated by the relation $v_c = P \cdot \frac{T^2 c}{Y}$, which is a simplified form of Equation 3 valid when $T/Y < 0.7$. The term P is a constant, Y is an average flow stress, and changes in sheet thickness are neglected. This relation for v_c is also valid for plane strain. These two assumptions require:

$$\frac{(T_1)^2 c_1}{Y_s} = \frac{(T_2)^2 c_2}{Y} \tag{C-3}$$

where T_1, c_1, and Y_s are values of stress, crack length, yield stress, the instant the crack begins to extend, while T_2, c_2, and Y are the relevant values for a crack traveling at a speed U. Defining $K_c \equiv T_1 \pi c$, and $K \equiv T_2 \sqrt{\pi c_2}$ leads to

$$\frac{K_c^2}{Y_s} = \frac{K^2}{Y} \tag{C-4}$$

or

$$\left(\frac{K}{Y}\right)^2 = \left(\frac{K_c}{K}\right)^2 \cdot \left(\frac{K_c}{Y_s}\right)^2. \tag{C-5}$$

Finally, combining Equations (C-1), (C-2), and (C-4) gives

$$U \approx \frac{Y_s}{F\left(\frac{\Delta\epsilon}{\Delta x}\right)} \left[\frac{K^2}{K_c^2} - 1\right]. \tag{C-6}$$

DYNAMIC MECHANICAL BEHAVIOR OF METAL AT THE TIP OF A PLANE STRAIN CRACK

J. M. KRAFFT

Naval Research Laboratory
Washington, D.C.

ABSTRACT

Although the stress intensity around a crack is continuously varying in the elastic singularity, in real metals the pattern breaks down at the tip across discretely sized regions of stress relaxation. Within these regions, conditions of stress, strain, and strain rate can be characterized as uniform and equal to the point value in the singularity at the distance equal to the region size. High strain rate within this zone can be attained by crack movement, or by rapid loading of a stationary crack. Fracture strength is largely governed by the triaxial plastic stability and size of these regions. Determinations of plastic stability from dynamic compression tests correspond nicely to speed variations and temperature-wise trends in fracture strength in two ferritic steels.

I. INTRODUCTION

The hypothesis which we attempt to explain and then to demonstrate in this paper is one of fracture-strength. It is not necessarily nor even usually the same as one of fracture-separation, both being but stages of the total fracture process. In most structural metals, the separation event commences only after substantial plastic deformation. At the tip of a tensile crack, this deformation results from tensile extension. The maximum strength obtainable in tensile extension is at the point of tensile instability; for example the ultimate strength of the ordinary tensile test. Correspondingly the maximum strength, or stress intensity factor, which a crack can sustain prior to fracture is just that required to produce local tensile instability. Large plastic strains may follow the establishment of tensile instability before the physical separation can commence. But this occurs without application of additional loading but rather by the contraction with load relaxation of the distended metal adjacent the region of tensile instability. It is this intermediate stage of deformation, uncoupled by instability from the measurable loading, which ordinarily separates the event which determines mechanical strength from subsequent fracture separation as the fractographer may observe it.

Identification of fracture strength with tensile instability means that the one is a measure of the other. In the context of this Symposium, one can, in principle, measure dynamic tensile instability with a dynamic fracture test. However, at the present state of this art we consider ourselves fortunate to make the converse connection, i.e., to measure dynamic tensile instability in a rapid plastic flow test and then from it predict the fracture behavior. High strain rate at the crack tip can be produced either by rapid loading of a stationary crack, or by rapid crack velocity with fixed crack-loading.

The connection between fracture strength and tensile instability has been eminently although infrequently proposed in past decades. Read, Marcus and McCaughey in 1947 [1] noted a correlation between impact fracture strength and the strain hardening exponent n evaluated in slow compression. Although they associated the effect with adiabatic shear instability proposed earlier by Zener and Hollomon [2], it might as well have been related to the tensile instability strain with which it can also be identified

In 1952 George and LaTorre [3] at NRL made observations of crack extension in very thin zinc foil. They observed crack extension by the growth and coalescence of holes nucleated in advance of the crack tip. Even with such obvious ductility, their cracks eventually became unstable and progressed rapidly across the sheet without increase of load. Irwin described this behavior to Orowan [4] who in turn reasoned that the fracture instability could be a result of tensile instability in the ligaments formed between the holes. The effect of specimen compliance, in analogue to a soft testing machine, he reasoned to account for the onset of instability close to the maximum strength point.

Fracture strength in the presence of a dominant flaw is generally directly related to the distance or size scale to which yielding, then instability, is sectioned off by obstructing free surfaces. In the case of thin sheets, the plate surfaces are these obstructions; the yielding and plastic flow tends to scale to the dimension of thickness. In thicker plates, where most of the crack tip is remote from the free surfaces, a homogeneous defect structure of the metal may provide the largest available free surface separations. The fracture strength K_{Ic} then becomes independent of plate thickness, a convenience utilized in experiments described here. It is conceivable that under such high constraint the fracture-separation might occur and thus set the fracture-strength before the tensile instability strain is reached to set it. But this does not appear to be the case. In rate sensitive steels under the most brittle conditions of low temperature and high strain rate, we observe [5] a direct correspondence between fracture-strength K_{Ic} and the strain hardening coefficient or tensile instability strain n.

This invariant proportionality between K_{Ic} and n was first interpreted as indicating an invariant process zone cell whose uniaxial tensile instability governs the plane strain fracture strength. But n is only the strain for tensile instability, not for rupture. As with Orowan ten years earlier, Williams and Turner [6] reminded NRL experimenters of 1964 that specimen compliance could reduce the instability strain requirement to that for maximum uniaxial strength n and even further—to about $n/2$—with the condition of maximum local triaxiality. Before it became apparent as to how the triaxiality might be assessed and used in these correlations, several and more extensive sets of data showed a fairly satisfactory correlation of K_{Ic} to n. Of particular interest here is my study with Eftis [7] of speed/temperature effects in a ship plate steel and more recently [8] of the detailed speed sensitivity of a gas transmission line-pipe steel. However, with recent attempts to interpret environmentally assisted crack extension, the degree of local triaxiality became apparent and a way of treating it in the analysis was developed.

The re-analysis of the data on mild steel to allow for local triaxiality is presented in this paper. The new results tend to support a literal interpretation of the presence of specimen-like ligaments at the crack tip, and the relevance of the fracture-strength to their dynamic mechanical properties.

II. CONDITIONS FOR TENSILE INSTABILITY

In crystalline metals, it is well known that gross plastic deformation is accommodated by the nucleation and movement of line and screw dislocations. Moving rather easily at first in annealed metals, their accumulation and interactions with the defect structure tends to repress their continued movement. The external manifestation of this is a work strengthening or hardening of the metal with strain ϵ_0. The rate of increase of hardening with strain, the strain hardening rate, $\frac{d\bar{\sigma}}{d\epsilon} = \bar{\theta}$ is a stabilizing mechanical influence. However, it diminishes with the strain.

If the straining affecting strain hardening is of tensile extension, a consequent contraction of area tends to cancel the stabilizing influence of strain hardening. When the strain hardening rate has decreased with strain to the point where it is just cancelled by geometric contraction, a condition of tensile instability occurs. Expressed as a rate process for the case of maximum lateral constraint [9], the instability condition is

$$A\frac{\partial\bar{\sigma}}{\partial\bar{\epsilon}}\frac{d\bar{\epsilon}}{dt} - 2\bar{\sigma}\frac{dA}{dt} = 0 \tag{1}$$

and the second derivative greater than zero,* where A is specimen area, $\bar{\sigma}$ and $\bar{\epsilon}$ its true (uniaxial) stress and strain, t is time, and the coefficient 2 is the maximum permissible plastic constraint factor derived by Williams and Turner [6]. Lateral contraction due to stretching reduces the area at a rate $\frac{dA}{dt} = 2\nu A\frac{d\bar{\epsilon}}{dt}$ where ν is the Poisson ratio.

A typical strain hardening behavior is diminution inversely as the strain, the parabolic law

$$\frac{\partial\bar{\sigma}}{\partial\bar{\epsilon}} \equiv \bar{\Theta} = \frac{n\bar{\sigma}}{\bar{\epsilon}}. \tag{2}$$

If parabolic strain hardening does occur, appropriate substitutions in (1) show that tensile instability occurs when the strain equals $n/2$.

III. TENSILE YIELDING AND INSTABILITY AT A CRACK TIP

It is now beyond reasonable doubt that Irwin's stress intensity factor K provides an extraordinarily successful characterization of the mechanical environment of an elastically stressed crack [10]. With it the pattern of y-stress (σ_y) redistribution around a crack is simply disclosed as an inverse square root of distance singularity

$$\sigma_y = \frac{K}{\sqrt{2\pi r}} \tag{3}$$

where r is distance directly in advance (x-direction) of the crack tip ($\theta = 0$). The K level is usually expressed as a function of nominal stress, σ, and (half) crack length, a,

$$K = Q\sigma\sqrt{\pi a} \tag{4}$$

where Q is a (near unity) constant of specimen and crack geometry for which many solutions are available [11].

It is necessary to determine the K level required to bring a ductile ligament along the crack front to the point of tensile instability. Assume it occupies a distance d_T ahead of the crack tip, as depicted in Fig. 1. Because of triaxiality the stress required to establish and main-

* Usually presumed, nonetheless necessary, as noted by P. Rees, Imperial College, London, in private communication.

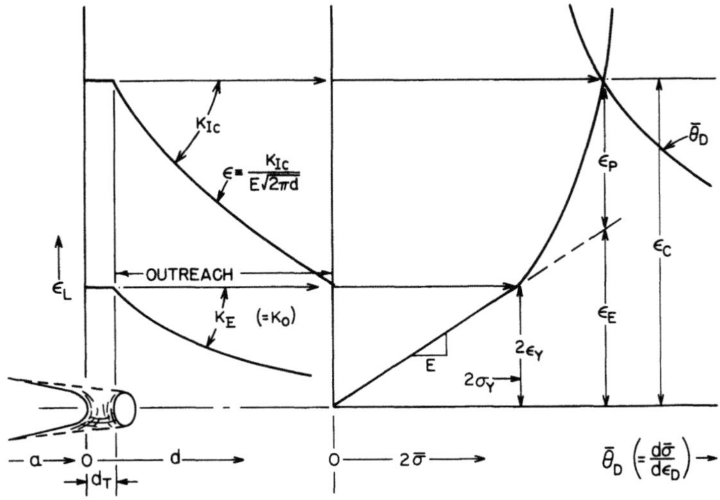

Fig. 1. Upon increases of the crack tip stress field K the process cell is formed by its plastic yielding and then brought to tensile instability with attainment of the fracture strength K_{Ic}. At K_{Ic} the potential yield zone has a substantial "outreach."

tain plastic yielding over the distance d_T is, after Williams and Turner [6], twice the uniaxial yield stress σ_Y. The K level required is then just twice what one might expect for uniaxial yielding, or a plane stress plastic radius of d_T.

$$\sigma_y = 2\sigma_Y = \frac{K_E}{\sqrt{2\pi d_T}}\left(= \frac{K_0}{\sqrt{2\pi d_T}}\right). \tag{5}$$

Plastic yielding levels both the stress and the strain singularity across d_T as depicted in Fig. 1. Thus conditions of strain and strain rate calculated for the boundary position $r = d_T$ should apply throughout the region closer than d_T.

As the K level is further increased above K_E, say by an increment K_P, plastic extension correspondingly increases across d_T. With this the yield zone advances, effecting a more extensive blunting of the y-stress singularity. Nonetheless the tensile strain across d_T continues to increase by at least

$$\epsilon_P = \frac{K_P}{E\sqrt{2\pi d_T}}. \tag{6}$$

We can say at least this value because plasticity solutions for analogous situations [12, 13] indicate stronger than inverse $\frac{1}{2}$ power singularities as a result of the plastic relaxation. Thus to raise ϵ_P to a level required for tensile instability ϵ_c, K_P no greater, and possibly less than that calculated by (6) for the elastic singularity would be required. Present experimental data indicates that assumption of the elastic strain singularity provides satisfactory correlations, although the subject requires further study. Thus for this paper we assume the elastic singularity, summing elastic and plastic K required to elastically stretch and then to plastically flow the ligament of size d_T.

$$\epsilon = \frac{2\sigma}{E} + \epsilon_P = \frac{K_E + K_P}{E\sqrt{2\pi d_T}} = \frac{K_I}{E\sqrt{2\pi d_T}}. \tag{7}$$

Here σ is the uniaxial flow stress after the plastic strain of interest, which would be the yield strength of the work hardened material rather than the initial value of (5). When K_I becomes large enough to establish the triaxial tensile instability strain, from (3), across d_T ($\epsilon = \epsilon_c$), then $K_I = K_{Ic}$, the point of plane strain fast fracture instability and maximum strength of the specimen.

As with the K_E point, it is consistent with this tensile ligament model to calculate the strain rate (as well as the strain) at the border of the process zone, the distance $r = d_T$ in the "K-strain" singularity.

Differentiating (7)

term → I → II

$$\dot{\epsilon} = \frac{d\epsilon_P}{dt} - \frac{2}{E}\frac{d\sigma}{dt} = \frac{1}{\sqrt{2\pi}E}\left(\underset{\text{I}}{d_T^{-\frac{1}{2}}\frac{dK}{dt}} + \underset{\text{II}}{\frac{K_I}{2}d_T^{-\frac{3}{2}}V_c}\right) \tag{8}$$

where V_c is crack velocity. Experiments described later are designed to bring into play separately and independently each of the two ingredients of strain rate at the crack tip: the first with a moving crack with fixed K_I (Sec. VI); the second with varied loading rate $\dot{K}$ on a fixed crack (Secs. VII and VIII).

IV. ORIGIN OF THE INSTABILITY PROCESS ZONE

Consider next, three hypotheses relevant to the attainment of the fracture strength:

1) In the path of the crack, any element which, by action of the stress field and isolated to a size by the defect structure, is brought to the point of complete tensile yielding must subsequently be extended to the point of tensile instability;

2) Control of the fracture strength is passed upward in size by the extension of yielding across larger scales of the defect structure, consequent of the drawing-out to tensile instability of a smaller scale yield zone; and

3) Fracture strength is the K field required to produce tensile instability in the largest yield zone which can be reached in the "passing upward" process.

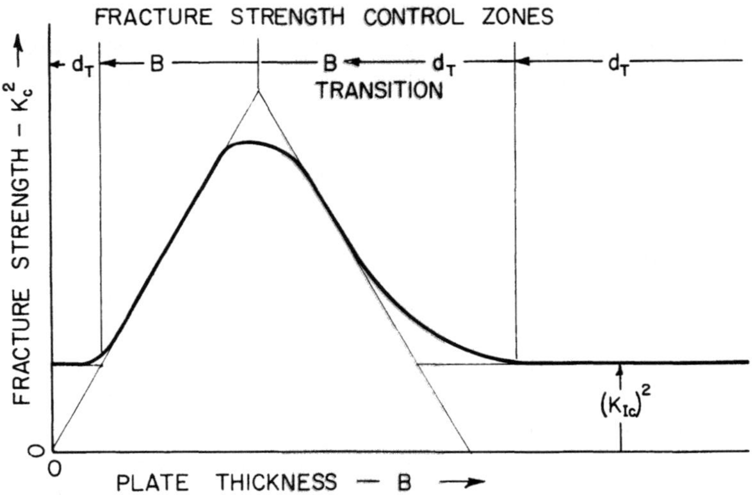

Fig. 2. When the "outreach" of the yield zone from d_T approaches specimen thickness, the fracture mode transition transfers control of the process cell size from d_T to the thickness B.

To illustrate these hypotheses consider the most readily observed "passing upward" process, the fracture mode transition from control of strength by d_T to control by plate thickness B (Fig. 2). To the existence of d_T in control of the plane strain "fracture strength" K_{Ic}, Birkle, Wei and Pellisier [14] have examined a series of Ni, Cr, Mo (4345) steels of widely varied sulfur-content. Measured in tensile tests on the metal tempered at 800° F, the tensile instability strain was not affected by sulfur content. However, K_{Ic} decreased with sulfur content so that the calculated instability distance (Eq. 7) * also decreased. Replicas of the fracture surface revealed that the spacing of the largest inclusions, later determined to be sulfides, was in direct correspondence with the calculated d_T; the greater the sulfur content the more there were. These inclusions centered on the largest dimples. However, the actual fracture separation was thought to have involved the coalescence of much smaller, carbide nucleated, dimples. Presumably these did not control the fracture strength, but passed it upward to the scale of the larger sulfide inclusions. Since these were in fact the largest, their instability set the strength.

The range or "outreach" of the zone of plastic yielding in the square root singularity is $(\epsilon_c/2\epsilon_Y)^2$, assuming the triaxial yield strength governs both defect scales. This is approximately $(n/4\epsilon_Y)^2$ for parabolic hardening. A typical outreach would be a factor of ten, one order of magnitude in the size scale of defects. It should be somewhat greater than (at most 4X) this in the transition from d_T to B, as full thickness yielding is

* Actually the uniaxial n was used for ϵ_c in (7) but with such a high yield strength the triaxial value of d_T is also in correlation with the measured inclusion spacing.

of reduced triaxiality. With given constraint, the outreach can be widely varied by conditions of temperature, and strain rate, and thermal condition, as these factors affect ϵ_c and ϵ_Y.

If the outreach of the (potential) yield zone from d_T becomes comparable to plate thickness, the familiar fracture mode transition is observed. With fixed thickness it may be caused by increased "outreach" with temperature. Or with fixed "outreach" in strain rate insensitive materials, it can result if the plate is made thinner. When the plate is made thinner than d_T times the outreach ratio, the fracture strength should decrease as the square root of the thickness which is then the process zone size or a $G_c(= K_c^2/E)$ as the first power (Fig. 2), the effect predicted by Bluhm [15] and by Hahn and Rosenfield [16] using a strip yield model. If thinned below d_T, the strength might again revert to control by d_T rather than approach zero with the thickness, although test measurements are rarely accomplished with such thin specimens.

V. THE DEVIOUS CHARACTER OF FRACTURE APPEARANCE

In fracture of a plate for which the outreach of the yield zone is only part of the (half) thickness, the center region of fracture will be controlled by d_T while near the surface a larger plastic enclave must be drawn out to tensile instability. In the elevated K field required to stretch the yield zone formed at the surface, the central (d_T) crack tunnels forward to a region of strain intensity equal to ϵ_c over the range d_T. Often the extent of the plastic enclave is revealed by the fracture appearance, where the separation follows the elastic plastic boundary, the familiar surface shear lips and center flat. However, it is important to remember that the final separation need not take this path. It often passes directly through the surface zone, leaving the appearance of a flat, more brittle fracture.

When fracture is controlled wholly on the d_T scale, there is no reason to hope that nature will be more kindly disposed to reveal its secrets. The fracture may follow the enclave of tensile instability and reveal the largest dimples. But on the other hand, it may go through them in any of a number of smaller scale, lower strength, processes. It may cleave through grains of no particular symmetry relative to the d_T enclave, or even pass between the grains.

To sum up then, the path of separation is the final stage in the total fracture process. It may, but it need not reveal a prior stage for which maximum external stress was required. Thus it need bear no unique relationship to the fracture strength. This should not discourage us from looking. There should be a tendency for the separation to pass through and thus reveal the largest inclusions. Or perhaps polished and etched sections would reveal their size and spacing in a useful way. But we must be wary of deceptive appearances in fractographic examinations.

VI. STRAIN RATE AT d_T WHEN CRACK VELOCITY IS INDUCED BY STRESS CORROSION

If the strength controlling ligamental cell is formed with establishment of plastic yielding over it ($r_Y = d_T$), then its instability can be induced thereafter by the rate of surface attack. Usually, the sole source of the areal diminution rate in degradation of tensile stability is the Poisson contraction. For corrosion assisted cracking we can assume that a surface degradation, such as actual dissolving of metal in a corrosive environment, also contributes to the areal diminution rate. If we assume ligaments of circular perimeter πd_T suffer a surface annihilation at a rate V_D, then their total rate of areal diminution will be

$$\frac{dA}{dt} = 2\nu A \frac{d\epsilon}{dt} + \pi d_T V_D. \tag{9}$$

Substituted in (2) and rearranged, the instability criterion becomes

$$\dot{\epsilon} d_T = 4 \bigg/ \left(\frac{\bar{\theta}}{2\bar{\sigma}} - 1\right) \tag{10}$$

where $\bar{\theta} = \frac{\partial \bar{\sigma}}{\partial \bar{\epsilon}}$, $A = (\pi/4)\, d_T^2$ and $\dot{\epsilon} = d\bar{\epsilon}/dt$.

If the crack loading is relatively constant, the strain rate can be expressed by using only the crack velocity term (II) of (8). Also d_T can be expressed in terms of K_{Ic} and ϵ_c, from (7), as well as direct a proportionality between ϵ and the instant level of K_I; whence

$$V_c = 8V_D \bigg/ \epsilon\left(\frac{\bar{\theta}}{2\sigma} - 1\right). \tag{11}$$

The right side of (11), save V_D, derives from the shape of an ordinary stress strain curve. Suitable procedures for this are detailed in [9].

The actual crack velocity and thus the strain rate across d_T will be directly proportional to the surface degradation velocity, V_D. Present evidence indicates that V_D is reasonably constant for a given combination of metal alloy and corrosive reagent. A typical example is shown in Fig. 3 for a Titanium alloy immersed in salt water (6 Al, 4 Zr, 2 Sn, 0.5 Mo, 0.5 V of about 125 Ksi yield strength, $K_{Ic} = 124$ Ksi $\sqrt{\text{in.}}$). Here the time required for extension of an initial fatigue crack to complete fracture was measured on cantilever bend specimens. The mechanical plastic flow properties were measured on compression plugs using a diametral strain gage. With a summation of the thus derived V_D/V_c

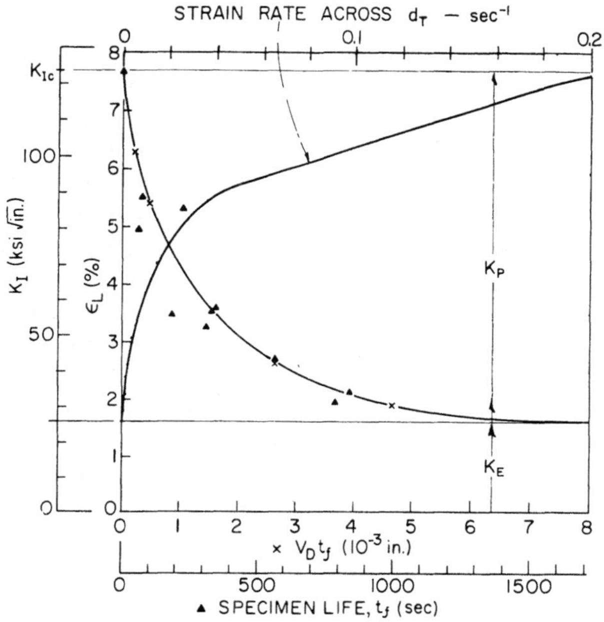

Fig. 3. The aqueous reagent 3.5% NaCl induces the crack growth below K_{Ic}, with corresponding "dynamic" strain rates within d_T of the order of 10^{-1}/sec in a Titanium alloy.

ratio over the crack path to K_{Ic} in the actual fracture specimen, the fracture time, t_f, multiplied by the unknown degradation velocity V_D was calculated. A cross plot indicated a constant value $V_D = 4.7$ μin./sec which provided a satisfactory match between measured and predicted specimen life.

The naturally occurring strain rate in the process zone is generally rather low; however, it is truly dynamic in the approach to K_{Ic}. Some shift in the sampling of strain rate could be obtained by using solutions of differing reactivity rate V_D. However, it has so far been difficult to reduce scatter in the subcritical crack velocity measurements to the degree required to precisely define mechanical properties of the process cell ligament.

VII. TEMPERATURE/RATE EFFECTS IN MILD STEEL

The fracture strength of mild steel has always been diabolically evasive of quantitative measurement in a form suitable for stress analysis. In the present context, this difficulty can be attributed to the extreme extent and variability, with temperature, strain rate, and thermal condition, of the outreach of the yield zone from the process zone. The outreach ratio from d_T to B should be roughly $(nE/2\sigma_Y)^2$ as discussed in Sec. IV. For typical low speed values around the notch impact test transition temperature of a mild steel ($n = 0.3, E = 3 \times 10^7$ psi, $\sigma_Y = 50$ Ksi) the outreach ratio is about 10^4. A typical d_T process zone size is 10^{-4} in. in such steel. The product, 1 inch, means that a plate twice this thickness will yield across its thickness. The fracture strength with an inch scale process zone size is too high to measure with reasonable specimens. At the same temperature, however, high strain rates would typically reduce n by $\frac{1}{2}$ and increase σ_Y by 2, reducing the outreach by a factor of $4^2 = 16$, to $\frac{1}{16}$ inch. Here a $\frac{1}{8}$ in. would be just through the d_T to B mode transition at this temperature and one 0.4 in., Charpy thickness somewhat in the middle or at its transition temperature, which is the result observed. The ratio of masses between specimens of linear dimension in the ratio of 16 can be out of all reason ($\simeq$4000). It is thus that the prediction of fracture strength from dynamic mechanical properties is sorely needed.

Our data for such an attempt on the U. of Illinois ship plate [7] is perhaps the most extensive collection available today. It is reanalyzed here for the fully triaxial instability model which was discovered subsequent to its publication in connection with the stress corrosion studies of the preceding section (VI).

The steel supplied by Prof. W. F. Hall from their wide plate specimen #39 was $\frac{3}{4}$ in. thick 20 points carbon 0.76% Mn semikilled. An extensive investigation of fast crack propagation, which need not concern us here, had already been carried out on this material. Our tests for plane strain fracture strength K_{Ic} were carried out on $\frac{1}{4}$ in. thick single edge notch tension specimens. They were prepared by fatigue precracking, then loaded over a wide range of speeds in the NRL dynamic universal testing machine.* Assuming a stationary crack ($V_c = 0$) prior to the instability point, the $\dot{K}$ needed to estimate strain rate within d_T from (8) is obtained directly from the measured loading rate just prior to fracture. These specimens were large enough for valid K_{Ic} measurement at −320° F (liq. N_2), −220° F and at high speeds at −120° F, as shown in Fig. 4A.

Compression tests for the instability strain, then regarded to be simply the (uniaxial) strain hardening exponent n, were also run at varied speeds. Here specimen size is no limitation so temperatures as high as desired, and certainly through the $d_T \rightarrow B$ transition temperature range, were employed. For clarity in the ratio matching of the strain for tensile instability with K_{Ic}, the data in Fig. 4B show only temperatures for which there is valid fracture-strength data for matching.

The estimates of n were all corrected to the isothermal condition of

* Now commercially available.

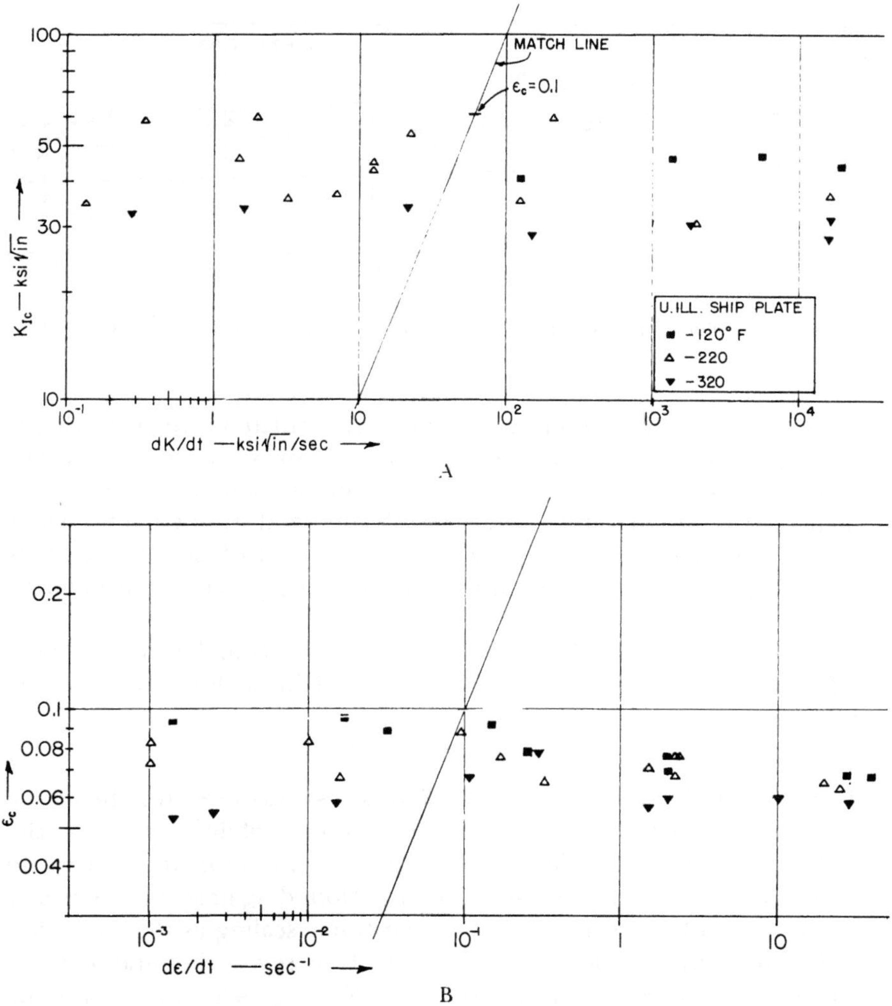

Fig. 4. For a ship steel, fracture strength K_{Ic} (4A) and the triaxial tensile instability strain ϵ_c (4B) are shown vs speed for three low temperatures, −120, −220 and −320° F.

deformation. The quenching time from a body as small as the process zone d_T is shorter than the briefest possible time for fracture [17]. Two methods were used and found in acceptable agreement as previously shown in [18]. First, records of continuous compression to about 10% strain were run. Those fast enough to be fully adiabatic were then measured at the 3% strain point. The temperature sensitivity of flow stress was also evaluated at this point. With this, and handbook values of the specific heat, the rate of thermal softening could be put in terms of an n value correction, which could then be added to the measured n value. This procedure is detailed in [7].

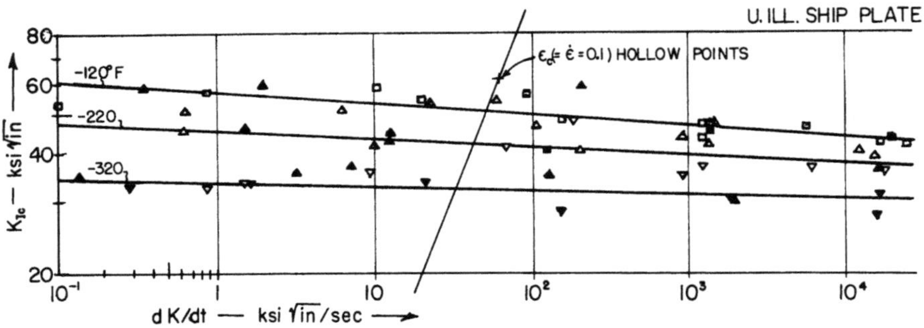

Fig. 5. The plots of Fig. 4 matched together indicate $K_{Ic}/\epsilon_c = 610$ Ksi $\sqrt{\text{in.}}$ and $d_T = 66$ μin.

Alternatively the isothermal n value was simulated directly by dividing the deformation into a series of short strokes, of about 2% each. Following each abrupt arrest, specimens were unloaded, cooled back to their starting temperature and then reloaded and strained. The "yield point" upon each reloading was taken as a point of isothermal stress for the strain accomplished in prior cycles. The procedure is detailed in [8].

In ordinary mild steels, the fit of the parabolic strain hardening law is fairly satisfactory. In such case the triaxial instability strain can be estimated as

$$\epsilon_c = 2\sigma/E + n/2. \tag{12}$$

In retrieval of these data, the 3% flow stress was used for the elastic stress (σ) rather than the stress at the point of instability. Since here the yield strain is small relative to n, this simplification introduces no appreciable error. The values of ϵ_c are plotted against the prevailing strain rate in Fig. 4B in the same logarithmic scaling as the K_{Ic} vs $\dot{K}$.

The matching of plots to test and hopefully reveal the ratio between K_{Ic} and ϵ_c involves superposition of the plots Figs. 4A and 4B with one special constraint: The strain rate in the process zone must be held identical with that of the compression test. This can be accomplished by matching while maintaining the quotient of K_{Ic} to $\dot{K}$ everywhere the same as that of ϵ_c to $\dot{\epsilon}$ (assuming E = constant). To do this on the log plots, we draw on each a line for which the ratio of ordinate to abscissa scales is a constant, e.g., 1:1 on the plots shown. These "match lines" are then maintained superimposed during the comparison, the only permissible adjustment is a translation along the lines. If this is done, any point common to both plots will meet the prescribed condition of proportionality. Matching is discussed more fully in [19].

The superposition of Figs. 4A and 4B according to the matching rule is shown in Fig. 5: the correspondence appears reasonably satisfactory. It is entirely satisfactory if one anticipates the effects to be

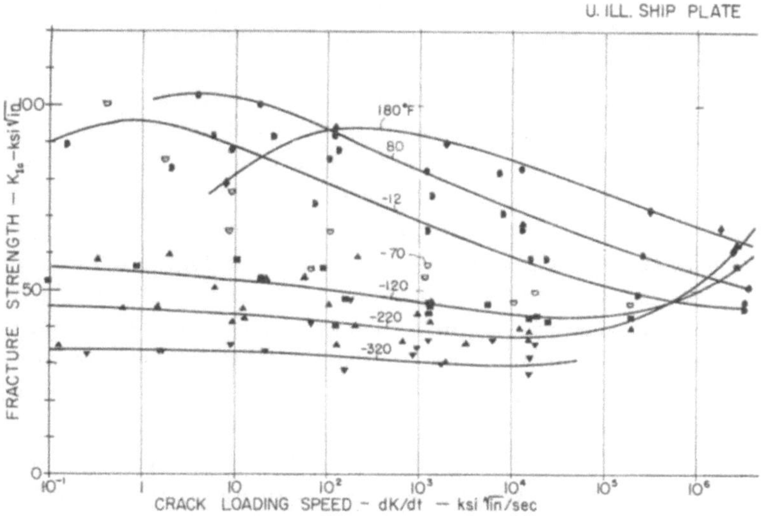

Fig. 6. Converted to K_{Ic} with the match ratio from Fig. 5, extensive ϵ_c data show the general pattern of speed/temperature sensitivity for this mild steel. The lines are roughly 100° F apart.

discussed in the next section: namely that there is a marked structure in strain rate sensitivity of the strain hardening rate in mild steel at lower temperatures. Though suggestive of it, the speed variation was not sampled finely enough here to clearly define it. The ratio of K_{Ic}/ϵ_c, and thus of $\dot{K}$ to $\dot{\epsilon}$, for the best match is 610 Ksi $\sqrt{\text{in.}}$ for which the process zone size d_T would be 66 μin. This is much smaller than in the titanium alloy of Sec. VI for which d_T was about 1600 μin.

If we dare now assume d_T invariant, a grand pattern of the speed-temperature variation of K_{Ic} emerges as displayed in Fig. 6. Augmenting the range of this prediction are some very high strain rate data obtained with Hopkinson-Kolsky pressure-bar loaders, as described in [20]. It will be noted that with increasing temperature the minimum fracture strength occurs at higher speeds. These correspond roughly to the strain rate for maximum flow stress, previously published in [7]. A peaking of fracture strength toward the lower speeds is suggestive of onset of a strain aging effect which should also depress the tensile instability strain and thus the fracture strength. Charpy impact $\dot{K}$, estimated at about 10^5 Ksi $\sqrt{\text{in.}}$/sec, is about as fast as one dare go before seeing an inversion in the temperature sensitivity of fracture strength K_{Ic} in Fig. 6.

In leaving this subject, it should be noted that still lower estimates of fracture strength result if the process zone strain hardening is assumed degraded by adiabatic heating. Freely running cracks should provide this condition across d_T. The appropriate comparison, which may be of interest to students of this Symposium, is available in [7].

VIII. SPEED-SPECTRUM OF STRAIN-HARDENING RATE

We began the last section by lamenting, for the practitioner of fracture mechanics, the appalling extent of the yield zone outreach in mild steel. But even if temperatures are low so this is not prohibitive, another baffling characteristic is observed. There appears to be a strong rate dependent variation in K_{Ic} and ϵ_c, superimposed on the general trends with speed and temperature as outlined in Fig. 6. With the usual large gaps in speed sampling, this effect in the past could only be interpreted as data scatter. We now have collected a set of data profuse enough to define some of the structure as distinct from scatter. An X-52 line pipe steel was supplied by Robert Eiber of Battelle, Columbus for this work. When these data were first matched [8] the uniaxial instability criterion ($\epsilon_c = n$) should have introduced a factor of two shift in the strain rate from what we now believe proper. The data have now been recalculated for the triaxial criterion with results described in this Section.

This structure or "spectrum" in K_{Ic} is seen as speed sensitive alone, superimposed on general trends with temperature and strain rate. It is apparently lost at room temperature, but is becoming evident at −120°, and is clearly present from −170° F to −320°. Accordingly, to display this structure we attempt to normalize out the background trends—for data collected at within ±100° F to a reference of −220° F. This is done by assuming that the strain hardening exponent should vary inversely as the flow strength level, and for these low yield strain materials, so should ϵ_c triaxial. Thus both K_{Ic} and ϵ_c values were corrected to −220° F by multiplying by the ratio of their 3% flow stress to that at the actual test temperature. Most of the data was actually collected at −220° F but this normalization procedure appears to bring the data from adjacent temperatures into line.

The fracture strength K_{Ic} data normalized to −220° F as shown in Fig. 7A as a function of the crack loading rate $\dot{K}$. Values of ϵ_c recalculated according to (11) are plotted against $\dot{\epsilon}$ in Fig. 7B, and normalized to −220° F. The plastic flow data at other temperatures are omitted for the sake of clarity. However, the trends are quite similar to those shown for the ship plate steel of Fig. 6.

The matching rules for process zone to (compression) plastic flow test consistency of strain and strain rate of Sec. VII are adhered to in the match of Fig. 8. The coordinates are those of strain rather than K as these should be the more basic and invariant from one steel to another. There is naturally some experimental uncertainty both in ordinates and abscissa matched. It is thought that this uncertainty is far less than the observed variation from the general trend. Moreover there are general coincidences throughout the pattern of both fracture

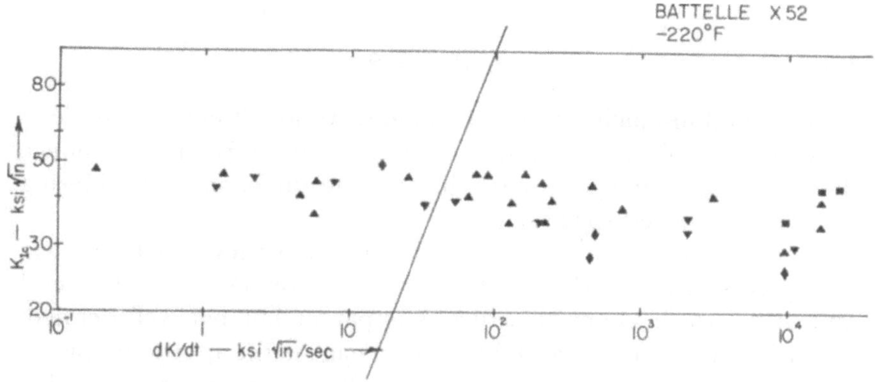

A

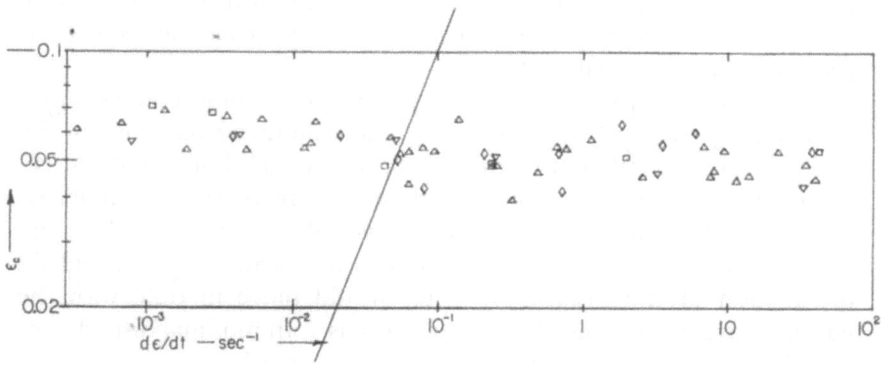

B

Fig. 7. The enormous variability of K_{Ic} (7A) and ϵ_c (7B) with speed is shown on line pipe steel at −220° F. Temperatures ± 100 from −220 are corrected by their 3% flow stress ratio.

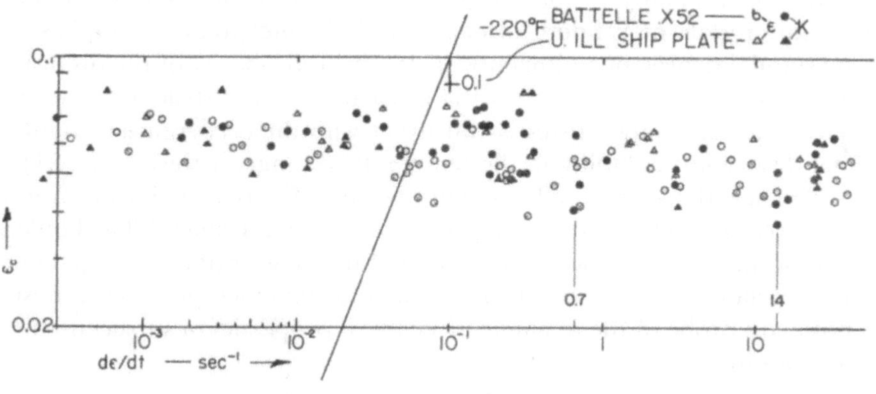

Fig. 8. The matching at $K_{Ic}/\epsilon_c = 710$ Ksi$\sqrt{\text{in.}}$ from Fig. 7 shows a marked pattern or rate spectrum of fracture-strength (solid points) and tensile instability strain (hollow). Also superimposed is the corresponding corrected data for the ship steel as triangles from Fig. 5.

strength and instability strain data points. Although not impossible, it is extremely unlikely that vibrations in the loading system could induce these. The machine is running in opposite directions and differing speeds for matched speed points.

Taking these variations seriously, a pattern emerges from the match. One perceives two fairly well-defined major troughs at 7×10^{-1} sec^{-1} and at 14 sec^{-1}. Other structure is apparent but not well defined. There is a suggestion of a subharmonic image in the approach upward in strain rate to these major troughs. With the present matching scheme, superharmonics are less clearly delineated than in the (improper) previous matching with n [8]. This is helpful as there seems to be no mechanism by which dislocations can be driven at supersonic speeds so as to rationalize these.

A plausible conjecture regarding the cause of these patterns, previously proposed in [8], involves the dislocations which move in the lattice in accommodation of the strain. These could accelerate in response to strain rate much more than they increase in number or density. As they approach the shear wave velocity for given direction in the crystal, one might expect a resonant interaction which would aid the motion of the dislocation. They would move in tune with the natural wave excursions of the lattice atoms with minimal disturbance and damping. At factors of 2, 3, etc., lower velocity (i.e., strain rate) corresponding subharmonic but weaker troughs could also be expected—as suggested by the data. But to attempt to drive beyond the fundamental resonance, the facilitating mechanism might be suddenly detuned, the rather abrupt walls of Fig. 8 suggesting such barrier in restoration of normal work hardening rate (i.e., its ϵ_c equivalent).

We are straining a polycrystalline aggregate composed largely of ferrite crystals of random orientation. This interpretation implies a substantial degree of uniformity of density and velocity of the dislocations accommodating the shearing of each. That the structure is seen only at lower temperatures is consistent with the occurrence of minimal dislocation damping in the temperature range of these tests [21].

The speed spectrum effects represent enormous variations in fracture strength. They are thus of great practical significance. This should help to justify as well as motivate their further scientific investigation. It is certainly an area in which the plane strain crack provides a most powerful probe of the dynamic mechanical properties of subject in this Symposium.

References

1. T. A. Read, H. Marcus and J. H. McCaughey, "Plastic Flow and Rupture of Steel at High Hardness Levels," Fracturing of Metals, p. 228, American Society Metals, Chicago, 1948.

2. C. Zener and J. H. Hollomon, "Plastic Flow and Rupture of Metals," Trans. Am. Soc. Metals, *33*, 163 (1944).
3. T. W. George, "Nucleation and Growth of Flow and Fracture Markings," Industrial and Engineering Chemistry, *44*, 1328 (1952).
4. E. Orowan, "Condition of High-Velocity Ductile Fracture," J. Appl. Physics, *26*, 900 (1955).
5. J. M. Krafft, "Correlation of Plane Strain Crack Toughness with Strain Hardening Characteristics of a Low, a Medium, and a High Strength Steel," Appl. Materials Research, *3*, 88 (1964).
6. J. G. Williams and C. E. Turner, "The Plastic Instability Viewpoint of Crack Propagation," Appl. Materials Research, *3*, 144 (1964).
7. J. Eftis and J. M. Krafft, "A Comparison of the Initiation with the Rapid Propagation in a Mild Steel Plate," J. Basic Eng., *87*, 257 (1965).
8. J. M. Krafft, "A Rate 'Spectrum' of Strain Hardenability and of Fracture Toughness," Rept. NRL Progress, (January 1966), pp. 6–16.
9. J. M. Krafft, "Role of Local Dissolution in Corrosion-Assisted Cracking of Titanium Alloys," Rept. NRL Progress, (March 1967), pp. 6–18.
10. G. R. Irwin, *Structural Mechanics,* p. 557, Pergamon, Oxford (1960).
11. P. C. Paris and G. C. Sih, "Stress Analysis of Cracks," ASTM STP 381, p. 30 (1965).
12. J. A. H. Hult and F. A. McClintock, "Elastic-Plastic Stress and Strain Distribution Around Sharp Notches Under Repeated Shear," IXth International Congress of Applied Mechanics, Vol. 8, Univ. of Brussels, p. 51 (1957).
13. J. L. Swedlow, M. L. Williams, and W. H. Wang, "Elasto-Plastic Stresses and Strains in Cracked Plates," Proc. First International Conf. on Fracture Sendai, Vol. 2, p. 259 (1965).
14. A. J. Birkle, R. P. Wei, and G. E. Pellisier, "Analysis of Plane Strain Fracture in a Series of 0.45C, Ni-Cr-Mo Steels with Different Sulfur Contents," Trans. Am. Soc. Metals, *59*, 981 (1966).
15. J. I. Bluhm, "Geometry Effect on Shear Sip and Fracture Toughness Transition Temperature for Bimodal Fracture," ASTM Preprint No. 75 (1962).
16. G. T. Hahn and A. R. Rosenfield, "Local Yielding and Extension of a Crack Under Plane Stress," Acta Metallurgica, *13*, 293 (1965).
17. J. G. Williams, "The Thermal Properties of a Plastic Zone," Appl. Material Research, *4*, 104 (1965).
18. J. M. Krafft, "Fracture Toughness of Metals," Rept. NRL Progress, p. 4 (November 1963).
19. J. M. Krafft and G. R. Irwin, "Crack-Velocity Considerations," ASTM STP 381, 114 (1965).
20. J. M. Krafft, "Instrumentation for High Speed Strain Measurement," *Response of Metals to High Velocity Deformation,* Proc. Metallurgical Society Conferences, Vol. 9, Interscience, New York, p. 9 (1961).
21. J. Weertman and J. M. Krafft, Agenda Discussion on High Speed Dislocations. Proceedings. Battelle Colloquium on Dislocation Dynamics, Seattle-Harrison (May 1967) (Publication forthcoming).

DYNAMICAL BEHAVIOR OF DISLOCATIONS *

J. J. Gilman
University of Illinois
Urbana, Illinois

The most important single property of a dislocation is its mobility. For a given applied stress, this can be very high as in a weakly bonded crystal like talc, or very low as in tightly bonded diamond. The large ratio (of order 10^8), as well as the sensitivity of dislocation mobility to impurities, temperature, other dislocations, etc., makes a simple universal description of it unlikely. However, certain limiting cases can be described in a simple way and intermediate cases can be interpolated. This will be attempted here.

Although it is often convenient to describe dislocations as if they were lines moving through space, this is quite misleading because their fields of action are not localized. The motion of a dislocation causes the external shape of a body to change so a large volume of material can be affected when the dislocation moves. This inevitably introduces changes in the body forces and surface tractions which increase the complexity of the behavior. In spite of these difficulties, however, some reproducible patterns of behavior have been found and used to increase both our knowledge of the physics of plastic flow, and our ability to predict the mechanical response of engineering materials [1].

Other papers at this symposium will consider macroscopic responses in some detail, so this one will concentrate on microscopic matters. The plan is to review the work that has been done on the measurement of dislocation mobilities, to present some interpretations of the observed behavior, and to outline some predictions based on present knowledge.

DISLOCATION MOBILITY MEASUREMENTS

There are two general methods for measuring dislocation mobilities. One uses direct observations of positions by means of selective etching

* This work has been supported by the Office of Naval Research (3985-06).

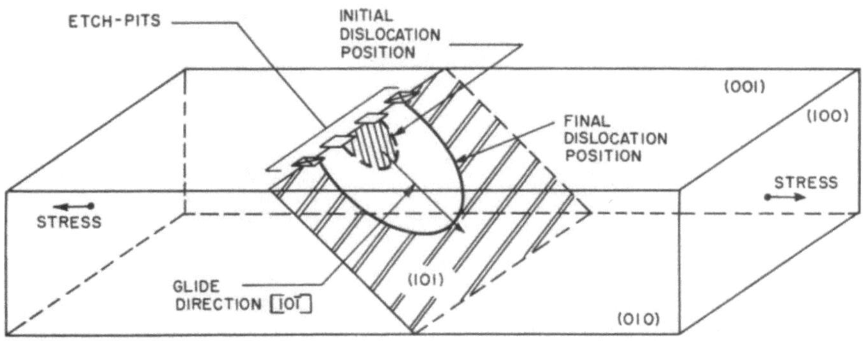

Fig. 1. Schematic drawing showing motion of an individual dislocation loop in a LiF crystal, as revealed by etch-pits.

[2] while the other uses indirect analysis of ultrasonic attenuation as a function of frequency at very-high-frequencies [3]. The former allows average velocities to be determined for large displacements (one micron and up), while the latter allows the damping coefficient to be determined for small displacements.

Methods for selectively etching dislocations will not be reviewed here, but they have been developed for most crystals of interest. Selective etching before and after a stress pulse has been applied to a crystal determines the initial and final positions of dislocations; thus the distance, Δx moved during a stress pulse of known duration, Δt is found. Figure 1 illustrates an actual observation. The average velocity is simply the Δx divided by the Δt, since experiments have shown that the time spent for acceleration is very short (much less than one microsecond).

Several means for producing stress pulses have been used. They include dead-weight loading (>10 sec); levers with rolling weights (>1 sec); electromagnets ($>10^{-2}$ sec); bouncing of spheres off specimens ($>20 \times 10^{-6}$ sec); and the use of a torsion shock bar ($>10^{-6}$ sec). All except the last of these methods were first used by Gilman and Johnston [4] and are described in their review.

For very short stress pulses the best present method is the torsion shock bar developed by Pope, Vreeland, and Wood [5]. It is illustrated in Fig. 2. A section of a bar is pretwisted to produce a certain maximum stress. At one end it is initially held by a metallic membrane, and when this membrane is quickly broken by means of an electric current pulse, a mechanical torsion pulse propagates along the bar toward the specimen. The torsion pulse stresses the specimen as it passes through it

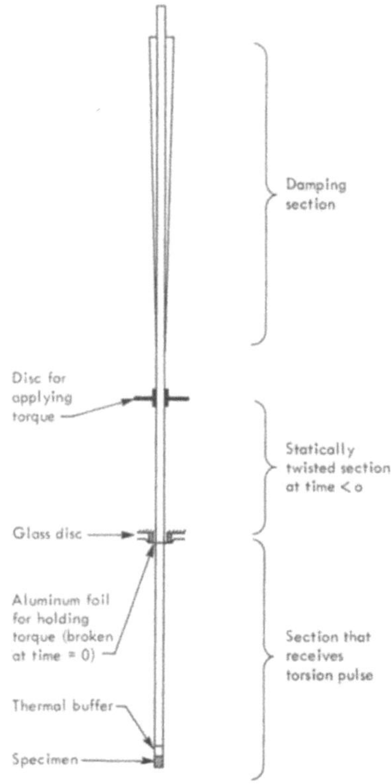

Fig. 2. Schematic drawing of Pope-Vreeland-Wood machine for producing microsecond torsional stress-pulses.

until it reflects at the far end. Then an unloading wave proceeds back through the specimen and is eventually absorbed by the attenuator at the opposite end of the assembly. A prime advantage of the method is that sharply rising wave-fronts remain sharp because there is no geometrical dispersion for torsional waves. Another is that the preload is uniform so the pulses are quite flat between their front and back edges.

The methods above have been applied to a variety of substances as listed in Table I. Several sets of representative data are shown in Fig. 3 for a variety of crystals. They will be interpreted in a later section.

The indirect method for determining dislocation mobility is based on a particular interpretive analysis of ultrasonic attenuation measurements. This analysis is substantiated by a large amount of circumstantial evidence and is generally thought to be a valid means for obtaining the viscosity coefficient for a dislocation. The analysis begins with the equation of motion for a dislocation line that lies parallel to the x-axis and moves in the xy-plane when acted upon by a shear stress σ_{zy}. Let

Table I

MEASUREMENTS OF INDIVIDUAL DISLOCATION VELOCITIES BY MEANS OF SELECTIVE ETCHING

CRYSTAL	REFERENCES
Salts	
LiF	a, b, o
NaCl	b
KBr	c
KCl	v
RbI	v
CaF_2	u
Covalents	
Ge	d, e, f, q, t
Si	d, h, g, r
GaSb	d
InSb	d
Metals	
Cu	m
Fe−3% Si	i, j, l
Nb	s
Ni	p
W	k
Zn	n

a. W. G. Johnston and J. J. Gilman—J. Appl. Phys. *30*, 129 (1959)
b. E. Y. Gutmanas, E. M. Nadgornyi, A. V. Stepanov—Soviet Phys.-Sol. St. *5*, 743 (1963)
c. V. B. Pariiskii, S. V. Lubenets, V. I. Startsev—Soviet Phys.-Sol. St. *8*, 976 (1960)
d. A. R. Chaudhuri, J. R. Patel, and L. F. Rubin—J. Appl. Phys. *33*, 2736 (1962)
e. M. N. Kabler—Phys. Rev. *131*, 54 (1963)
f. O. W. Johnson—J. Appl. Phys. *36*, 3247 (1965)
g. T. Suzuki and H. Kojima—Acta Met. *14*, 913 (1966)
h. J. R. Patel—Discuss. Farad. Soc. #38, p. 201 (1964)
i. D. F. Stein and J. R. Low—J. Appl. Phys. *31*, 362 (1960)
j. J. S. Erickson—J. Appl. Phys. *33*, 2499 (1962)
k. H. W. Schadler—Acta Met. *12*, 861 (1964)
l. D. F. Stein and R. P. Laforce—J. Appl. Phys. *36*, 661 (1965)
m. W. F. Greenman, T. Vreeland, D. S. Wood—J. Appl. Phys. (August 1967)
n. T. Vreeland—J. Appl. Phys. (September 1967)
o. J. Cotner and J. Weertman—Discuss. Farad. Soc. #38, p. 225 (1964)
p. R. W. Rohde and C. H. Pitt—J. Appl. Phys. *38*, 876 (1967)
q. S. Schafer—Phys. Stat. Sol. *19*, 297 (1967)
r. J. R. Patel and P. E. Freeland—Phys. Rev. Lett. *18*, 833 (1967)
s. H. Guberman—submitted to Acta Met. (1967)
t. J. R. Patel and A. R. Chaudhuri—Phys. Rev. *143*, 601 (1966)
u. G. A. Keig and R. L. Coble—Amer. Cer. Soc. Bull. *46*, 359 (1967)
v. E. M. Nadgornyi, E. Y. Gutmanas—*Proc. Conf. on Yield and Fracture*, Oxford, Sept. (1966)

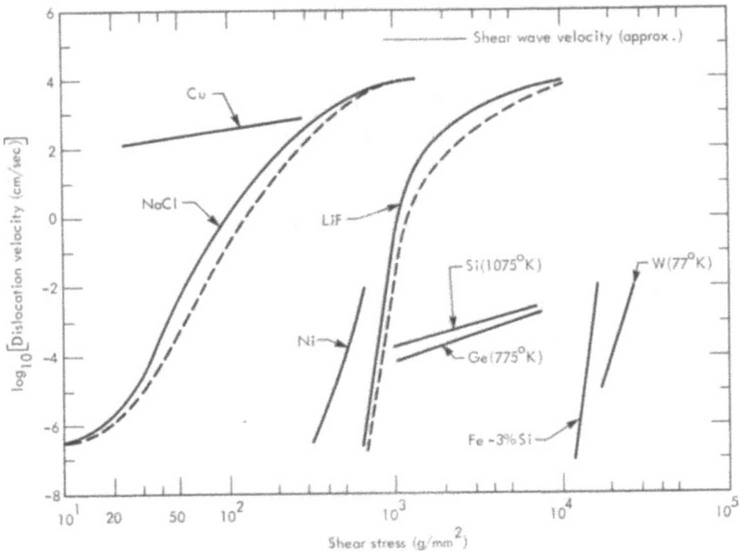

Fig. 3. Velocity-stress behavior for dislocations in several representative materials (solid lines—edge type; dashed—screw type).

m be the effective mass of the dislocation per unit length; T its effective line tension; and b its Burgers displacement. Also, assume that the viscous drag on it is proportional to its velocity, $\partial y/\partial t$. Then an equation of motion may be written [6]:

$$m\frac{\partial^2 y}{\partial t^2} + B\frac{\partial y}{\partial t} - T\frac{\partial^2 y}{\partial x^2} = b\sigma_{zy} \tag{1}$$

where B is the viscous damping constant that relates the driving force, σb, and the drag force, Bv:

$$B = \frac{\sigma b}{v}. \tag{2}$$

For a line segment of length, L, and an applied stress of the form: $\sigma_{zy} = \sigma_0 \sin \omega t$ where ω is the angular frequency, resonance occurs at the frequency:

$$\omega_0 = \frac{\pi}{L}\left(\frac{T}{m}\right)^{1/2} \tag{3}$$

in the absence of damping. For the case of large damping the logarithmic decrement displays a maximum at a frequency [7]:

$$\omega_m = \frac{\pi^2 T}{BL^2}. \tag{4}$$

Table II

DAMPING CONSTANTS FOR DISLOCATION MOTIONS

	Damping Constant = $B(10^{-4}$ d-sec/cm$^2)$			
Crystal	From vhf. ultrasonic attenuation	Ref.	From slope of linear velocity-stress curve	Ref.
LiF	16.0	b	7.0	a
	2.5	c	—	
	2.4	e	—	
NaCl	(2.5–10.5)	d	—	
	1.6	e	—	
KCl	3.2	b	—	
KBr	—		~20	f
Al	~10.0	l	—	
Cu	8.0	g	7.0	i
	6.5	h	—	
	0.8	b	—	
Zn	—		7.6	j
Pb	~3.7 (300° K)	k	—	
	~1.1 (60° K)		—	

a. W. P. Mason—J. Acoust. Soc. Am. *32*, 458 (1960)
b. T. Suzuki, A. Ikushima, M. Aoki—Acta Met. *12*, 1231 (1964)
c. O. M. Mitchell—J. Appl. Phys. *36*, 2083 (1965)
d. R. A. Moog—Ph.D. Thesis, Cornell Univ. (1965)
e. F. Fanti, J. Holder, and A. Granato—to be published
f. V. B. Pariiskii, S. V. Lubenets, V. I. Startsev—Sov. Phys.-Sol. St. *8*, 976 (1966)
g. G. A. Alers and D. O. Thompson—J. Appl. Phys. *32*, 283 (1961)
h. R. M. Stern and A. V. Granato—Acta Met. *10*, 358 (1962)
i. W. F. Greenman, T. Vreeland, D. S. Wood—J. Appl. Phys. (August 1967)
j. T. Vreeland et al.—J. Appl. Phys. (September 1967)
k. W. P. Mason and A. Rosenberg—J. Appl. Phys. *38*, 1929 (1967)
l. W. P. Mason and A. Rosenberg—Phys. Rev. *151*, 434 (1966)

At frequencies higher than this (typically in the range 10–100 *m-Htz*), the dislocations no longer vibrate like strings, but move more like rigid rods; and the ultrasonic attenuation, α, approaches a limiting value:

$$\alpha_\infty = \frac{8\varphi T\rho}{\pi^2 B} \tag{5}$$

where φ is an orientation factor that relates the coordinates of the dislocations to the coordinates of the sound wave, and ρ is the dislocation density.

If the dislocation density is known, and T is estimated to be $Gb^2/2$ (G is the shear modulus), then B can be estimated. Numerical values are listed in Table II. For more details of this method, the reader is referred to reviews by Stern and Granato [3] and Granato [8]. Values of *B* can also be obtained from linear velocity-stress curves or tangents at high velocities; and some values are listed in the Table for comparison with the ultrasonic results. It may be seen that there is approximate agreement.

VELOCITY-STRESS RELATIONS

A glance at Fig. 3 indicates that no general relation between dislocation velocity and stress can be written. In some cases, such as copper and germanium, a linear relation is observed; whereas in others, such as lithium fluoride and iron-silicon, the relation is very non-linear. There are some limiting cases, however, according to the type of crystal; and to whether the motion is induced by stress alone, or is assisted by thermal fluctuations. These cases are classified schematically in Table III.

In crystals with non-local binding (metals, salts, molecular) the energy of a straight dislocation is very nearly independent of its position in the perfect structure. This is because the cohesive energies of such crystals depend predominantly on the atomic volume; being rather independent of the atomic packing pattern. The mobility of such a dislocation is very high and the main source of drag that acts on it is the viscosity caused by thermal oscillations. The most direct evidence of this is that the drag seems to increase with increasing temperature. Copper is an example [9] and the high mobility of dislocations in it is indicated in Fig. 3.

In covalently bonded crystals, bond angles as well as bond lengths strongly affect the cohesive energy. Therefore, the energy of a dislocation tends to depend strongly on its position in the structure. This makes its mobility low, and the viscous drag on it decreases with increasing temperature because its motion is aided by thermal oscillations.

Table III

CLASSIFICATION OF DISLOCATION BEHAVIOR

		Bonding	
		Local (covalent)	Non-Local (metals and salts)
Stress	Low (<G/100)	Thermally Activated	Phonon and/or Free-Electron Drag Limited
	High (>G/100)	Stress Activated	–

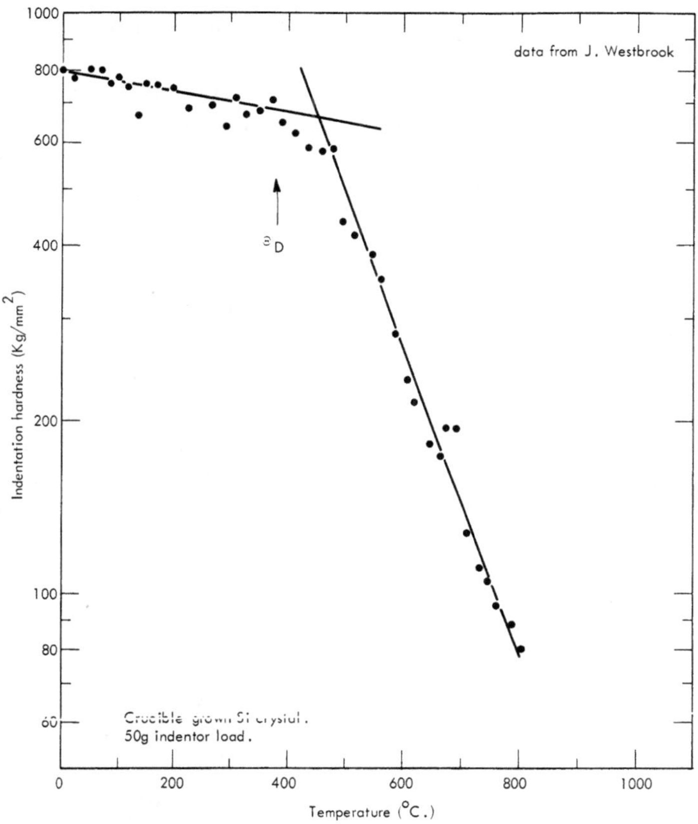

Fig. 4. Showing change in temperature dependence of hardness of silicon below its Debye temperature.

At low temperatures the thermal energy per vibrational mode ($\frac{1}{2}kT$) becomes too small to assist the stress in moving dislocations and their motion can only be caused by high applied stresses. This is clearly shown by the dependence of the indentation hardness on the temperature as illustrated by data for silicon in Fig. 4. In the 400–800° C range the hardness depends strongly on the temperature, but in the range below the Debye temperature, $\theta_D \simeq 390°$ C, the hardness is high and nearly independent of the temperature.

The existence of stress activation at low temperatures in covalent crystals is also indicated by the sharp onset of yielding that is observed when a plane shock wave passes through germanium (or silicon) at room temperature. When the pressure reaches 44 ± 4 kbar for a $\langle 111 \rangle$ wave in germanium, flow begins [10]. This corresponds to a hardness number of 440 kg/mm², so the two kinds of measurement agree.

The two limiting bond types are often mixed in impure metal and salt crystals to form a heterogeneous system that can have a complex behavior.

A common feature for any crystal type is that the dislocation velocity cannot increase without limit. It must saturate below the velocity of sound if the stress field is continuous (supersonic motion is possible for discontinuous stress fields). Therefore, if the limiting velocity is called v^*, the velocity can be quite generally written as:

$$v = v^* P_m(\sigma_s) \tag{6}$$

where P_m is the average probability for it to have the velocity, v^* at a given instant. P_m is a function of the stress and may also depend on the prior plastic strain, the temperature, impurity concentration, etc.

Any general functional form that is proposed for P_m must satisfy at least two conditions at the limits of high and low stress. Namely, the value must be zero when $\sigma_s = 0$ and it must approach unity asymptotically for high values of σ_s. Of the functions that have been shown to fit at least some of the data, only the one proposed by Gilman [11] satisfies both of these conditions. Its form is:

$$e^{-D/\sigma_s} \tag{7}$$

where D is a constant called the drag stress, and it is known to fit much of the data, especially at medium to high stress levels. However, its form at low stress levels is poor because its derivatives are complicated. Also, it does not even approximate the linear velocity-stress relations that have recently been observed for Cu, Ge, and Zn. Therefore, its applicability is limited. Nevertheless, it is believed that it is consistent with the "stress-activated" category of Table III, and can be derived from a physical model to be discussed in a later section.

Two functions that satisfy the conditions mentioned above, and also have linear behavior for small stresses are:

$$1 - e^{-A\sigma} \tag{8a}$$

$$\tanh A\sigma. \tag{8b}$$

The second of these has the largest range of approximate linearity and therefore is suggested as a possible form for describing experimental data. The following combination of (2), (6), and (8b) is suggested:

$$V = V^* \tanh (b\sigma_s/V^*B) \tag{9}$$

which reduces to $V = b\sigma_s/B$ when σ_s is small.

The forms of (7) and (8b) are compared in Fig. 5, and it may be seen that (7) has the effect of shifting the transition between small and large velocities to a finite stress level. This is why it is useful for describing

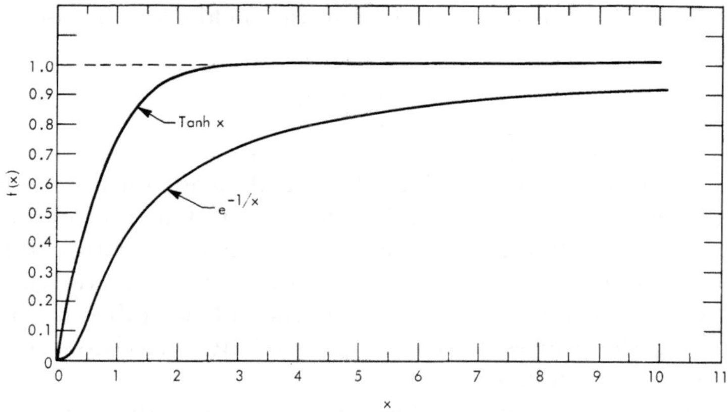

Fig. 5. Comparison of two possible probability functions for describing the stress dependence of dislocation velocities.

materials with distinct yield stresses. On the other hand, (9) gives a means for describing visco-elastic materials.

It seems likely that the behavior of almost any material can be described by mixing (7) and (8b) into (6). This will give a three or four constant equation that is probably the most simple form with adequate generality. The mixed equation would be:

$$\bar{V} = \bar{V}_c^* \tanh (C\sigma) + \bar{V}_d^* e^{-D/\sigma}. \tag{10}$$

This procedure may seem arbitrary, but it is consistent with the fact that real materials are heterogeneous so that more than one mechanism controls their behavior.

VISCOUS RESISTANCE TO DISLOCATION MOTION

When a straight dislocation moves along a glide plane, a velocity gradient exists in the direction perpendicular to the plane. The gradient is largest at the center of the dislocation and decreases rapidly away from the center. If a means exists in the medium for transferring momentum from the higher velocity regions to the ones with lower velocities, then viscous losses occur which create a dragging resistance to the motion. These losses can be described in various ways, but the most familiar is the use of a viscosity (or fluidity) coefficient. Mason [12] has considered the losses in the strain field of a screw dislocation from this point of view. The rate of energy loss, dW, in a small volume element, dV, is dependent on the strain rate, $\dot{\epsilon}$, within it and on the viscosity coefficient, η:

$$dW = \eta(\dot{\epsilon})^2\, dV. \tag{11}$$

Integration of this loss over the strain-rate field gives the viscous drag force $= W/v$:

$$f_m = b^2\eta v/8\pi r_0^2 \tag{12}$$

where r_0 is the lower limit of the integration.

This result has two difficulties. First, it depends sensitively on r_0 as might be expected since the strain-rate is greatest near the dislocation's core. This is not desirable because r_0 is a somewhat arbitrary quantity. Second, and most important, it neglects the effect of the velocity gradient associated with the "sliding" on the glide plane at the core. Yet, it is this sliding that causes a dominant part of the total viscous loss. This may be shown as follows.

The velocity gradient at the glide plane of a dislocation can be evaluated from the displacements given by the solution of Peierls' integral equation [13]. The x-component of the relative displacement across the glide plane (for an edge dislocation) is:

$$u(x) = -\frac{b}{\pi}\tan^{-1}\left(\frac{2x}{w}\right) \tag{13}$$

where w is the dislocation's width, and $x = vt$. This relation does not depend on the particular atomic force law which acts indirectly to determine w.

The expression for the velocity gradient at the glide plane is:

$$\dot{\epsilon} = \frac{1}{a}\left(\frac{du}{dt}\right) = \frac{v}{a}\left(\frac{\partial u}{\partial x}\right) \tag{14}$$

where a is the separation distance of the atomic layers.

If (13) and (14) are substituted into (11) and this is integrated:

$$P_g = \frac{4\eta v^2 b^2}{\pi^2 w^2 a}\int_{-\infty}^{+\infty}\left[1+\left(\frac{2x}{w}\right)^2\right]^{-2} dx = \eta v^2 b^2/\pi a w \tag{15}$$

which may be compared with Mason's result by setting $a \simeq w \simeq b \simeq r_0$ to obtain:

$$f_g \simeq 8 f_m$$

so the power loss associated with the sliding along the glide plane is large compared with the losses in the surrounding elastic field.

An expression for the damping constant, B, can be obtained from (15) since Bv equals F_d [neglecting the small contribution from (12)]:

$$B = \left(\frac{b^2}{\pi a w}\right)\eta. \tag{16}$$

A more difficult problem is the calculation of the viscosity coefficient.

At high temperatures (above the Debye temperature) it is valid to treat η as the viscosity of a uniform fluid because the mean free path for phonons is of the order of atomic dimensions. At low temperatures the detailed structure at the glide plane must be considered. The limiting cases of localized and non-localized bonding will be discussed here in turn.

For germanium (localized bonding), high temperature data are available [14] from which values for η can be obtained through (15). Near the melting point these values might be expected to correspond to the directly measurable viscosity of liquid germanium. Schafer measured dislocation velocities as a function of stress and temperature and found:

$$V = \beta_0 \sigma_s e^{-Q/kT} \tag{17}$$

where $\beta_0 = 0.05 - 0.15$ cm^3/d-sec and $Q = 1.62$ e.v. At $T_m = 1230°$ K then, $\beta = (3.1 - 9.4) \times 10^{-10}$ cm^3/d-sec. For a dislocation line, $\beta = \dfrac{a}{\eta}$, but kink motion controls the dislocation velocity in germanium and if the kink concentration is C_k the kink velocity is v_d/C_k. Thus β becomes aC_k/η. If the energy of formation of a kink is U_k, the kink concentration will be: $e^{-U_k/kT}$. Equating η to the liquid viscosity, η_0, a value for U_k can be obtained:

$$U_k = kT_m \ln (a/\beta\eta_0). \tag{18}$$

The liquid viscosity at the melting point as estimated by means of Andrade's equation is: 2.8×10^{-2} poise which yields a value for U_k of: 0.68 e.v. This is quite reasonable since it is approximately one-half the activation energy in (17), or the single-bond energy. Therefore it may be concluded that the glide plane at the core of a dislocation in Ge behaves approximately like the liquid.

The analog of the damping constant, B, is bC_k/β which has a value of $\sim 4.4 \times 10^{-2}$ for Ge at T_m. Note that this is much greater than the values listed in Table II for metals and that it increases rapidly with decreasing temperature.

Metals may behave differently, but this is not yet clear. For example, the viscosity of copper at its melting point is 4.1×10^{-2} poise, and with $1.2a \simeq b \simeq \frac{1}{4}w$ (16) would yield a value of $B \simeq 4 \times 10^{-3}$ d-sec/cm^2 if the dislocation core behaved like liquid copper. The values of B observed at room temperature are less than about 7×10^{-4} d-sec/cm^2, or an order of magnitude smaller than the liquid value. Furthermore, B appears to increase with increasing temperature which contradicts the temperature dependence of η for liquids.

There is considerable doubt about the exact cause of the glide plane viscosity in pure metals (and pure salts), but the best proposal to date is that of Mason [12] who attributes it to phonon viscosity. Thermal oscillations in the solid are considered to be quasi-particles (phonons) that can be formally described in terms of gas dynamics. A suddenly applied strain temporarily leaves the gas in a non-equilibrium state which relaxes into equilibrium with the strained crystal in a time period given by the thermal relaxation time [15]:

$$\tau = \frac{3K}{\rho C_v V_0^2} \tag{19}$$

where K = lattice thermal conductivity
ρ = mass density
C_v = specific heat at constant volume
V_0 = average Debye sound velocity.

The product of this relaxation time and the modulus defect between the relaxed and unrelaxed state equals the effective viscosity.

Mason's value for the phonon gas viscosity in copper at room temperature is: $\eta_p = 7 \times 10^{-3}$ poise. In (16), again let $b/a = 1.2$ and $w = 4b$; so $B \sim 7 \times 10^{-4}$ poise which compares favorably with the observed values of $1–8 \times 10^{-4}$ poise. Thus Mason's theory correctly estimates the viscosity.

STRESS ACTIVATION

Dislocation motion in Ge is clearly thermally activated at high temperatures [14] but dislocations can also move in covalent crystals at low temperatures if the stress is high enough (Fig. 4). Electronic conduction in solids provides a good analogy. At high temperatures, small electric fields can induce the motion of thermally excited carriers in pure crystals. However, conduction can still occur at low temperatures provided a large electric field is applied. It is proposed here that low temperature plastic flow occurs because kinks can undergo quantum mechanical tunneling through the potential barriers that block them. This is analogous to electron tunneling in the conduction case.

In order to account for the creep of metals at very low temperatures, Mott [16] proposed that dislocations that are pinned by other dislocations which cross their glide planes can tunnel through the energy barrier associated with creating a jog in order to move past the pinning point. It is proposed here that this is a much more general phenomenon; in particular that the tunneling rate for kinks controls the flow of covalently bonded crystals at low temperatures [17].

There is considerable evidence that in covalent crystals (such as Ge and Si) dislocation lines move as a result of the motion of sharply defined kinks along their lengths. That is, the velocity of a line depends on the velocity of kinks along it, and the concentration of kinks (C_k = fraction of atomic sites occupied by kinks):

$$v_d = C_k v_k. \tag{20}$$

The evidence is the strong temperature dependence of the velocity at high temperatures where the phonons have atomic wavelengths, the crystallographic shapes of dislocation loops, the close correlation between chemical binding energy and glide activation energy, and a detailed calculation of the kink configuration by Labusch [18] which has yielded 0.46b as the kink width.

Other authors [14] have argued that the rate at which kinks are created (in pairs) is the rate controlling step, but kink creation should be quite easy at a free surface (where image forces cause the line tension to disappear); yet dislocation lines that move near free surfaces in silicon show no marked change of shape. It is therefore concluded that kink motion is the rate determining process.

At low temperatures, the concentration of kinks is determined by a balance between creation and annihilation rates [19]. If N is the number of kinks per unit length, and α is the number of pairs created per unit length per second, then since the pair annihilation rate will be: $V_k N^2$, the kink density changes at the rate (assuming a straight line):

$$dN/dt = 2(\alpha - V_k N^2). \tag{21}$$

The kink density saturates when $dN/dt = 0$, so the saturation density is:

$$n^* = \left(\frac{\alpha}{V_k}\right)^{1/2}$$

and the dislocation velocity is (provided $\alpha < V_k/b^2$):

$$v_d = b(\alpha V_k)^{1/2}. \tag{22}$$

The quantities α and V_k can be further resolved. The pair creation rate equals the product of the number of possible creation sites (which is $1/d$ where d is the kink length), and the pair creation rate at any site. This latter equals the creation attempt frequency, ν_c, times the success probability, P_c. Thus α is given by:

$$\alpha = \frac{\nu_c P_c}{d}. \tag{23}$$

The kink velocity depends on the jump rate (which is the jump attempt frequency, ν_m, times the success probability, P_m, and also on the jump distance, d. Hence:

$$v_k = \nu_m p_m d. \qquad (24)$$

Substitution of (23) and (24) into (22) yields:

$$v_d = b(\nu_c \nu_m)^{1/2}(P_c P_m)^{1/2}. \qquad (25)$$

Since atomic dimensions are involved, ν_c and ν_m must approximately equal the Debye cut-off frequency. Then the first two terms of (25) can be replaced by a terminal kink velocity, V_k^*, which equals the elastic shear-wave velocity, approximately.

The production of a kink pair and the motion of a kink is very nearly the same process in the diamond structure. To create a kink pair, two adjacent atoms must be separated by shear. To move a kink, the same thing happens (but the configuration is asymmetric). Since the processes are similar, the probabilities P_c and P_m should be approximately equal and (25) becomes:

$$V_d = V_k^* P_m. \qquad (26)$$

Then the next step is to calculate P_m.

In the absence of an applied stress (and other external fields) there is no net force acting on a kink. It is located at a minimum of potential energy, which lies a depth, ϵ_p, below the free-kink level. Experimental evidence indicates that this binding energy is substantial, being of the order of one electron-volt in Ge and Si. For small values of ϵ_p, the zero-point energy ($\sim k\theta/2$, where $k =$ Boltzmann's constant and $\theta =$ Debye temperature) will excite a kink at all temperatures. However, for silicon, $k\theta/2$ is about 3×10^{-2} e.v. so thermal excitations are important at high temperatures only.

Figure 6 shows a schematic energy diagram for a kink. The force ex-

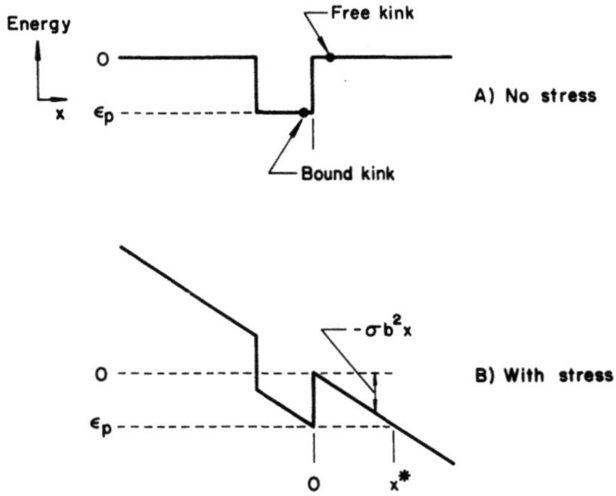

Fig. 6. Schematic potential energy diagrams for tightly bound kinks along dislocation lines.

erted by an applied stress is σb^2 so the potential energy is $-\sigma b^2 x$ and the tunneling distance, x^* (where the total energy of a free kink becomes equal to that of a bound one), is:

$$x^* = \epsilon_p/\sigma b^2. \tag{27}$$

Since the configuration relaxes after each jump, the energy diagram is aperiodic.

The WKB approximation for the tunneling probability is:

$$\exp\left\{-\frac{\sqrt{8m}}{h}\int_0^{x^*}[V(x)-W]^{1/2}\,dx\right\} \tag{28}$$

where: m = kink effective mass; h = Planck's constant $/2\pi$; and $V(x) - W = \epsilon_p - \sigma b^2 x$. The kink has a wavelength associated with it, so:

$$\lambda = \frac{h}{\sqrt{2_m \epsilon_p}} \simeq d \tag{29}$$

and this relation plus (28) leads to:

$$P_m = \exp\left[\frac{-4}{3db^2}(\epsilon_p/\sigma)\right]. \tag{30}$$

In the diamond structure, $d = b$ for a 60° dislocation, and $b^3 = \sqrt{8}\ V$ where V is the atomic volume. To a sufficient approximation then, (30) becomes:

$$P_m = \exp\left(\frac{-\epsilon_p}{2V\sigma}\right) \tag{31}$$

where $\epsilon_p/2V = D$ = characteristic drag stress.

For rapid kink motion, the exponent in (31) must approach unity so the applied shear stress must approach:

$$\sigma_s^* \simeq \epsilon_p/2V. \tag{32}$$

These two equations just above give the principal results of the quantum tunneling theory. First, kink motion may be stress activated at low temperatures, and is a very sensitive function of the applied stress. Second, a simple condition for rapid flow results that depends only on the binding energy density.

The binding energy, ϵ_p, is a critical quantity; it is also an elusive one because there is no means at present for measuring it directly, and its calculation is difficult. One method for estimating it is to recognize that the binding energy must be proportional to the shear cohesive energy and write:

$$\epsilon_p \sim Gb^3$$

then, from (31):

$$P_m = \exp(-\Phi G/\sigma) \tag{33}$$

where Φ is a proportionality coefficient to be determined by experiment. Its magnitude is expected to be about ten. The appropriate shear modulus, G, is the one referred to the glide plane $\{111\}$ and glide direction $\langle 110 \rangle$; designated,

$$C_{gd} = 3[S_{44} + 4(S_{11} - S_{12})]^{-1}.$$

Two approximate measures of low-temperature flow stresses are available: indentation hardness numbers and the elastic limits observed for plane shock waves. Figure 7 shows that hardness numbers do indeed correlate with C_{gd} as suggested by (33). Furthermore, the elastic limit from shock wave data is about 44 kbars = 440 kg/mm^2 [10] in approximate agreement with the hardness number.

Another measure of ϵ_p is the energy gap because excitation of one electron of a bond across the gap destroys the bond. This measure was

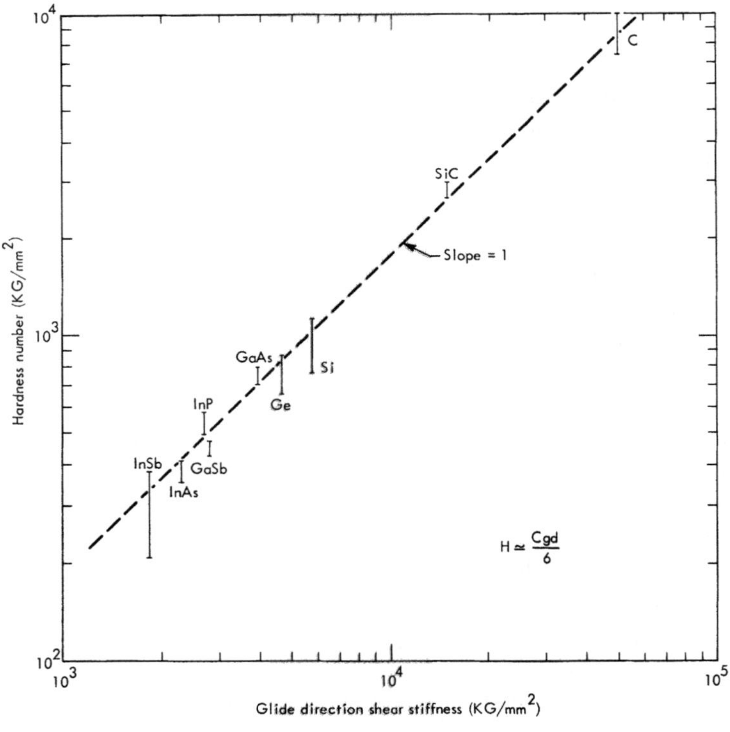

Fig. 7. Correlation of flow stress (hardness) and shear stiffness for tetrahedrally bonded crystals.

Table IV

COMPARISON OF FLOW STRESSES CALCULATED FROM ENERGY GAPS AND OBSERVED HARDNESSES

Crystal	Eg (e.v.)	b (A)	V (10^{-24}cm^3)	Flow Stress (10^{10}d/cm^2)	Hardness Number (10^{10}d/cm^2)
Ge	0.80	3.99	22.7	5.6	6.6
Si	1.14	3.83	20.0	9.1	8.0

first applied to dislocations by Suzuki [20] who used it to set a sharp upper limit on the allowable strain for a covalent bond. Here it is assumed that it represents the energy needed to excite a kink into its mobile state. Table IV compares calculated flow stresses (compressive) with hardness numbers and it may be seen that the numerical agreement is good.

A consequence of the idea that the energy gap measures the limit of shear resistance is that large pressure, magnetic, and electric fields might influence kink motion by affecting the gap.

BEHAVIOR OF HETEROGENEOUS CRYSTALS

Most crystals consist of mixtures of regions in which the bonding is either localized, non-localized, or intermediate. This gives great variety to their mechanical responses and makes a general interpretation difficult. The most detailed study of the heterogeneous behavior was that of Pariiskii, Landau, and Startsev [21] who determined distribution curves for dislocation "jumps" in LiF crystals of two different purities (Fig. 8). In such crystals, dislocations move rapidly for short distances and then stop momentarily before moving rapidly again. In an impure (10^4 ppm) crystal the authors found the most frequent "jump" distance to be about 1.8 microns, whereas the most frequent distance for a higher purity (7.6 ppm) was about 4.5 microns. This is consistent with the idea that dislocations move easily through pure salts and simple metals, but become stopped by impurities and other defects when these are present.

Figure 9 shows a possible interpretation of the macroscopic behavior of LiF. The velocity-stress curve on the left shows a hypothetical be-

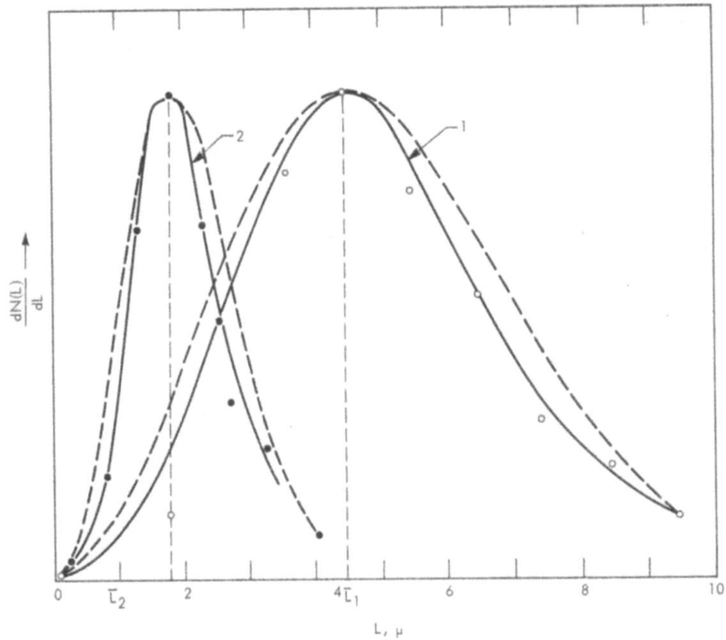

Fig. 8. Distributions of dislocation "jumps" relative to the length of the jump in LiF. N(L) = number of jumps of length less than L. Curve 1 is for crystal with 7.6 ppm impurities; Curve 2 for one with 10^4 ppm impurities. Dotted curves correspond to Gaussian distributions (after Pariiskii, Landau, and Startsev).

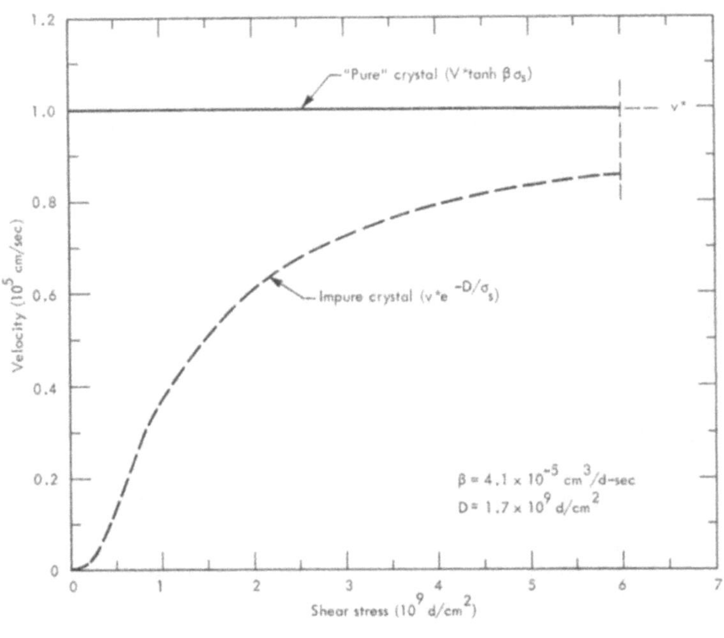

Fig. 9. Comparison of velocity-stress curve deduced from ultrasonic attenuation data for "pure" LiF with a curve measured for an impure crystal by means of selective etching.

havior for the pure matrix, while the curve on right shows how typical experimental data behave. The difference between the two curves measures the losses caused by impurities. The initial slope of the curve for the pure crystal is determined by the damping constant, $B = 7 \times 10^{-4}$ *d*-sec/cm^2, taken from ultrasonic measurements. On the scale of Fig. 9 this gives negligible power loss and the stress required to bring the velocity up to its terminal value is quite small. Impurities produce substantial losses, however, as indicated by the shading on the drawing. Most real materials behave like the impure crystal.

To interpret the data of Fig. 9, it is suggested that the regions adjacent to impurities behave like small sections of covalent crystals (at low temperatures) and that when short segments of dislocations get stopped at these places, they are strongly bound and must tunnel through their potential barriers in order to move on. This accounts for the stress activated form of the velocity-stress relation. There are two major complications though. One is that the local stress is concentrated at impurities so the applied stress is much smaller than the unknown actual stress. The other is that because of the impurities there are long-wavelength internal stress-fields present that amplify the effective viscous drag as a result of the mechanism proposed by Chen, Gilman, and Head [22].

Most real substances a) are heterogeneous, b) contain short-range concentrated stresses, and c) contain long-range stress fluctuations. Therefore, it is unlikely that detailed physical models of them can be constructed. It is more realistic to be satisfied with adequate phenomenological descriptions such as (10).

EFFECT OF MOTION ON DISLOCATION CORE STRUCTURE (DYNAMIC ANISOTROPY)

It is well known that when the velocity of a dislocation approaches the velocity of sound, the "shape" of its stress field changes because of the Fitzgerald-Lorentz contraction. However, drag forces can also be expected to change the shape of a moving dislocation, primarily at the core where the viscous losses are greatest. It was suggested some time ago that this might influence plastic behavior [23] and recently a specific model for f.c.c. crystals has been proposed and studied by Copley and Kear [24].

The above authors have considered the motion of glide dislocations in f.c.c. crystals subjected to a uniaxial stress. They assume that the steady-state velocity of each Shockley partial is a function of the total force acting on it per unit length. Then the velocity of an $a/2\ \langle 110 \rangle$ glide dislocation is a function of the average Schmid factor for its

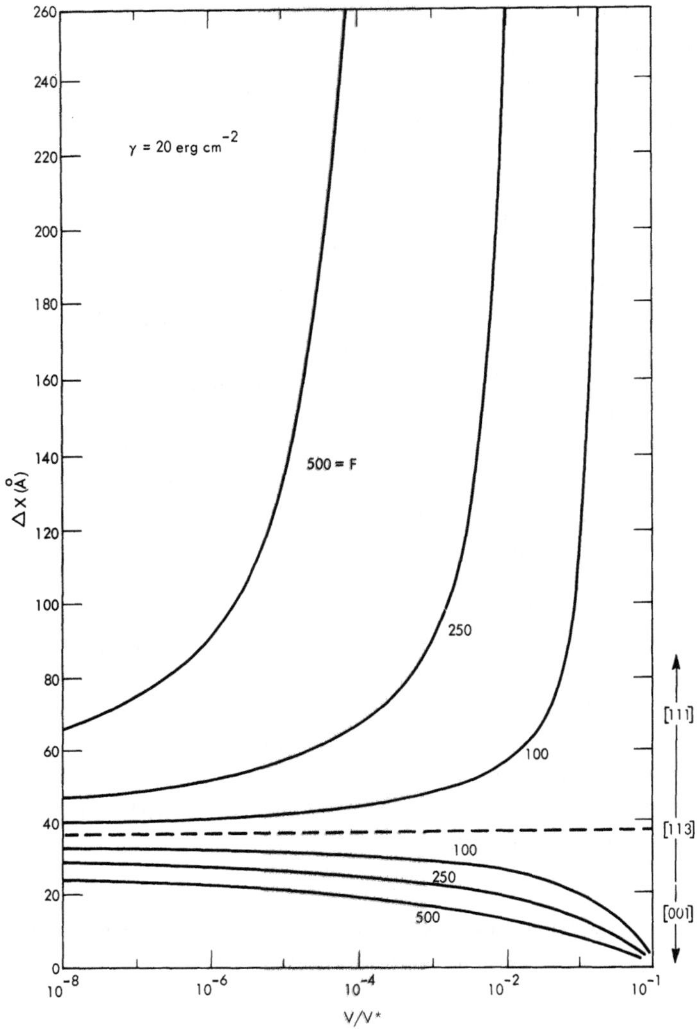

Fig. 10. Results of numerical calculations of the widths of extended dislocations in f.c.c crystals as a function of relative velocity. Curves are for various values of the drag force, $F(d/\mathrm{cm})$ and for three orientations of the axis of applied tension (after Copley and Kear).

Shockley partials, and its degree of dissociation depends on its velocity and the direction and sense of an applied uniaxial stress. For tension applied near a {111} direction, glide dislocations are completely dissociated at moderately high velocities (~10 cm/sec) in alloys with high drag-stresses and low stacking-fault energies. For tension applied near {001} on the other hand, glide dislocations constrict at high velocities in alloys with high frictional drag-stresses and high stacking-fault energies. These effects should exert an influence on the strain hardening characteristics of such crystals at high strain rates through their effect on the probability of cross-glide.

The dynamics were studied by means of the velocity-force relation:

$$V = V^* e^{-F/\Sigma f_i} \tag{33}$$

where V^* is the terminal velocity, F is the drag force acting on a given dislocation line, and Σf_i is the sum of forces caused by the resolved shear stress, the stacking fault, and others. Some of Copley and Kear's numerically calculated results are shown in Fig. 10, where the width of an extended glide dislocation is given as a function of its relative velocity (log scale). Several curves are shown with various values of the drag force, $F(d/\text{cm})$; a stacking fault energy of 20 erg/cm^2; and three orientations of the stress-axis (111, 113, and 001). It may be seen that extension occurs more readily than contraction because the attraction of the partials is constant, whereas the repulsion depends strongly on their separation distance. Also, extension begins at very low relative velocities, so these core structure changes are not limited to very fast dislocations. If the signs of the applied stresses are reversed so are the signs of the width changes.

The dynamical anisotropic behavior of extended dislocations may be expected to influence the macroscopic flow properties because wide dislocations will tend not to cross-glide as much as narrow ones. Also, although the present theory neglects changes in the total viscous drag with the width, such changes are likely to occur.

SUMMARY

Several aspects of dislocation motion in viscous or quasi-viscous media are considered. The discussion begins with a review of two methods for measuring dislocation mobilities as a function of stress; one is the selective etching method and the other is ultrasonic attenuation measurements at very high frequencies. Available damping constant data are summarized.

It is pointed out that no simple analytic function (two constants) can describe all of the observed velocity-stress behavior. A classification scheme is given for the four general behavioral types. It is suggested that two broad classes of crystals need to be distinguished: first, those with localized chemical bonds (covalents); second, those with non-local bonding (metals and salts). In addition, it is important to distinguish stress activation from thermal activation for the localized bonding case.

Several authors have discussed damping mechanisms for dislocations. Apparently they have all underestimated the dominant role played by the viscous forces along the glide plane. It is shown here that this is so large that the viscous drag produced by the integrated strain-rate field away from the glide plane is small enough to be neglected. As a result some previous numerical calculations are incorrect by an order of magnitude.

Various elementary sources of glide plane viscosity are discussed.

It is emphasized that stress alone can activate dislocation motion at low temperatures, and a quantum-mechanical description of the behavior is proposed and shown to be semi-quantitatively consistent with available hardness and plane shock-wave data.

Real crystals are mixtures of regions of the limiting types. Therefore, they are poorly suited for detailed analysis and are best described in terms of phenomenological parameters.

The effect of motion through a viscous medium on the structure of dislocation cores is discussed. Depending on the direction of an applied tension or compression relative to the crystal axes, extended dislocations may be expected to extend further or contract. Thus crystals may possess "dynamical anisotropy."

References

1. J. J. Gilman, Proc. Fifth U.S. National Congress of Applied Mechanics, p. 385, ASME, New York (1966).
2. J. J. Gilman and W. G. Johnston, Dislocation and Mechanical Properties of Crystals, Ed. by J. C. Fisher, et al., p. 116, J. Wiley and Sons, New York (1957).
3. R. M. Stern and A. V. Granato, Acta. Met. *10,* 358 (1962).
4. J. J. Gilman and W. G. Johnston, Sol. St. Phys. *13,* 327 (1960).
5. D. P. Pope, T. Vreeland and D. S. Wood, "Machine for Producing Square Torsion Pulses of Microsecond Duration," Rev. Sci. Inst. *35,* 1351 (1964).
6. J. S. Koehler, Imperfections in Nearly Perfect Crystals, Ed. by W. Shockley, et al., p. 197, J. Wiley and Sons, New York (1952).
7. A. Granato and K. Lucke, J. Appl. Phys. *27,* 583 (1956).
8. A. V. Granato, Proc. Battelle Colloquium on Dislocation Dynamics, to be published (1967).
9. G. A. Alers and D. O. Thompson, J. Appl. Phys. *32,* 283 (1961).
10. R. A. Graham, O. E. Jones and J. R. Holland, J. Appl. Phys. *36,* 3955 (1965).
11. J. J. Gilman, Australian J. Phys. *13,* 327 (1960).
12. W. P. Mason, J. Acoust. Soc. Amer. *32,* 458 (1960).
13. R. Peierls, Proc. Phys. Soc. *52,* 34 (1940).
14. S. Schafer, Phys. Stat. Sol. *19,* 297 (1967).
15. A. Akheiser, J. Phys. USSR *1,* 277 (1939).
16. N. F. Mott, Phil. Mag. *1,* 568 (1956).

17. J. J. Gilman, J. Metals *18*, 1171 (1966).
18. R. Labusch, Phys. Stat. Sol. *10*, 645 (1965).
19. V. Celli, M. Kabler, T. Ninomiya, R. Thomson, Phys. Rev. *131*, 58 (1963).
20. H. Suzuki, Proc. Int. Conf. Cryst. Latt. Def., J. Phys. Soc. Japan *18*, Suppl. I, 182 (1963).
21. V. B. Pariiskii, A. I. Landau and V. I. Startsev, Sov. Phys.-Sol. State *6*, 1002 (1963).
22. H. S. Chen, J. J. Gilman and A. K. Head, J. Appl. Phys. *35*, 2502 (1964).
23. J. J. Gilman, Proc. Cranfield Fracture Symposium (1962).
24. S. M. Copley and B. H. Kear, "The Dependence of the Width of a Dissociated Dislocation on Dislocation Velocity," Report No. 67-007, AMRAD Laboratory, Pratt and Whitney Aircraft.

CONSTITUTIVE EQUATIONS FOR DYNAMIC MATERIAL BEHAVIOR

S. R. BODNER

Technion—Israel Institute of Technology
Haifa, Israel

ABSTRACT

The constitutive equations that have been developed for the dynamic behavior of materials presuppose the existence of a reference "static" yield criterion. An alternative formulation motivated by the work on dislocation dynamics considers the total deformation to consist of elastic and plastic components throughout the deformation history. This procedure permits the consideration of large deformations (finite strains) in a direct manner. The present paper outlines an elastic-viscoplastic theory based on this approach and includes numerical results for an internally pressurized thick walled sphere.

INTRODUCTION

The problem of developing constitutive equations for the dynamic behavior of metals resolves mainly in the determination of expressions for strain rate sensitivity under multiaxial loading. The extent to which the stress-strain relations of metals are dependent upon the strain rate is still not a settled matter. Bell has shown [1, 2] that the stress-strain curves of a number of pure, or almost pure, fully annealed FCC metals are independent of strain rate over a large dynamic range (from 1 to 10^4 sec^{-1}). On the other hand, it has been shown that these materials do show time effects in the low strain rate region ($\dot{\epsilon} \sim 10^{-4}$ sec^{-1}) [3]. Although the magnitudes of these time effects are small, they can lead to important consequences such as steps or serrations of the stress-strain curve. These irregularities of the stress-strain curve are due to a combination of negative strain rate dependence of the flow stress in the range of testing, and the method of load or strain application [4].

The FCC metals studied by Bell are characterized by smooth stress-strain curves in the dynamic range, i.e., they do not exhibit a sharp yield point. There is, however, ample experimental evidence to show that metals with sharp yield points are fairly rate sensitive over a large range of strain rates. Most technological metallic alloys have this property. For example, the yield stress of mild steel increases by a factor of 2.5 over a strain rate span from 10^{-6} to 10^{2} sec^{-1} [5].

There has been considerable interest in formulating constitutive laws for rate sensitive materials in order to analyze dynamic loading problems. Problems of elastic-plastic wave propagation and the permanent deformation of structures due to blast loading depend upon such laws. There is also a class of problems for which strain rate effects can be important but inertia terms are still small.

Only relatively few general forms of constitutive laws have been proposed. The best known is that due to Malvern [6, 7] which hypothesizes the existence of a reference, "static," stress-strain function and then relates an increase in the stress from this datum to the imposed strain rate. This law has shown some success in predicting certain experimental observations. The best known is the result that a small amplitude stress pulse would propagate at the elastic velocity in a plastically prestrained medium. Another result of a special form of this law is the prediction of permanent deformation values for blast loaded structures. These values were found to be in good agreement with experiments [8, 9].

A modification of Malvern's treatment to include a limiting maximum stress-strain curve was proposed by Lubliner [10]. A multiaxial version of Malvern's hypothesis was formulated by Perzyna [11] as a generalization of a more restricted viscoplastic law by Hohenemser and Prager [12]. Most of these constitutive laws have been summarized and discussed in papers by Perzyna [13] and by Lindholm [14].

Perzyna and Wojno have recently extended the multiaxial formulation to include finite strains [15]. The basic assumption of that paper is the additivity of the elastic and inelastic parts of the rate of deformation tensor. The paper retains, however, the requirement of completely elastic response for stress states below a yield criterion. The inelastic part of the rate of deformation tensor is taken to be a function of excess stresses above the static yield criterion.

The common feature of all of these constitutive laws is the hypothesis of a reference "static" stress-strain relation. For multiaxial loading, this corresponds to a reference "static" yield surface which grows in some prescribed manner with strain rate. The results of multiaxial strain rate experiments are usually represented in a similar manner. The dynamic yield surface, expressed in terms of stress and stress deviator invariants, has been considered to be expressible as a function of invariants of the elastic strain rate tensor, e.g. [14].

Although this formulation can serve to predict important physical observations, it is not entirely satisfactory as a general law of material behavior. The question of the reference "static" yield surface is of more than academic interest since yield surfaces do vary in the low, so-called "quasi-static," range. For example, a small negative strain rate dependence in this range was shown to be the underlying cause of steps or serrations in the stress-strain curves [4]. The constitutive laws for this range should include time effects such as those due to thermally activated yielding mechanisms and to hardening resulting from impurity and vacancy diffusion. In addition, the requirement to refer dynamic stress-strain relations to a reference "static" case does not lend itself readily to account for dynamic strain hardening. The manner in which the stress at high strain levels depends on the strain rate may differ appreciably from that of the yield stress, e.g. [5].

CONSTITUTIVE LAWS BASED ON "DISLOCATION DYNAMICS"

An alternative formulation of elastic-viscoplastic behavior of materials has been developed based on considerations of microscopic deformation mechanisms [16–18]. The motivation for this theory is the experimentally observed properties of dislocation motion and multiplication. The theory is, however, a macroscopic continuum one in that it relates macroscopic quantities such as stress, strain, and strain rate.

In terms of continuum mechanics, the theory considers the deformation from the initial state to have both elastic and plastic (irreversible) components. The plastic strain rate $\dot{\epsilon}^p$ is related to a function of the strain (the mobile dislocation density ρ) and a function of the stress (the dislocation velocity v). That is

$$\dot{\epsilon}^p = b\rho(\epsilon^p)v(\sigma) \tag{1}$$

where b is a constant. The measured elastic strain rate $\dot{\epsilon}^e$ is given by

$$\dot{\epsilon}^e = (1/M)\dot{\sigma} \tag{2}$$

where M is the combined elastic modulus of the specimen and the loading system. The total strain rate imposed by the loading device ϵ is then the sum: $\dot{\epsilon}^e + \dot{\epsilon}^p$.

Strain hardening can be introduced by considering v to be a function of ϵ^p as well as of σ. Forms for v that have been suggested from direct measurements and from theoretical considerations are

$$v = \left[\frac{\sigma - q\epsilon^p}{\sigma_0}\right]^n \tag{3}$$

and

$$v = A \exp\left[-B/(\sigma - q\epsilon^p)\right] \tag{4}$$

where σ_0, q, n, A and B are material constants. The first form (3) is somewhat easier to interpret and use mathematically, but (4) seems to hold over a greater range and has more physical meaning [19].

These relatively simple equations can be combined to give an elastic-viscoplastic equation for the material response. The equation can be integrated from the initial state (with an initial dislocation density) for a variety of loading conditions. These include straining in a flexible machine, constant stress (creep), and rapid stressing to high level to obtain yield delay times. The results for imposed straining show that sharp yield points and post yield stress drops would result from suitable initial parameter values, and that the results depend upon the rate of straining.

There is accumulating evidence that this description of material behavior is a good approximation to the actual behavior of a number of metals. The repeated discontinuous yielding effect can also be adequately described by this approach [20].

If (3) is used, then it can be shown that both sharp yield points and strong strain rate sensitivity result from small values of the exponent n. It is noted that the derived value of n for steel, which has a sharp yield point and is rate sensitive, is about 30 [18]. The value of n for FCC metals, such as Cu, which have smooth stress-strain curves and are relatively rate insensitive, is about 200 [18]. The theory accounts, in this manner, for the experimental observation of an association between the sharp yield point and strain rate sensitivity. Another interpretation of this result is that the plastic strain rate component is small for the lower values of n so that the elastic effect is more pronounced.

An important consequence of this formulation from the viewpoint of continuum mechanics is that yielding is not a separate and independent criterion but is a consequence of a general constitutive law of the material behavior. Elastic and plastic deformations are not separate entities bounded by a yield criterion, but contribute to the total deformation throughout the loading history. Recent experimental work tends to indicate that this is a more realistic formulation of the actual behavior of metals. Accurate measurements in the microstrain region have shown that irreversible strains do occur at very low strain values.

FORMULATION OF ELASTIC-VISCOPLASTIC THEORY

There does not seem to be a unique method of generalization of the preceding equations based on dislocation kinetics to the multiaxial case. Direct interchange of stress components by tensor invariants is arbitrary and, by itself, may not lead to a consistent formulation. It seems that a systematic development of a viscoplastic theory motivated by "dislocation dynamics" could be obtained within the framework of classical plasticity theory. This has been carried out by Partom and Bodner [21] and applied to the case of a thick walled sphere under internal pressure. Aspects of the theoretical development and some of the results are described in this paper. The principal point for the present discussion is that it is possible to formulate a general elastic-viscoplastic theory without an independent "static" yield criterion.

The theory [21] accounts for large deformations and finite strains. It therefore proved more convenient to use particle velocities and deformation rates rather than strains as the basic variables. The total deformation rate d_{ij} is the symmetric part of the velocity gradient $v_{i,j}$:

$$d_{ij} = (\tfrac{1}{2})(v_{i,j} + v_{j,i}) \tag{5}$$

and is considered to be the sum of elastic and plastic (irreversible) components

$$d_{ij} = d_{ij}^{e} + d_{ij}^{p}. \tag{6}$$

This is also the basic assumption in the work of Perzyna and Wojno [15]. Thermal effects are not considered in the present discussion.

The large deformation viscoplastic theory has been developed in complete generality [21], but for the present purpose it will be directly applied to the case of spherical symmetry. The particle location (Lagrangian coordinate) is designated "a" and the spatial location (Eulerian coordinate) by "r."

The total radial and circumferential deformation rates d_r and d_c can be expressed in terms of the radial velocity v:

$$\begin{aligned} d_r &= \frac{\partial v}{\partial r} \\ d_c &= v/r. \end{aligned} \tag{7}$$

The strain rate expressions depend on both the deformation rates and the strains. For the Almansi strain measure with spherical symmetry, these are

$$\begin{aligned} \dot{\epsilon}_r^{y} &= (1 - 2\epsilon_r^{y})d_r^{y} \\ \dot{\epsilon}_c^{y} &= (1 - 2\epsilon_c^{y})d_c^{y} \end{aligned} \tag{8}$$

where the superscript y refers to either the elastic or plastic component. The strain components can be obtained by integrating (8). It is noted that in view of the nonlinearity of (8), simple superposition of strain rate components cannot be applied.

The total strains can also be expressed in terms of the particle location. For the Almansi strain measure these relations are

$$\epsilon_r = \frac{1}{2}\left[1 - \left(\frac{\partial a}{\partial r}\right)^2\right]$$
$$\epsilon_c = \tfrac{1}{2}[1 - (a/r)^2]. \tag{9}$$

The preceding equations constitute the essential kinematics of the problem. The equilibrium equation, for spherical symmetry, is

$$\frac{\partial \sigma_r}{\partial r} + \frac{2}{r}(\sigma_r - \sigma_c) = 0. \tag{10}$$

The remaining equations in the formulation of the problem are the constitutive equations of the material behavior, and the boundary and initial conditions. For the particular numerical examples to be described, the boundary and initial conditions consisted simply of a stress free outer surface and a constant radial velocity imposed on the inner surface commencing at $t = 0$.

The constitutive laws relate the elastic and plastic strain rate components to the stress. The total stress is a state variable since it is directly related to the elastic strain component. Additional stresses that are not state variables, such as would be generated by internal viscosity, are not considered in the development. The rate at which work is done on a unit volume element is therefore

$$\dot{W} = \sigma_{ij}(d_{ij}^e + d_{ij}^p). \tag{11}$$

In finite strain analyses it is convenient to relate the elastic strain to the stress through the strain energy function. The stress-strain relation for spherical symmetry is

$$\sigma_r = (\rho/\rho_0)[F_0 + F_1\epsilon_r^e + F_2(\epsilon_r^e)^2]$$
$$\sigma_c = (\rho/\rho_0)[F_0 + F_1\epsilon_c^e + F_2(\epsilon_c^e)^2] \tag{12}$$

where ρ/ρ_0 is the density ratio referred to the initial state and is a function of the strain invariants, E_1, E_2, E_3:

$$(\rho/\rho_0)^2 = 1 - 2E_1 + 4E_2 - 8E_3. \tag{13}$$

The coefficients F_0, F_1, F_2 can be obtained from the strain energy function ψ and the strain invariants,

$$F_0 = \frac{\partial \psi}{\partial E_1} + E_1\frac{\partial \psi}{\partial E_2} + (E_2 - 2E_3)\frac{\partial \psi}{\partial E_3}$$
$$-F_1 = 2\frac{\partial \psi}{\partial E_1} + (2E_1 + 1)\frac{\partial \psi}{\partial E_2} + E_1\frac{\partial \psi}{\partial E_3} \tag{14}$$
$$F_2 = 2\frac{\partial \psi}{\partial E_2} + \frac{\partial \psi}{\partial E_3}.$$

An uncoupled form of the strain energy function was used, that is, one for which the distortional and dilational components are distinct [22]:

$$\psi=\frac{9}{4}\frac{K_0}{\alpha(\beta-\alpha)}\left\{\left[1-\left(\frac{\rho}{\rho_0}\right)^{2/3\alpha}\right]-\frac{\alpha}{\beta}\left[1-\left(\frac{\rho}{\rho_0}\right)^{2/3\beta}\right]\right\}+6\mu_0\left\{\frac{E_1{}^2-3E_2}{(3-2E_1)^2}\right\} \quad (15)$$

where K_0 and μ_0 are the bulk and shear moduli of the initial state. The parameters α and β determine the change of volume with pressure: $\alpha=1, \beta=2$ are suggested as reasonable values. The stresses σ_r and σ_c could therefore be computed from the elastic strains and the invariants of the strain tensor.

The plastic deformation rate component can be expressed as a function of the stress by means of a viscoplastic flow law. For a systematic development of these relations, it is convenient to work with the flow law of classical plasticity theory. A requirement on the plastic stress-strain relation is that positive work is done during plastic deformation, that is

$$\sigma_{ij}d_{ij}{}^p \geqslant 0. \quad (16)$$

The flow law of classical plasticity satisfies this condition by stipulating that

$$d_{ij}{}^p=\underline{d}_{ij}{}^p=\lambda\underline{\sigma}_{ij} \quad (17)$$

where the bar symbol refers to the deviator. This relation also satisfies the conditions for material stability (the condition of normality to the yield surface is not directly relevant here). The deviatoric character of the flow relation means that the plastic deformation is incompressible. A compressibility term does appear, however, in the elastic strain energy function (15).

Equation (17) can be squared to obtain λ,

$$\lambda^2=\underline{D}_2{}^p/\underline{J}_2 \quad (18)$$

where $\underline{J}_2$ is the second invariant of the stress deviator and $\underline{D}_2{}^p$ is the second invariant of the plastic deformation rate.

In classical plasticity, plastic flow takes place when a yield condition is satisfied. The von Mises yield condition is, in the present notation,

$$\underline{J}_2=-Y^2/3 \quad (19)$$

where Y is the yield stress in tension. The flow law then can be written as

$$\underline{d}_{ij}{}^p=[-\underline{D}_2{}^p/(Y^2/3)]^{1/2}\underline{\sigma}_{ij} \quad (20)$$

which is the classical von Mises law of plastic flow. Equation (20) applies, of course, only when (19) is satisfied. If $-\underline{J}_2 < Y^2/3$, then λ and $\underline{d}_{ij}{}^p$ are zero.

A viscoplastic constitutive law could be obtained from (17) by assuming that a relationship exists between $\underline{D_2}^p$ and $\underline{J_2}$. Forms for this relation are suggested by the work on dislocation dynamics, i.e. (3) and (4). The proposed multiaxial generalizations of those equations are

$$\underline{D_2}^p = -C_0^2[-\underline{J_2}/(X^2/3)]^n \tag{21}$$

and

$$\underline{D_2}^p = -D_0^2 \exp\,[(Z^2/3)/\underline{J_2}]. \tag{22}$$

In these equations D_0, n, X and Z are material constants. The coefficient C_0 in (21) is a scale factor rather than an independent material parameter. The quantities X and Z cannot be directly related to the yield stress Y of the perfectly plastic material. If (21) is used, then perfectly plastic behavior would be obtained in the limit as $n \to \infty$. For this special case X would be the same as Y. Equation (22) does not admit such an interpretation since D_0^2 is the upper bound value of $\underline{D_2}^p$. Equations (21) and (22) do not include strain hardening, but they can be modified to do so.

When (3) or (4) are combined with (1) to give the relation between $\dot{\epsilon}^p$ and σ, a term $\rho(\epsilon^p)$, the density of mobile dislocations, appears in the stress-strain rate relation. However, the plastic strain ϵ^p is not a state function, and this representation is only valid for special loading histories for which $\rho(\epsilon^p)$ is known. Generalizing (21) or (22) in this manner would require C_0 or D_0 to be a function of state variables such as the stress, temperature, and others that may be introduced.

The general method of numerical solution of these equations requires working with a space and time network. At each time interval, a new velocity distribution is assumed based on the previous state. The deformation rates, total strain rates, and strains are calculated at each point from the kinematic relations. The total deformation rate is considered to be divided into elastic and plastic components, and the stress σ_{ij} is calculated from the respective constitutive law. The correct stress at all points must, of course, satisfy the equilibrium equation and the constitutive laws. A successive approximation procedure can be developed to solve the equations over the space and time network.

In this manner it is possible to solve problems for a variety of loading conditions and different constitutive laws. The procedure does not require prior determination whether the material behavior is elastic or plastic since the formulation accounts for both components at all stages.

Numerical analyses were performed for the sphere problem for a number of loading histories and for elastic, elastic-perfectly plastic, and elastic-viscoplastic material behavior. Inertia terms were not considered in these examples. It does not appear that the method of numerical solution used for the quasi-static problems could be readily adopted to include inertia terms. Wave propagation problems could be formulated in a similar manner, but would require a different numerical technique.

NUMERICAL EXAMPLES

Calculations were performed for the deformation of a thick walled sphere subjected to an imposed velocity distributed uniformly around the internal surface. Specifying the velocity rather than the pressure is a more convenient numerical procedure since an instability is experienced by the system. After the instability, the load becomes multi-valued whereas the radial deformation increases monotonically.

Results for the pressure-radial displacement relation of an elastic-viscoplastic sphere are shown in Fig. 1. The initial inner and outer radii of the sphere are 10 and 20 cm respectively, and viscoplastic behavior according to (22) is assumed. The quantities used in the strain energy function, (15), are $K_0 = 10^4$ Kg/cm^2, $\mu_0 = 10^3$ Kg/cm^2, $\alpha = 1$, $\beta = 2$. The results in Fig. 1 are for a prescribed stress parameter Z(100 Kg/cm^2) and imposed velocity v_i(0.05 cm/sec) on the inner surface. Higher curves are obtained for smaller values of the parameter D_0 where D_0^2 is the maximum value of the second invariant of the

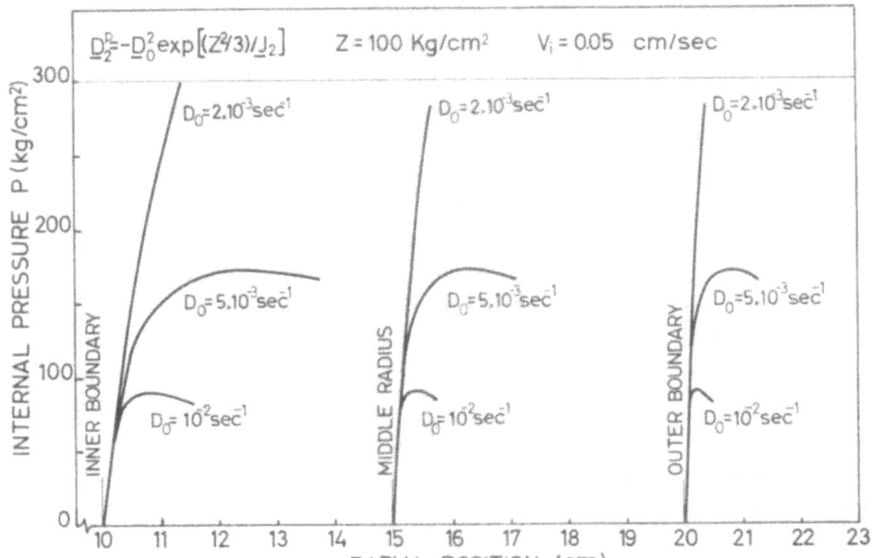

Fig. 1. Internal pressure – radial displacement relations for various material parameter values of elastic-viscoplastic sphere.

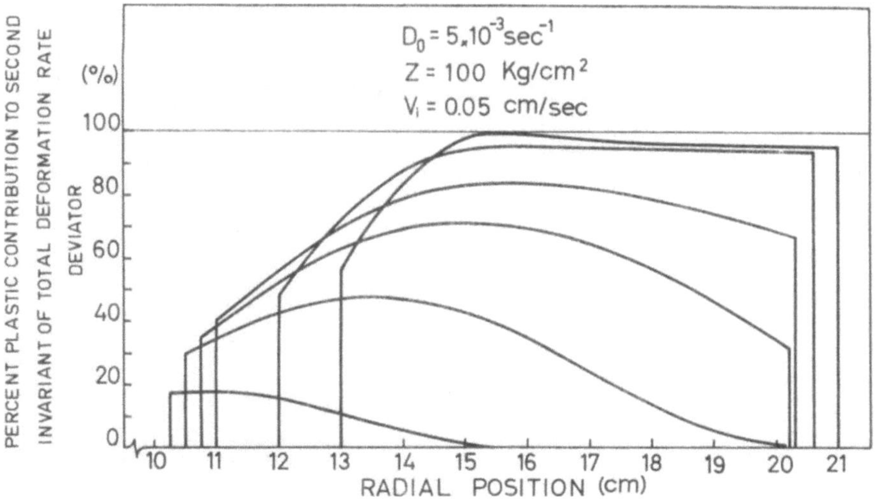

Fig. 2. Amount of plastic deformation throughout the wall thickness for elastic-visco-plastic sphere.

plastic deformation rate. This is a consequence of a smaller plastic deformation rate contribution to the total deformation rate. A larger elastic component would correspond to higher pressure values for the same deformation.

The instabilities in the curves in Fig. 1 are a combination of both material and geometrical effects. An instability occurs even for the fully elastic case, but plastic strain contributions tend to reduce the level of the curves and sharpen the instability.

The amount of plastic straining throughout the sphere thickness for maintained velocity on the inner surface is shown in Fig. 2 for a particular set of parameter values. It is noted that the inner region initially has the largest plastic deformation, whereas the reverse occurs for large deformations. The fully plastic condition is approached at the larger deformations.

The effect of different values of the imposed velocity on the pressure-displacement relation is shown in Fig. 3. The parameters used in the calculations correspond to highly rate sensitive materials since a five-fold increase in the imposed velocity appreciably alters the curves for the same value of D_0. The curves in Fig. 3 indicate that rate sensitivity tends to increase for smaller values of D_0 (corresponding to a smaller inelastic contribution to the total deformation rate). That is, the curves for the different velocities would have a greater spread, e.g., the velocity effect is greater for $D_0 = 5 \times 10^3$ sec^{-1} than for $D_0 = 10^{-2}$ sec^{-1}. An appraisal of rate sensitivity effects could have been more readily performed had (21) been used in the calculations. The ex-

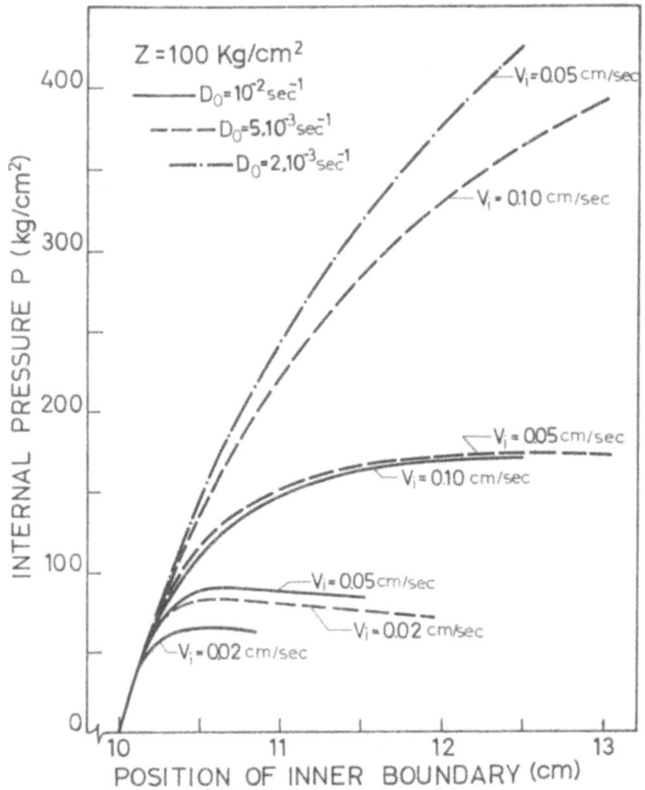

Fig. 3. Effect of imposed radial velocity on pressure-inner boundary displacement relation of elastic-viscoplastic sphere.

ponent n in that equation is more directly associated with rate sensitivity effects [18, 23].

Pressure-inner boundary displacement curves for elastic-perfectly plastic behavior are shown in Fig. 4. These curves correspond to the large deformation analysis and are somewhat below the corresponding curves of the classical infinitesimal theory. The fully elastic case could be obtained from this analysis as Y becomes increasingly large.

The effect of increasing plastic deformation in reducing the maximum pressure and sharpening the instability is well illustrated in Fig. 4. Lower values of the yield stress Y correspond to increased plastic deformation at the same radial displacement.

From Figs. 1 and 4 there appears to be a correlation between decreasing D_0 and increasing Y. Both correspond to smaller plastic contributions to the total deformation and therefore to higher pressure-displacement curves. From Fig. 3 it is seen that increasing the imposed

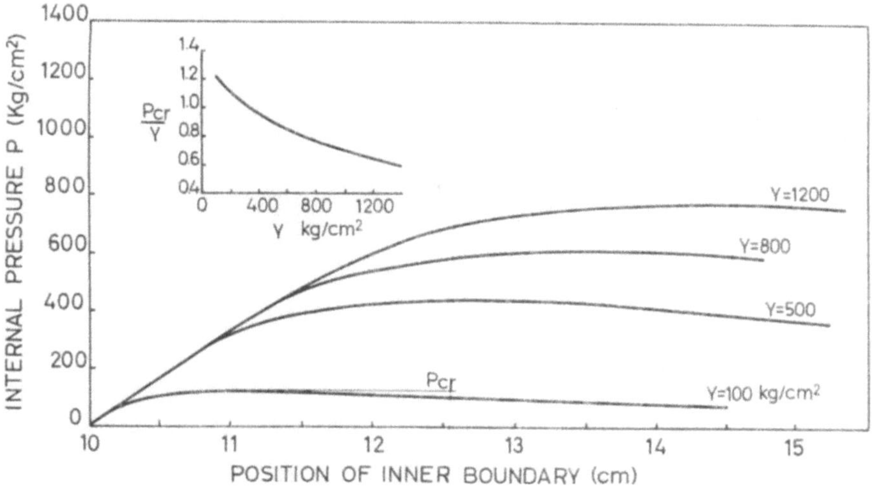

Fig. 4. Pressure-inner boundary displacement relations for various values of the yield stress of elastic-perfectly plastic sphere.

velocity for an elastic-viscoplastic material also increases the pressure-displacement curve in the same manner as increasing Y for a perfectly plastic material.

The viscoplastic laws (21) and (22) are represented in Fig. 5 for a number of parameter values. Equation (22) leads to a limiting value for the second invariant of the plastic deformation rate $\underline{D}_2''$, whereas $\underline{D}_2''$ is unbounded for (21). Perfectly plastic behavior could be obtained from (21) in the limit as $n \to \infty$. The corresponding curve in Fig. 5 would be a horizontal line at $\underline{D}_2'' = 0$ until $-\underline{J}_2 = X^2/3$, and then a vertical line.

Pressure displacement curves based on (21) are shown in Fig. 6 for various values of the exponent n. The results approach that of the perfectly plastic material for increasing n. The curves are lower than that of the perfectly plastic case since lower values of n correspond to larger plastic deformation rate contributions when $-\underline{J}_2 < X^2/3$ (Fig. 5). These results are similar to those of Fig. 1 where larger values of D_0 correspond to larger plastic deformation rate contributions. No plastic deformation occurs in the perfectly plastic case until the yield stress is exceeded. The instability in the curves in Fig. 6 is due to both material and geometry effects, but plastic instability appears to be the principal factor for the cases shown.

The present analysis could be readily modified to include an "over-stress" viscoplastic law of the kind formulated in [15]. This would correspond in Fig. 5 to a translation of the origin of the curves along the horizontal axis to $-\underline{J}_2 = X^2/3$. The subsequent analysis and numeri-

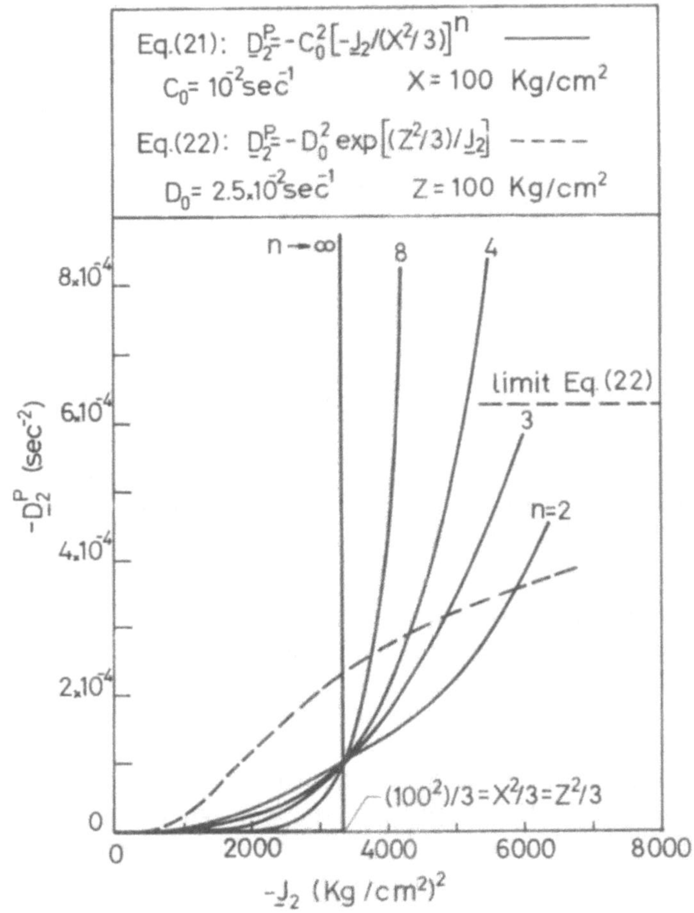

Fig. 5. Representation of viscoplastic flow laws (variation of second invariant of plastic deformation rate with second invariant of the stress deviator).

cal procedure would be essentially the same. Although this would reduce the formulation to the more classical forms, it would eliminate the primary physical motivation of the theory and many important physical considerations. The formulation of viscoplastic laws in terms of the "overstress" due to strain rate does not seem to be a fundamental requirement. On the contrary, the inclusion of elastic and inelastic components to the total deformation rate appears to have a sounder physical basis and is simpler to treat analytically.

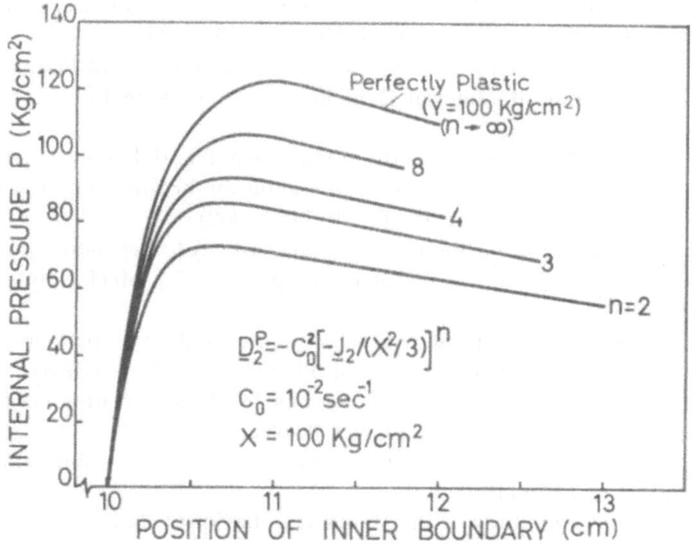

Fig. 6. Pressure-inner boundary displacement relations for various parameter values of viscoplastic and perfectly plastic sphere.

ACKNOWLEDGMENTS

The author would like to express his appreciation to Mr. Y. Partom for providing the numerical examples (based on his technique of numerical solution) and for helpful discussions in the preparation of this paper.

Part of the research reported in this paper was supported by the Air Force Office of Scientific Research under Contract AF 61(052)-951, through the European Office of Aerospace Research (OAR), United States Air Force.

References

1. J. F. Bell, "Single, Temperature-Dependent Stress-Strain Law for the Dynamic Plastic Deformation of Annealed FCC Metals," J. Appl. Phys., *34*, 134 (1963).
2. J. F. Bell, "Generalized Large Deformation Behavior for FCC Solids," Phil. Mag., *11*, 1135 (1965).
3. A. Rosen and S. R. Bodner, "The Influence of Strain Rate and Strain Ageing on the Flow Stress of Commercially-Pure Aluminum," J. Mech. Phys. Solids, *15*, 47 (1967).
4. S. R. Bodner and A. Rosen, "Discontinuous Yielding of Commercially-Pure Aluminum," J. Mech. Phys. Solids, *15*, 63 (1967).
5. M. J. Manjoine, "Influence of Rate of Strain and Temperature on Yield Stresses of Mild Steel," Trans. ASME, *66*, A211 (1944).

6. L. E. Malvern, "Plastic Wave Propagation in a Bar of Material Exhibiting a Strain Rate Effect," Quart. Appl. Math., *8*, 405 (1950).
7. L. E. Malvern, "The Propagation of Longitudinal Waves of Plastic Deformation in a Bar of Material Exhibiting a Strain Rate Effect," J. Appl. Mech., *18*, 203 (1951).
8. S. R. Bodner and P. S. Symonds, "Experimental and Theoretical Investigation of the Plastic Deformation of Cantilever Beams Subjected to Impulsive Loading," J. Appl. Mech., *29*, 719 (1962).
9. S. R. Bodner and W. G. Speirs, "Dynamic Plasticity Experiments on Aluminum Cantilever Beams at Elevated Temperature," J. Mech. Phys. Solids, *11*, 65 (1963).
10. J. Lubliner, "A Generalized Theory of Strain Rate Dependent Plastic Wave Propagation in Bars," J. Mech. Phys. Solids, *12*, 59 (1964).
11. P. Perzyna, "The Constitutive Equations for Rate Sensitive Plastic Materials," Quart. Appl. Math., *20*, 321 (1963).
12. See: W. Prager, *Introduction to Mechanics of Continua*, Ginn and Co., Boston, Mass., 136 (1961).
13. P. Perzyna, "The Study of the Dynamic Behavior of Rate Sensitive Plastic Materials," Arch. Mech. Stos., *15*, 113 (1963).
14. U. S. Lindholm, "Dynamic Deformation of Metals," in *Behavior of Materials Under Dynamic Loading*, ed. by N. J. Huffington, Jr., ASME, New York (1965).
15. P. Perzyna and W. Wojno, "On the Constitutive Equations of Elastic/Viscoplastic Materials at Finite Strain," Arch. Mech. Stos., *18*, 85 (1966).
16. W. G. Johnston and J. J. Gilman, "Dislocation Velocities, Dislocation Densities, and Plastic Flow in Lithium Fluoride Crystals," J. Appl. Phys., *30*, 129 (1959).
17. W. G. Johnston, "Yield Points and Delay Times in Single Crystals," J. Appl. Phys., *33*, 2716 (1962).
18. G. T. Hahn, "A Model for Yielding with Special Reference to the Yield Point Phenomena of Iron and Related BCC Metals," Acta Met., *10*, 727 (1962).
19. J. J. Gilman, "Dislocation Mobility in Crystals," J. Appl. Phys., *36*, 3195 (1965).
20. S. R. Bodner, "The Mechanics of Repeated Discontinuous Yielding of Metals," Mat. Science and Engin. *2*, 213 (1967).
21. Y. Partom and S. R. Bodner, "Large Deformations of an Elastic-Viscoplastic Sphere" (in preparation).
22. Y. Partom, "The Dilatational and Distortional Partition of the Energy Function in Finite Strain," Israel J. Tech., *4*, 218 (1966).
23. G. T. Hahn, C. N. Reid, and A. Gilbert, "Dislocation Dynamics of Yielding and Fracture," Proc. of Symposium on the Role of Substructure in the Mechanical Behavior of Metals, Report No. ASD-TR-63-324, U.S. Air Force Systems Command, 253 (1963).

EFFECT OF STRAIN RATE ON THE DISLOCATION SUBSTRUCTURE IN DEFORMED NIOBIUM SINGLE CRYSTALS

J. W. EDINGTON
Battelle Memorial Institute
Columbus, Ohio

ABSTRACT

This paper describes an investigation of the relationship between mechanical properties and dislocation substructure in the strain rate range 10^{-4} to 10^{4} sec^{-1}. The relationship between lower yield stress τ_{yield} and strain rate ϵ follows two consecutive semilogarithmic relationships of the form $\tau_{yield} = \alpha + \beta \log \epsilon$ with a break at a strain rate of approximately 10 sec^{-1}. The dislocation configurations produced by deformation in these two regions have been investigated at strain rates of 1.2×10^{-4} sec^{-1} and 1.5×10^{3} sec^{-1} using transmission electron microscopy techniques. The relationship between dislocation density, flow stress, and plastic strain has been determined and is discussed in terms of the dislocation structures observed.

INTRODUCTION

Up to the present time a number of studies have been made of the behavior of metals deformed at high strain rates. Considerable interest has been centered on the measurement of mechanical properties, largely because of the difficulty of instrumentation at high strain rates. Although rather less data exist for b.c.c. metals than for f.c.c. metals, the information at present available indicates that the effect of strain rate on the flow stress of b.c.c. metals is quite complicated. For example, in iron the strain-rate dependence of the room temperature lower yield stress τ_{yield} in the strain-rate range 10^{-4} to 10^{3} sec^{-1} can be described by two consecutive semilogarithmic relationships [1] of the form

$$\tau_{yield} = \alpha + \beta \log \dot{\epsilon}$$

where ϵ is the applied strain rate and α and β are constant over particular strain-rate ranges. These have been interpreted in terms of two rate-controlling mechanisms [1–3]. In polycrystalline molybdenum Orava [4] has found that while the same form of relationship applies in the strain-rate range 10^{-6} to 10^{-1}, several linear regions exist, the details of which are strikingly sensitive to grain size. Davidson, et al. [5] have also shown several linear stages for single crystals of molybdenum deformed in the strain-rate range 10^{-4} to 10^{3} sec^{-1} although the details of these linear regions do not agree with those of Orava. In polycrystalline niobium three linear regions have been reported by Sargent, et al. [6], in the strain-rate range 10^{-5} to 10^{-1} sec^{-1}, but Fourdeux and Wronski [7] report only two. In contrast, Mitchell, et al. [8] report two regions for single crystals of niobium. Four regions have been found for polycrystalline tantalum [6] deformed in the strain-rate range 10^{-5} to 10^{-1} sec^{-1}, and three ranges have been found for single crystals of tantalum [9] deformed in the strain-rate range 10^{-5} to 1 sec^{-1}.

None of these studies were supported by structural studies of the dislocation substructure generated at different strain rates except those of Fourdeux and Wronski on niobium which was deformed over a relatively restricted strain-rate range. This paper presents a detailed study of the mechanical properties of single crystals of niobium deformed in the strain-rate range 10^{-4} to 10^{3} sec^{-1}, together with extensive investigations of the dislocation configurations and densities generated by deformation at the extremes of the strain-rate range investigated.

EXPERIMENTAL PROCEDURE

General

Single crystals of niobium of the orientation 42 shown in Fig. 1 were grown by electron beam melting. The starting material was $\frac{1}{2}$-inch-diameter "beam melted" niobium supplied by Fansteel Corporation. The rods were given three zone passes from bottom to top in a vacuum of 10^{-5} torr and at a speed of $\frac{1}{4}$ inch per hour. The analysis of the impurities present after melting is shown in Table 1.

Specimens were deformed in compression because this is the simplest mode of operation of the Hopkinson bar testing device which was used to achieve the high strain rates. In addition, relatively short specimens were used to prevent stress gradients in the specimen during testing with the Hopkinson bar. In order to reach strain rates of the order of 10^{3} sec^{-1} one series of specimens 0.25 inch long and approximately 0.19 inch in diameter was produced by spark machining. The ends of the compression specimens were machined parallel and flat to within 0.00025 inch. Approximately 0.015 inch was polished from

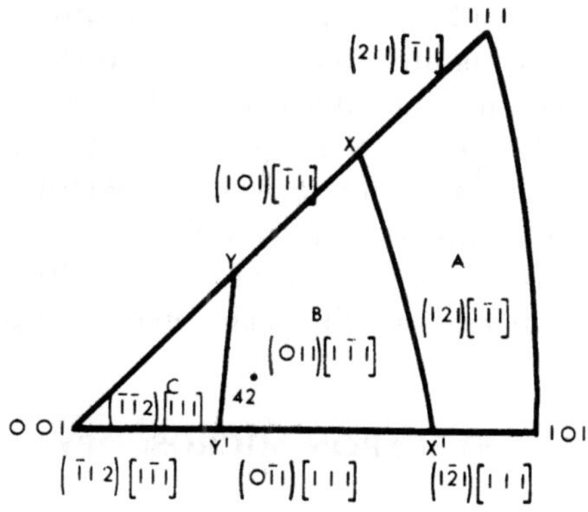

{101}{211} slip systems in Niobium single crystals.

Fig. 1. The orientation (at 42) of the single crystals and the division of the unit triangle into most highly stressed slip systems.

the surface using a chemical polishing solution of 30% HF 40% HNO_3 in order to remove the damage caused by the spark machining process. In addition to these specimens a series of shorter specimens were also produced in the same way in order to reach strain rates close to 10^4 sec^{-1}. The dimensions of the short specimens were, length 0.080 inch, diameter 0.170 inch, with the same tolerances as the long specimens for the compression faces. It was not possible in either series of specimens to increase the axial ratio of the specimens by cutting down the diameter of the specimens in order to minimize end effects. To do so would introduce too great a mismatch between the specimen and the anvils of

Table 1

IMPURITY CONCENTRATIONS IN NIOBIUM AFTER MELTING

Element	Concentration, ppm	Element	Concentration, ppm
C	9	W	50
O	95	Zr	20
N	39	Mn	15
H	0.5	Al	15
Fe	15	Cu	15
Si	15	Sn	15
Ta	350	Cr	15
Ti	15	V	15
Mo	15	Pb	15
Ni	15		

the Hopkinson bar resulting in possible departure of the loading surfaces from a plane, thus introducing errors in the measurements.

The niobium crystals were deformed at room temperature in the strain-rate range 10^{-4} to 6.5×10^3 sec^{-1} using three different testing machines. The low strain rate tests were performed on a standard "Instron" testing machine, the intermediate strain rate tests were performed on a "Krafft" dynamic loading device [10], and the high strain rate tests were performed on a "Hopkinson bar" impact loading device.

ELECTRON MICROSCOPY

General

Transmission electron microscopy experiments were performed on thin foils taken from long specimens deformed at strain rates of 1.2×10^{-4} sec^{-1} and 1.5×10^3 sec^{-1} to a series of strains up to 30% in shear. The strain on the Hopkinson bar tests was limited by use of a series of thick steel collars of different lengths, shorter than the specimen and separated from the specimen by loose cotton wool packing. The action of the collar is to prevent further deformation of the specimen past a certain strain by taking up the stress applied to the specimen. In this way different strains were achieved at the same strain rate, with the same striker bar velocities. Thin sheets 0.030 inch thick were spark cut parallel to the primary slip plane (011) and discs 2.3 mm in diameter were spark trepanned from these sheets to obtain discs from the center, and as close as possible to the compression faces, of the test specimen. In practice thin areas were obtained within 2 mm of the compression face. These discs were indented with the solution developed by Stickler and Engle [11], which consists of 2% HF, 5% HNO_3, and 93% CH_3OH used at $-60°$ C, 250 volts, and 60 ma, and produces a bright indented surface. Finally, the discs were chemically thinned using a solution of 30% HF and 70% HNO_3 at $-50°$ C. The final procedure was controlled by the usual light beam and microscope method. After washing in alcohol, specimens were observed in a Siemens Elmiskop 1A electron microscope using a Swann-type double tilt goniometer.

Burgers Vector Determination

The multiplicity of slip systems in b.c.c. crystals complicates the unambiguous determination of the Burgers vector using the g.b. = 0 criterion. The possible Burgers vectors to be considered include

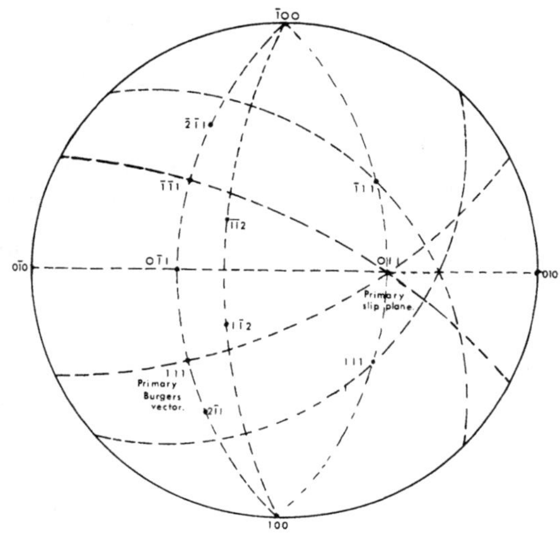

Fig. 2. A schematic diagram showing the 200, $2\bar{1}1$, $0\bar{1}1$, $\overline{21}1$, $1\bar{1}2$, $\overline{11}2$ reflections which can be used in the Burgers vector analysis, for a thin foil parallel to the primary slip plane (011). Note that all $\frac{a}{2}\langle 111\rangle$ or a $\langle 001\rangle$ Burgers vectors are invisible under at least one reflection.

$a\langle 100\rangle$ as well as $\frac{a}{2}\langle 111\rangle$ since both are stable [12]. Figure 2 shows the important reflections which are available, i.e., within the tilting range of the goniometer stage, for Burgers vector determination. These reflections include 100, $0\bar{1}1$, $2\bar{1}1$, $\overline{21}1$, $\overline{11}2$, $1\bar{1}2$, and the directions of various Burgers vectors which are invisible for these reflections are shown in Table 2.

Table 2

INVISIBILITY CRITERIA FOR VARIOUS BURGERS VECTORS IN NIOBIUM

Reflection	Burgers Vector Directions for Which Dislocation Is Invisible
100	[010] [001] [011]
$0\bar{1}1$	[100] $[\bar{1}11]$ [111] [011]
$2\bar{1}1$	$[\overline{11}1]$ [011]
$\overline{21}1$	$[1\bar{1}1]$ [011]
$\overline{11}2$	[111] $[1\bar{1}0]$
$1\bar{1}2$	$[\bar{1}11]$ [110]

It may be seen from this table that, by use of all the available reflections, all $\frac{a}{2}\langle 111\rangle$ Burgers vectors and $a\langle 001\rangle$ Burgers vectors can be identified (assuming no $a\langle 011\rangle$ Burgers vectors are present) and that some $a\langle 011\rangle$ Burgers vectors can also be determined. However, because of the symmetry of the [011] axis, it is not possible to distinguish between primary dislocations with Burgers vector $\frac{a}{2}[1\bar{1}1]$ and secondary dislocations on the primary slip plane with Burgers vector $\frac{a}{2}[\bar{1}\bar{1}1]$ unless the specimen is placed in the microscope in a known orientation. For similar reasons it is not possible to differentiate between the Burgers vectors $\frac{a}{2}[111]$ and $\frac{a}{2}[\bar{1}11]$. However, in all cases the orientation of the specimen was known in the electron microscope so that these distinctions could be made.

A further problem associated with Burgers vector determination in niobium is the effect of lattice anisotropy on the dislocation image. Head, et al. [13] have computed dislocation images of screw dislocations lying parallel to $\langle 111\rangle$ directions in α-iron for the 011 and 112 reflections when g.b. = 0. For the 011 reflection no significant contrast is predicted but for the 112 reflection a weak alternating double image is predicted. Similar effects are found in niobium but it is still possible to determine Burgers vectors knowing that dislocations will show weak alternating double images for 112-type reflections when g.b. = 0.

Dislocation Density Determination

Dislocation density measurements were made on plates taken from 15 different areas of each foil examined. Three foils were examined from each specimen, one from the center, one from near the compression face, and one from between the previous two. In the absence of a cell structure the method used was that given by Keh [14] in which the average dislocation density N is given by

$$N=\left\{\frac{n_1}{L_1}+\frac{n_2}{L_2}\right\}\frac{1}{t}$$

where n_1 and n_2 are the average number of intersections between the dislocations and two orthogonal sets of grid lines lengths L_1 and L_2, and t is the foil thickness. However, in the presence of a cell structure or when marked dislocation tangling occurs, the dislocation density was determined using the method outlined by Ham [15] where $N = 2n/Lt$, where n is the number of intersections between the dislocations and a random set of lines length L and t is the foil thickness. The thickness of the foil for most samples was determined from observations of dotted contrast on steeply inclined dislocations which were usually present on secondary slip planes in each foil. It was found that t was generally close to 2500 Å.

There is no single reflection which gives a sharp image that does not have some dislocations out of contrast (see Table 2). Consequently, for every thin foil, dislocation density counts were made on four different areas with each reflection in Table 2 operating on each used. In this way the fractions of dislocations invisible under each reflection could be determined. Subsequently, one reflection, usually 200, was chosen and 15 plates for dislocation density counts were taken. After counting, the dislocation density counts with the 200 reflection operative were multiplied by the necessary factor obtained from the previous four areas to determine the densities of primary and secondary dislocations in each foil.

RESULTS

Mechanical Properties

Figure 3 shows typical shear stress shear strain curves for (a) the long specimens and (b) the short specimens at various strain rates. X-ray measurements carried out on the crystals after deformation showed that the crystal axis remained within two degrees of the initial orientation. Consequently, the stress/strain curves were resolved along the initial orientation of the primary slip plane. The long specimens all showed yield points, the size of which increased with increasing strain rate. The absence of a yield point for the Hopkinson bar tests is a result of the occurrence of Pochhammer-Chree oscillations in the initial portion of the oscilloscope trace which mask the yield point. Consequently, the initial portion of the stress/strain curve was obtained by averaging these oscillations. The short specimens also show a yield point at high strain rates 250 sec^{-1} but not at very high strain rates for the same reason as described for the long specimens, and not at the low strain rates, probably because of end effects. No stage I hardening is observed for either long or short specimens, although the crystals are favorably oriented for single slip. This absence of stage I is probably a result of the occurrence of multiple slip from the beginning of deformation because of end effects. This explanation is strengthened by the lack of rotation of the specimen axis during deformation as described previously. It is noticeable that the work-hardening rate is higher in the short specimens compared with the long specimens (see Table 3). This

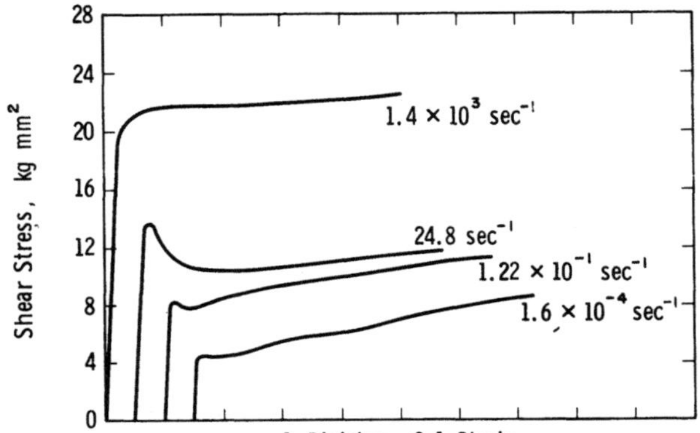

(a)

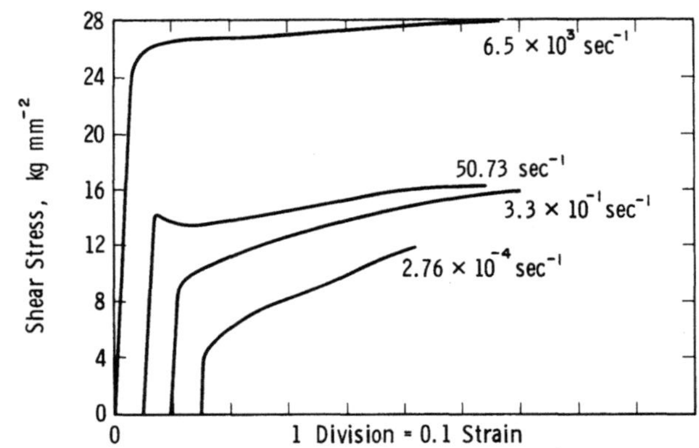

(b)

Fig. 3. Typical resolved shear stress/resolved shear strain curves for (a) long (b) short specimens.

Table 3

THE EFFECT OF STRAIN RATE ON WORK-HARDENING RATE AT 10% STRAIN

Short Specimens		Long Specimens	
Strain Rate, sec^{-1}	Work-Hardening Rate, Kg mm^{-2}	Strain Rate, sec^{-1}	Work-Hardening Rate, Kg mm^{-2}
1.6×10^{-4}	18.25	2.46×10^{-4}	7.7
1.22×10^{-1}	10.05	3.3×10^{-1}	5.9
24.8	7.75	50.73	4.1
1.4×10^{3}	3.5	6.5×10^{3}	2.0

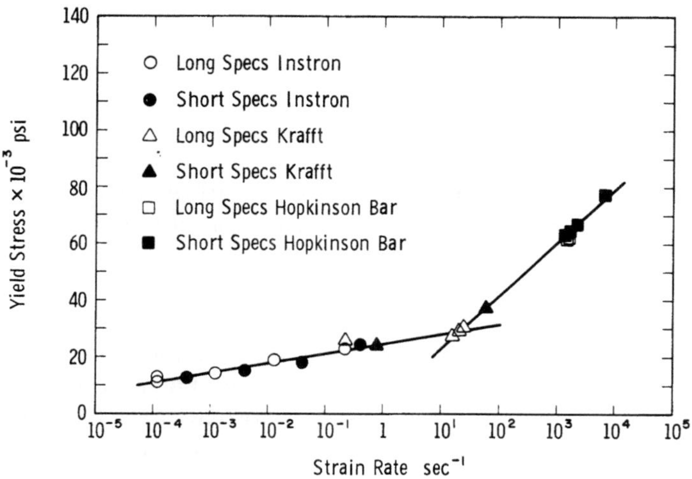

Fig. 4. The variation of yield stress vs. the logarithm of the strain rate for niobium single crystals.

effect is probably caused by an increased dislocation-multiplication rate with strain in short compared with long specimens because of end effects although this has not been verified at present. Both the long and short specimens show the same trend, namely a reduction in work-hardening rate with increasing strain rate. This is similar to the effect of lowering the deformation temperature observed in tantalum [9] and niobium [8], and similar to the effect of strain rate on work-hardening rate observed in iron [5].

The variation of lower yield stress with strain rate for the present specimens is shown in Fig. 4. It is immediately apparent that this material exhibits two regions with different strain-rate sensitivity of the yield stress like those found in *a*-iron in contrast to the several regions found by other workers for niobium [6–8]. The difference between the present work and that of previous workers [6–8] may be a result of impurity differences because the results of Sargent, et al. [6], and Fourdeux and Wronski [7] on different purity polycrystalline material did not agree with each other, and the results of Mitchell, et al. [8] on very pure single crystals of niobium are different from the present results on less pure single crystals. Note that both the long and the short specimens fall on the line and that in the region of overlap of the "Instron" and "Krafft" machines the results agree well. Note also that the change from Region 1 to Region 2 occurs within the "Krafft" testing range and is not related to a change in testing technique.

ELECTRON MICROSCOPY

Strain Rate 1.2×10^{-4} sec^{-1}

Specimens were deformed to shear strains of 3.7%, 5.4%, 12.64%, 19.88%, and 29.7%. The characteristic dislocation configurations of each strain are described separately.

(a) 3.7% Shear Strain

At this low strain the dislocation configurations consisted of a few tangles such as that shown in Fig. 5 separated by relatively large areas of low dislocation density which were approximately 10 μ in diameter. The dislocation configurations were independent of the position of the foil within the compression specimen up to 2 mm from the compression faces. Most of the dislocations were either mixed or screw in nature, and those edge dislocations visible were present in the form of short dipoles. The majority of the dislocations present are primary dislocations with Burgers vector $\frac{a}{2}[1\bar{1}1]$, see Fig. 6 at *A*. However, short dislocation segments with Burgers vector $\frac{a}{2}[\bar{1}11]$ are visible at *B* lying on (101) plane. These two sets of dislocations frequently interact to produce short segments of dislocation with Burgers vector $a\,[001]$, see Fig. 6 at *C*.

(b) 5.4% Shear Strain

The dislocation substructure is very similar to that shown in Fig. 5 for a shear strain of 3.7%. However, dislocation tangles are more frequent in this specimen and are separated by regions of low dislocation density which are approximately 7 μ in diameter. In addition, Fig. 7 at *A* and *B* shows that dislocation dipoles are present lying perpendicular to $[1\bar{1}1]$ and $[\bar{1}\bar{1}1]$, the active slip directions in the (011) slip plane indicating that slip on the primary slip plane is not confined to the most favorably oriented slip direction even at this early stage in deformation. Again, the dislocation distribution is constant throughout the compression specimen up to 2 mm from the compression faces. The Burgers vectors of dislocations are $\frac{a}{2}[1\bar{1}1]$, $\frac{a}{2}[\bar{1}11]$, $\frac{a}{2}[\bar{1}\bar{1}1]$, and $a[001]$. The latter dislocations with Burgers vector $a[001]$ are formed in short segments by interaction of dislocations with Burgers vectors $\frac{a}{2}[\bar{1}11]$ and $\frac{a}{2}[1\bar{1}1]$ as described previously.

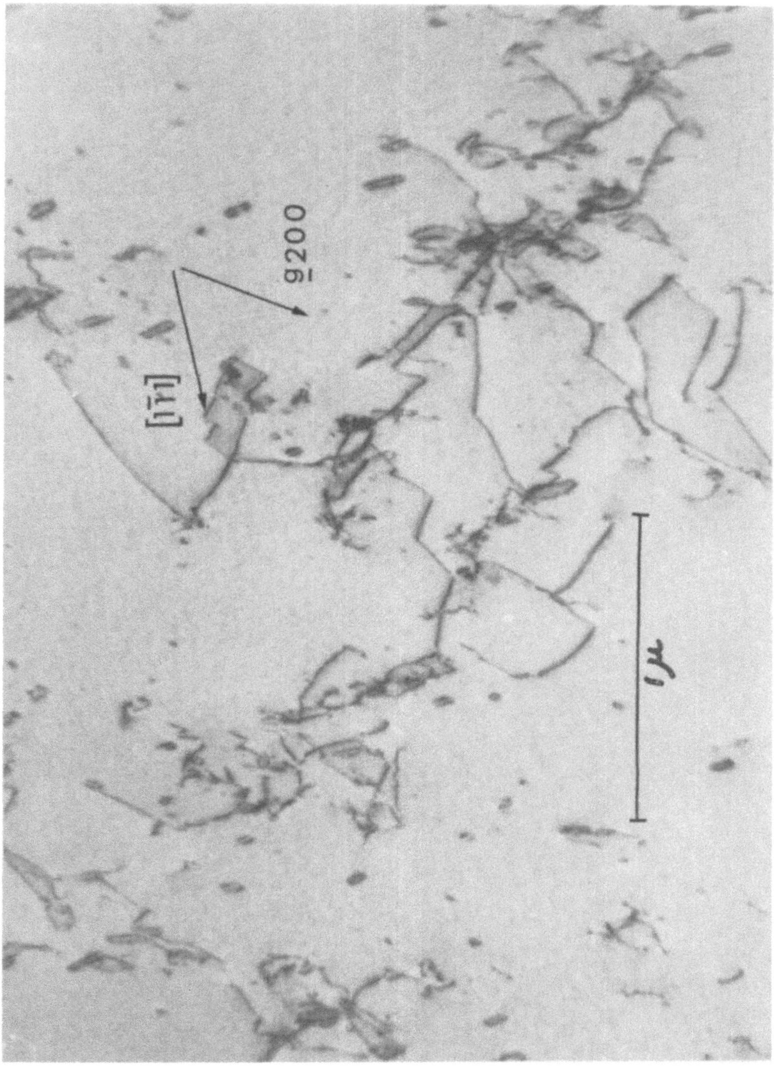

Fig. 5. $\dot{\epsilon} = 1.2 \times 10^{-4}$ sec^{-1}, 3.7% shear strain (011) foil. A typical dislocation tangle. $[1\bar{1}1]$ is parallel to primary Burgers vector.

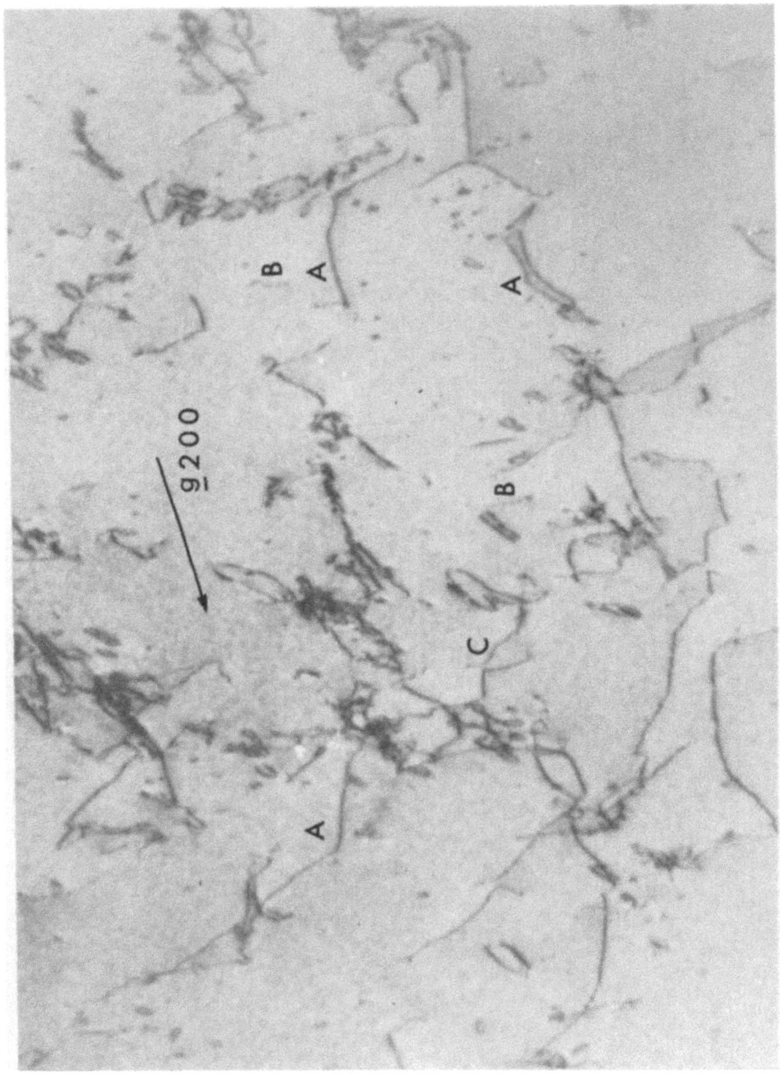

(a)

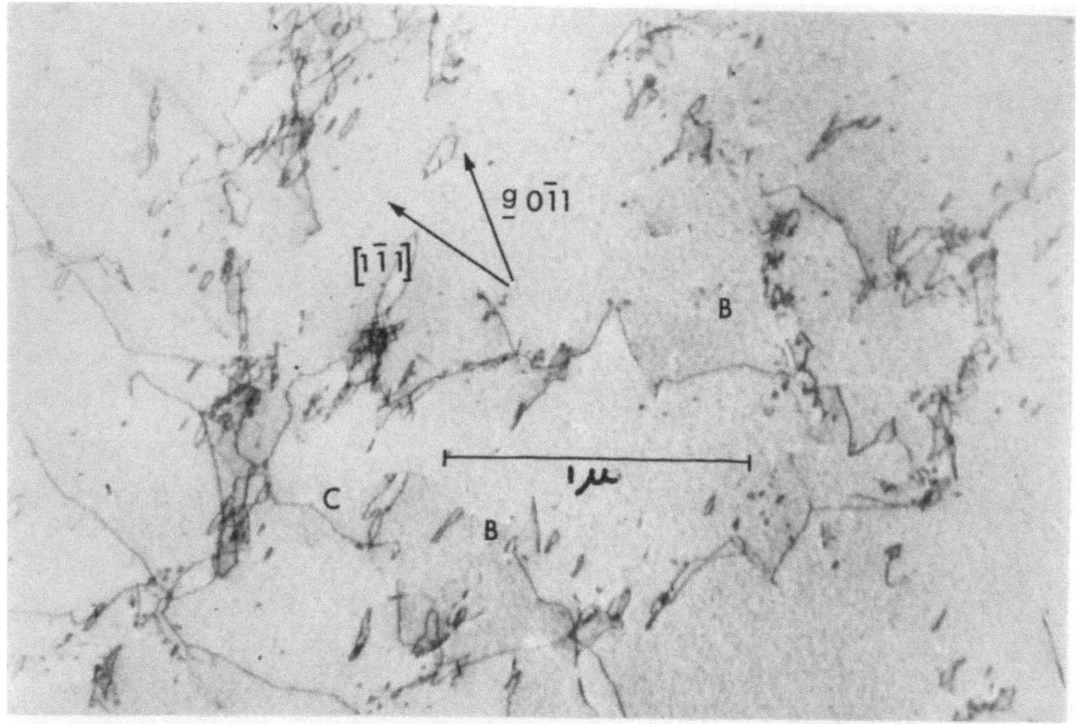

(b)

Fig. 6. $\dot{\epsilon} = 1.2 \times 10^{-4}$ sec^{-1}, 3.7% shear strain (011) foil. Burgers vector analysis. Primary dislocation $b = \frac{a}{2}$ $[1\bar{1}1]$ at A are invisible under a $\overline{21}1$ reflection. Secondary dislocations $b = \frac{a}{2}[\bar{1}11]$ at B are invisible under 011 and $1\bar{1}2$ reflections. Note interaction product at C with Burgers vector a [001] invisible under a 200 reflection.

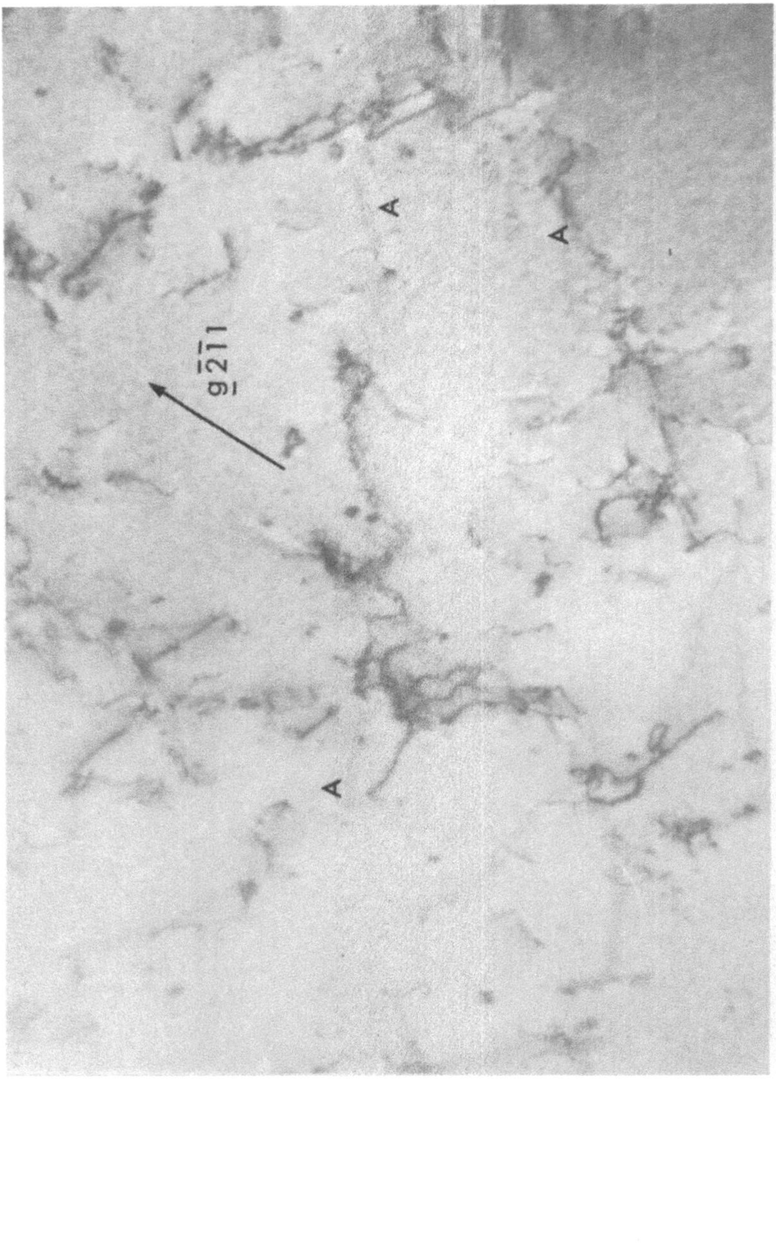

(c)

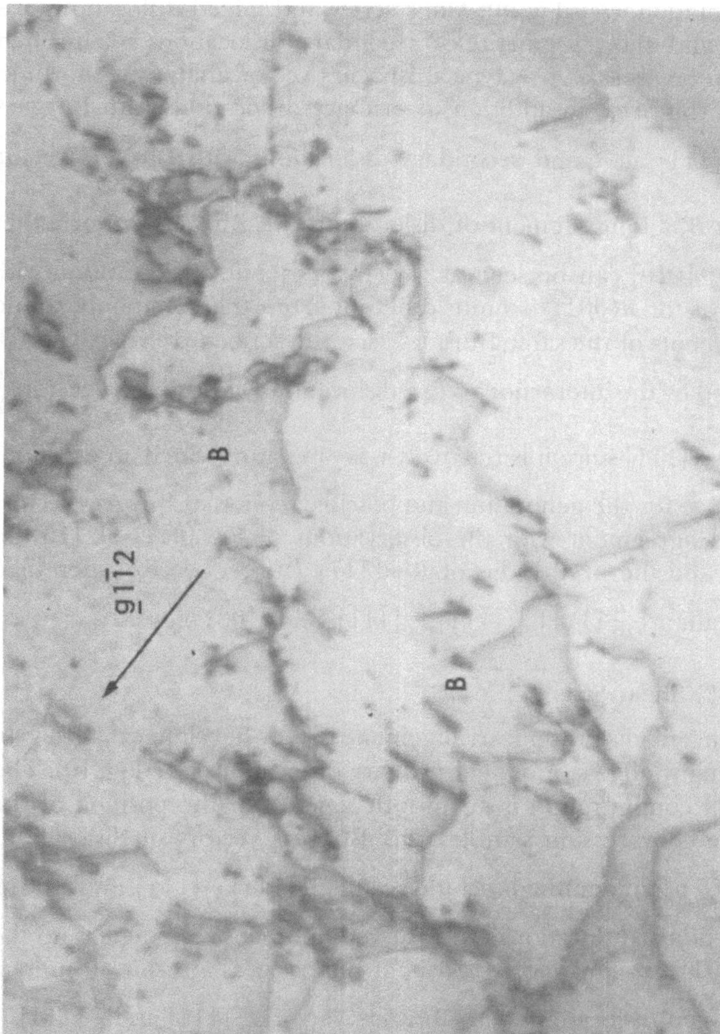

(d)

Fig. 6. *(Continued)*

(c) 12.64% Shear Strain

At this stage in the deformation process dislocation tangles are well developed and contain large numbers of primary dislocations with edge orientation, (see Fig. 8). Many screw and mixed dislocations are observed and short segments of secondary dislocations on inclined planes can be seen at *A*. A typical Burgers vector analysis is shown in Fig. 9. In this area, in addition to primary dislocations with Burgers vector $\frac{a}{2}[1\bar{1}1]$ at *A* and secondary dislocations with Burgers vector $\frac{a}{2}[\bar{1}11]$ at *B*, a long segment of dislocation with Burgers vector either $a[001]$ or $a[010]$ can be seen at *C*. This segment of dislocation with Burgers vector $a\langle 001\rangle$ is quite different from the commonly found short segments of the same Burgers vector described previously which are formed by the interaction of two dislocations with Burgers vectors $\frac{a}{2}[1\bar{1}1]$ and $\frac{a}{2}[\bar{1}11]$ since it is relatively wavy in nature. Such observations are evidence for the generation and possible motion of Burgers vectors $a\langle 001\rangle$ in agreement with the observations of Foxall, et al. [16] on niobium, and the predictions of Reid [17]. Burgers vectors identified at this strain are $\frac{a}{2}[1\bar{1}1]$, $\frac{a}{2}[\bar{1}11]$, $\frac{a}{2}[\overline{111}]$, and $a[001]$.

(d) 19.88% Shear Strain

The tangled dislocation arrangement is well developed, and again contains many edge component primary dislocations (see Fig. 10). The dislocation configuration is again independent of the position of the foil in the compression sample. The Burgers vectors of dislocations which have been identified include $\frac{a}{2}[1\bar{1}1]$, $\frac{a}{2}[\bar{1}11]$, $\frac{a}{2}[\overline{111}]$, and $a[001]$. Most secondary dislocations are observed in or close to tangles and the latter $a[001]$ Burgers vector is usually observed as a result of interactions between dislocations with Burgers vectors $\frac{a}{2}[1\bar{1}1]$ and $\frac{a}{2}[\bar{1}11]$.

(e) 29.7% Shear Strain

At this stage, a rudimentary cell structure is formed (see Fig. 11) and again the dislocation configurations do not vary with the position of the foil in the compression specimen. Figure 12 shows a tangle under different reflections. It can be seen that, while many primary edge dislocations are present in the tangle, large numbers of secondary dislocations are also present. Burgers vectors determined at this strain include $\frac{a}{2}[1\bar{1}1]$ at A in Fig. 12, $\frac{a}{2}[\bar{1}11]$, $\frac{a}{2}[\overline{11}1]$, and $a[001]$. Most secondary dislocations including those with Burgers vector $\frac{a}{2}[\bar{1}11]$ which were observed between tangles at small strains seem to be associated with the tangles at this strain. The majority of dislocations with Burgers vector $a[001]$ are produced by intersection interaction as described previously.

Strain Rate 1.5×10^3 sec^{-1}

Specimens were deformed to shear strains of 2.8%, 6%, 12%, and 30%. The characteristic dislocation substructure corresponding to each strain is described separately.

(a) 2.8% Shear Strain

A typical area is shown in Fig. 13. Long, heavily jogged screw dislocations are present lying parallel to $[1\bar{1}1]$ and $[\overline{11}1]$ directions and in addition small dipoles are visible at A. In addition there are large areas 10–15 μ in diameter free of dislocations. Figure 14 shows a Burgers vector analysis. Dislocation A has Burgers vector $\frac{a}{2}[1\bar{1}1]$, dislocation B, $\frac{a}{2}[\overline{11}1]$, and dislocation C, $\frac{a}{2}[\bar{1}11]$. An example of a common interaction is visible at C' to produce a short dislocation segment with Burgers vector $a[001]$. The general dislocation configurations are the same throughout most of the length of the specimen although the dislocation density fluctuates somewhat. Within 2 mm of the compression face the dislocation density is higher, and we have observed one case where the dislocations are of unusual character (see Fig. 15). In this case long dislocations with Burgers vector $a[100]$ are present at A. Such long segments are unlikely to form from dislocation interactions on the basis of the results described previously, and it is concluded that friction effects near the compression face promote slip with $a\langle 001\rangle$ Burgers vectors at high strain rates (i.e., stress levels). Thus even at this early stage in the deformation process dislocations with Burgers vectors $\frac{a}{2}[\bar{1}11]$, $\frac{a}{2}[1\bar{1}1]$, $\frac{a}{2}[\overline{11}1]$, and $a[001]$ are observed.

(b) 6% Shear Strain

A typical area is shown in Fig. 16 which shows mainly primary screw dislocations lying parallel to the $[1\bar{1}1]$ direction together with short inclined dislocations with Burgers vector $\frac{a}{2}[\bar{1}11]$. Short dislocation seg-

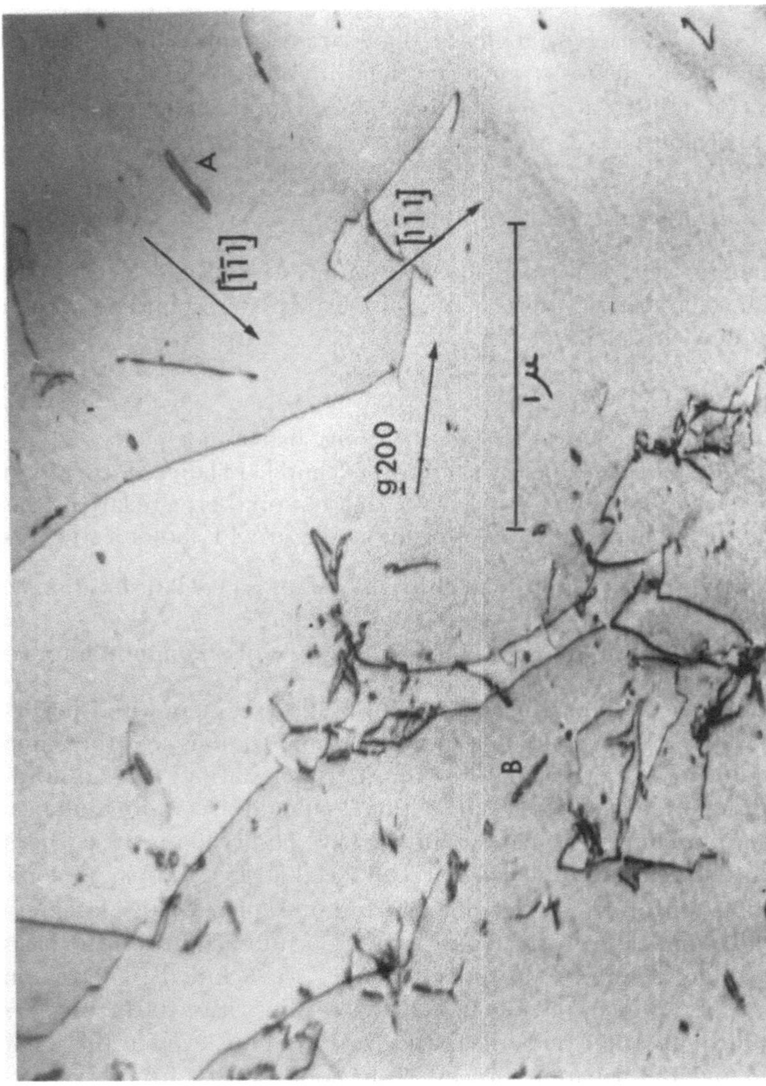

Fig. 7. $\dot{\epsilon} = 1.2 \times 10^{-4}$ sec^{-1}, 5.4% shear strain (011) foil. A typical dislocation tangle. Note dipoles *A* and *B* perpendicular to the primary slip direction $[1\bar{1}1]$ and the secondary slip direction $[\overline{11}1]$ in the primary slip plane.

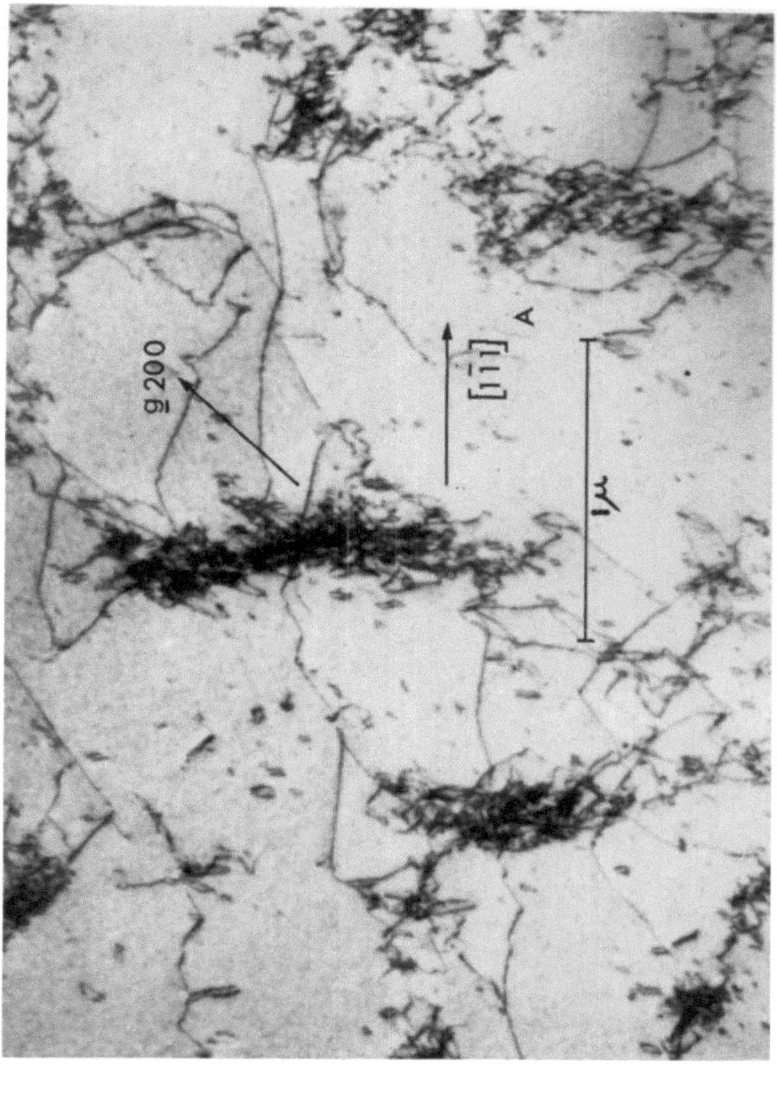

Fig. 8. $\dot{\epsilon} = 1.2 \times 10^{-4}$ sec^{-1}, 12.64% shear strain (011) foil. A typical area showing long primary screws and many edge dislocations in tangles. Short secondary dislocations are visible at *A*.

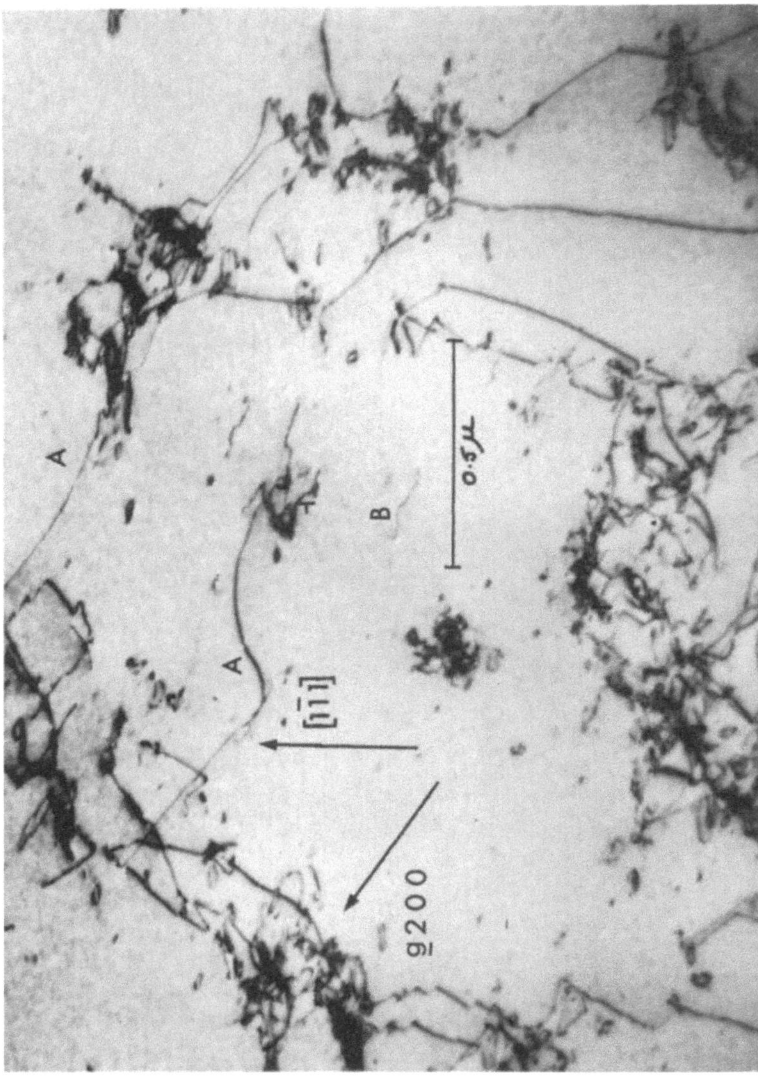

(a)

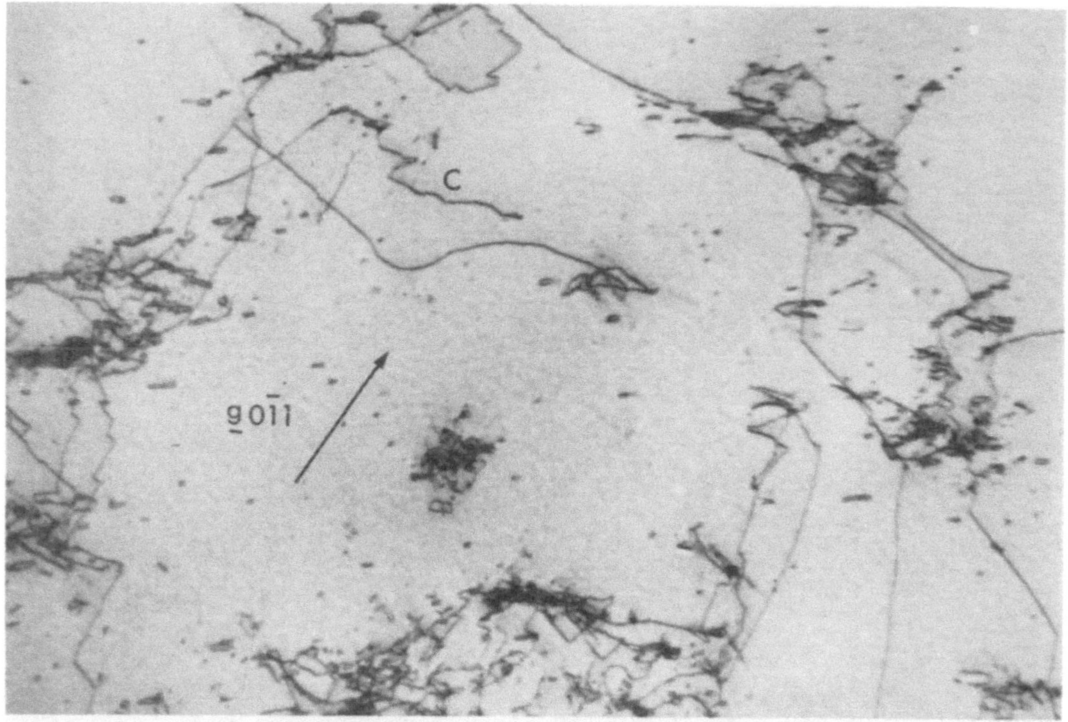

(b)

Fig. 9. $\dot{\epsilon} = 1.2 \times 10^{-4}$ sec^{-1}, 12.64% shear strain (011) foil. Note primary dislocation at A, $b = \frac{a}{2}[\bar{1}11]$, invisible under a $\overline{2}\overline{1}1$ reflection and dislocation with $b = \frac{a}{2}[\bar{1}11]$ at B, invisible under a $1\bar{1}2$ reflection with $b = a$ [010] at a [001] at C invisible under 200 reflection.

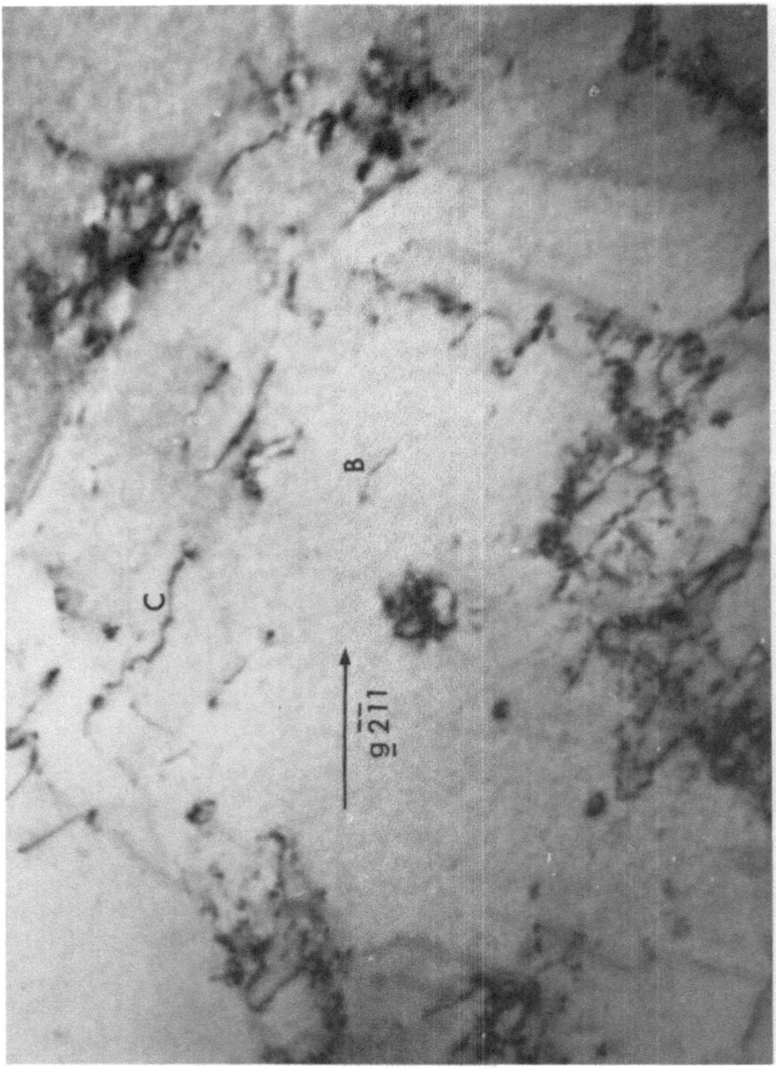

(c)

Fig. 9. *(Continued)*

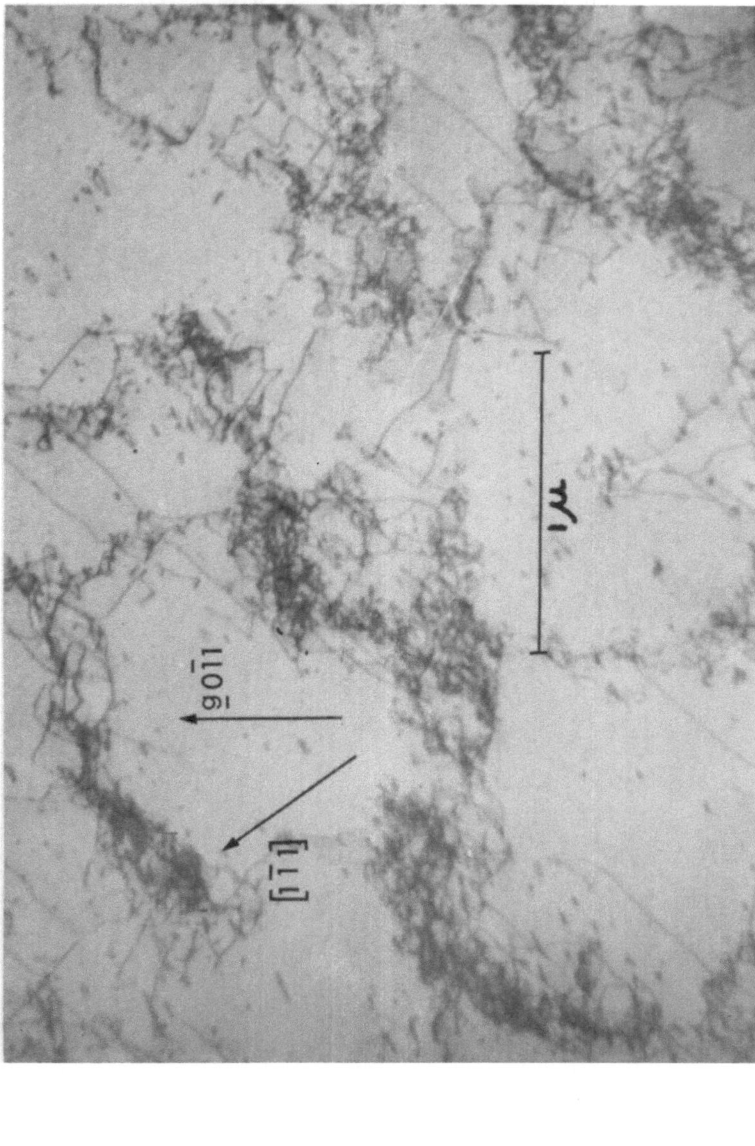

Fig. 10. $\dot{\epsilon} = 1.2 \times 10^{-4}$ sec^{-1}, 19.88% shear strain (011) foil. A typical area showing a partially developed cell structure.

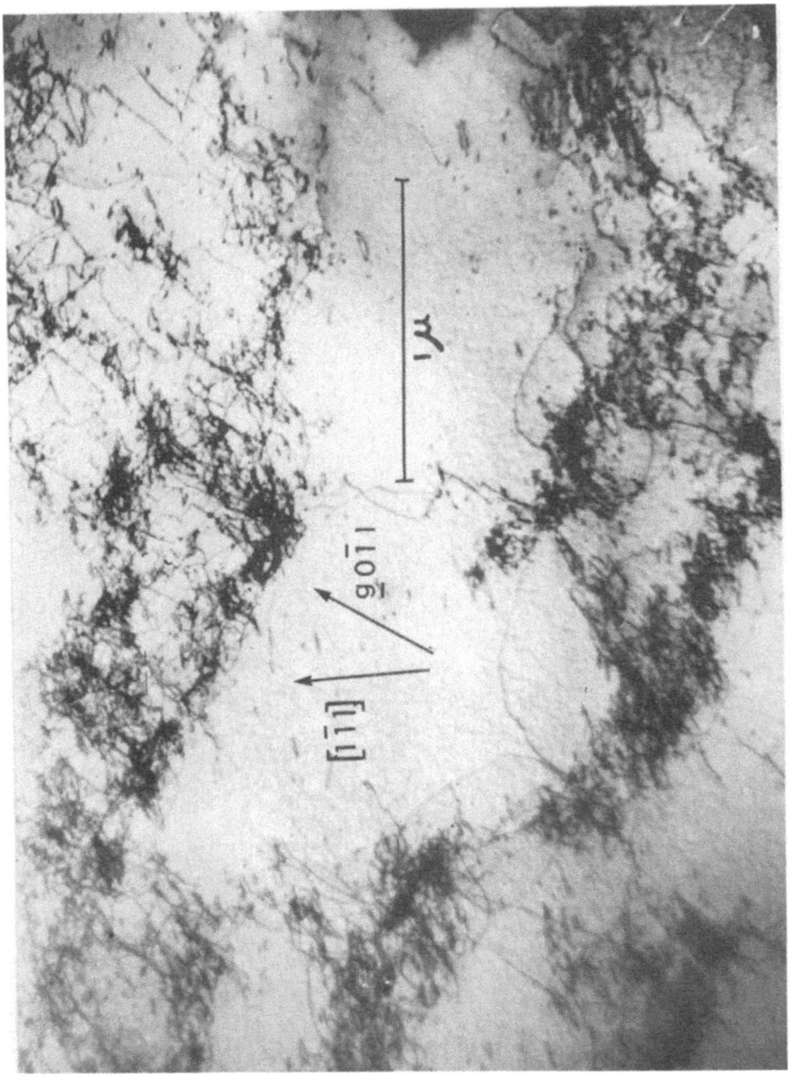

Fig. 11. $\dot{\epsilon} = 1.2 \times 10^{-4}$ sec^{-1}, 29.7% shear strain (011) foil. A typical area showing a rudimentary cell structure with variable mesh size.

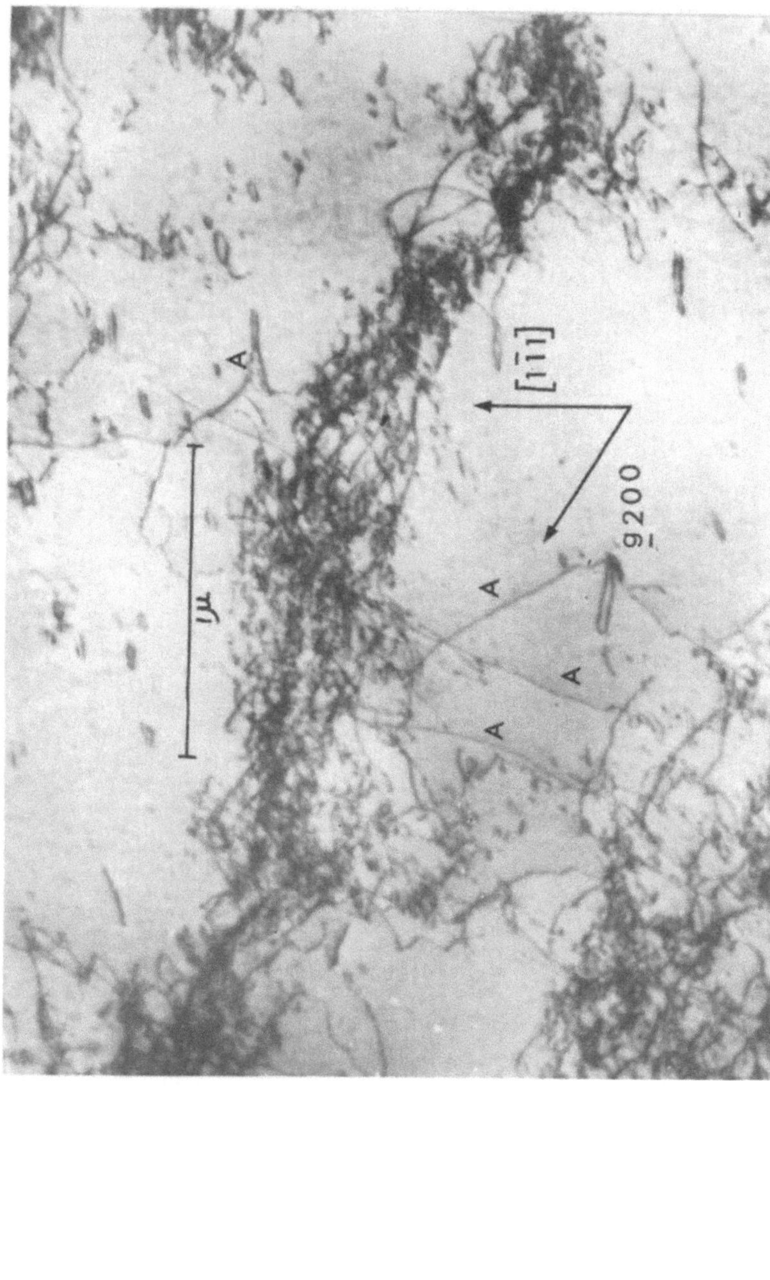

(a)

Fig. 12. $\dot{\epsilon} = 1.2 \times 10^{-4}$ sec^{-1}, 29.7% shear strain (011) foil. A typical dislocation tangle. Note high density of secondary dislocations when the primary dislocations are invisible under a $\bar{2}11$ reflection.

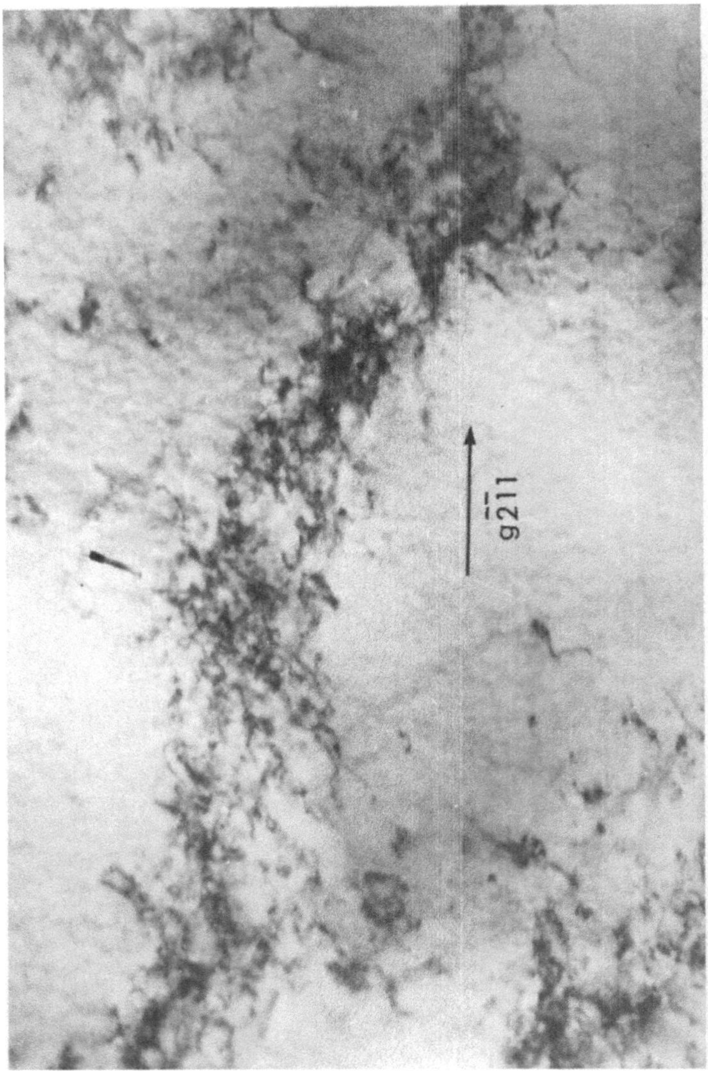

(b)

Fig. 12. (*Continued*)

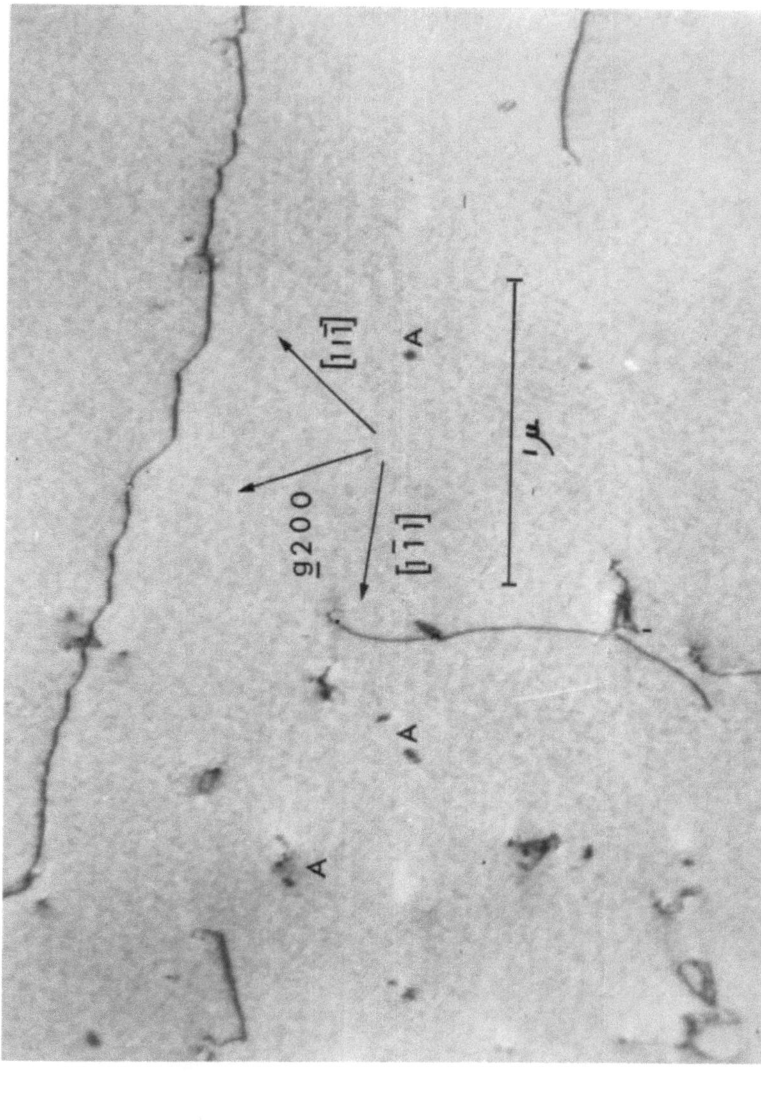

Fig. 13. $\dot{\epsilon} = 1.5 \times 10^3$ sec^{-1}, 2.8% shear strain, (011) foil. Long screw dislocations parallel to the primary slip direction [1$\bar{1}$1] and the secondary slip direction [$\bar{1}\bar{1}$1] in the (011) slip plane. Note small dipoles at *A*.

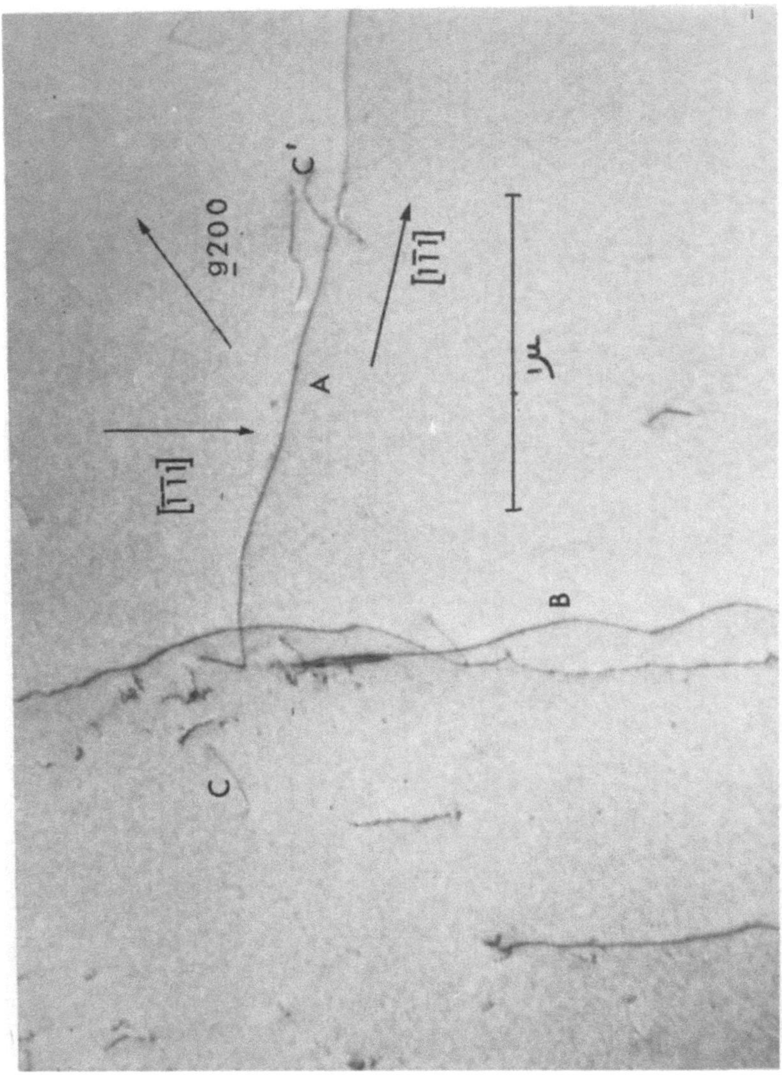

(a)

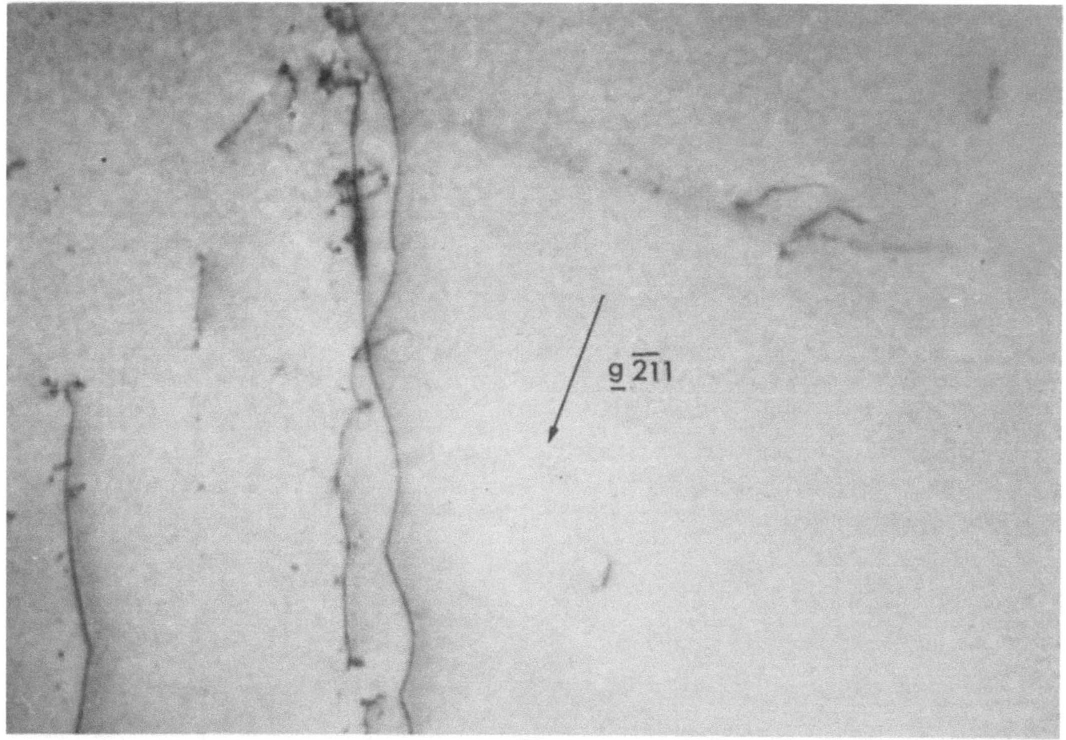

(b)

Fig. 14. $\dot{\epsilon} = 1.5 \times 10^3\ sec^{-1}$, 2.8% shear strain, (011) foil. Dislocation *A* has $b = \frac{a}{2}[1\overline{1}1]$ and is invisible under the $\overline{21}1$ reflection, dislocation *B* has $b = \frac{a}{2}[\overline{11}1]$ and is invisible under $2\overline{1}1$ reflection, dislocation has $b = \frac{a}{2}$ $[\overline{1}11]$ and is invisible under a $\overline{11}2$ reflection. Note interaction product at *C* with Burgers vector a [001] is invisible under a 200 reflection.

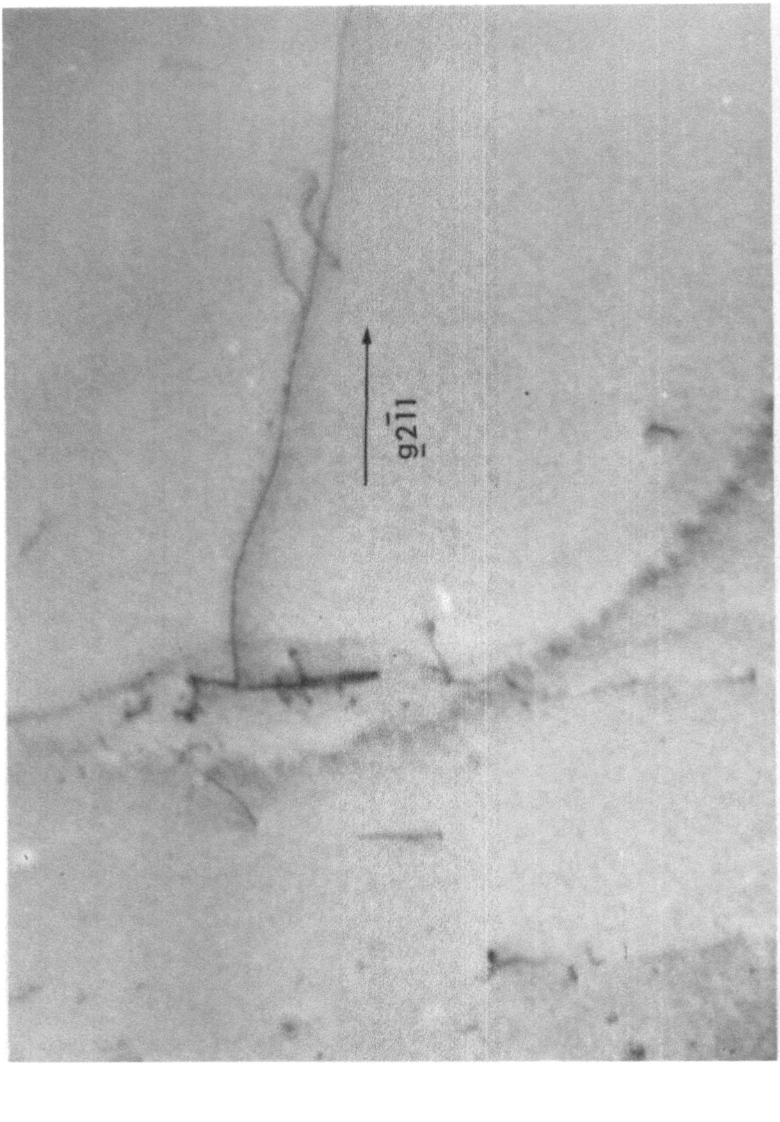

(c)

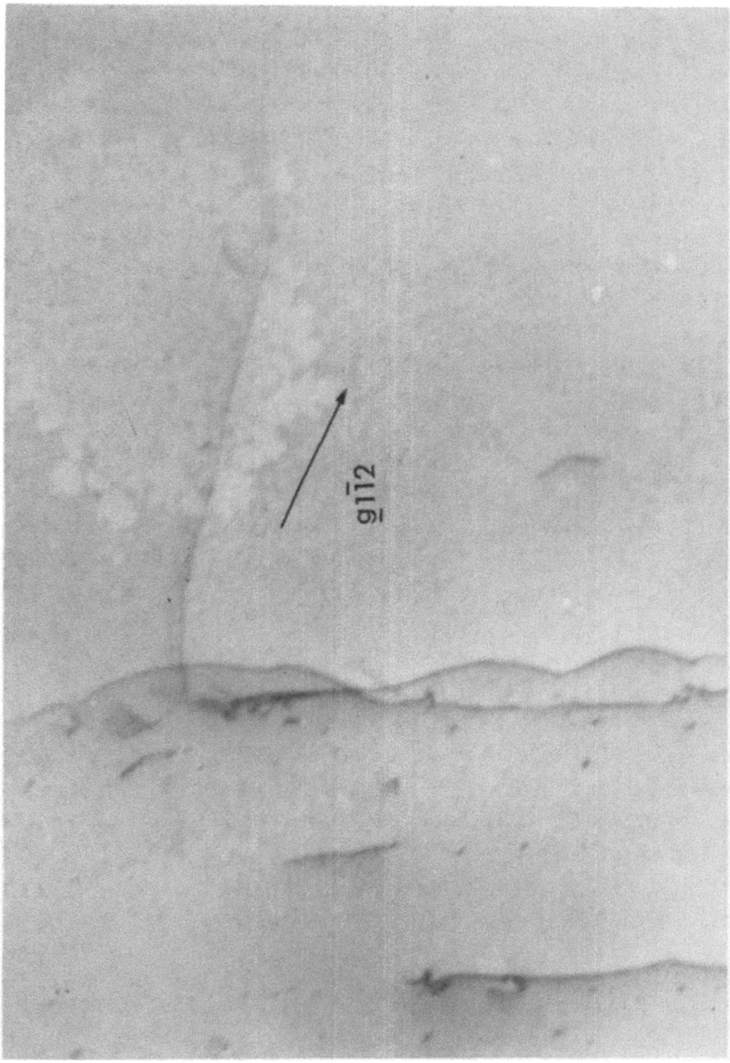

(d)

Fig. 14. *(Continued)*

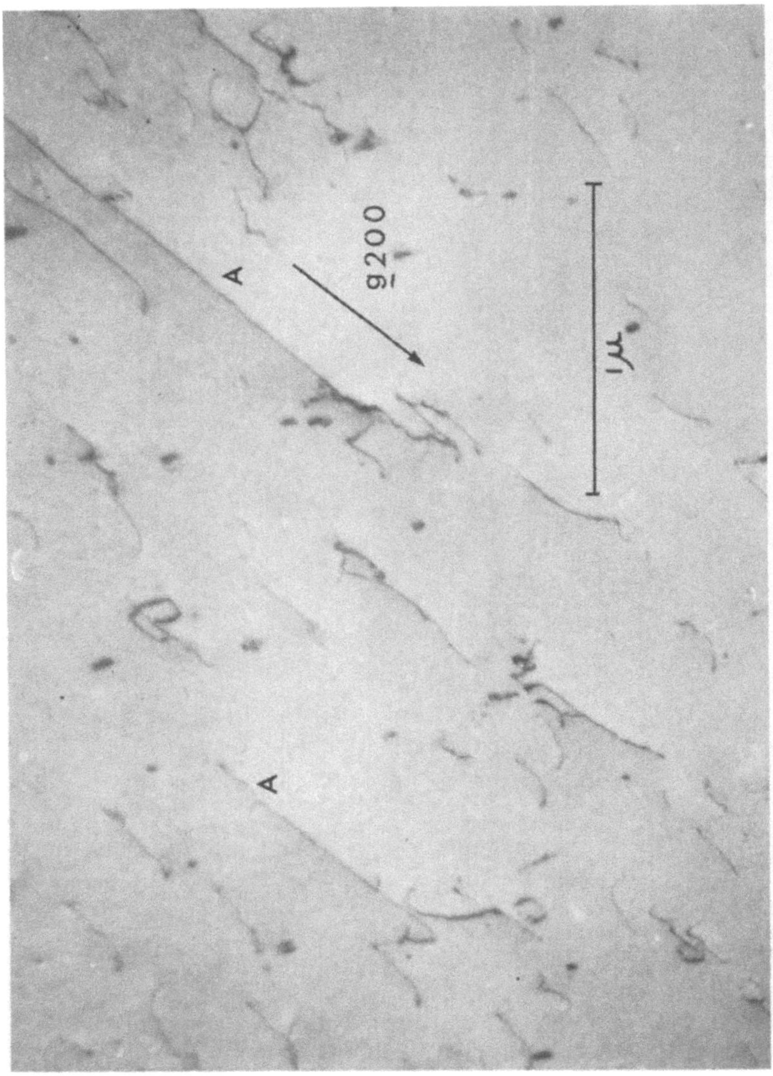

(a)

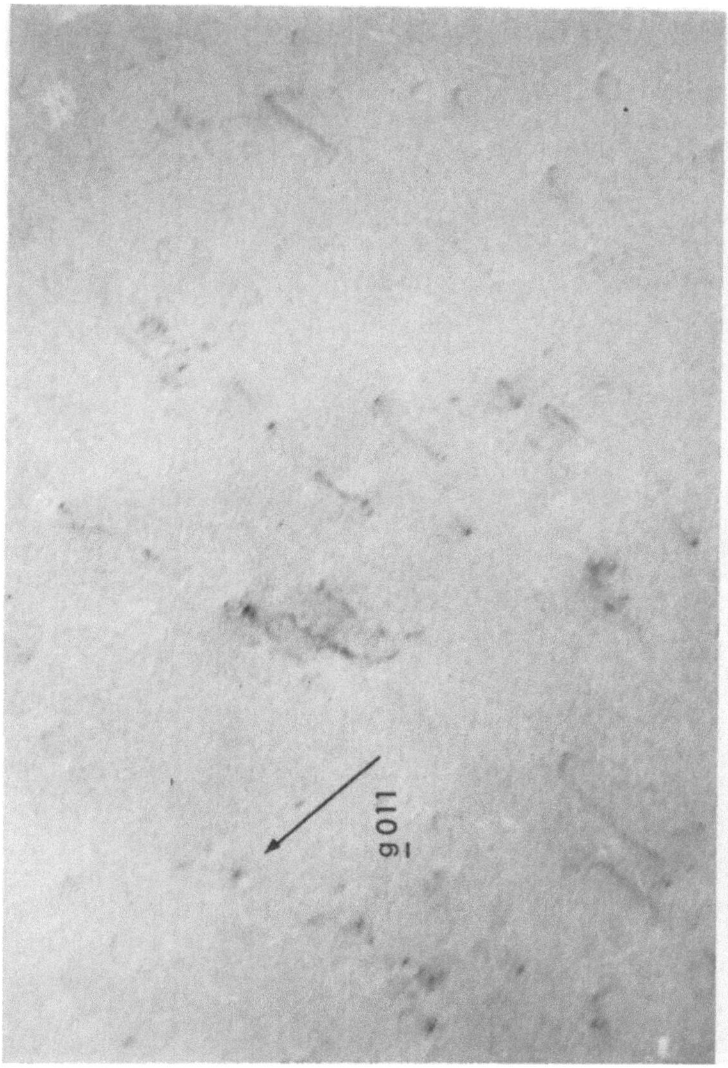

(b)

Fig. 15. $\dot{\epsilon} = 1.5 \times 10^{-3}$ sec^{-1}, 2.8% shear strain, (011) foil. Note dislocation at *A* with $b = \frac{a}{2}[100]$ is invisible under an 011 reflection only.

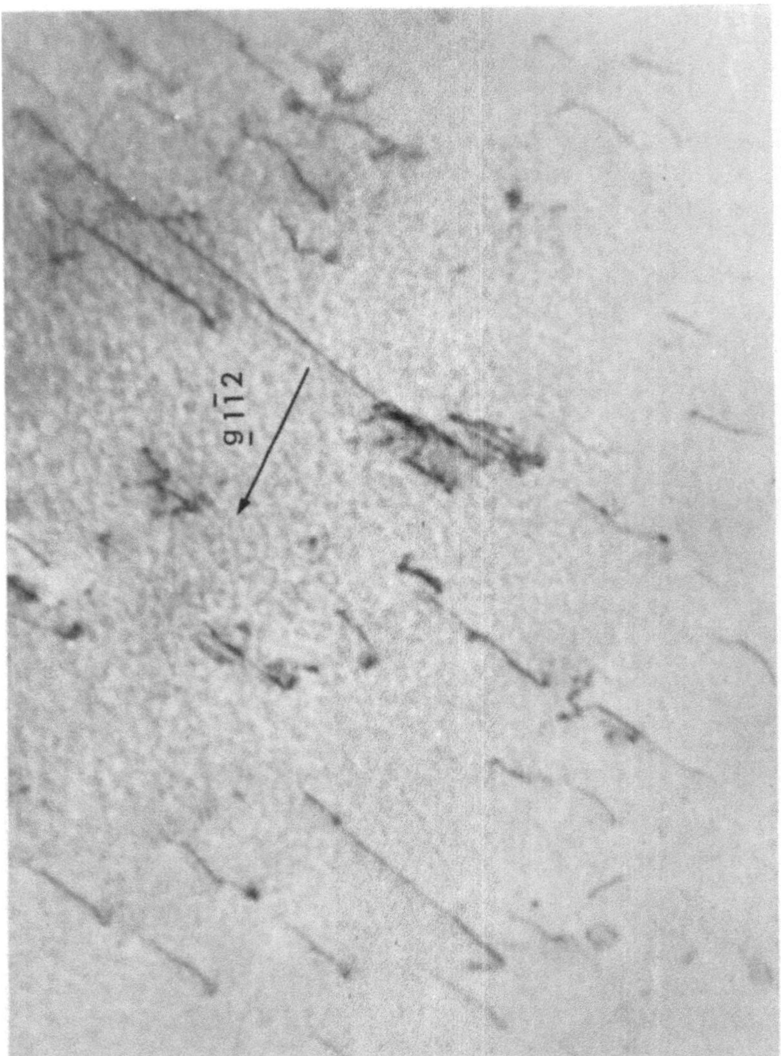

(c)

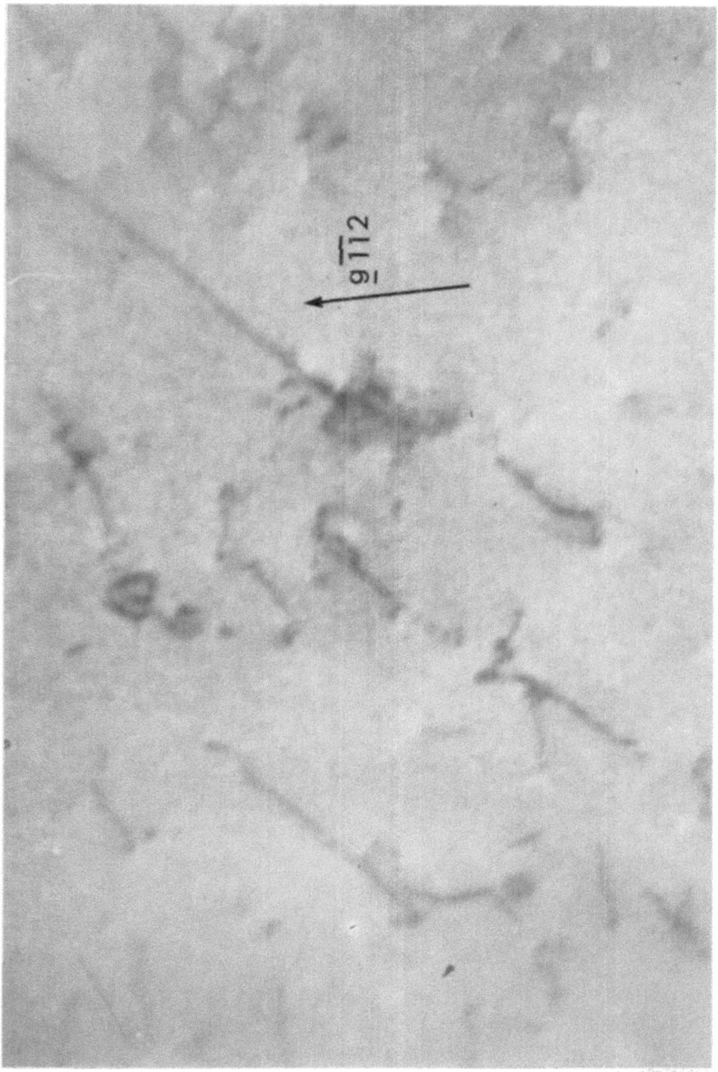

(d)

Fig. 15. (*Continued*)

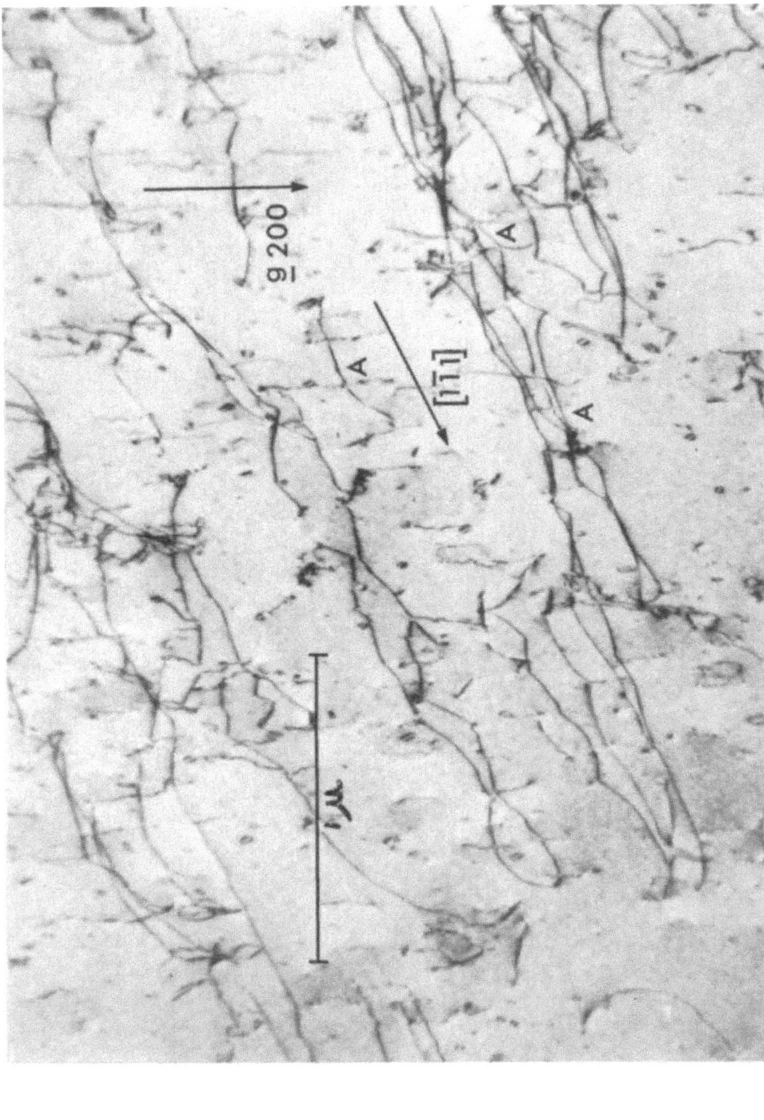

Fig. 16. $\dot{\epsilon} = 1.5 \times 10^3$ sec^{-1}, 6% shear strain (011) foil. Note long screw dislocations parallel to the primary slip direction $[1\bar{1}1]$ and the short secondary slip dislocations $b = \frac{a}{2}[\bar{1}11]$. Dislocations with $b = a\,[001]$ formed by interaction at *C*.

ments with Burgers vector $a[001]$ are formed at interaction points such as A. There is a high density of small dislocation edge loops present. There is some fluctuation of dislocation density with position of the foil in the specimen although the general features of the dislocation configuration remain constant. Burgers vectors identified are the same as those described for 2.8% strain at this strain rate.

(c) 12% Shear Strain

At this stage of deformation the primary dislocations still remain largely screw in character, as shown in Fig. 17a, and there is a high density of primary edge dislocation loops. However, the dislocation density still varies from point to point in the specimen. For example, Figs. 17b, c show typical areas in other parts of the compression specimen. Figure 17b shows an area of lower dislocation density but with essentially the same dislocation configuration as Fig. 17a. However, Fig. 17c is taken from another area and shows a slip band cutting the slip plane. Here the dislocation density and configuration are quite different from those shown in Figs. 17a, b. Dislocation Burgers vectors identified at this strain are $\frac{a}{2}[1\bar{1}1]$, $\frac{a}{2}[\bar{1}11]$, $\frac{a}{2}[\bar{1}\bar{1}1]$, and $a[001]$. Dislocations with the latter Burgers vector are formed by dislocation interaction.

(d) 20% Shear Strain

A typical area is shown in Fig. 18. The dislocation distribution is more uniform from point to point in the compression specimen except for occasional areas like that in Fig. 17c where secondary slip bands cross the foil plane. There is a high density of primary dislocation dipoles. Dislocation Burgers vectors identified at this strain are $\frac{a}{2}[1\bar{1}1]$, $\frac{a}{2}[\bar{1}11]$, $\frac{a}{2}[\bar{1}\bar{1}1]$, and $a[001]$. Again dislocations with Burgers vector $a[001]$ are generally formed by interaction as described previously.

(e) 30% Shear Strain

Thin foils obtained at this strain show a high density of primary screw dislocations and dipoles (see Fig. 19). There is still no dislocation tangling. Again it is found that the dislocation configurations are constant from point to point in the compression specimen except in areas where secondary slip bands cut the primary slip plane. Burgers vectors identified are $\frac{a}{2}[1\bar{1}1]$, $\frac{a}{2}[\bar{1}11]$, $\frac{a}{2}[\bar{1}\bar{1}1]$, and $a[001]$. Again the latter Burgers vectors are produced by dislocation interaction.

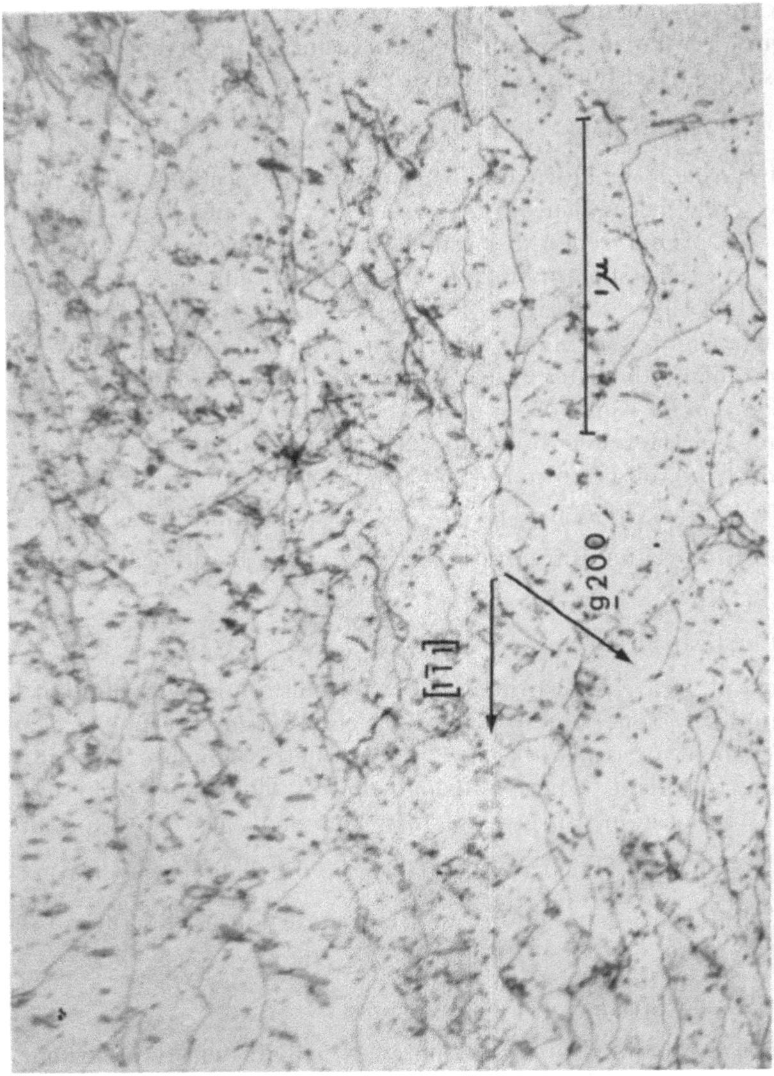

(a)

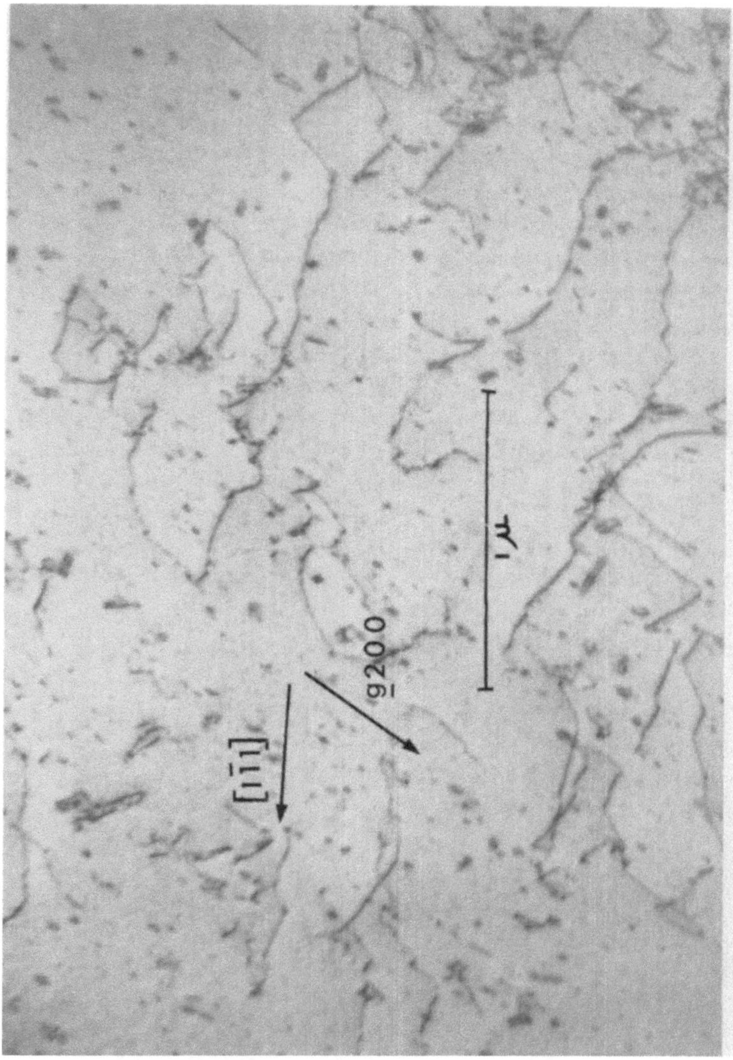

Fig. 17. $\dot{\epsilon} = 1.5 \times 10^3$ sec^{-1}, 12% shear strain, (011) foil. Examples of the variation of dislocation density variation within the specimen.

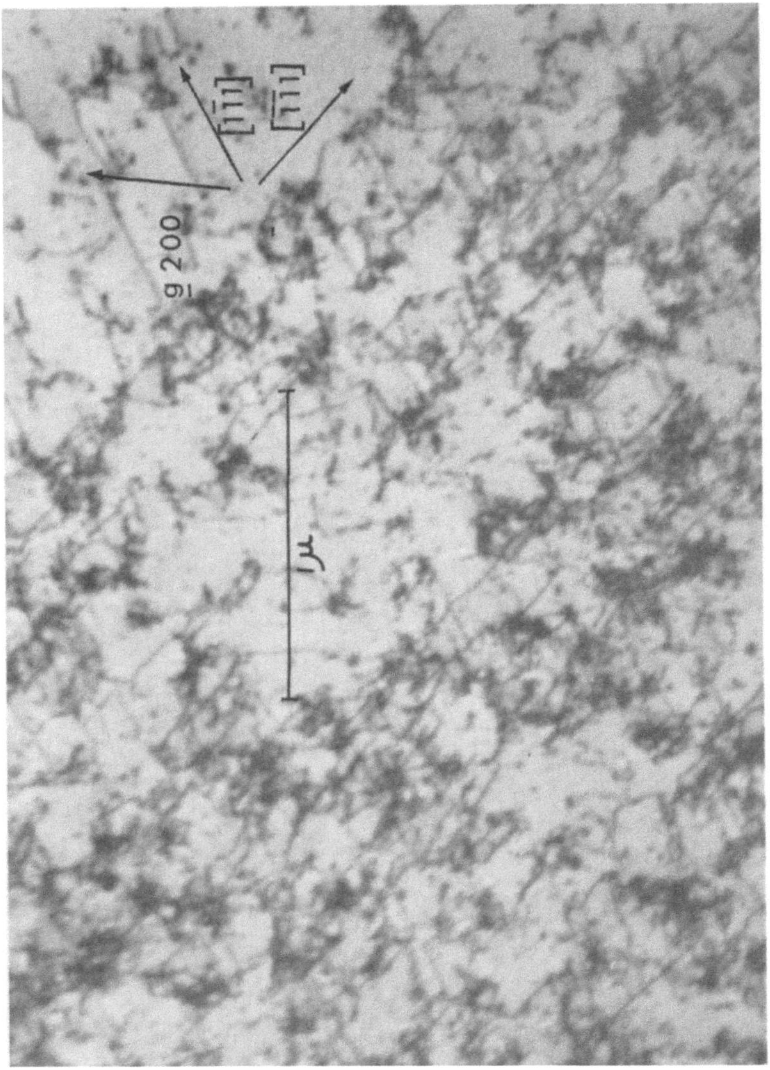

(c)

Fig. 17. *(Continued)*

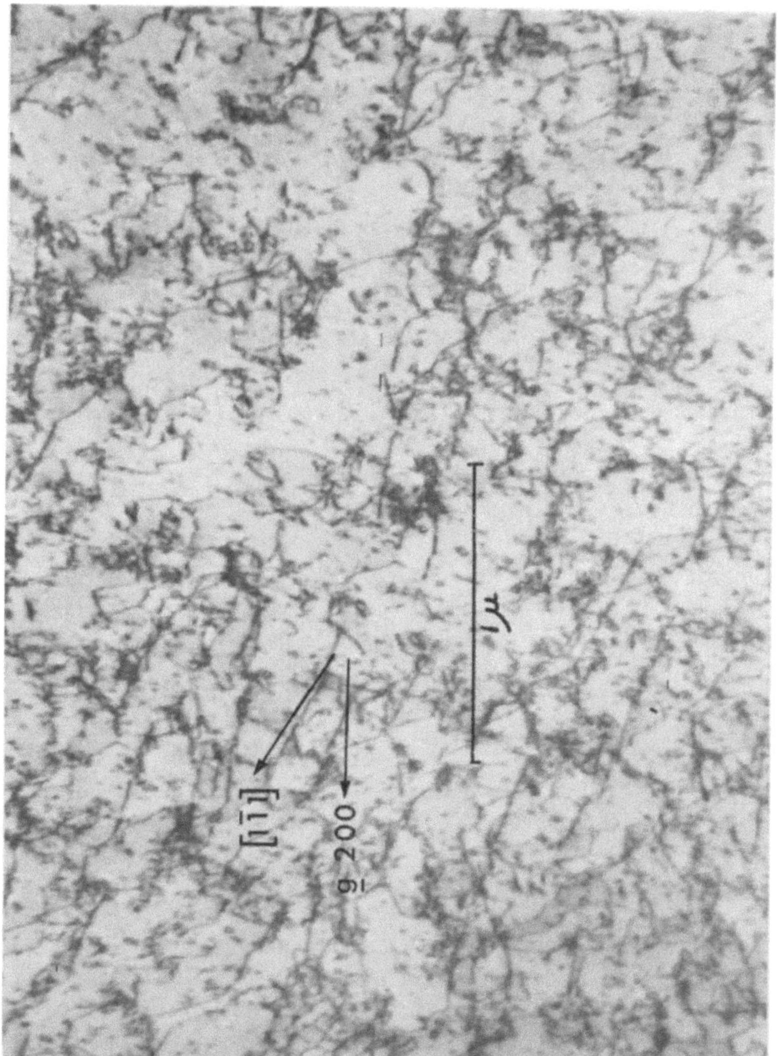

Fig. 18. $\dot{\epsilon} = 1.5 \times 10^3$ sec^{-1}, 30% shear strain, (011) foil. A typical area.

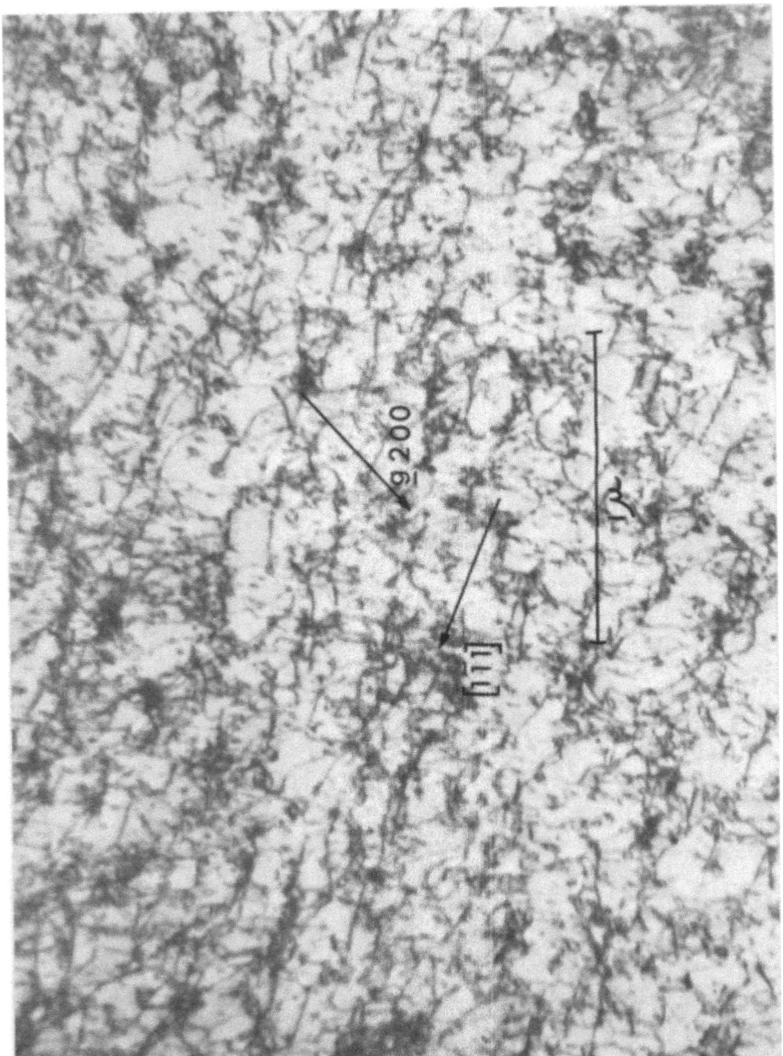

Fig. 19. $\dot{\epsilon} = 1.5 \times 10^3$ sec^{-1}, 30% shear strain, (011) foil. A typical area.

Dislocation Densities

Both the total dislocation density and the primary dislocation density are related to the flow stress at both the high and the low strain rate by an equation of the form (see Fig. 20),

$$\tau = \tau_0 + \alpha\mu b\sqrt{N}$$

where τ is the flow stress, α is a constant, μ is the shear modulus, b the Burgers vector, and N the dislocation density. It is not clear what significance can be attached to the term τ_0. For example it has been termed the friction stress [18–21] or the stress necessary to move one dislocation in the absence of others [22, 23]. However, it has also been pointed out [24] that τ_0 can be considered to be simply a result of the linear interpretation of the data relating τ and $\sqrt{N}$ and has no intrinsic significance. Values for α and τ_0 obtained from Fig. 18 are $\alpha_{total} = 0.134$, $\alpha_{primary} = 0.198$, τ_0 3.45 kg mm^{-2} at a strain rate of 1.2×10^{-4} sec^{-1} and $\alpha_{total} = 0.0375$, $\alpha_{primary}$ 0.05, $\tau_0 = 21.35$ kg mm^{-2} at a strain rate of 1.5×10^3 sec^{-1}.

The dislocation density is a linear function of the plastic strain with slope 1.25×10^{-11} cm^{-2} as shown in Fig. 21. Thus it can be concluded that the dislocation multiplication rates are essentially the same at both strain rates.

Finally, for comparison purposes the density of dislocation dipoles has been measured as a function of strain at both the high and the low strain rates. The results are summarized in Table 4.

Table 4

Dipole Density vs. Plastic Shear Strain

$\dot{\epsilon} = 1.2 \times 10^{-4}$ sec^{-1}		$\dot{\epsilon} = 1.5 \times 10^{3}$ sec^{-1}	
Shear Strain, %	Dipole Density, /cc $\times 10^{14}$	Shear Strain, %	Dipole Density, /cc $\times 10^{14}$
3.7	7.5	2.8	2.25
5.4	9.0	6.0	15.5
12.64	14.75	12.0	35.0
19.88	22.75	20.0	58.0
29.7	32.75	30.0	88.5

It can be seen that the density of dipoles at the high strain rate is approximately twice that at the low strain rate for a given strain. This result is in general agreement with the qualitative estimate of Gilbert,

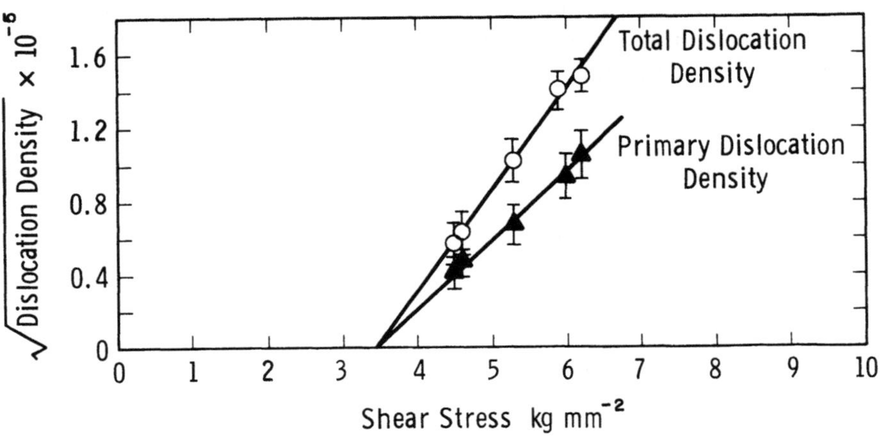

(a)

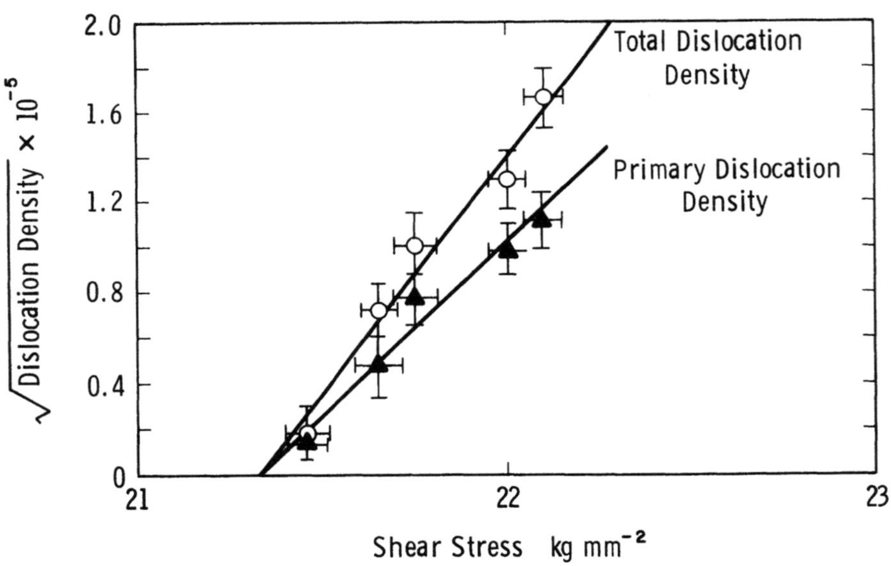

(b)

Fig. 20. The relationship between the flow stress and the square root of the dislocation density (a) $\dot{\epsilon} = 1.2 \times 10^{-4}$ sec^{-1}, (b) 1.5×10^{3} sec^{-1}.

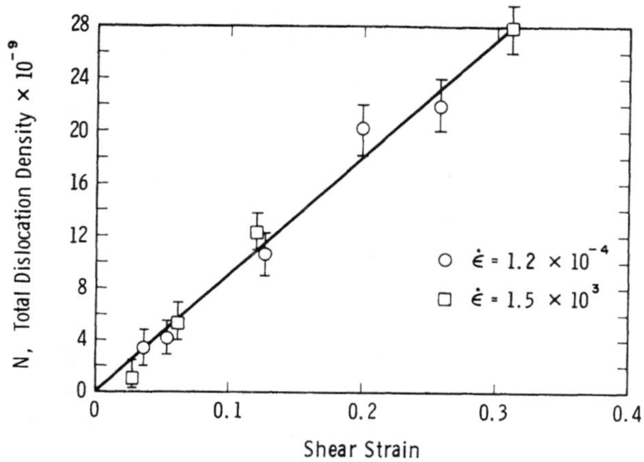

Fig. 21. The relationship between the plastic strain and the dislocation density.

et al. [25], on deformed molybdenum. However, in the present work measurements of dipole density at the low strain rate were underestimates because dipoles in tangles were not included because they could not be resolved. Consequently, it is concluded that the actual dipole density is probably not too different at the two strain rates.

DISCUSSION

One of the most interesting features of the mechanical property measurements is the two-stage relationship between the lower yield stress and the logarithm of the strain rate shown in Fig. 4. It is possible to analyze this relationship in terms of an activation volume, V^*, and an associated rate-controlling deformation process for each linear stage using current theories of thermally activated flow [3, 26]. With this approach one obtains $V^* \simeq 26b^3$ for the low strain rate range, subsequently designated range 1, and $V^* \simeq 5b^3$ for the high strain rate range, subsequently designated range 2. Thus range 1 corresponds generally to the high range reported by Mitchell, et al. [8], for single crystals of niobium and range B reported by Fourdeux and Wronski [7] for polycrystalline niobium. The present range 2 corresponds reasonably well with range 2 for iron [1, 3].

It would be dangerous to draw any definite conclusion as to the operative rate-controlling mechanisms for ranges 1 and 2 from the values of activation volume given above. In the present work we have plotted lower yield stress for our specimens, which is, in practice, the stress necessary to produce the Luders strain in the specimen. Figure 3 shows that the shape of the yield point, and consequently the strain at the end of the Luders elongation, varies with strain rate. As a result of this our flow stress values are not produced for constant strain. Furthermore, the more accurate method for determination of activation volumes involves rapid strain-rate changes during the test, and the analysis of the results requires that the dislocation substructure remains unchanged during the strain-rate change. However, as pointed out previously, the lower yield stress values in Fig. 4 are measured at different strains, i.e., different dislocation densities and configurations, and consequently the above values of activation volume cannot be considered to be accurate. In this it is somewhat surprising that there is such a well-defined linear relationship between lower yield stress and the logarithm of the strain rate. It should be further pointed out that, for the reasons mentioned previously, the presence of a yield point engenders difficulties in making distinctions between stages with similar slopes for the lower yield stress vs. logarithm strain rate relationship. In consequence, some of the small changes in slope of the lower yield stress vs. log $\dot{\epsilon}$ graph which have been observed in polycrystalline materials [4, 6, 7] cannot be regarded as significant evidence for the operation of different rate-controlling deformation mechanisms. In other work on single crystals [5, 8] which do not exhibit a yield point, small changes in slope of the τ_0 vs. log $\dot{\epsilon}$ curve have been observed that appear to be more reliable than the data on polycrystalline materials. However there is some evidence [8, 16, 27, 28] that the value of τ_0 in b.c.c. single crystals is sensitive to work hardening in the microstrain region. Consequently, small changes in slope of the τ_0 vs. log $\dot{\epsilon}$ curve may reflect the different microstrains at which τ_0 is measured through the strain-rate dependence of the microstrain work-hardening rate, rather than the operation of different rate-controlling mechanisms.

The large differences in slope of Regions 1 and 2 probably reflect a difference in deformation mechanism, particularly since the transmission electron micrographs indicate that the dislocation substructure is very different after deformation in ranges 1 and 2. Rosenfield and Hahn [1] have suggested that range 1 reflects the stress dependence of the edge dislocation velocity. However, in the present case the stress dependence of the dislocation velocity measured by Guberman [29] is much lower than the strain-rate dependence of the lower yield stress in range 1, although Guberman has shown that the dislocation velocity vs. stress relationship is dependent on specimen purity and so may be, in fact, parallel to the lower yield stress vs. log $\dot{\epsilon}$ for our particular material. However, both edge and screw dislocations are present in approximately equal densities in thin foils taken from the specimens deformed at a strain rate of 1.2×10^{-4} sec^{-1} and in the specimens of

Foxall, et al. [16]. It is concluded therefore that the screw and edge components are behaving in essentially the same way in range 1 and that it is improbable that the behavior of edge components is rate controlling. At present it is not clear what the rate-controlling mechanism for deformation is in range 1. Range 2 is considered by Conrad [3], Dorn and Rajnak [2], and Arsenault [30] to reflect a rate-controlling process which consists of overcoming the Peierls stress by the nucleation of double kinks on screw dislocations. This mechanism is consistent with the long heavily jogged dislocations observed after deformation at a strain rate of 1.5×10^3 sec^{-1}, and the absence of edge dislocations in these specimens.

It is notable that there is no stage 1 work hardening observed in these specimens, despite being oriented for single slip. As pointed out previously this would be expected if multiple slip occurred from the beginning of deformation, as evidenced by the constant orientation of the compression axis throughout deformation. The transmission electron micrographs support this contention since dislocations with Burgers vectors $\frac{a}{2}[\bar{1}11]$, $\frac{a}{2}[1\bar{1}1]$, $\frac{a}{2}[\overline{11}1]$, and $a[001]$ are observed even in the initial stage of deformation $\simeq 3\%$ shear strain at both the high and the low strain rates.

The work-hardening rate is low compared with that observed by other workers [16, 27, 31] for stage 2 in niobium for our specimens. However, this is probably caused by the high impurity content of our specimens since Mitchell, et al. [8] have shown that the work-hardening rate in stage 2 increases with increasing number of zone passes in the electron zone refining apparatus. The dislocation configurations observed in the specimens deformed at low strain rates are similar to those found on the primary slip plane in stage 2 by Foxall, et al. [16], Bowen, et al. [27], and Taylor and Christian [39], although there are some differences. For example, the present specimens show more evidence of secondary slip, i.e., secondary dislocations are not confined to dislocation tangles, see for example Fig. 6. This effect is probably due to the strong end effects exerted on the present specimens. A further difference is that dislocation multipoles have not been observed in the present specimens, neither have dislocation networks. It is not clear why dislocation networks do not occur; however the absence of multipoles is probably directly related to the absence of stage 1 hardening in the present tests. Foxall, et al. [16] have pointed out that the multipoles are similar to the loose arrangements of wide dipoles observed in stage 1, and consequently their occurrence is probably related to a process occurring during the transition from stage 1 to stage 2.

The dislocation arrangements in the specimens deformed at the high strain rate are similar to those observed at low temperatures in niobium [16, 31] except that again there is more secondary slip for the reason outlined above. As pointed out by Arsenault and Lawley [32] long straight dislocations and the absence of tangling and cell formation appears to be associated with a high effective stress τ^* and a low work-hardening rate, in agreement with the present results.

At both the high and the low strain rates the flow stress is a square root function of the total dislocation density and the primary dislocation density. However, the values for the slope of the line α are very different. Indeed, the value of α at low strain rates agrees reasonably well with that found by Taylor and Christian for niobium in compression but is only $\frac{1}{3}$ of that found in tension [33]. Thus, it appears that low α values are associated with compression tests.

Most work-hardening theories predict a square root relationship between the total and primary dislocation density and the flow stress [34] as found in the present work. Consequently, it is not possible to differentiate between work-hardening theories on this basis. The greater rate of work hardening at low strain rates is not caused by a higher dislocation density at the low strain rate because the dislocation density at a given strain is independent of strain rate, see Fig. 21. Rather it appears that the dislocation cell walls are more effective hardeners than the more homogeneous dislocation distribution encountered at the high strain rate.

It is interesting to note that the dislocation density at a given strain is independent of strain rate. Thus, it can be inferred that the multiplication rate with strain is independent of strain rate, and consequently the number of mobile dislocations is probably the same over the strain-rate range investigated, assuming that the same cross-slip multiplication processes occur at the high and low strain rates. This assumption is consistent with the substructure observed in the electron microscope.

CONCLUSIONS

1. The strain-rate dependence of the room temperature lower yield stress of single crystals of niobium in the strain-rate range 10^{-4} to 6.5×10^3 sec^{-1} can be described by two consecutive semilogarithmic relationships of the form

$$\tau_{\text{yield}} = \alpha + \beta \log \dot{\epsilon}$$

where ϵ is the applied strain rate and α and β are constant over each strain-rate range.

2. The dislocation arrangement after deformation at strain rates of 1.2×10^{-4} sec^{-1} and 1.5×10^3 sec^{-1} indicates that there is a different rate-controlling mechanism operative at low and high strain rates.

3. It is not clear what rate-controlling mechanism operates at low strain rates. At high strain rates the dislocation configurations are consistent with a Peierls mechanism.
4. The tangles and cell structure formed by deformation at a strain rate of 1.2×10^{-4} sec^{-1} is a more effective hardener than the uniform dislocation distribution formed at a strain rate of 1.5×10^3 sec^{-1}.
5. The relationship of the total, or primary, dislocation density N to the flow stress τ has the form

$$\tau = \tau_0 + \alpha\mu b\sqrt{N}$$

at both the high and the low strain rates.
6. There is a linear relationship between the total dislocation density and the plastic strain at high and low strain rates which is independent of strain rate.
7. The mobile dislocation density is independent of strain rate.

References

1. A. R. Rosenfield and G. T. Hahn, Trans. ASM, *56*, 962 (1966).
2. J. E. Dorn and S. Rajnak, Trans. AIME, *230*, 1052 (1964).
3. H. Conrad, J.I.S.I., *198*, 364 (1961).
4. R. N. Orava, Trans. AIME, *230*, 1614 (1964).
5. D. L. Davidson, U. S. Lindholm and L. M. Yeakley, Acta Met., *14*, 703 (1966).
6. G. A. Sargent, P. J. Sherwood and A. A. Johnson, Nature, London, *195*, 374 (1962).
7. A. Fourdeux and A. Wronski, Acta Met., *11*, 1270 (1963).
8. T. E. Mitchell, R. A. Foxall and P. B. Hirsch, Phil. Mag., *8*, 1895 (1963).
9. T. E. Mitchell and W. A. Spitzig, Acta Met., *13*, 1169 (1965).
10. J. M. Krafft and J. C. Hahn, U.S. Patent No. 3,194,062; 1965.
11. R. Stickler and R. J. Engle, J. Sci. Inst., *40*, 518 (1963).
12. F. C. Frank and J. F. Nicholas, Phil. Mag., *44*, 1213 (1953).
13. A. K. Head, Phys. Stat. Sol., *20*, 505 (1967).
14. A. S. Keh, "Direct Observations of Lattice Defects in Crystals," Interscience, New York, p. 231 (1961).
15. R. K. Ham, Phil. Mag., *6*, 1183 (1961).
16. R. A. Foxall, M. S. Duesbery and P. B. Hirsch, Canadian Journal of Physics, *45*, Part 2, 607 (1967).
17. C. N. Reid, Acta Met., *14*, 13 (1966).
18. A. S. Keh, "Direct Observations of Imperfections in Crystals," Interscience, New York, p. 213 (1962).
19. A. S. Keh and S. Weissman, "Electron Microscopy and Strength of Crystals," Interscience, New York, p. 231 (1963).
20. A. Lawley and H. L. Gaigher, Phil. Mag., *10*, 15 (1964).
21. J. W. Edington and R. E. Smallman, Acta Met., *12*, 1313 (1964).

22. H. Wiedersich, AIME J. Met., *16*, 425 (1964).
23. G. E. Hollox and R. E. Smallman, J.A.P., *37*, 818 (1966).
24. D. J. Bailey and W. F. Flanagan, Phil. Mag., *15*, 43 (1967).
25. A. Gilbert, B. A. Wilcox and G. T. Hahn, Phil. Mag., *12*, 649 (1965).
26. H. Conrad, "High Strength Materials," Ed. Zackay, Wiley, New York (1965).
27. D. K. Bowen, J. W. Christian and G. Taylor, Canadian Journal of Physics, *45*, 903 (1967).
28. A. S. Keh, Phil. Mag., *12*, 9 (1965).
29. H. D. Guberman, Report ORNL 4020, Oak Ridge National Laboratory, p. 41 (1966).
30. R. J. Arsenault, Acta Met., *14*, 831 (1966).
31. G. Taylor and J. W. Christian, Phil. Mag., *15*, 893 (1967).
32. R. J. Arsenault and A. Lawley, Phil. Mag., *15*, 549 (1967).
33. R. A. Foxall, Private Communication.
34. F. R. N. Nabarro, Z. S. Basinski and D. B. Holt, Adv. Phys., *13*, 193 (1964).

CONSTITUTIVE RELATIONSHIPS FROM IMPACT STUDIES

WILLIAM J. GILLICH

U.S. Army Ballistic Research Laboratories
Aberdeen Proving Ground, Maryland

ABSTRACT

It was discovered through free flight impact tests using specimens traveling at a constant velocity prior to impact that one form of the deformation curve for high purity aluminum and copper single crystals could be represented as a $\frac{1}{2}$ power law. This particular deformation law as found to be predictable using the stress and strain ratios of the aggregate theory of Taylor. These predictions were made on the basis of the deformation of polycrystalline metals.

A generalization of the $\frac{1}{2}$ power law constitutive relation is shown to account for uniaxial deformation that occurs for both single crystal and polycrystalline specimens subjected to static or dynamic deformations. This formulation is related to the parabolic type of constitutive development proposed by Bell.

INTRODUCTION

The phenomenological science of mechanics rests upon certain field postulates which take the form of balance or conservation equations. Conservation laws for linear momentum, moment of momentum, mass, and energy supply a rational basis for all continuum theories. These conservation laws, either as field equations or in the jump condition form, are, however, insufficient to yield specific answers to problems and we are left with an undetermined situation. The necessity of obtaining a determinate mathematical system leads to the so-called constitutive assumption. A functional relation between stress σ_{ij} and the motion $x_i = x_i(X_i, t)$ of the body is usually supplied in tensorial form. These equations, nine of which are needed in general, define a material. The postulates of mechanics including both the field equations or jump conditions and the constitutive assumption rest on a foundation of physical experience and experimental data. Forms of the constitutive function are usually assumed and a coherent mathematical theory is then developed from the balance conditions.

For example consider the motions permissible in the theory of linear infinitesimal elasticity where Hooke's law is the postulated constitutive assumption. For certain of these motions the field equations give a symmetric stress tensor and together with the constitutive equation give a hyperbolic equation of motion. D'Alembert's solution to this equation of motion $U_i = F_i(X_i\nu_i \pm Ct)$, where F_i are arbitrary functions and the ν_i are direction cosines, gives a wave solution of unchanging shape and of constant velocity C. Experiments in the propagation of elastic waves have shown this solution to be accurate to the extent that general agreement is obtained as to the form of Hooke's law. The dynamical portion of the theory of linear elasticity is now used as one of the main sources of establishing constitutive data where ultrasonic methods are employed. Wave velocities are measured and elastic constants are inferred.

Another example where a logical mathematical model consistent with and developable from classical field theory is the deformation of materials at very high pressures. A material is considered as a perfect fluid defined by a constitutive assumption known as the caloric equation of state which in practice is represented by so-called Hugoniot curves. Since very high pressures are most conveniently generated by the propagation of steady-state shock waves the conservation equations are now given as jump conditions instead of field equations. The nature of these jump conditions allows the determination of the Hugoniot curves through shock and particle velocity measurements. Constitutive data for solids under very high pressures developed from shock wave studies are found to be generally in good agreement throughout the literature and are considered representative of materials under the given loading conditions.

These examples are given to point out the historical fact that only through self-consistent dynamic theories have reproducible constitutive data been developed.

The alternative methods of developing constitutive data from direct measurements in homogeneous unidirectional deformation fields, both static and dynamic, are found to give results that in general disagree [1]. One of the difficulties with these approaches is that little theory exists to serve as a guide. As a consequence agreement on the proper variables to describe the deformation does not exist. Secondly the methods of measuring the proper variables directly, at least in the case of dynamic deformation, are difficult if not impossible. For instance in the case of dynamic plasticity there is no way to accurately measure stress and only a few accurate methods of strain measurement. Lack of standardization of materials and their state before the desired deformation, and lack of control of ambient conditions add to the confusion.

Fortunately, in the case of dynamic plasticity, there is recourse to a mathematical model which supplies a convenient way to substantiate the theory and to supply constitutive data. For deformation under a uniaxial stress condition the field equations supply a quasilinear hyperbolic equation of motion if the constitutive assumption is taken such that the engineering or 1st Piola Kirchhoff [2] stress is a single valued function of some deformation measure. This deformation measure can be interpreted as a finite longitudinal contraction or extension. To avoid discontinuities the slope of the deformation curve must be greater than zero, and the second derivative must be less than zero. The equation of motion has a simple wave solution $\dot{U} = f(\epsilon)$ (where $\dot{U}$ is the particle velocity and ϵ the finite strain) which becomes a centered simple wave for impact at constant velocity. This solution allows the determination of wave speeds from experimentally measured strain-time data obtained at various material points along specimens as well as the prediction of maximum values of deformation for given impact velocities. These predictions are easily tested and once they are established a specific constitutive equation can be determined. This one-dimensional theory was first devised by von Karman [3], Taylor [4] and Rakhmatulin [5] for semi-infinite specimens impacting at a constant velocity. This work shall be referred to as the finite amplitude wave theory of solids when used in its material coordinate form.

The work to be presented here originated from impact studies where the semi-infinite specimens were single crystals of high purity aluminum and where the finite amplitude wave theory was found to be the appropriate description of the deformation [6]. Experimentally substantiation of the theory was accomplished with the use of diffraction grating strain gauges [7]. Strain-time information obtained during passage of the wave front at various gauge positions supplied the necessary data. The constitutive equation that was found to govern the predominant type of deformation was a $\frac{1}{4}$ power law. This $\frac{1}{4}$ power deformation law was found to be related to the static polycrystalline tensile deformation using the stress and strain ratios of the aggregate theory of Taylor [8] and Bishop and Hill [9]. Another $\frac{1}{4}$ power deformation curve was developed from these impact studies which could not be accounted for through the aggregate ratios.

Studies of finite amplitude waves in high purity copper single crystals of the [111] orientation also revealed a $\frac{1}{4}$ power constitutive equation which could be related to polycrystalline tensile data. However the $\frac{1}{4}$ power type deformation was not found to be the predominant form of deformation law for copper.

The occurrence of $\frac{1}{4}$ power deformation curves from these two impact studies and similarity to a parabolic form of a constitutive equation proposed by Bell [10–12] lead to a general form of the $\frac{1}{4}$ power law.

This representation is shown to categorize certain aspects of uniaxial deformation for six different metals. The form of this law for uniaxial polycrystalline deformation is:

$$\sigma = \beta_0 \left(1 - \frac{T}{T_m}\right)^2 \epsilon^{1/4} \tag{1}$$

where σ is the nominal engineering stress, ϵ the nominal strain, T is the test temperature and T_m the melting temperature. Both temperatures are in degrees Kelvin. The constant β_0 takes the following form:

$$\beta_0 = (\tfrac{2}{3})^{r/4}\mu(0)M_0$$
$$M_0 = .0212, \quad r = 1, 2, 3 \cdots. \tag{2}$$

Here $\mu(0)$ is the isotropic elastic shear modulus $\mu(T)$ at $T = 0$, M_0 is a dimensionless constant and r is an integral index. The aggregate relation in the following form:

$$\frac{\sigma}{\tau} = \bar{m} = \frac{\gamma}{\epsilon}$$

where (3)

$$\bar{m} = 3.06$$

used in conjunction with (1) yields a deformation law in terms of resolved shear stress, τ and strain, γ for single crystals:

$$\tau = \beta_{r0} \left(1 - \frac{T}{T_m}\right)^2 \gamma^{1/4} \tag{4}$$

where

$$\beta_{r0} = \left(\frac{2}{3}\right)^{r/4} \mu(0) \frac{M_0}{\bar{m}^{5/4}}. \tag{5}$$

It was first discovered that the coefficient given by (5) of the $\frac{1}{4}$ power law representation took this form from the observation:

$$\beta_{r0} \text{ (parabola)} = \beta_{r0} \text{ } (\tfrac{1}{4} \text{ power law}). \tag{6}$$

The β_{r0} (parabola) is the coefficient used in Bell's [12] development of a general constitutive relation. Equation (6) first appeared in the single crystal work with aluminum [6]. The form of (5) was developed through this comparison.

The data that supports the form of the constitutive equations given as (1) and (4) are shown in Table 1. Table 2 supplies additional information for the tests referred to in Table 1. Figures 1 and 2 show two different methods employed to obtain the data of Table 1. Figure 1 gives a log-log determination of a power law of the form $\sigma = \beta\epsilon^a$. Both

Table 1

Material	T(°K)	r	$\mu(0)$(kg/mm^2)	β_{r0}(kg/mm^2) (Predicted)	β_{r0}(kg/mm^2) (Experimental)	Ref.
Al (99.99%)	300	3	3,110	12.01	11.86	a
	300	12		4.82	4.83	a
	300	12		4.82	4.88	b
	140, 195, 300	12		4.82	4.79	c
	300	16		3.21	3.35	d
Ag (99.97%)	299, 473	4	3,170	11.06	10.97	e
Cu (99.999%)	300	6	5,086	14.49	14.12	f
	273, 300, 473, 523	6		14.49	14.35	g
Mo (99.95%)	724, 814	13	15,400	21.59	21.65	h
Mild Steel (≈ .38% C)	300	2	8,420	36.07	36.46	i
Ta (99.84%)	591, 701, 813	2	7,010	29.95	29.55	j

a. W. J. Gillich, Phil. Mag. *15*, 659 (1967).
b. U. S. Lindholm and L. M. Yeakley, J. Mech. Phys. Solids *13*, 41 (1964).
c. R. P. Carreker and W. R. Hibbard, J. Metals *9*, 1157 (1957).
d. W. N. Sharpe, Ph.D. Thesis, The Johns Hopkins Univ., Baltimore, Md. (1956).
e. R. P. Carreker, J. Metals *9*, 112 (1957).
f. W. J. Gillich, unpublished (1967).
g. R. P. Carreker and W. R. Hibbard, Acta Metallurgica *1*, 654 (1953).
h. R. P. Carreker and R. W. Guard, J. Metals *8*, 178 (1956).
i. D. Tabor, Proc. Roy. Soc. A *192*, 247 (1948).
j. J. W. Pugh, Trans. of the ASM *48*, 677 (1956).

β and "a" may be determined from such plots. The static tensile data of Carreker and Hibbard [13] for copper are used as an example of this method. The stress and strain have been changed from the true stress and true strain originally given. Figure 2 is a σ^4 vs. ϵ plot used for the same purpose. The molybdenum data from static tensile tests of Carreker and Guard [14] changed to nominal stress and strain are plotted in Fig. 2. In practice these plots served only as a guide to determine which regions of the stress strain data were of the $\frac{1}{4}$ power law form. The constants of the equations were calculated by statistical methods over the predetermined regions. The predicted values in Table 1 are

Table 2

Material and Structure	Tm(°K)	Annealing Temperature (°K)	Grain Size (mm)	Specimen Diameter (mm)	Type of Test	Ref.
Al (99.99%) Fcc	933	–	Single Crystal	25.4	Impact	a
		–	Single Crystal	25.4	Impact	a
		–	Single Crystal	12.7	Split Hopkinson Bar	b
		724	.065	.762	Static Tension	c
		Chilled Cast	4.76	12.7	Static Tension	d
Ag (99.97%) Fcc	1234	974	.017	.508	Static Tension	c
Cu (99.999%) Fcc	1356	–	Single Crystal	25.4	Impact	f
		524	.012	.762	Static Tension	g
Mo (99.95%) Bcc	2898	1374	.18	.762	Static Tension	h
Mild Steel (≈ .38% C) Bcc	1808	–	–	–	Static Compression or Tension	i
Ta (99.84%) Bcc	3269	–	.036 to .051	.254 × 5.08	Static Tension	j

given by (5). Table 2 also supplies the values of the index r and the shear modulus $\mu(0)$.

Since the index r is allowed to take on the values 1, 2, 3, etc., it offers a series of constants, β_{r0}, for a given material. This distribution of β_{r0} in this series is such that as r increases, the values of β_{r0} get closer together. Consequently as a measure of error and to point out that the series of β_{r0} are not sufficiently dense so that any $\frac{1}{4}$ power law may fit this scheme, consider the aluminum data of Sharpe [15] of Table 2. The error from the predicted value is 4.3% while the difference from the values that would occur for $r = 15$ and 17 are 10.90% and 9.34% respectively. This example was chosen since the r of 16 is the largest that occurs in Table 1 and the error of 4.3% is also the greatest. The error in the predictions in all other cases is less than 2% where the distribution is less dense.

The type of test from which the deformation data of Table 1 were developed is indicated in Table 2. Single crystal deformation and polycrystalline deformation are related in the case of copper and alumi-

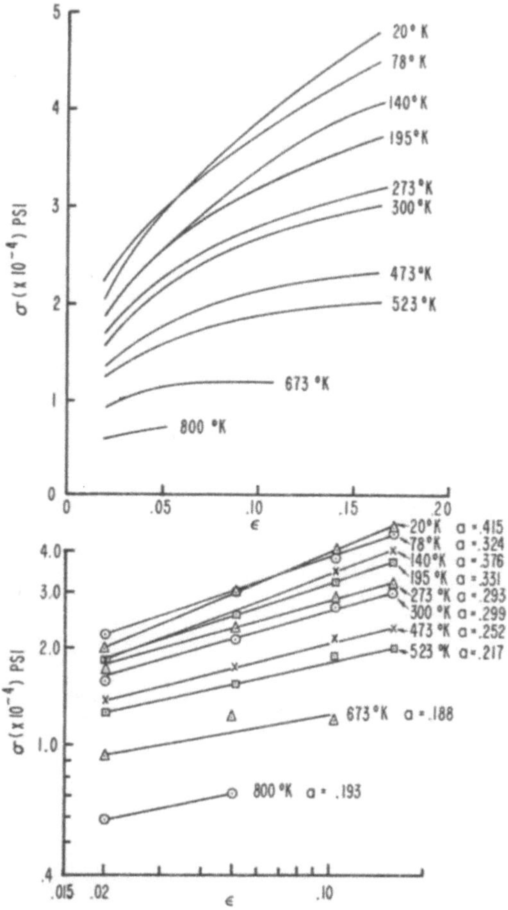

Fig. 1. Static tensile polycrystalline deformation curves for copper (Carraker and Hibbard).

num. Polycrystalline data alone are given for silver, molybdenum, mild steel and tantalum.

The $\frac{1}{4}$ power law developed as the predominant type of deformation from finite amplitude wave studies in high purity aluminum single crystals is represented by the $r = 12$ entry of Table 1. The $r = 3$ entry represents another $\frac{1}{4}$ power law that also occurred [6]. The split Hopkinson bar experiments of Lindholm and Yeakley [16] for high purity single crystals gives the $r = 12$ law over a limited portion of the imposed deformation. Table 3 gives the resolved data from the tests of Lindholm and Yeakley [16] and the range of nominal strain where agreement is obtained. Test number 1 agrees throughout the total

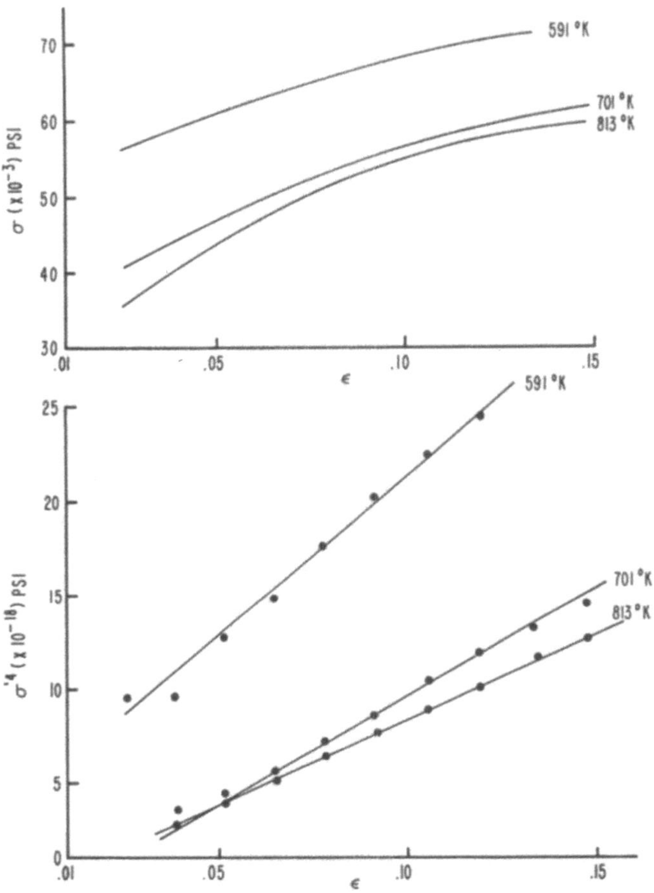

Fig. 2. Static tensile polycrystalline deformation curves for tantalum (Pugh).

deformation. The lack of correlation in these tests, throughout the indicated range, with the impact tests could be due to a short specimen size (12.7 mm × 12.7 mm) which would cause constrained lattice rotation. Fleischer [17] showed that two types of lattice rotations (Fig. 3) can occur during compression of single crystals. One rotation which occurs is opposite in sense from the one that occurs in tension where the specimen axes rotate on a great circle connecting the [101] direction and the specimen axes. The other rotation is where the specimen axes rotate towards the normal to the slip plane (the [111] direction). Fleischer's [17] experiments (Fig. 4) in static compression showed that one type rotation occurred near the constrained ends while the other occurred near the center of 40 mm long single crystals

Table 3

POWER LAW ($\tau = \beta_s \gamma^a$) FIT TO SPLIT HOPKINSON BAR ALUMINUM SINGLE CRYSTAL TESTS (LINDHOLM & YEAKLEY)

Test No.	a	$\Delta\epsilon(\%)$	$\beta_s(\mathrm{kg/mm^2})$
1	0.25	0.05–9.6	21.10
2		No Power Law Fit	
3	0.25	0.05–2.18	23.07
4	0.5	0.05–0.66	–
	0.25	0.81–9.1	26.23
5	0.25	0.04–0.60	19.13
6	0.5	0.35–1.2	–
		Ave.	22.38
		β_{r0} =	4.88 kg/mm²

of copper and aluminum that were 12 mm square. The single crystals used in the impact study were 254 mm long and no measurement was made within 25.4 mm or one diameter of the impact face.

Deformations that are described by $\frac{1}{4}$ power laws occur within a specific temperature range. This type of deformation is also sensitive to the initial state of the material. The studies cited for the polycrystalline tensile tests in Table 1 for aluminum, silver, copper and molybdenum were conducted for a variety of grain sizes and annealing temperatures. The $\frac{1}{4}$ power law is the appropriate description for the deformation of aluminum, tantalum, and molybdenum for the grain size and annealing temperatures given in Table 2 while its applicability is less obvious in the cases of copper and silver. Figures 5 and 6 show graphically the region where the power law exponent is $\frac{1}{4}$ for the grain

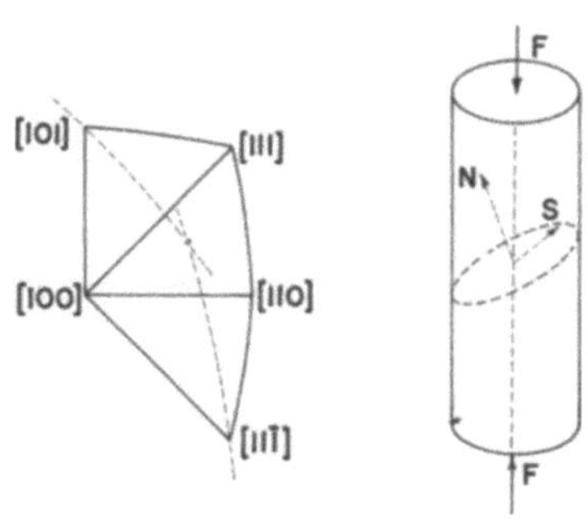

Fig. 3. Lattice rotation during compression.

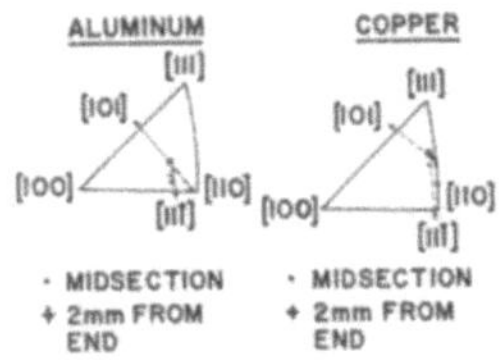

LATTICE ROTATION DURING COMPRESSION FOR COPPER AND ALUMINUM SINGLE CRYSTALS (FLEISCHER 1957) SPECIMAN SIZE 1.2cm X 4.0cm.

Fig. 4

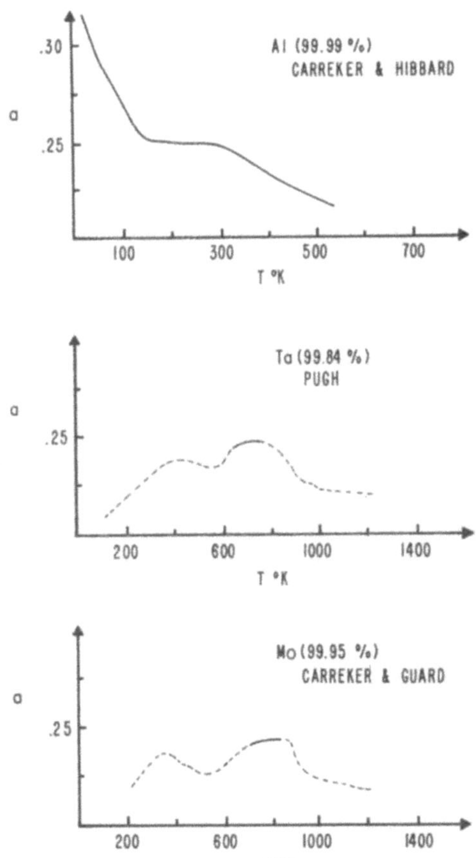

POWER LAW EXPONENT AS A FUNCTION OF TEMPERATURE

Fig. 5

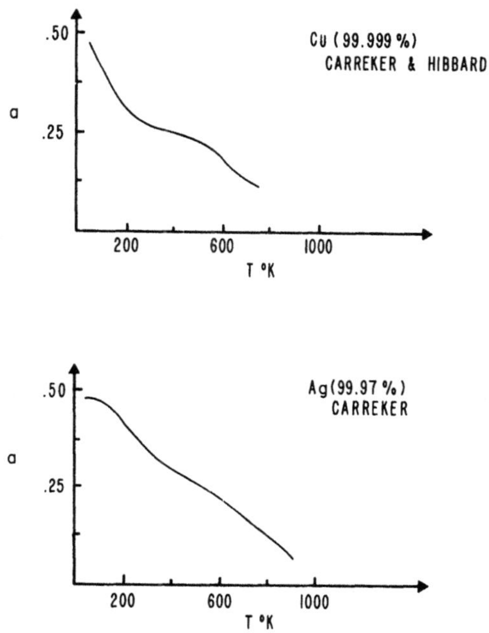

POWER LAW EXPONENT AS A FUNCTION OF TEMPERATURE

Fig. 6

size and annealing temperatures given in Table 2. The distinct plateau for aluminum at $a = 0.25$ and the maximum at this value for tantalum and molybdenum supply impetus for categorizing this type of polycrystalline deformation. The regions where $a = 0.25$ are not as distinct for copper and silver (Fig. 6). The solid line portions of the curves in Figs. 5 and 6 have been recalculated using nominal stress and strain while the broken line portions of these curves are displaced forms of the original data.

Instead of interpreting the $\frac{1}{4}$ power law as a deformation law it may be considered as a locus of a series of parabolic deformation curves where each parabolic segment is terminated by a transition of physical significance. The generalized treatment Bell [12] gives of deformation curves has the capability of handling forms of deformation which were previously described by power laws with a variety of exponents. For example, for the materials described in Table 1, Figs. 5 and 6 show a continuous variation of power law exponent with temperature. Bell [12] has shown that a formulation, with predictable transitions, and a single power law exponent ($a = 0.5$) can be used to describe the situation. Bell postulates that material stability is the rationale that supports this formulation. Considerable order is obtained when this description is applied in conjunction with

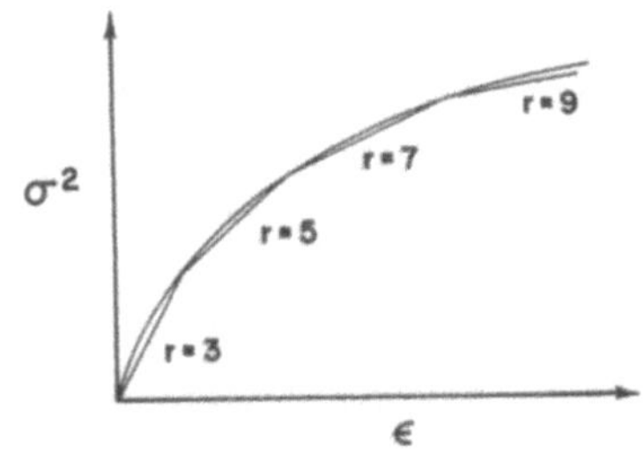

Fig. 7. Correspondence with parabolic law.

the $\frac{1}{4}$ power law. For the polycrystalline tensile data given in Table 1 Bell [12] gives a series of parabolic segments defined by the equations:

$$\sigma = \beta_0 \left(1 - \frac{T}{T_m}\right) \epsilon^{1/2} \tag{7}$$

$$\begin{aligned} \beta_0 &= (\tfrac{2}{3})^{r/2}\mu(0)B_0 \\ B_0 &= 0.0280, \quad r = 1, 2, 3 \cdots \end{aligned} \tag{8}$$

with the exception that r takes on only odd values. This situation is shown in Fig. 7 where squared stress is plotted against the strain. The parabolas of (7) appear as straight line segments and the $\frac{1}{4}$ power law of (1) appears as a limiting parabolic curve. The segments always meet at Bell's transition strains of $\epsilon = 1.45\%$, 4.25%, 7.5%, 12.5%, etc. Consequently the $\frac{1}{4}$ power law's usefulness in classifying deformation may also be extended to a study of the behavior of transitions from different deformation states as suggested from the work of Bell.

SUMMARY AND CONCLUSION

The constitutive equations of mechanics which are most universally accepted and from which general consistency is obtained have as a basis a theory of wave propagation. The theory offers a mathematical model formulated from either the field equations or jump condition form of balance laws for momentum, moment of momentum, mass and energy and a general constitutive assumption. Constitutive assumptions of the form of Hooke's law and the caloric equation of state were taken as examples where linearity and a specified functional dependence gave specific constitutive equations for a great number of real materials. These equations are established through inference from particle velocity and wave speed measurements.

The finite amplitude wave theory offers an analogous model for the development of constitutive equations for plastic deformation. This theory is used here to establish a form of a deformation law which is found to be related to tensile polycrystalline deformation. The discovery that the coefficient of one form of the proposed law is numerically equal to the coefficient of a law suggested by Bell allowed a more general form of the proposed law. This more general form of the law, which is a $\frac{1}{4}$ power law, was found to describe the deformation within a limited temperature range and for a given annealed state for six different metals. The $\frac{1}{4}$ power law not only offers an analytic description for certain types of deformation, but has a potential use in the study of transitions from various deformation states as proposed by Bell. Specifically the $\frac{1}{4}$ power law can be considered the bounding curve for tensile polycrystalline deformation which proceeds in a piecewise parabolic manner from different states with predictable transitions.

References

1. J. F. Bell, J. Mech. Phys. Solids *14*, 309 (1966).
2. C. Truesdell and W. Noll, Handbuch der Physik III/3, P124 (1964).
3. T. von Karman, N.D.R.C. Report A29 O.S.R.O., 365 (1942).
4. G. I. Taylor, British Official Report R.C., 329 (1942).
5. K. Rakhmatulin, Prikl. Mat. Mekh. *9*, 19 (1945).
6. W. J. Gillich, Phil. Mag. *15*, 659 (1967).
7. J. F. Bell, Proc. of the 3rd U.S. Natn. Congr. Appl. Mech., p. 489 (1958).
8. G. I. Taylor, J. Inst. Metals *62*, 30 (1938).
9. J. F. W. Bishop and R. Hill, Phil. Mag. *42*, 414 (1951).
10. J. F. Bell, Phil. Mag. *10*, 107 (1964).
11. J. F. Bell, Phil. Mag. *11*, 1135 (1965).
12. J. F. Bell (in preparation), Physics of Large Deformation of Crystalline Solids (Springer-Berlin).
13. R. P. Carreker and W. R. Hibbard, Acta Metallurgica *1*, 654 (1953).
14. R. P. Carreker and R. W. Guard, J. Metals *8*, 179 (1956).
15. W. Sharpe, Ph.D. Thesis, The Johns Hopkins Univ., Baltimore, Md. (1956).
16. U. S. Lindholm and L. M. Yeakley, J. Mech. Phys. Solids *13*, 41 (1964).
17. R. L. Fleischer, J. Mech. Phys. Solids *6*, 301 (1958).

PREDICTION OF ELASTIC-PLASTIC WAVE PROFILES IN ALUMINUM 1060-0 UNDER UNIAXIAL STRAIN LOADING

A. H. JONES, C. J. MAIDEN, S. J. GREEN, AND H. CHIN

General Motors Technical Center
Warren, Michigan

INTRODUCTION

In an ideally elastic-perfectly plastic material, in which the elastic moduli are constant, a high intensity wave propagating from the impact interface of two flat plates has a two-wave structure as shown in Fig. 1. The elastic wave propagates at velocity

$$\sqrt{\frac{\lambda + 2\mu}{\rho}} \tag{1}$$

with intensity

$$\frac{\lambda + 2\mu}{2\mu} Y \tag{2}$$

where Y is the yield stress of the material in a uniaxial stress test, λ and μ are Lame's constants, and ρ is the material density. This is followed by the higher intensity plastic wave travelling at a slower velocity

$$\sqrt{\frac{\lambda + (\frac{2}{3})\mu}{\rho}}. \tag{3}$$

In such a material the two-wave structure will prevail at all pressures. However, for a real material, the two-wave structure is modified by the

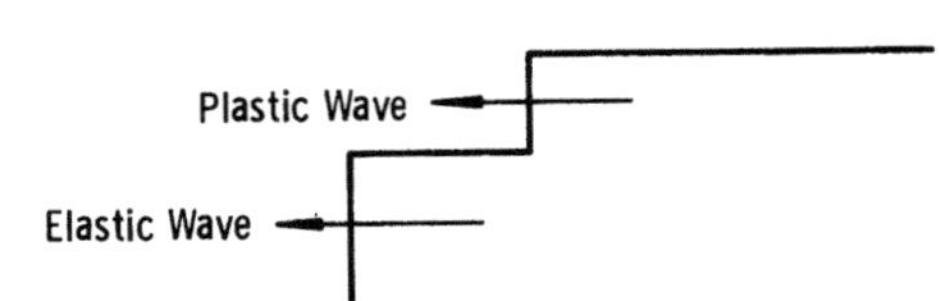

Fig. 1. Elastic plastic wave propagation.

increased stiffness with pressure. The increased stiffness will mean a higher plastic wave velocity for higher intensity waves and, eventually, the plastic wave will travel faster than the elastic wave.

In this paper we will consider the effect strain rate has in modifying the structure of the two waves shown in Fig. 1 in a pressure region where we can assume that the elastic moduli are constant. It is obvious, from inspection, that at the elastic and plastic wave fronts there exist high strain-rate regions. It is in these and associated relaxation regions that strain rate effects will influence the loading wave profile. Most dramatic, and the most easily analyzed effects lie at the elastic wave front. Specifically, we will consider wave profiles in aluminum alloy 1060-0 which has been tested at strain rates from 10^{-3} to 10^{3} sec^{-1} at the Material and Structures Laboratory of Manufacturing Development, General Motors Corporation [1]. The results of these tests and others will be used to predict the wave profiles from flat plate experiments, and these predictions will be compared with the experimental results published by Sandia Corporation [2].

UNIAXIAL STRESS EXPERIMENTS

In these experiments, aluminum 1060-0 specimens, $\frac{3}{8}$ inch in diameter, were machined from plate stock. A photomicrograph of the material, shown in Fig. 2, indicates it to be composed of large grains with preferred orientation normal to the direction of loading. Chemical composition and measured material properties are listed in Table 1.

Table 1

Chemical Analysis	
Aluminum	99.60 percent
Silicon	0.11
Iron	0.28
Copper	<0.01
Magnesium	<0.01
Zinc	<0.01
Density	= 2.706 gm/cm^3
Longitudinal Wave Velocity	= 6.33 km/sec
Shear Wave Velocity	= 3.14 km/sec

These specimens were tested in two separate instruments. A Medium Strain Rate Machine (a gas-operated device, similar to machines used by Clark and Wood [3] and Campbell and Marsh [4]), in which the specimen is compressed by a piston travelling at a controlled velocity, was used for determining the stress-strain curves at strain rates be-

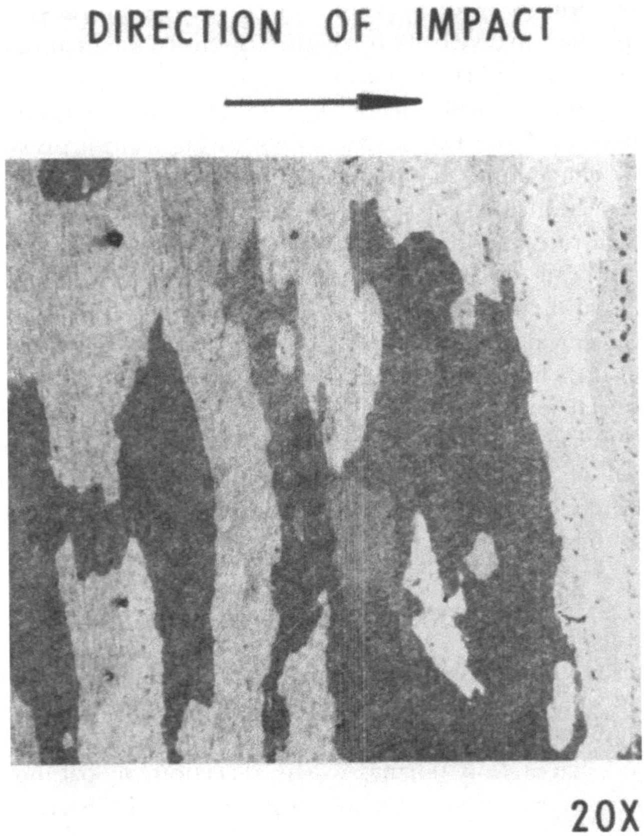

Fig. 2. Photomicrograph of the aluminum 1060-0 target material.

tween 0.001 sec^{-1} and 40 sec^{-1}. Strain was measured by an optical extensometer and stress by strain gages attached to an anvil bar. During a test, plastic strain rate varies at most by a factor of 3, and it is the average strain rate that is referred to in the results. A full description of the machine, the corrections made for machine stretch, the method of data reduction and specimen lubrication is given elsewhere [5].

A split Hopkinson bar apparatus was used to obtain stress-strain curves at strain rates between 10^2 sec^{-1} and 10^3 sec^{-1}. In this technique, stress-strain curves are obtained by considering the transmission of a stress wave through the specimen, which is sandwiched between two elastic bars [6, 7]. The method of analysis and discussion of variations in strain rate, stress and strain throughout the specimen are given in Reference 5.

Results from both pieces of equipment are plotted in Fig. 3 as engineering stress-strain curves at different average plastic strain rates.

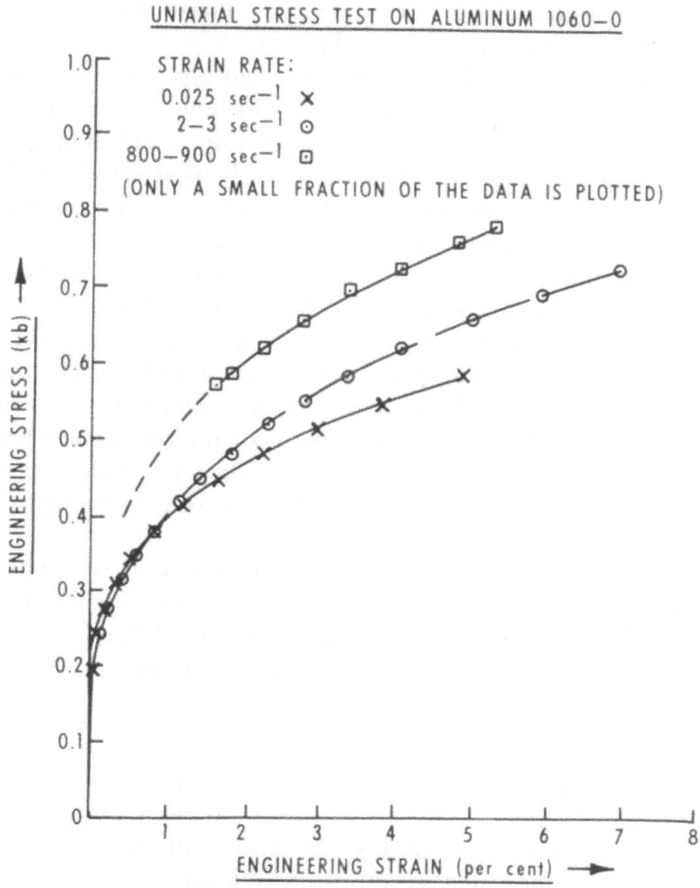

Fig. 3. Uniaxial stress strain curves at various strain rates for aluminum 1060-0.

Repeatability was within $\pm 1\frac{1}{2}\%$ for the Medium Strain Rate Machine and $\pm 5\%$ for the split Hopkinson bar.

For f.c.c. crystal metals, such as aluminum 1060-0, Seeger [8] has demonstrated that plastic flow is controlled by the thermally activated intersection of glide dislocations with forest dislocations, as long as the shear stress is not great enough to effect intersection instantaneously. Under these circumstances the applied shear stress alone is insufficient to allow continuous motion of the dislocations. The dislocations are momentarily arrested at barriers and their mean velocity is determined by the frequency of the thermal activation over such barriers. If the obstacles are localized and their energy is less than $50kT$, thermal activation assists the applied stress in moving dislocations over the obstacles and allows them to continue to sweep across the glide plane.

For activation energies greater than $50kT$, plastic flow is insensitive to temperature and does not depend on strain rate.

In the case of very high shear stresses, dislocations can cut through barriers instantaneously and continuously without the aid of thermal fluctuations, since these do not have enough time to aid the jog formation. As a result, the jogs are formed instantly by stress alone. In this region, Kumar [9] has suggested that the flow stress in aluminum is the stress required to overcome the force exerted by forest dislocation barriers. He visualizes this force as being composed of two parts, one is dislocation velocity independent and the other is proportional to the dislocation velocity. The velocity independent part is the force required to overcome long-range back stresses and cut through junctions of the forest dislocations. This force increases with increasing strain or dislocation density. The other component is a retarding force which is proportional to the velocity of the moving dislocation and arises due to various energy dissipative phonon and electron damping mechanisms [9].

Data, shown in Fig. 3 for 1060-0 aluminum, as well as results from other tests, have been compared with the Seeger [8] thermally activated model for describing plastic flow. From this model, the stress at constant strain can be presented by a logarithmic relation:

$$\begin{aligned} \sigma &= \sigma_A(\epsilon) + C(\epsilon) \log \dot{\epsilon}^p & \dot{\epsilon}^p &\geqslant \dot{\epsilon}_{\text{crit}} \\ \sigma &= \sigma_A(\epsilon) & \dot{\epsilon}^p &< \dot{\epsilon}_{\text{crit}} \end{aligned} \tag{4}$$

where σ is the flow stress and $\dot{\epsilon}^p$ the plastic strain rate. Since the theoretical foundation of this approach indicates that σ_A and C are sensitive to dislocation density and arrangement, the logarithmic dependence of flow stress on strain rate is expected only for a well-specified (cold-worked) structure. Also, where it is found that flow stress at constant strain follows these equations, the inference is that dislocation arrangement and density are insensitive to strain rate but depend on strain. This condition is known not to hold for the very high strain rates [10].

The comparison of theory and experiment is given in Fig. 4. It is seen that the experimental stress-strain dependence is consistent with a thermally activated process for strain rates in the range 0.03 sec^{-1} to over 40 sec^{-1}. At the higher plastic strain rate, the sensitivity increases as indicated by the higher slopes of the lines. This behavior is consistent with experiments reported for pure aluminum by Karnes and Ripperger [11], Ferguson, Kumar and Dorn [12], and Kumar [9]. Also, it is consistent with the concept described earlier, that at very high stresses plastic flow is influenced by viscous dislocation motion [9]. Unfor-

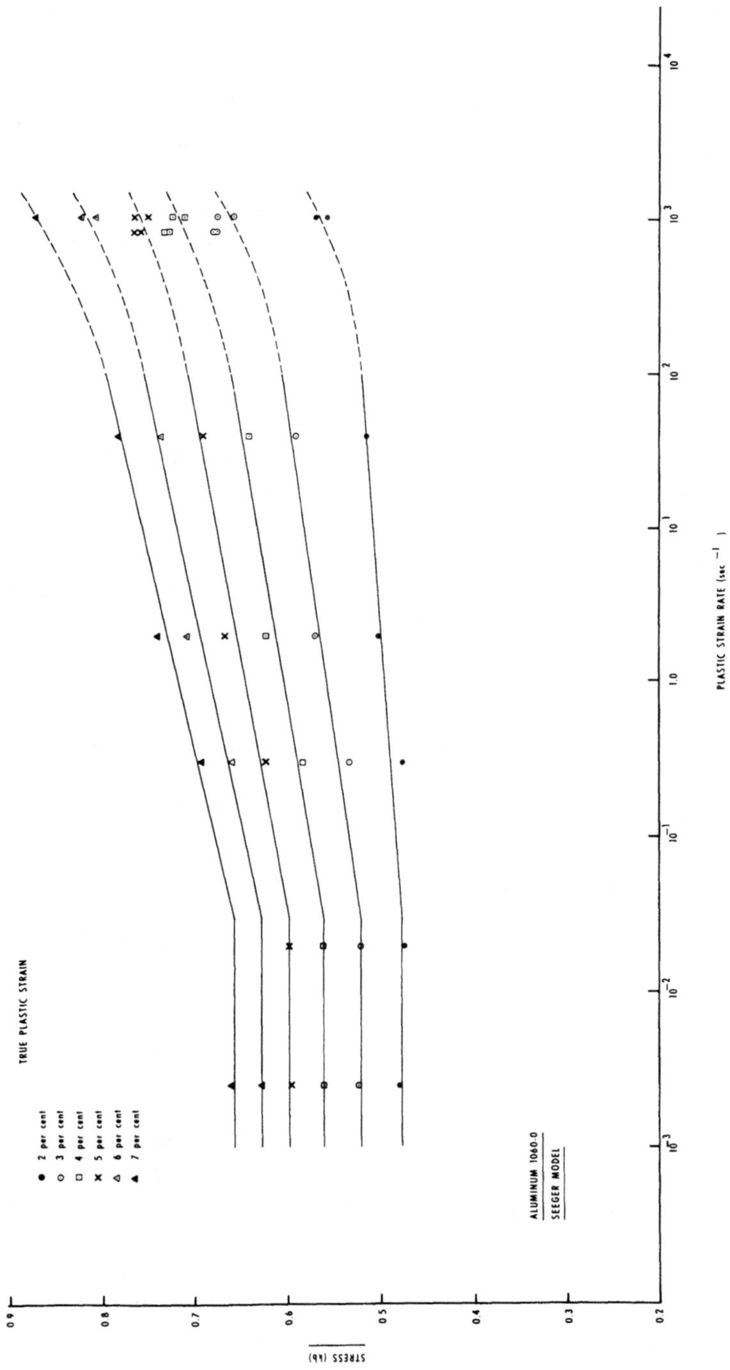

Fig. 4. Analysis of the uniaxial stress-strain-strain rate data for aluminum 1060-0.

tunately for the 1060-0 aluminum insufficient data exist to allow a more complete characterization of the linear-behavior region.

UNIAXIAL STRAIN ANALYSIS

Details of this analysis are as follows: Wave profiles are determined analytically by the approximate method of solving the non-linear partial differential equations using the finite difference technique. The equations for one-dimensional strain in rectangular coordinates are

$$\rho \frac{\partial u}{\partial t} = \frac{\partial \sigma}{\partial x} \tag{5}$$

$$\dot{\epsilon}_x = \frac{\partial u}{\partial x} = \frac{\dot{\rho}}{\rho} \tag{6}$$

where ρ is the density, u the particle velocity, σ the axial stress and the dot indicates differentiation with respect to time. These, combined with a constitutive equation and initial and boundary conditions, allow a calculation of the wave propagation.

The constitutive equations are based on Hookean elastic response, and incompressible plastic flow

$$\dot{\sigma} = (\lambda + 2\mu)\dot{\epsilon} - 2\mu\dot{\epsilon}^p \tag{7}$$

the stresses and strain being measured in the axial direction, together with a relation describing plastic strain rate in terms of stress and strain. Generalization of the uniaxial stress data to a uniaxial strain condition is based on the equivalence of plastic work. Thus, appropriate stresses and strains are obtained from the equivalent stress and equivalent plastic strain rate, $\bar{\sigma}$ and $\overline{d\alpha^p}$, respectively.

$$\bar{\sigma} = \sqrt{\tfrac{1}{2}[(\sigma_1 - \sigma_2)^2 + (\sigma_2 - \sigma_3)^2 + (\sigma_3 - \sigma_1)^2]}$$
$$\overline{d\alpha^p} = \tfrac{1}{3}\sqrt{2[(d\epsilon_1{}^p - d\epsilon_2{}^p)^2 + (d\epsilon_2{}^p - d\epsilon_3{}^p)^2 + (d\epsilon_3{}^p - d\epsilon_1{}^p)^2]} \tag{8}$$

where σ_1, σ_2, σ_3 are the principal stresses and $d\epsilon_1{}^p$, $d\epsilon_2{}^p$, $d\epsilon_3{}^p$ the associated plastic strain rates.

The finite difference form for equations 5 and 6 has been adequately described in the open literature [13]. The degree of approximation depends on the mesh size and artificial viscosity coefficient which are needed. The finer mesh size leads to better representation. Unfortunately, refining the mesh size increases computer times which can easily become prohibitive. Introducing artificial viscosity in the program is essential for simplicity since it eliminates shock wave discontinuities which complicate finite difference formulations. Visco-plastic behavior of the material does contribute some viscosity and there is a contribution inherent in the finite difference scheme. However, in the present calculations these were insufficient to dampen the oscillations to an acceptable level. For this reason, a small artificial viscosity has been introduced. Although this does not completely eliminate unwanted oscillations immediately behind the elastic precursor, they are reduced to an acceptable level. It should be pointed out that care must be exercised in the choice of viscous terms as too large an artificial viscosity would predict premature attenuation of the elastic precursor. Thus, as an independent check, the elastic precursor has been studied separately.

Following the method outlined by Duval [14] the precursor decay is governed by the equation

$$\frac{\partial \sigma}{\partial t} = -\mu \dot{\epsilon}^{p} \tag{9}$$

where μ is the modulus of rigidity and $\dot{\epsilon}^{p}$ the plastic rate of strain. We have relied on this method quite heavily to study the effect of various terms in the strain rate representation. However, for a more complete study of the effect of the rate term on the plastic flow, it is necessary to obtain the wave profiles in the plastic wave front.

DISCUSSION

Using the uniaxial strain analysis described in the previous section, together with various constitutive equations, wave profiles in 1060-0 aluminum have been calculated to compare with the experimental results presented in Reference 2. These results are for an initial impact stress of 2.6 kbars in the uniaxial strain configuration.

Using (4) to describe strain rate behavior of 1060-0 aluminum (for which σ_A and C were evaluated at yield by extrapolation of the analyzed test data), the elastic precursor attenuation is shown in Fig. 5. Attenuation is seen to be very rapid and is in disagreement with the experimental data reported by Karnes [2]. Thus, the Seeger [8] model predicts an almost instantaneous relaxation of the elastic overstrain to plastic strain, as might be expected from the uniaxial stress-strain rate behavior of this model (depicted in Fig. 6) which shows very little increase in flow stress with strain rate. Due to the rapid relaxation, the computed wave profiles shown in Fig. 7 are similar to that of an elastic-perfectly plastic material. Thus, apart from the small viscous effect introduced by the finite difference technique and artificial viscosity, the profile is the same as that shown schematically in Fig. 1.

Improved agreement with experimental results can be obtained with a model that predicts an increased rate sensitivity. In the higher strain-

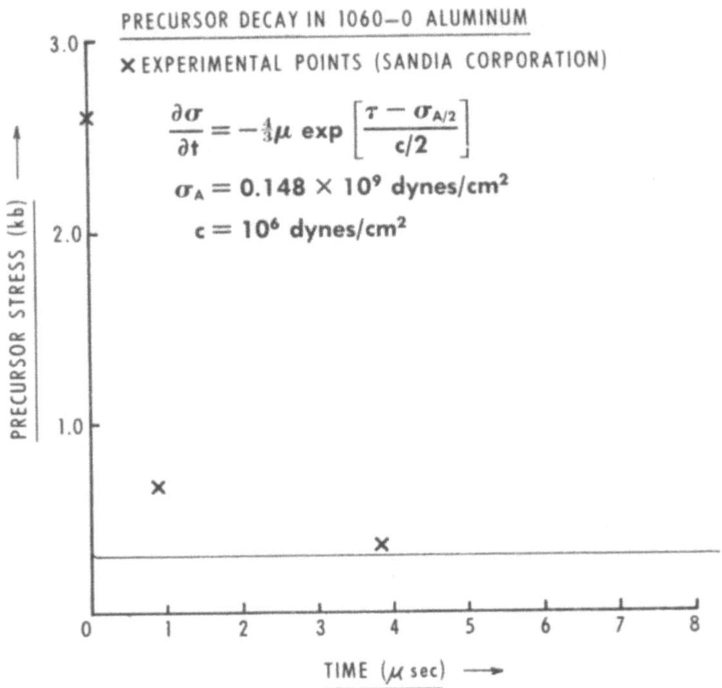

Fig. 5. Attenuation of the elastic precursor for the Seeger model.

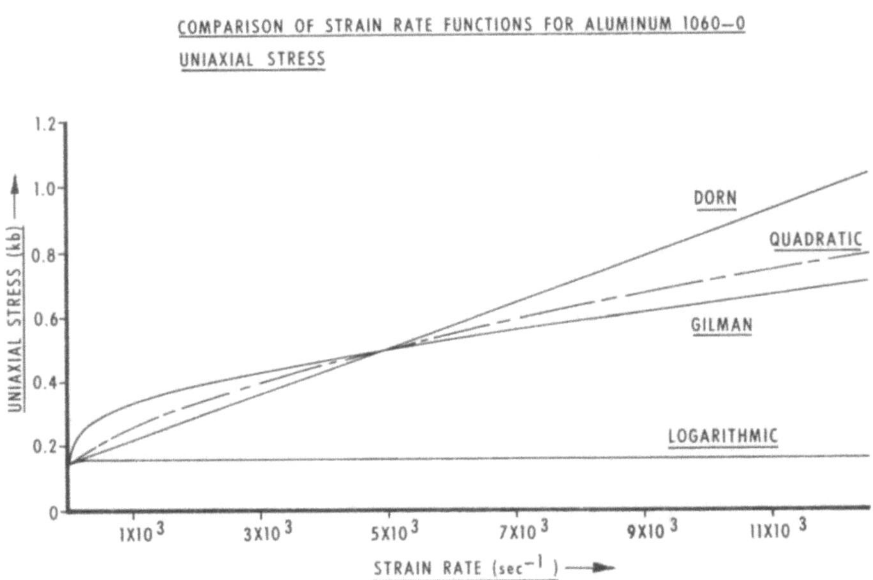

Fig. 6. Variation of the flow stress with strain rate for the various strain rate functions.

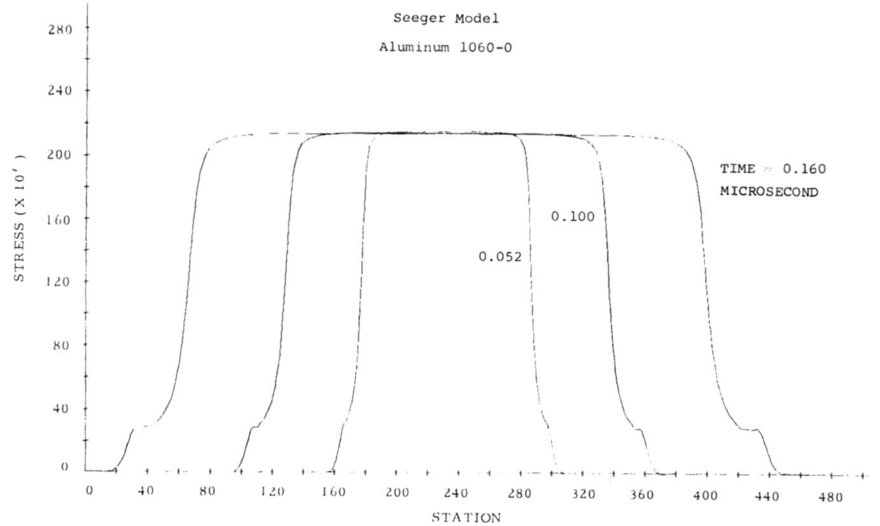

Fig. 7. Wave profile for flat plate impact test using the Seeger model.

rate region (above 10^3 sec^{-1}) where increased sensitivity was observed, we rely on data presented by Ferguson et al. [12]. For strain rates from 10^3 sec^{-1} to 2.6×10^4 sec^{-1} the empirical fit to their data is a relation of the form

$$\tau = \tau_0 + \alpha \dot{\gamma}^p \tag{10}$$

where τ is the applied shear stress and $\dot{\gamma}^p$ the plastic shear strain rate. Combining with (9), the attenuation of the elastic precursor is given by

$$\sigma = \left(\sigma_0 - \frac{Y}{D}\right) \exp\left[-\frac{4}{3}\mu \frac{Dt}{\alpha}\right] + \frac{Y}{D} \tag{11}$$

where

$$D = \frac{\mu}{\lambda + 2\mu}$$

and σ_0 is the initial stress assuming elastic impact. Using this solution, the predicted attenuation is shown in Fig. 8. The values of α used to fit the flat plate experimental results are 0.983×10^5 dynes/cm^2 and 0.472×10^5 dynes/cm^2 which are of the order of the value reported by Ferguson et al. [12], namely 10^4 to 10^5 dynes/cm^2. For the initial stress of $2.60kb$ a plastic strain rate of the order of 10^4 sec^{-1} is expected, i.e., in the range reported by Ferguson. Therefore, since the values of α are approximately the same, it can be assumed that a similar mechanism for plastic flow is occurring in both the uniaxial stress and uniaxial strain conditions. However, a single value of α will not fit the two data

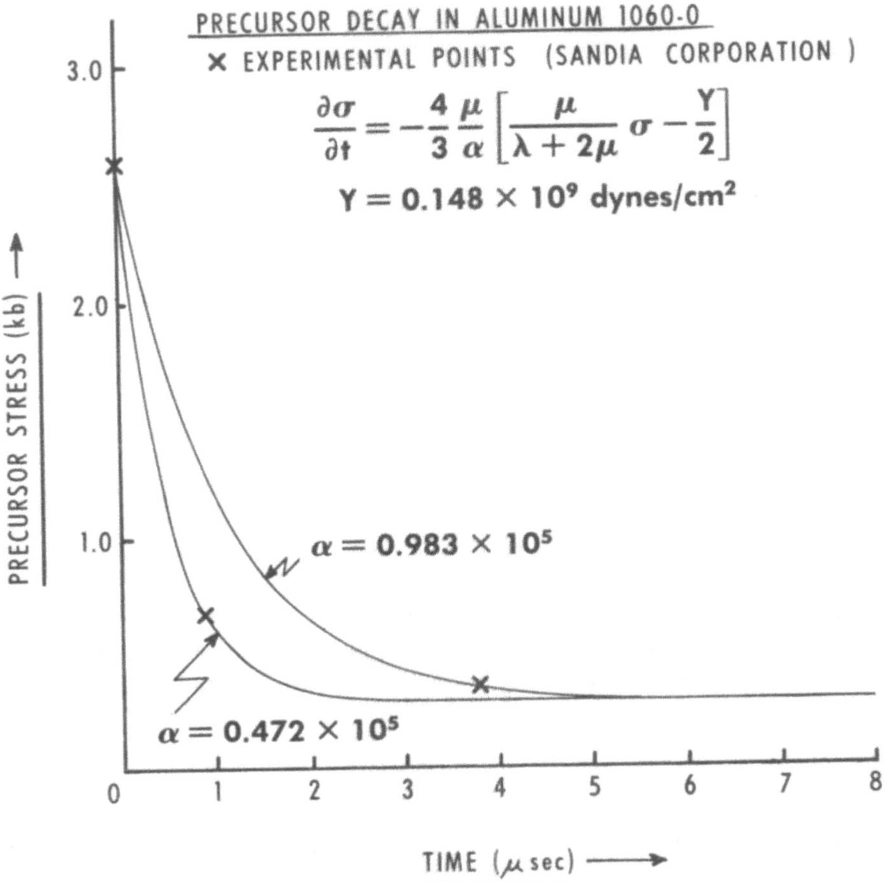

Fig. 8. Attenuation of the elastic precursor for the dislocation viscous damping model.

points reported. Note that the uniaxial stress-strain rate representation for the relation given by (10) is shown in Fig. 6 for $\alpha = 0.983 \times 10^5$ dynes/cm^2.

Better agreement can be obtained with the flat plate experiments for a model that predicts an even higher rate of sensitivity at lower strain rates. To incorporate this higher sensitivity, a possible stress-strain rate relation is given by

$$\dot{\gamma}^p = \frac{1}{\alpha}(\tau - \tau_0) + \frac{1}{\beta}(\tau - \tau_0)^2 \qquad (12)$$

where τ_0 is the static flow stress in shear.

Combining this equation with (9) leads to the solution of the elastic precursor attenuation

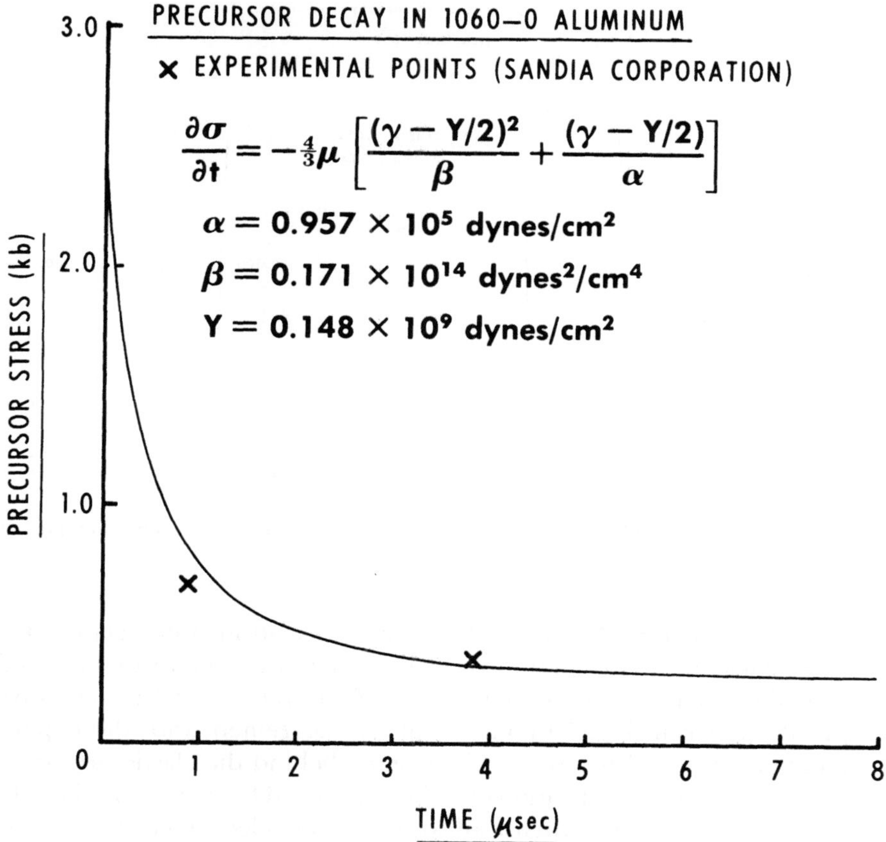

Fig. 9. Attenuation of the elastic precursor for the quadratic representation of the strain rate effects.

$$\frac{4}{3}\mu t = \frac{1}{\sqrt{b^2 - 4ac}}\left[\ln\left|\frac{2a\sigma + b - \sqrt{b^2 - 4ac}}{2a\sigma + b + \sqrt{b^2 - 4ac}}\right|\right]_{\sigma_0}^{\sigma} \tag{13}$$

for

$$b^2 - 4ac > 0$$

where

$$a = \frac{D^2}{\beta} \qquad b = \frac{2D\tau_0}{\beta} + \frac{D}{\alpha}$$

$$D = \frac{\mu}{\lambda + 2\mu} \qquad c = \frac{\tau_0^2}{\beta} - \frac{2\tau_0}{\alpha}.$$

Figure 9 depicts the best fit obtained for the flat plate impact data reported by Karnes [2] for $\alpha = 0.957 \times 10^5$ dynes/cm² and $\beta = 0.171 \times$

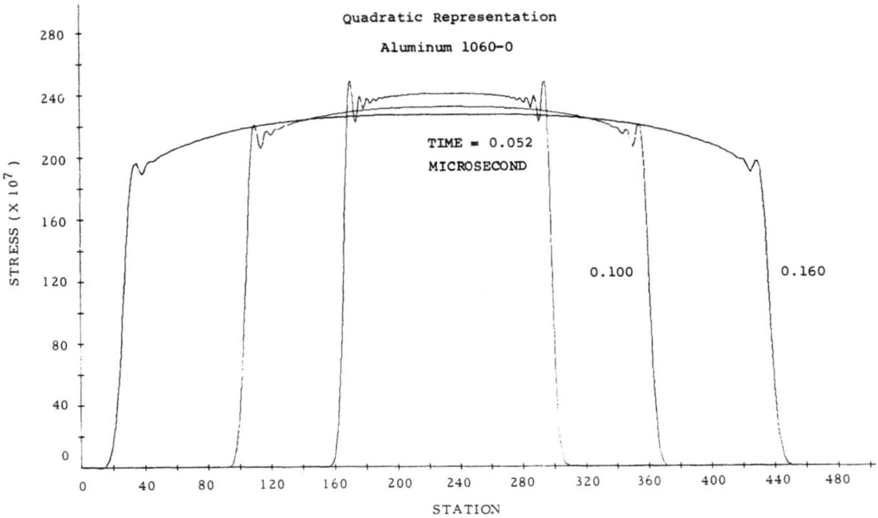

Fig. 10. Wave profile for flat impact testing using the quadratic representation of the strain rate effects.

10^{14} (dynes/cm^2)^2. (Figure 6 shows the stress-strain rate relation for these constants.) It is an improved agreement over the result obtained with the dislocation damping model. Moreover, the calculated wave profile shown in Fig. 10 gives qualitative agreement with the experimentally observed behavior. The ringing behind the elastic precursor is physically unrealistic and could be eliminated by increasing the artificial viscosity at the expense of losing the detailed description of the real profile. At later times the ringing disappears and we are left with a profile that momentarily supports a delayed yield which is in contrast to the behavior at lower strain rates (Fig. 3), but which is observed in the flat plate experiments at 1.9 μsec and 3.8 μsec after impact [2].

A further strain-rate sensitive constitutive relation due to Gilman [15] has been successfully used to analyze the decay of the elastic precursor in iron by Taylor [16]. This formulation stems from dislocation dynamics in which the dislocation velocity is given by

$$v = v_\infty \exp\left(-\frac{\tau_0}{\tau}\right) \tag{14}$$

where τ_0 is a constant and v_∞ is the shear wave velocity. Taking the mobile dislocation density to be constant at 10^7 cm^{-2}, the predicted elastic precursor decay is shown in Fig. 11 for $\tau_0 = 0.80 \times 10^9$ dynes/cm^2 and $\tau_0 = 0.65 \times 10^9$ dynes/cm^2. Also, the predicted wave profile derived with the Gilman model is shown in Fig. 12. Initially the model supports a delayed yield but after a short time it appears that it would

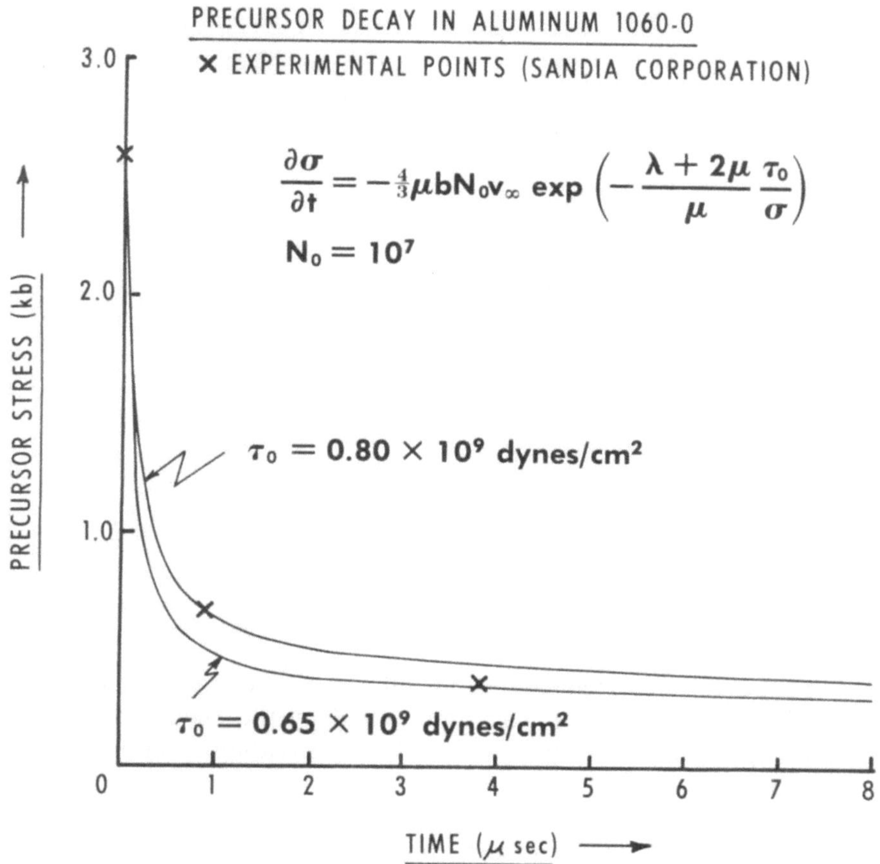

Fig. 11. Attenuation of the elastic precursor for the Gilman model.

no longer be present, and would therefore be in contradiction to experimental observation. Note that Fig. 6 shows the Gilman stress-strain rate relation for $\tau_0 = 0.80 \times 10^9$ dynes/cm^2.

In summary, of the constitutive equations considered, the best agreement with the experimental data for 1060-0 aluminum is obtained with (12). Also, it is evident that we have only skimmed the surface in the study of dynamic plastic deformation. Further work is planned, both experimental and theoretical, to improve the understanding of the plastic deformation process at high strain rates. Theoretical efforts will be directed to obtain wave profiles at later times that can be compared directly with experimental results, while further experiments will be carried out at higher strain rates and for different thickness targets.

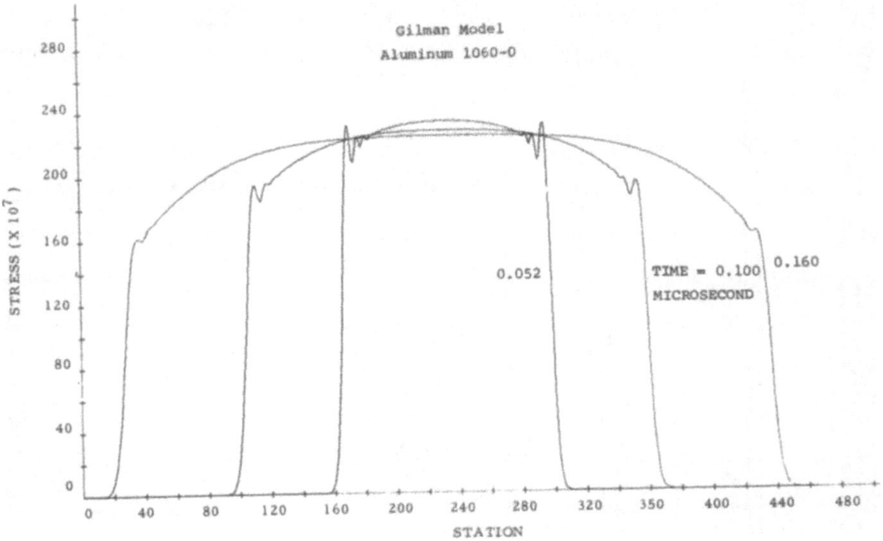

Fig. 12. Wave profile for flat plate impact using the Gilman model.

CONCLUSIONS

1. For strain rates from 0.03 sec^{-1} to 40 sec^{-1} the behavior of aluminum alloy 1060-0 is reasonably represented by the Seeger model. At higher strain rates, the material shows increasing strain-rate sensitivity.
2. The best representation of the flat plate impact results for an impact stress of 2.6 kb in the uniaxial strain configuration are obtained by assuming the plastic strain rate is given by

$$\dot{\gamma}^{p} = \frac{1}{\alpha}(\tau - \tau_0) + \frac{1}{\beta}(\tau - \tau_0)^2.$$

ACKNOWLEDGMENT

The authors wish to express their appreciation to Dr. A. Kumar for the discussions and criticism on the work presented and to Mr. R. Perkins and F. Schierloh for carrying out extra strain-rate tests.

References

1. D. L. Holt, S. G. Babcock, S. J. Green and C. J. Maiden, Trans. ASM, *60* (1967).
2. C. H. Karnes, "Symposium on the Mechanical Behavior of Materials Under Dynamic Loads," San Antonio, Texas, September 1967.
3. D. S. Clark and D. S. Wood, Proc. A.S.T.M., *49* (1949).

4. J. D. Campbell and K. J. Marsh, Phil. Mag., *7* (1962).
5. C. J. Maiden and S. J. Green, J. Appl. Mech., *33* (1966).
6. C. J. Maiden and J. D. Campbell, Phil. Mag., *3* (1958).
7. H. Kolsky, Proc. Phys. Soc., Series 3, *62* (1949).
8. A. Seeger, in *Dislocations and Mechanical Properties of Crystals*, Wiley and Sons, N.Y. (1956).
9. A. Kumar, Ph.D. Thesis, College of Engineering, University of California, Berkeley (1967).
10. S. Yoshida and N. Nagata, Trans. J.I.M., *8* (1967).
11. C. H. Karnes and E. A. Ripperger, J. Mech. Phys. Solids, *14* (1966).
12. W. G. Ferguson, A. Kumar and J. E. Dorn, J. Appl. Phys., *38* (1967).
13. M. L. Wilkins, *Methods of Computational Physics*, Academic Press, Inc., New York (1964).
14. G. E. Duvall, IUTAM Symposium on Stress Waves in Anelastic Solids, held at Brown University, April 1963, Springer-Verlag (1964).
15. J. J. Gilman and N. G. Johnston, in *Dislocation and Mechanical Properties of Crystals*, John Wiley and Sons, Inc., New York (1957).
16. J. W. Taylor, J. Appl. Phys., *36* (1965).

THE PLATE IMPACT CONFIGURATION FOR DETERMINING MECHANICAL PROPERTIES OF MATERIALS AT HIGH STRAIN RATES *

CHARLES H. KARNES

Sandia Corporation
Albuquerque, New Mexico

ABSTRACT

A review of the latest free surface motion and stress measuring instrumentation is presented along with a description of the plate impact one-dimensional strain configuration for determining mechanical properties under stress wave propagation conditions. A discussion is included of results of some other investigators who have used this configuration to determine dynamic yield stress and the degree of strain rate sensitivity.

INTRODUCTION

Most of the effort being expended to determine the dynamic mechanical properties of materials is concentrated in the area of impact and subsequent plastic wave propagation in long rods and wires or in configurations such as the split Hopkinson pressure bar. These configurations can be made to approximate the condition of one-dimensional stress when the times of observations are tens or hundreds of microseconds, or even milliseconds. Many investigators have not considered the use of the one-dimensional strain conditions of plate impact experiments to investigate the dynamic mechanical properties of materials. That type of experiment has been used primarily by solid state physicists to study electrical properties or to determine high pressure thermodynamic equations of state. The plate impact experiment is a very powerful method of studying the mechanical properties of solids under wave propagation conditions.

The advantages of this experiment for determining mechanical properties are: 1) measurements are unaffected by geometry; 2) an

* This work was supported by the Atomic Energy Commission.

exact, well-defined and non-changing state of strain is achieved; 3) very short time observations are made; 4) extremely high strain rates are achieved; and 5) a large amount of information is obtained from each experiment.

The disadvantages of this experiment are: 1) very high precision measurements are required; 2) one must make measurements of indirect quantities; 3) observation times are limited; 4) large samples are usually required; and 5) the cost per experiment is high.

The purposes of this paper are: 1) to describe the details of the experimental and analytical techniques that are involved in determining dynamic mechanical properties from plate impact experiments and 2) to present some of the results of previous investigators who have used the technique to study dynamic yielding and strain rate effects in metals. It is not intended to be a complete review article.

STATE OF STRESS AND STRAIN

In using the plate impact experiment, it is more convenient and more accurate to make use of the symmetric impact conditions that exist when the projectile and target plates are of the same material. For this situation, the impact conditions are well defined since the particle velocity at the impact surface is precisely one-half the projectile velocity.

Figure 1 is a cross-section of a projectile and target plate a short time after impact. The laboratory coordinate system is defined such that the stress wave resulting from impact propagates in the $+x$ direction, and the y and z directions are perpendicular to it and oriented arbitrarily. The dimensions of the plate are chosen so that the rarefaction waves from the lateral edges do not arrive at the interior of the target plate until after the measurements are complete.

Figure 1 shows qualitatively that the motion of a particle on a macroscopic scale is in the x-direction until the rarefaction waves have had time to propagate from the edges to the particle in question and produce a component of velocity in the y or z directions. Before that occurs, the strain components in the y and z directions are exactly zero in a homogeneous material, and the x component of strain is equal to the volumetric strain, i.e.,

$$\epsilon_{xx} = \frac{l_0 - l}{l_0} = \frac{V_0 - V}{V_0} = 1 - \frac{\rho_0}{\rho} \tag{1}$$

where l, V, and ρ represent length, specific volume, and density of an element, respectively, and the zero subscripts represent initial values.

Although the state of strain is very simple, the stress state is three di-

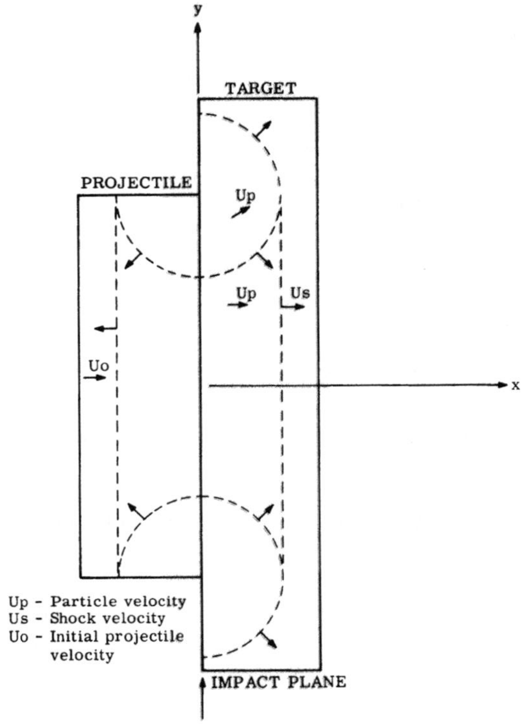

Fig. 1. Schematic showing location of compression and rarefaction waves after impact.

mensional. The lateral stress components are equal for a material exhibiting transverse isotropy. In order to describe the stress state in more detail, it is convenient to consider the stress-strain behavior as determined by a conventional one-dimensional stress compressive test. In the one-dimensional stress state, the stress and strain are denoted by Y and α, respectively. If Y is a known function of the plastic work W_p, the question is, How does the material yield and what is the resulting stress-strain path in one-dimensional strain? In 1952, Wood [1] showed for one-dimensional strain that the stress in the direction of propagation, when broken into spherical and deviatoric components, is of the form

$$\sigma_{xx} = K\epsilon_{xx} + \tfrac{2}{3}Y(W_p) \tag{2}$$

where $K\epsilon_{xx}$ is the spherical or hydrostatic component * and $\frac{2}{3}Y(W_p)$ is the deviatoric component.† The derivation of (2) involves the following assumptions: 1) the total strain is the elastic plus the plastic com-

* K is the adiabatic bulk modulus.

† Wood actually expressed Y as a function of plastic strain, but for one-dimensional strain, that is equivalent to expressing it in terms of plastic work [2].

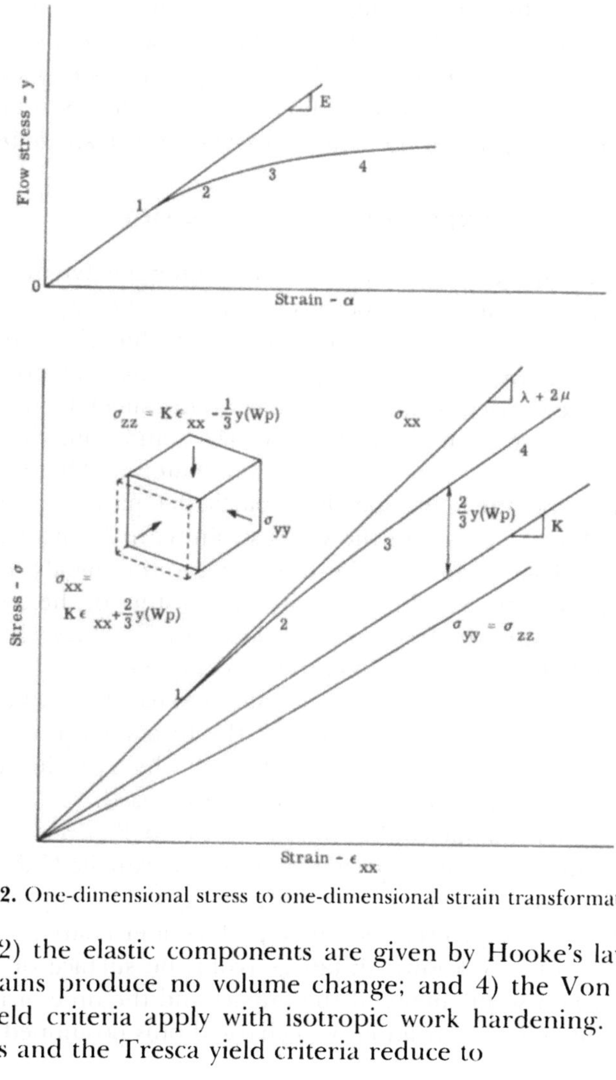

Fig. 2. One-dimensional stress to one-dimensional strain transformation.

ponents; 2) the elastic components are given by Hooke's law; 3) the plastic strains produce no volume change; and 4) the Von Mises or Tresca yield criteria apply with isotropic work hardening. Both the Von Mises and the Tresca yield criteria reduce to

$$\sigma_{xx} - \sigma_{yy} = \pm Y(W_p) \tag{3}$$

for the one-dimensional strain state.

In order to complete the transformation of the stress-strain path from the one-dimensional stress state to the one-dimensional strain state, one must know the relationship between ϵ_{xx} and α for "equivalent" conditions. If one assumes that equal plastic work defines equivalent conditions and that the bulk modulus K is independent of mean pressure, then Fowles [2] has shown that the total strain ϵ_{xx} is given by

$$\epsilon_{xx} = \tfrac{3}{2}\alpha - \frac{Y(W_p)}{6K}. \tag{4}$$

The procedure in transforming the stress-strain path is to select an α and a corresponding $Y(W)_p$, calculate the equivalent ϵ_{xx} from (4), and compute the corresponding σ_{xx} from (2). Typical stress-strain paths for the two states with corresponding points are shown in Fig. 2 along with an element showing the resultant stress and strain states.

EXPERIMENTAL TECHNIQUE

The plate impact experiment for determining the dynamic mechanical properties of materials proceeds as follows. The impact of two flat, smooth plates made of identical material produces a change in the particle velocity of the stationary plate which is instantaneous and uniform over the impacting surface. The resulting shock front is unstable if the stress-strain path is concave downward and immediately begins to spread as the wave propagates into the material. The details of the wave shape and the wave speeds of particular parts of the wave are determined by the details of the stress-strain path. As the wave reflects off the back free surface of the target plate, the details of the wave shape determine the resulting velocity-time history of the free surface. If one can measure the free surface velocity-time history with sufficient precision relative to impact time, then a complete analysis of the wave propagation problem, including the interactions which take place as a result of the waves being reflected off the free surface back into the plate, results in the average stress-strain path which produced the wave shape for that particular thickness of material used.

In the experimental configuration shown in Fig. 3, the projectile plate is bonded to a long projectile body and is propelled into the target plate at the desired velocity by compressed gas. The impact velocity is determined by the projectile's shorting of slender charged pins which protrude accurately known distances from the surface of the target plate. The non-simultaneity of the impact and the time of impact are both determined by four additional charged pins positioned flush with the impact surface.

The required precision is obtained only when exceptional care is used in preparing the targets. The target and the projectile nose surfaces have a roughness of from 0.02 to 0.1 micron (1 to 4 μin.) rms and are flat to within 0.3 micron (12 μin.) over the diameter of the impacting surface for impact velocities of the order of 100 fps. The finish and flatness requirement can be relaxed somewhat for higher velocities. The minimum tilt or non-simultaneity of impact that can be obtained averages about 10^{-4} rad. for 100 fps to 5×10^{-4} rad. for 2000 fps projectile velocities. The projectile nose plates are supported in such a way that they are not distorted by being accelerated down the barrel.

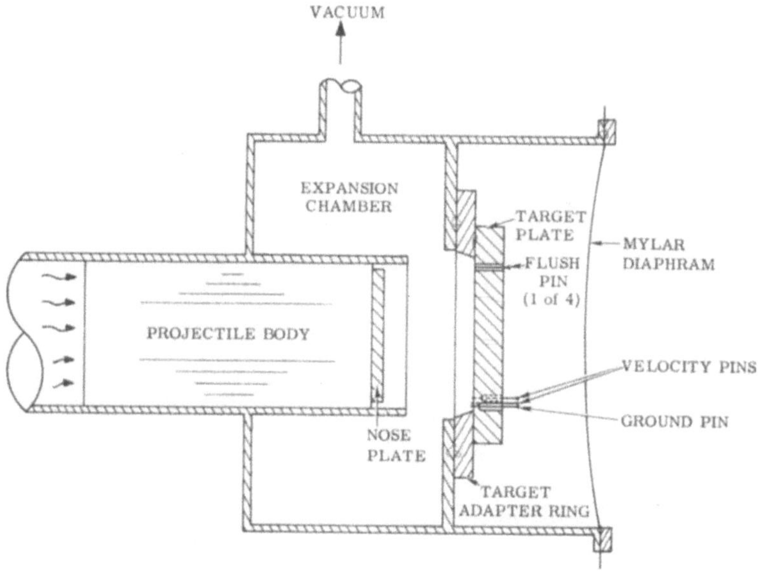

Fig. 3. Schematic showing orientation of target plate and projectile just before impact.

The latest developments in instrumentation to measure accurately the motion of the free surface as a result of its interaction with the oncoming wave profile are the standard Michelson displacement interferometer adapted for this purpose by Barker and Hollenbach [3] and the velocity interferometer also developed by Barker [4] and Hollenbach. A schematic of the Michelson interferometer is shown in Fig. 4. The free surface of the target plate is one mirror of the interferometer. Each time the surface moves one-half wave length of the source light, the photomultiplier observes one complete cycle from constructive to destructive to constructive interference. The observed frequency of the fringes is proportional to the free surface velocity and is approximately $10^6 HZ$ for each foot per second using a 6328 Å light source. The present limit on free surface velocity amplitude is about 700 fps due to photomultiplier frequency response.

The velocity interferometer (Fig. 5) overcomes the limitation on maximum free surface velocity by optically differentiating the free surface motion so that each fringe represents a change in velocity rather than a change in position. The photomultiplier views the surface by two light paths: path (A) is directly through the beam splitters and path (B) is around the delay leg. The beam passing directly through the beam splitters is of a wave length determined by the velocity of the free surface (by the Doppler shift) at time t. The beam just returning to the beam splitter from the delay leg is of a wave length determined by the free surface velocity at time $t-\tau$, where τ is the time

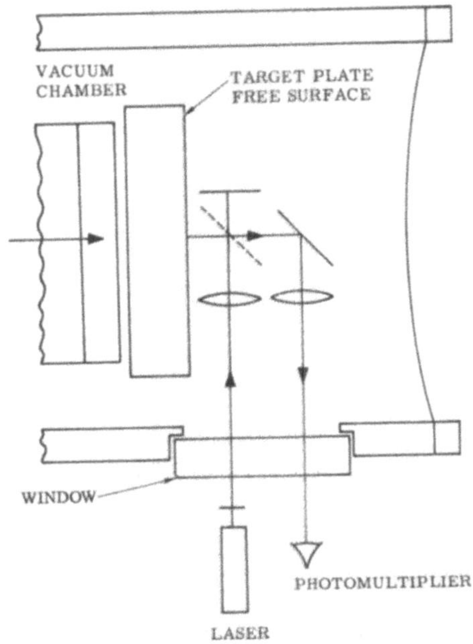

Fig. 4. Schematic of Michelson interferometer (from ref. 3).

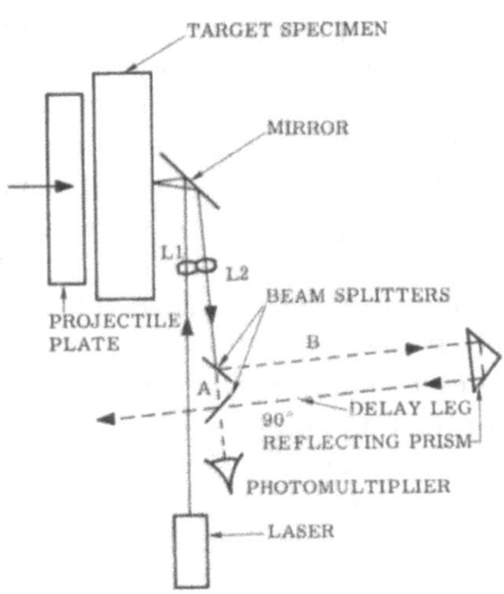

Fig. 5. Schematic of velocity interferometer (from ref. 4).

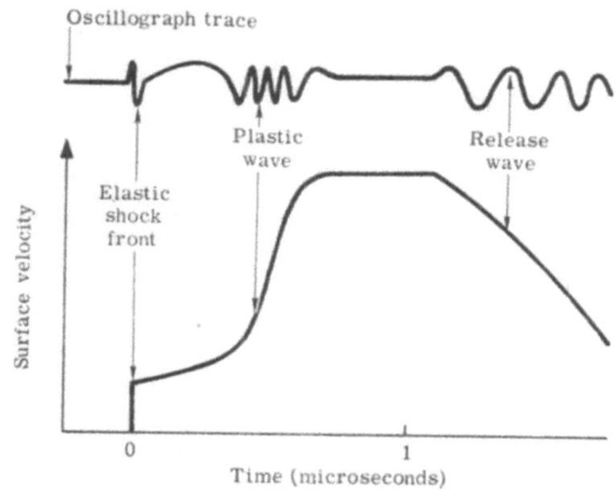

Fig. 6. Free surface velocity history and corresponding oscillograph trace of photomultiplier output (from ref. 4).

required for light to travel around the delay leg. The two beams mix, and the resulting beat frequency is proportional to the free surface acceleration averaged over the delay time τ. The time integral of the beat frequency is just the number of fringes observed and is proportional to the change in free surface velocity. Barker [4] has shown that

$$v(t) = \frac{\lambda}{2\tau} N(t) \tag{5}$$

where $v(t)$ is the free surface velocity at time t (averaged over time τ), λ is the light source wave length, and $N(t)$ is the number of fringes counted at time t. The coefficient, $\lambda/2\tau$, which is the change in velocity corresponding to one complete fringe, can be varied at will. For most current applications, τ is of the order of 0.010 μsec, this gives a value for $\lambda/2\tau$ of about 110 fps which is known to within 0.2 percent. The time resolution of both interferometer systems is approximately 0.002 μsec. Schematic of a free surface velocity-time history and the corresponding photomultiplier output from the velocity interferometer are shown in Fig. 6.

Another transducer, the "Sandia quartz gage," which was developed in the last five years by Graham, Neilson, and Benedick [5], enables one to measure stress-time history directly with a time resolution of the order of 0.010 μsec. As shown in Fig. 7, the Sandia quartz gage is used to monitor the stress-time history at the interface between the quartz and a target plate after the stress wave has traversed the target.* As the stress wave enters the quartz, the electrical current generated by the quartz is proportional to the interface stress and is given by the relationship

* The Sandia quartz gage can also be used to monitor the stress-time history at the impact surface.

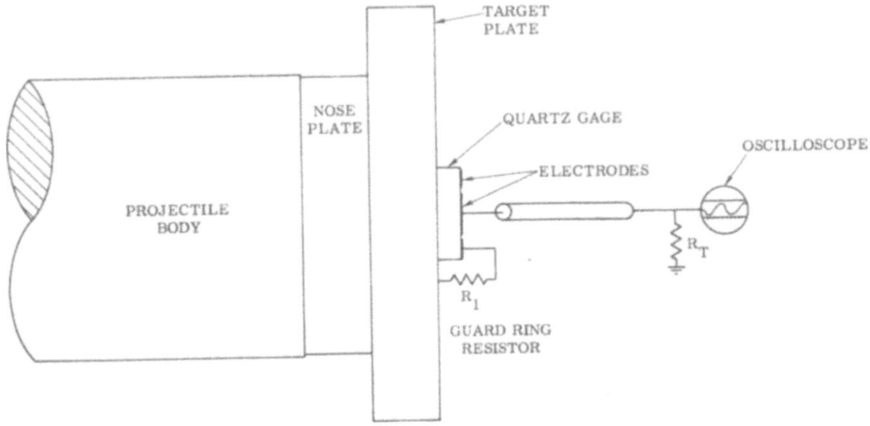

Fig. 7. Schematic of quartz gage as used in monitoring transmitted stress wave.

$$i(t) = \frac{AKU_s}{l}\,\sigma(t) \tag{6}$$

where A is the area of the "active" portion of the gage, K is the piezoelectric current coefficient, U_s is the shock velocity in the quartz, l is the gage thickness, and $\sigma(t)$ is the quartz-target interface stress. The recording time is limited by the arrival of the initial transmitted stress wave at the rear surface of the gage. The portion of the quartz which is generating the current $i(t)$ is subjected to one-dimensional strain and electric fields until after the measurements are complete. Figure 8 shows the stress-time history at an invar-quartz interface after the invar target was symmetrically impacted at a velocity of 616 fps. The maximum stress in the invar was 30.8 kbar [6].*

ANALYSIS

Since strain cannot be measured directly in a plate impact experiment, one must have a theoretical foundation on which to base an analysis of the complete wave propagation problem in order to compute the stresses and strains involved. An analysis commonly used by investigators at Sandia Laboratories follows after Barker, Lundergan, and Herrmann [7] in that it is assumed that a continuous stress profile can be represented by a series of stress increments or jumps, each propagating with a constant velocity appropriate for that stress level. The assumption of constant propagational velocity implies that the material is not strain rate dependent. It is also assumed that the Hugoniot jump equations apply for the interaction of the stress incre-

* One kbar is 14,504 psi.

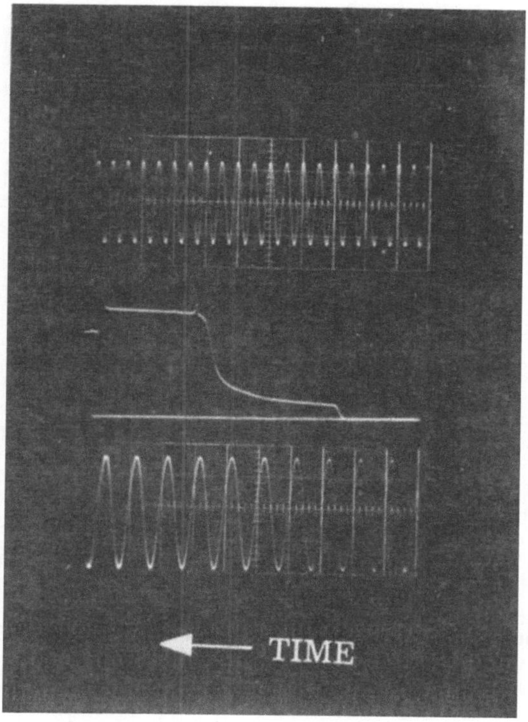

Fig. 8. Stress-time history at invar-quartz interface after transmission through 0.3387″ of invar showing arrival of elastic wave and slower moving plastic wave. Upper timing wave is 10 mc and lower trace is 7.0v (from Graham [6]).

ments with each other and with material interfaces, i.e.,

$$\Delta\sigma_i = \rho_{i-1} U_{si} \Delta U_{pi} \tag{7}$$

$$\frac{\rho_i}{\rho_{i-1}} = \frac{U_{si}}{U_{si} - \Delta U_{pi}} \tag{8}$$

where $\Delta\sigma_i$ is the change in stress across the ith jump which is propagating with a velocity U_{si}; this causes a jump in particle velocity of ΔU_{pi} and a final density of ρ_i. The density ahead of the jump is ρ_{i-1}. The assumption of constant U_{si} and, hence, strain rate independence, results in a propagational velocity of

$$U_{si} = \frac{1}{\rho_{i-1}} \sqrt{\rho_0 \frac{\Delta\sigma_i}{\Delta\epsilon_i}} \tag{9}$$

where ρ_0 is the density corresponding to zero strain, and $\Delta\epsilon_i$ is the jump in strain corresponding to $\Delta\sigma_i$. These assumptions imply that the stress-strain path is represented by a series of straight line segments, with the

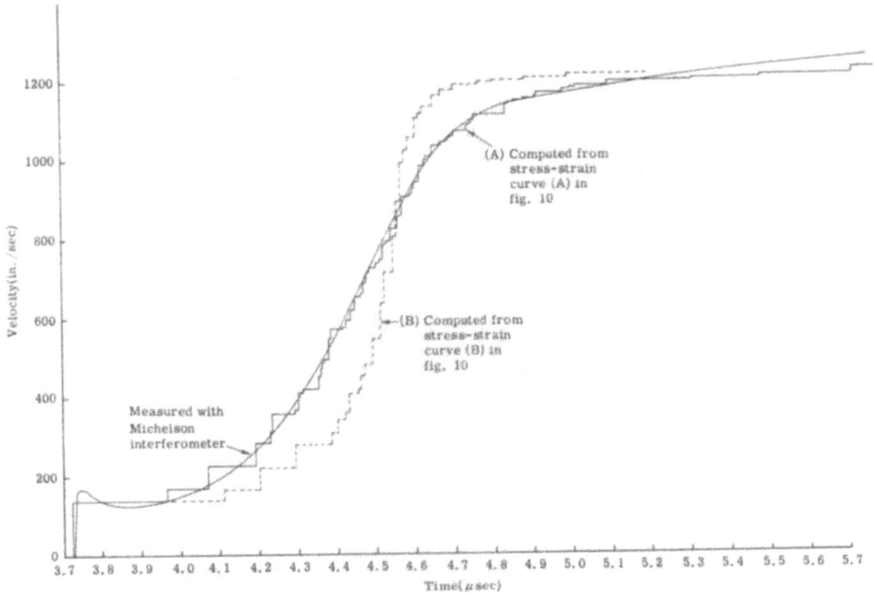

Fig. 9. Measured and computed free surface velocity history for 0.945″ target impacted at 105.4 FPS.

number of segments chosen determining the number of jumps used in representing the stress profile.

The positions and magnitudes of all the interactions which take place are calculated with a high speed digital computer which essentially completes an *x-t* diagram showing the positions of all the shocks as a function of time. The magnitudes of all the variables behind each shock, such as stress, strain, particle velocity, and internal energy, are also computed. Obviously, the computer solution to the complete wave propagation problem cannot be obtained unless the stress-strain relation in loading and unloading is known, but it is this relationship which is being sought. In order to obtain it, the problem is actually solved in reverse order by a process of iteration [7].

The free surface velocity history is used as a guide to obtain a first try as to the stress-strain path which the material had to follow in order to produce that free surface motion. With the first try stress-strain path as an input to the computer program, the complete solution is obtained and the computed free surface velocity-time history is compared with the measured history. Generally, a good fit is not obtained after one try, so the stress-strain path is adjusted appropriately until the computed and measured free surface velocity history compare within experimental and numerical error.

The free surface velocity history shown in Fig. 9 compares measured and computed values for a 0.945 inch thick target of annealed 1060 aluminum (99.6% Al) impacted by an identical projectile nose moving with a velocity of 105.4 fps. The free surface motion was monitored with the Michelson interferometer instrumentation and the computed free surface velocity histories were obtained from two different stress-strain paths. Stepped curve *A* is the final computed velocity history obtained after several of the iterations described above. In order to indicate the sensitivity of this technique in obtaining a final stress-strain path, another slightly different stress-strain path (curve *B* in Fig. 10) was used in the program to calculate another free surface velocity history which is shown by stepped curve *B* in Fig. 9. The region of the measured curve in Fig. 9 immediately after the arrival of the elastic wave cannot be reproduced by the computed curves because the relaxation in velocity, as first reported by Barker, Butcher, and Karnes [8], is not compatible with the assumption of strain rate independence used in the analysis.

The same stress-strain path which best reproduces the free surface velocity history for a 0.945 inch thick target was used to predict the free surface motion for 0.482 and 0.229 inch targets and is compared with

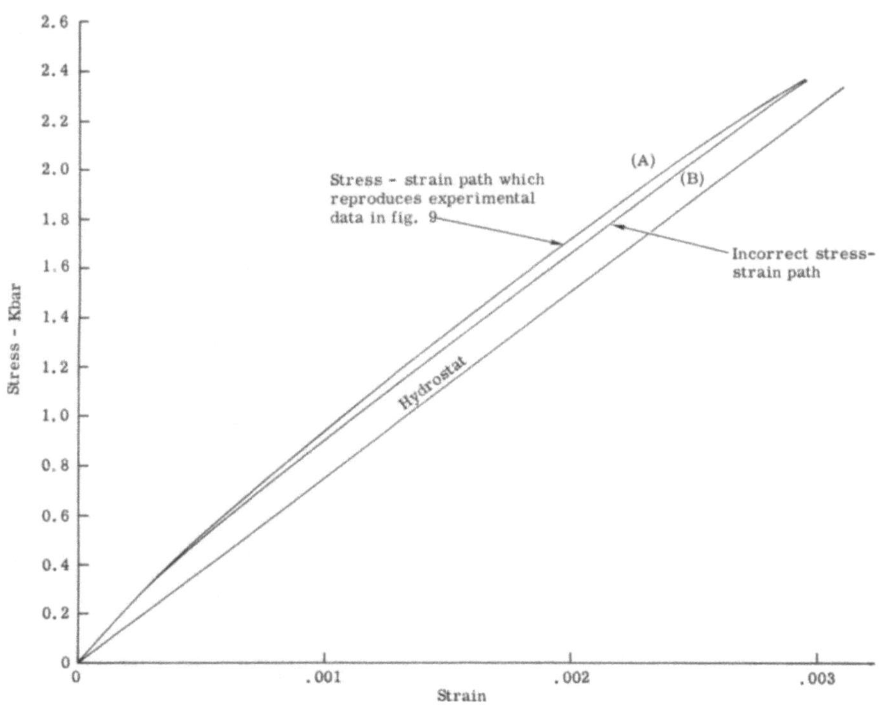

Fig. 10. Stress-strain paths which result in free surface velocity histories of Fig. 9.

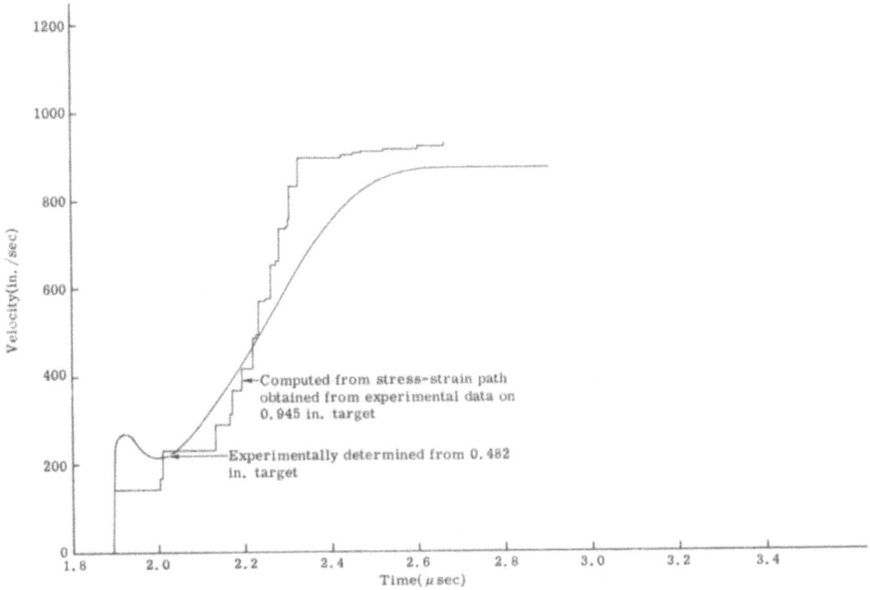

Fig. 11. Free surface velocity history for 0.482 in. thick 1060 aluminum target.

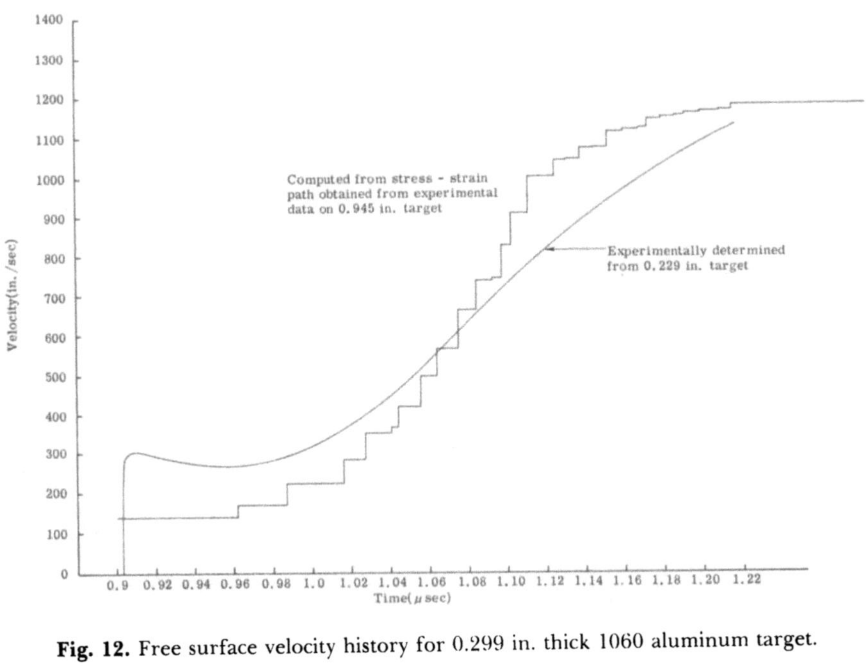

Fig. 12. Free surface velocity history for 0.299 in. thick 1060 aluminum target.

measured behavior for those thicknesses in Figs. 11 and 12, respectively. If this material were strain rate independent as assumed, i.e., if the propagational velocity associated with each particle velocity level were constant with distance of propagation, excellent agreement should have been obtained for the other thicknesses. Such is not the case.

The target thickness divided by the arrival time of a particular velocity level on the free surface velocity-time curve is the *average* propagational velocity, U_{si}, through the thickness of the target. The resulting stress-strain path which best reproduces the free surface motion is then an *average* stress-strain path for that particular target thickness and strain rate history. For propagation through the thin targets, the average strain rates are higher, and, if the material properties are rate or time dependent, the corresponding stress-strain paths should lie above those for the thick targets, as indicated in Fig. 13 where the stress-strain paths for the 0.229, 0.482, and 0.945 inch thick targets are shown. All curves bend toward the hydrostat at the peak stresses since the strain rates at the peak stresses are very low for all target thicknesses and at that time the material is approaching a state of

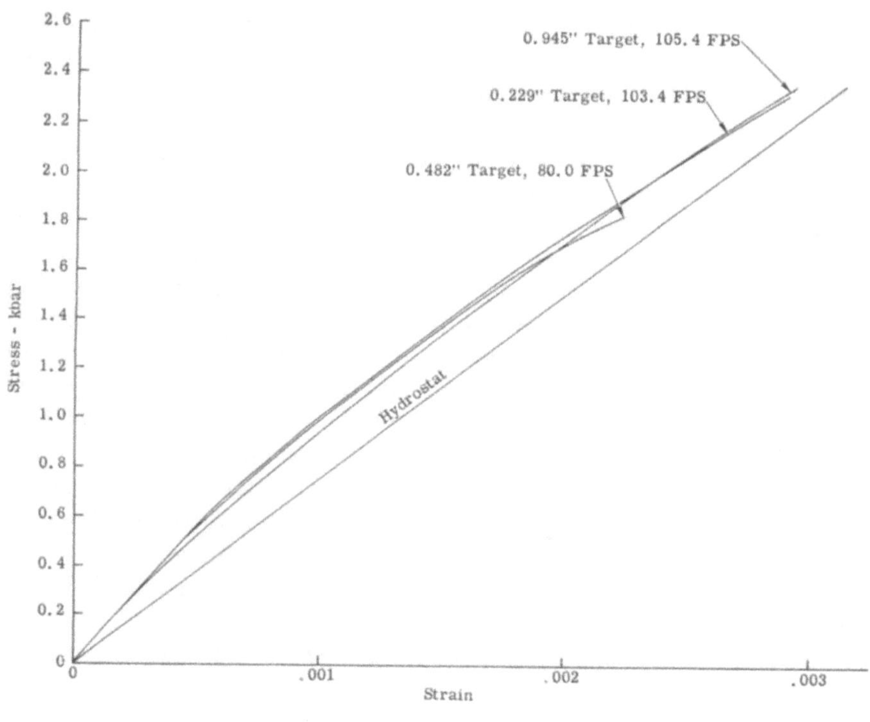

Fig. 13. Variation in average stress-strain path with target thickness and impact velocity.

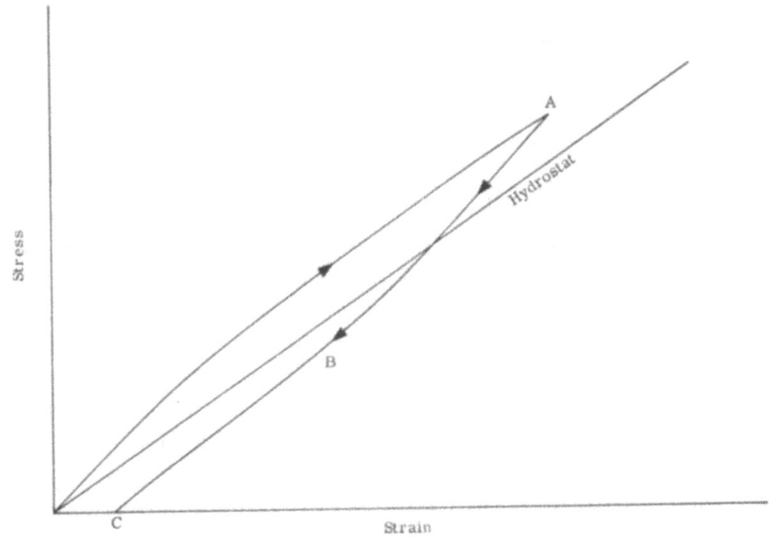

Fig. 14. Qualitative compression and unloading path in one-dimensional strain.

equilibrium. The curve for the 0.945 inch target crosses the other two curves because it is for a higher impact velocity and is associated with a higher average strain rate at stress levels where the thinner, lower velocity targets are approaching states of equilibrium.

A stress-strain path that best reproduces the experimentally determined free surface motion is not determined without uncertainties. If a material is loaded due to impact along a stress-strain path (such as in Fig. 14) to the maximum stress at point A, the stress and strain histories at a point within the interior of the material are dependent only on the details of the stress-strain path followed in reaching point A (assuming the path is unique). However, once the compressive wave reflects off a free surface, the material behind the reflected wave must unload to a zero stress condition. The details of the free surface motion, therefore, depend on the knowledge or assumption concerning the unloading path as the material unloads elastically and then undergoes reverse yielding along path B to point C in Fig. 14. The details of the unloading path are determined by the deviatoric stress in reverse yielding, the location of the hydrostat, and the yield criterion. Although the free surface motion cannot be calculated exactly without precise knowledge of the details of the unloading path, fortunately the calculated free surface motion depends only very weakly on the unloading path used [7].

VERIFICATION OF REFLECTION CALCULATIONS

In order to eliminate the uncertainty due to the unknown details of the unloading path an experiment was performed using the same conditions as before except that a quartz gage was placed on the back surface of the 1060 Al target. The mechanical impedances in one-dimensional strain of aluminum and quartz are so nearly the same whether or not the aluminum has yielded that any unloading which takes place as a result of the interface reflections is completely elastic. Therefore, the uncertainties associated with the unloading path are eliminated, and the only approximations remaining in the analysis are: 1) the representation of the continuous stress distribution with a series of steps, and 2) the assumption that each step propagates with a constant velocity characteristic of the stress increment it represents. The steps approach the continuous curve if a large number of small ones are used, and the second assumption results in an average stress-strain path based on the average propagation velocities of the steps.

The smooth line in Fig. 15 shows the stress history at the aluminum-quartz interface as determined by the quartz gage. A computer analysis was made for this configuration: the analysis considered the inter-

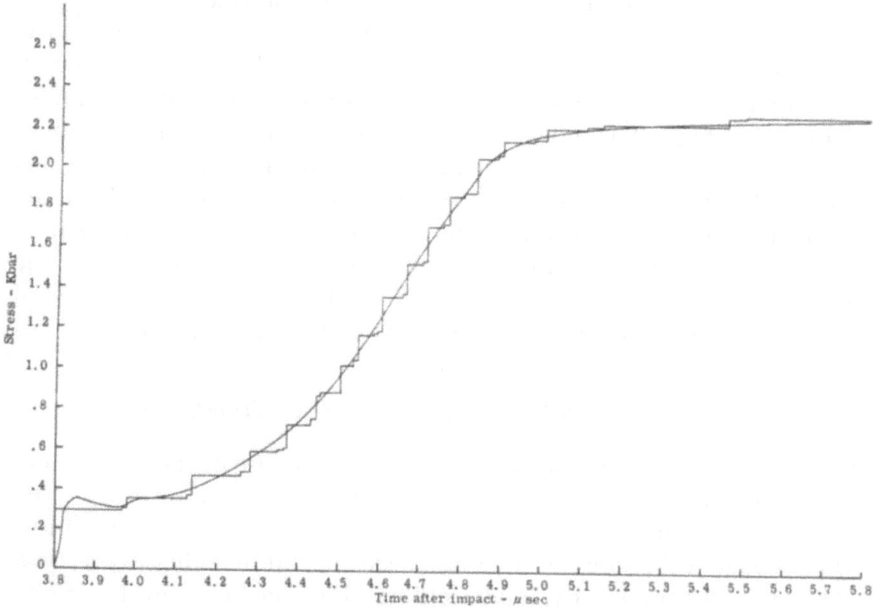

Fig. 15. Experimental and computed stress history at interface between 0.945 in. thick aluminum target and quartz gage.

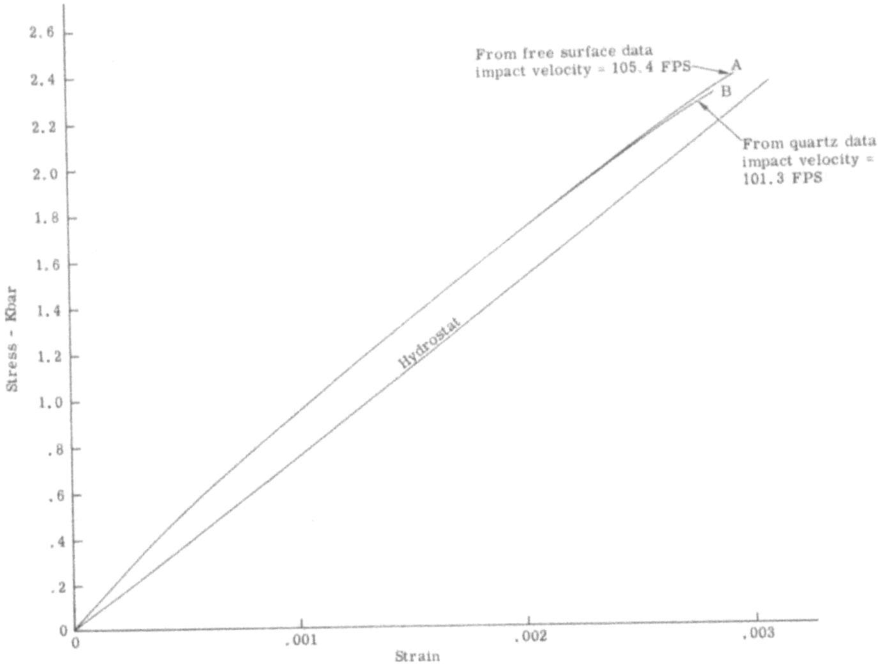

Fig. 16. Stress-strain paths obtained from quartz and free surface data.

actions as a result of the aluminum-quartz interface. The stepped curve in Fig. 15 is the computed stress history at the aluminum-quartz interface based on stress-strain path *B* in Fig. 16. The stress-strain path obtained from the free surface measurements with the interferometer (curve *A* in Fig. 10) is shown for comparison. The excellent agreement at the lower stress levels, even though the degree of unloading was drastically different, attests to the negligible effect the details of the unloading path have on the loading portion of the stress-strain curve. The curves diverge slightly near the maximum stress because of the slightly different impact velocities. Note that the difference in final stress levels is the same percentage as the difference in velocities.

RESULTS OF OTHER INVESTIGATORS

There are some investigators who have made use of the one-dimensional strain conditions of plate impact experiments to investigate dynamic mechanical properties. The first reported use of the Sandia quartz gage technique was by Jones, Neilson, and Benedick [9] in 1962. They investigated the dynamic yielding phenomena of duraluminum, iron, and steel of various compositions and heat treatments when shocked with high explosive plane wave generators. Some of their re-

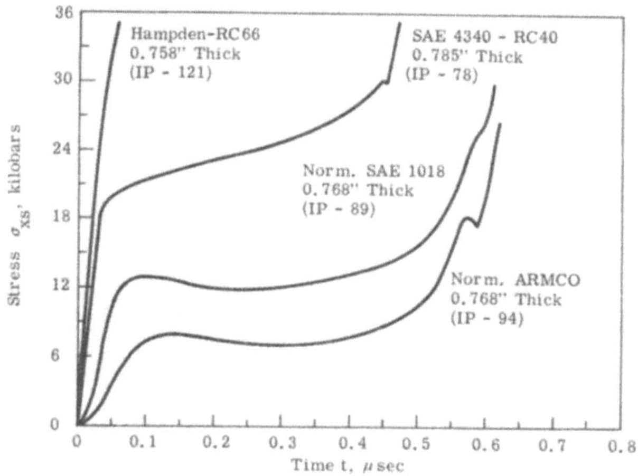

Fig. 17. Transmitted wave shapes through iron and iron alloys obtained with quartz gage (from ref. 9).

sults are shown in Fig. 17 from which the dynamic yield stress in one-dimensional strain (Hugoniot elastic limit) can be obtained for all the materials except Hampden tool steel whose yield stress is higher than the linear range of the quartz.

In 1963, Taylor and Rice [10] reported the use of a parallel plate capacitor free surface motion transducer to correlate the variation in amplitude of the elastic wave in Armco iron with the equation of state of a linearly relaxing solid. These same data were further analyzed and reported in 1965 by Taylor [11]. The Hugoniot elastic limit of Armco iron attenuates in amplitude with distance of propagation. Taylor showed that both the amplitude and its decay with distance of propagation (shown in Fig. 18) can be explained in terms of the dislocation theory of Gilman and Johnston [12]. He used the relation

$$\dot{\gamma} = bNv \tag{10}$$

to describe the plastic shear strain rate γ in terms of the Burgers vector b, the dislocation density N, and the average dislocation velocity v. The dislocation velocity was expressed in terms of the resolved shear stress by

$$v = v_\infty \exp\left(-\frac{\tau_0}{\tau}\right) \tag{11}$$

where v_∞ and τ_0 are constants and τ is the resolved shear stress. The dislocation density was expressed as

$$N = (N_0 + \beta\gamma) \exp\left(-\frac{\phi\gamma}{\tau}\right) \tag{12}$$

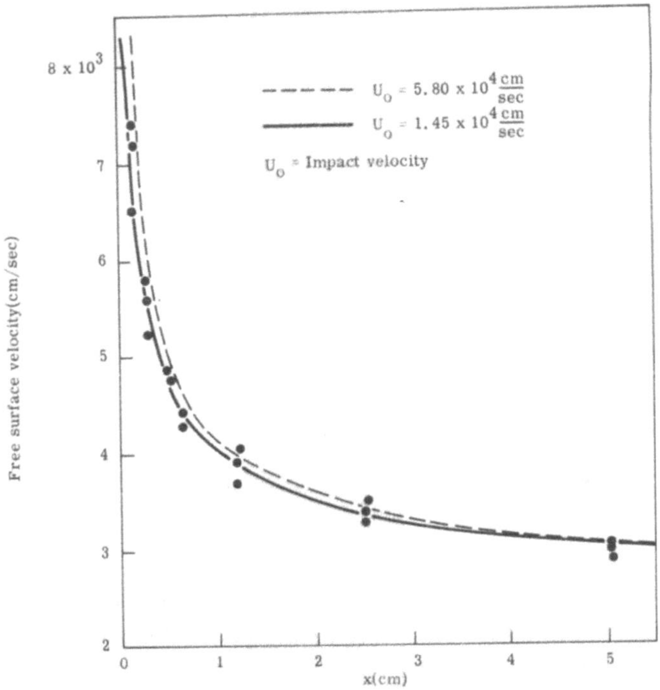

Fig. 18. Initial free surface velocity as result of elastic wave as function of distance of propagation. Points are from experiments. Lines are predictions from dislocation model (from ref. 11).

where N_0 is the initial dislocation density, and β is the multiplication rate. The term $\phi\gamma/\tau$ represents a work hardening function. By adjusting some of the constants, Taylor was able to fit the data as shown in Fig. 18.

In 1964, Barker, Lundergan, and Herrmann [7] published the first evidence of strain rate effects in aluminum in the one-dimensional strain configuration. They made use of the slant resistor free surface motion instrumentation system developed by Barker and Hollenbach [13]. The resultant stress-strain curves for 6061-T6 aluminum obtained from the same analysis technique previously described in this paper are presented in Fig. 19 along with the curve predicted from quasi-static data for the same material. The individual dynamic curves bend toward the static curve near the ends. This results in stress-strain curves which lie above the quasi-static curve and depend on the impact velocity or maximum stress obtained in the experiment. Although the increase in stress above the static curve is only about five percent of the total stress, the increase in the deviatoric component is about thirty percent.

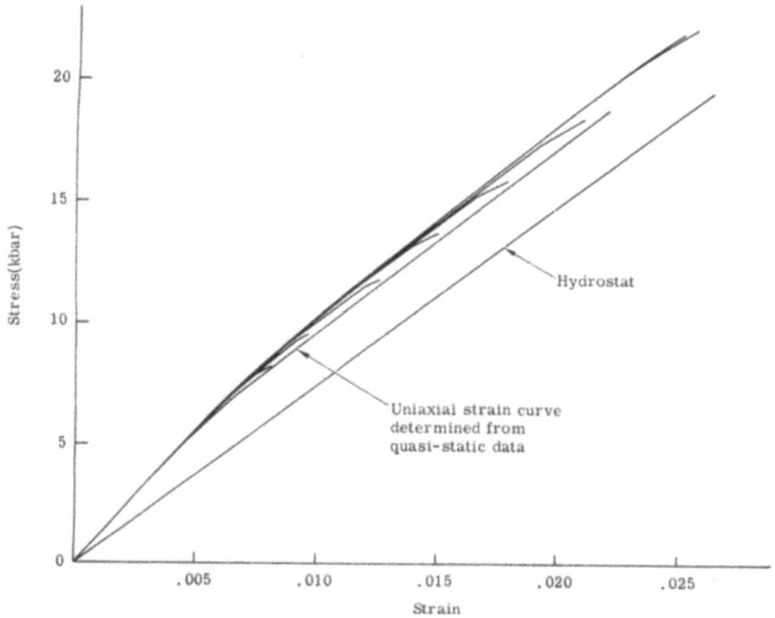

Fig. 19. Stress-strain paths for 6061-T6 aluminum obtained with slant resistor instrumentation (from ref. 7).

In 1966, Butcher and Karnes [14] showed that the strain rate behavior of 6061-T6 can be explained by a Malvern [15] type analysis using a constitutive relation for one-dimensional stress of the form

$$Y = Y_0 + A(\alpha - \alpha_0) + B'(1 - e^{-F(\alpha-\alpha_0)}) \tag{13}$$

where

$$B' = B \log \dot{\alpha}^p + D \qquad \text{for} \qquad \dot{\alpha}^p \leqslant \delta$$

$$= B[\log(\delta) - 1] + D + (B/\delta)\dot{\alpha}^p \qquad \text{for} \qquad \dot{\alpha}^p > \delta$$

where δ defines the boundary separating the logarithmic from the linear dependence on plastic strain rate and B and D are constants which determine the degree of strain rate dependence. It was found experimentally that the stress-strain path (based on the same analysis previously discussed) is not unique, is dependent on the distance of propagation (and hence, time), and is predicted accurately by (13) when transformed to one-dimensional strain. The computed average stress-strain paths for propagation through 0.5 and 2.5 in. targets along with the experimental data points and the curve predicted from static behavior are shown in Fig. 20.

Although strain rate dependence definitely exists in 6061-T6 alumi-

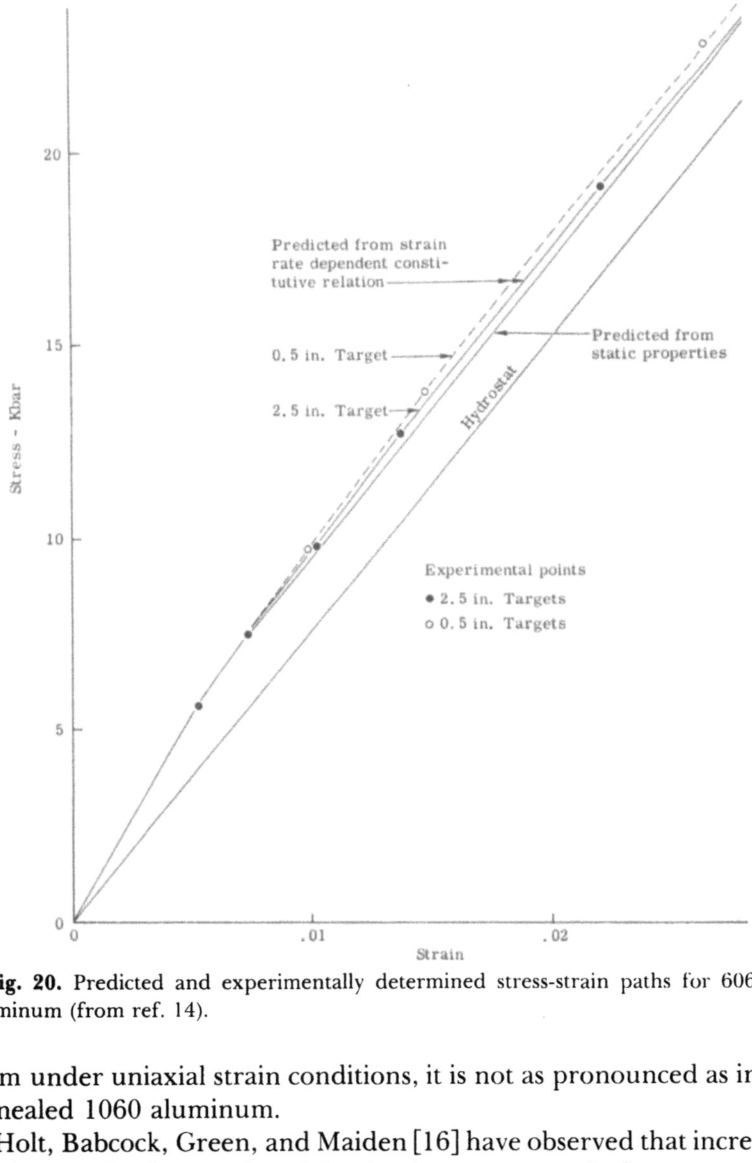

Fig. 20. Predicted and experimentally determined stress-strain paths for 6061-T6 aluminum (from ref. 14).

num under uniaxial strain conditions, it is not as pronounced as in the annealed 1060 aluminum.

Holt, Babcock, Green, and Maiden [16] have observed that increased hardness, whether produced by alloying content or by heat treatment, reduces the strain rate sensitivity of aluminum. Their results are shown in Fig. 21 where the relative increase in flow stress due to a six order of magnitude increase in strain rate is plotted against the flow stress corresponding to the lowest strain rate. Their results were obtained on systems which rapidly deform short cylindrical specimens, with the measurements being made after several reverberations of the stress waves have produced equilibrium in the specimens. Figure 21 also includes

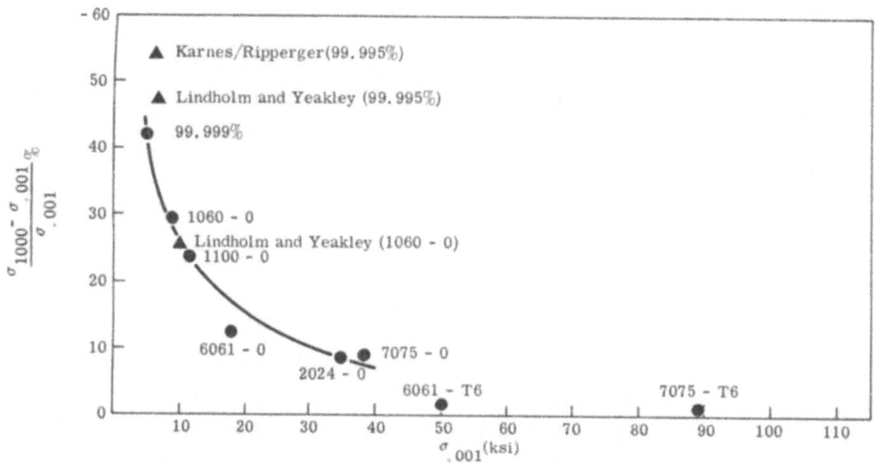

Fig. 21. Dependence of rate sensitivity on flow stress in aluminum alloys. Flow stress at 10^3in./in./sec, σ_{1000} and at 10^{-3}in./in./sec, σ_{001} are values at an arbitrary strain of 6% (from ref. 16).

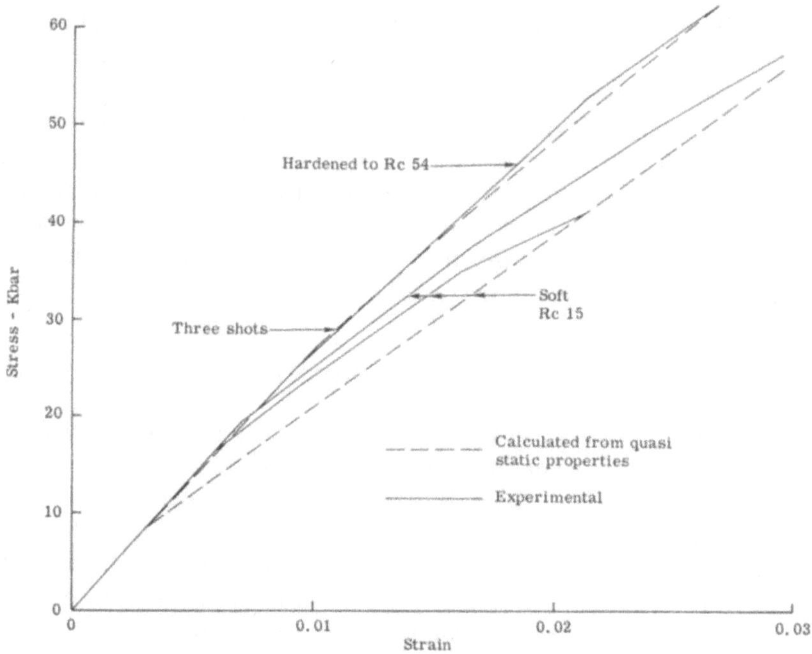

Fig. 22. Stress-strain paths for hard and soft 4340 steel (from ref. 19).

data points obtained by Lindholm and Yeakley [17] and by Karnes and Ripperger [18].

Butcher and Canon [19] have observed the same phenomena in plate impact experiments on 4340 steel. Their results are shown in Fig. 22 where the resulting stress-strain paths for hard and soft 4340 are included along with the corresponding curves based on quasi-static behavior. Jones, et al. [9] observed the same phenomena of rate sensitivity being reduced by hardness.

SUMMARY

It has been shown that the interactions which take place just inside the free surface of a target plate can be calculated with confidence using a series of steps to represent the continuous stress profile and the Hugoniot jump conditions to describe the interaction of the steps. Such a technique enables one to construct an average stress-strain path which reproduces experimental data.

Plate impact experiments on 1060 and 6061-T6 aluminum agree with the findings of Holt, Babcock, Green, and Maiden [16], using entirely different experimental techniques, that the mechanical properties of aluminum are rate dependent and that the extent of the effect is increased by either purifying or annealing, or both.

The plate impact experiments enable one to study mechanical properties of materials in a well-defined state of strain that is not affected by geometry or boundaries even though the stress state is different from the one usually considered. One can make meaningful measurements during strain rates which are several orders of magnitude higher than in rod or cylinder impact studies.

ACKNOWLEDGMENTS

The author wishes to thank L. M. Barker for his constructive comments during the preparation of the manuscript and also for allowing the inclusion of the velocity interferometer information prior to his presentation in Paris. He is also indebted to R. A. Graham for many discussions concerning the use of the quartz gage and also for inclusion of the unpublished invar-quartz record. Without the assistance of Mrs. Emily Young, the dozens of computer runs could not have been made in the time allowed. C. D. Lundergan contributed many helpful comments concerning the manuscript and the author is indebted to B. M. Butcher for his comments and the use of the partially unpublished interferometer data on the 0.482 in. 1060 aluminum target. The author is especially indebted to R. E. Hollenbach, T. C. Looby, and R. G.

Newman for making most of the experiments reported in this paper possible.

Thanks is also given to the American Institute of Physics, the American Institute of Aeronautics and Astronautics and the American Society of Metals for granting permission to republish some of the figures.

References

1. D. S. Wood, J. Appl. Mech., *19*, 521 (1952).
2. G. R. Fowles, J. Appl. Phys., *32*, 1475 (1961).
3. L. M. Barker and R. E. Hollenbach, Rev. Sci. Inst., *36*, 1617 (1965).
4. L. M. Barker (To be presented at IUTAM Symposium H.D.P.), Paris, Sept. 1967.
5. R. A. Graham, F. W. Neilson and N. B. Benedick, J. Appl. Phys., *36*, 1775 (1965).
6. R. A. Graham, Private Communication.
7. L. M. Barker, C. D. Lundergan and W. Herrmann, J. Appl. Phys., *35*, 1203 (1964).
8. L. M. Barker, B. M. Butcher and C. H. Karnes, J. Appl. Phys., *37*, 1989 (1966).
9. O. E. Jones, F. W. Neilson and N. B. Benedick, J. Appl. Phys., *33*, 3224 (1962).
10. J. W. Taylor and M. H. Rice, J. Appl. Phys., *34*, 364 (1963).
11. J. W. Taylor, J. Appl. Phys., *36*, No. 10 (1965).
12. J. J. Gilman and W. G. Johnston, *Dislocation and Mechanical Properties of Crystals*, John Wiley and Sons, Inc., New York (1957).
13. L. M. Barker and R. E. Hollenbach, Rev. Sci. Inst., *35*, 742 (1964).
14. B. M. Butcher and C. H. Karnes, J. Appl. Phys., *37*, 402 (1966).
15. L. E. Malvern, Quart. Appl. Math., *7–8*, 405 (1949–51).
16. D. L. Holt, S. G. Babcock, S. J. Green and C. J. Maiden, Trans. ASM, *60* (1967).
17. U. S. Lindholm and L. M. Yeakley, J. Mech. Phys. Solids, *13*, 41 (1965).
18. C. H. Karnes and E. A. Ripperger, J. Mech. Phys. Solids, *14*, 75 (1966).
19. B. M. Butcher and J. R. Canon, AIAA Journal, *2*, 2174 (1964).

THE RELATIONSHIP BETWEEN THE CONSTITUTIVE EQUATION AND ONE-DIMENSIONAL WAVE PROPAGATION

E. A. RIPPERGER
The University of Texas
Austin, Texas

and

HAL WATSON, JR.
Southern Methodist University
Dallas, Texas

ABSTRACT

A uniform normal stress is suddenly applied at the end of a semi-infinite rod. The stress then remains constant. The bar is made of a hypothetical material with a yield stress significantly lower than the applied stress. Various constitutive relationships are assigned to the bar material. Wave front profiles, in terms of the axial strain, are computed for each constitutive relationship, and for different values of the constants that appear in the constitutive equations. Profiles obtained in this way are compared to determine the sensitivity of the characteristic features of the profiles to the form of the constitutive equation. Special computation techniques for minimizing error are described. Results of the computations are discussed in terms of their relevance to the interpretation of experimental measurements.

INTRODUCTION

Measurements from which the constitutive relationships applicable at high rates of strain are deduced, inevitably involve plastic wave propagation [1, 2]. The measurements are usually made in either a one-dimensional stress or a one-dimensional strain situation. In this paper, attention will be focused on the one-dimensional stress problem.

The most general first-order quasi-linear form of the constitutive equation is [3, 4]

$$\dot{\epsilon} = f(\sigma, \epsilon)\dot{\sigma} + g(\sigma, \epsilon). \tag{1}$$

The functions f and g depend upon temperature, but temperature is assumed to be constant here. Two special forms of this equation have been used in analytical studies and have been investigated experimentally. In one case, the function $g(\sigma, \epsilon)$ is assumed to be negligible [5]. This is the rate independent relationship expressed simply as

$$\sigma = \sigma(\epsilon). \tag{2}$$

In the second case, the function $f(\sigma, \epsilon)$ is assumed to be a constant, $1/E$, and the function $g(\sigma, \epsilon)$ is taken in various forms [6]. Thus

$$\dot{\epsilon} = \frac{1}{E}\dot{\sigma} + g(\sigma, \epsilon). \tag{3}$$

In some of the experimental investigations intended to reveal the relationship which best describes the behavior of materials, various features of the profile of a plastic wave front propagating in a rod have been studied [7, 8]. These investigations have, in general, not clearly resolved the issue of which form of the constitutive relationship is most nearly correct. Direct measurements of stress and strain have indicated that the function $g(\sigma, \epsilon)$ is not negligible [9, 10]. This contradiction suggests that measurements of the wave profile characteristics are not particularly sensitive as indicators of the form of the constitutive relationship. With this thought in mind, the authors have made a computer investigation of one-dimensional plastic stress wave propagation assuming different forms of the constitutive equation and different values of the constants involved. The results of these computations are presented and discussed in terms of their relevance to the interpretation of experimental measurements.

In these discussions, no attempt is made to compare experimental measurements and computed results although ample experimental measurements are available. The emphasis here is on the similarities and differences in strains which result when different constitutive equations are assumed. The experimentation is done with the computer.

For $g(\sigma, \epsilon)$ the following forms have been investigated

$$g(\sigma, \epsilon) = \frac{1}{\tau}\left(\frac{\sigma - \bar{\sigma}}{\bar{\sigma}}\right)^{m} \tag{4}$$

$$g(\sigma, \epsilon) = K(\sigma - \bar{\sigma}) \tag{5}$$

$$g(\sigma, \epsilon) = K(\sigma - \sigma_y). \tag{6}$$

In each case, it is assumed that $\sigma > \bar{\sigma}$, and $\sigma > \sigma_y$.

Equation (4) is the form proposed by Chalupnik [10]. It differs from one of the special forms discussed by Perzyna [11] only in the substitu-

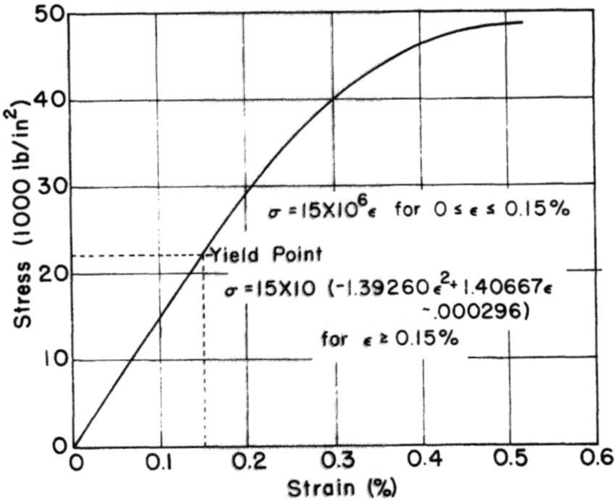

Fig. 1. Static stress-strain curve.

tion of $\bar{\sigma}$ for σ_y. The latter, σ_y, is the yield stress in simple tension and $\bar{\sigma}$ is the stress given by the static stress-strain relationship for the strain that exists when the stress is σ.

Thus, the quantity $\sigma - \bar{\sigma}$ is the dynamic overstress used by Malvern [6]. Equation (4) is also the form used by Bodner and Symonds [12] with $\bar{\sigma}$ replaced by σ_y, in a strain-rate dependent rigid-plastic analysis of the impact loading of beams. Equation (5) will be recognized as the linearized version of the exponential form of the "g" function suggested by Malvern. Equation (6) is the form used by Sokolovskii [13].

The meaning of constitutive relationships such as these is made somewhat clearer perhaps by examining a few of the stress-strain curves plotted from these "g" functions and comparing them to the static stress-strain curve in Fig. 1. The function $\dot{\epsilon}'' = g(\sigma, \epsilon)$ represents a surface. To examine some of the features of that surface, we will take a cross section in which stress appears as a function of strain with strain rate $\dot{\epsilon}''$ as a parameter. Some of these cross sections are shown in Fig. 2 for a function of the form given in (4). In this plot, five different combinations of the two parameters τ and m are represented. These curves were prepared by first calculating the value of $\dfrac{\sigma - \bar{\sigma}}{\bar{\sigma}}$ corresponding to a given set of parameters and a strain rate of 250 sec^{-1}. Then for selected values of $\bar{\sigma}$ the corresponding values of σ are readily computed from the relationship

$$\sigma = \bar{\sigma}[1 + (\tau\dot{\epsilon})^{1/m}].$$

From curves 2, 3 and 4, it may be seen that for a fixed value of m the strain rate effect increases with increasing τ. The extent to which the

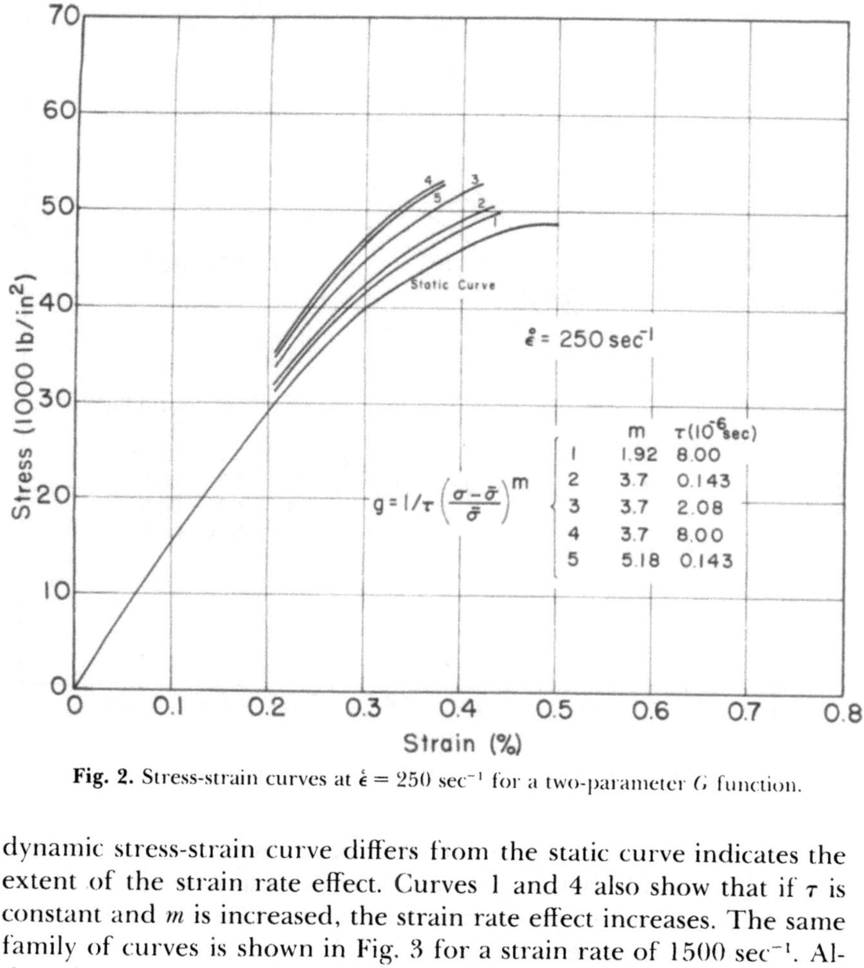

Fig. 2. Stress-strain curves at $\dot{\epsilon} = 250\ \text{sec}^{-1}$ for a two-parameter G function.

dynamic stress-strain curve differs from the static curve indicates the extent of the strain rate effect. Curves 1 and 4 also show that if τ is constant and m is increased, the strain rate effect increases. The same family of curves is shown in Fig. 3 for a strain rate of 1500 sec^{-1}. Although the relative positions of the curves are unchanged, it is obvious that the effects are somewhat greater than those produced by a strain rate of 250 sec^{-1}.

Stress-strain curves for the linearized flow law of (5) are shown in Fig. 4 for three different values of the constant K and two different strain rates. These curves show that as K increases, the rate effect decreases, or in other words, the smaller K is, the greater the dynamic overstress becomes at a given strain rate.

For the constants τ and m which appear in (4), the values selected for study are

m	τ
5.81	0.143×10^{-6} sec
3.7	2.08×10^{-6} sec
1.92	8.0×10^{-6} sec

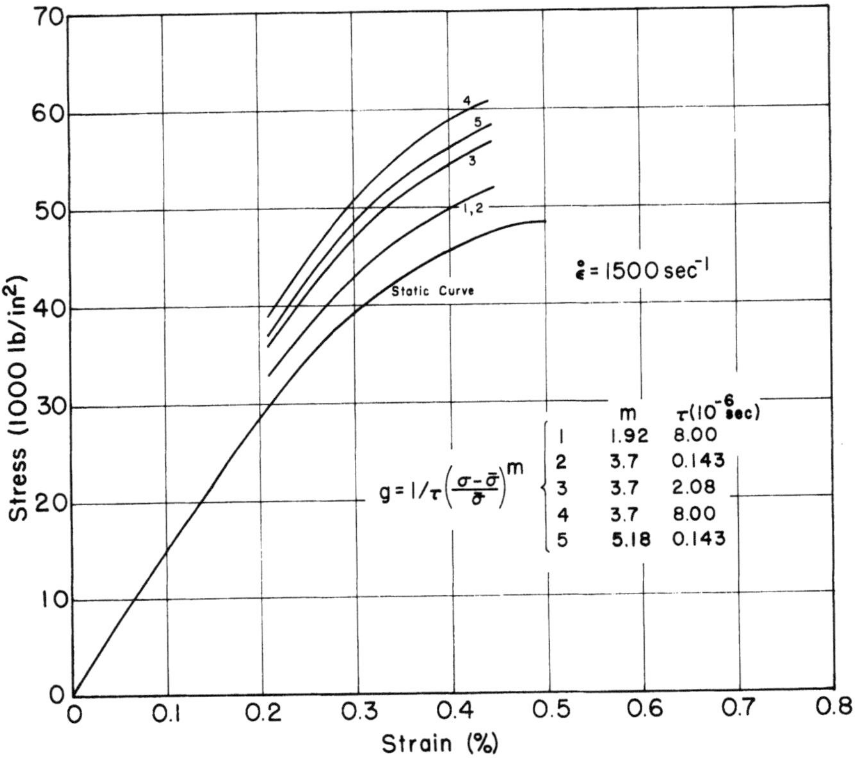

Fig. 3. Stress-strain curves at $\dot{\epsilon} = 1500\ \text{sec}^{-1}$ for a two-parameter G function.

These values were suggested to some extent by earlier experimental results [10, 14, 15], but, in general, the choice was based primarily on a need for a moderately wide range of realistic values rather than on a need to simulate some particular material.

The values for the constant K in (5) were chosen as follows:

$$K = 1.125 \times 10^6 \quad \text{and} \quad 4.50 \times 10^6\ \text{sec}^{-1}.$$

These values also were selected, not to match the particular data available, but to give a realistic range of values suitable for the type of comparisons to be made.

Equation (6) indicates that plastic flow will tend to bring the stress back to the yield stress level whereas, in the relationship given by (5), the stress decays toward the static-stress-strain curve. For an elastic-plastic material, the behavior predicted by the two equations would be the same.

Since a static stress-strain curve is required for (2), (4), and (5), a hypothetical curve was constructed. This curve shown in Fig. 1 resembles a copper stress-strain curve, but it must be emphasized again that

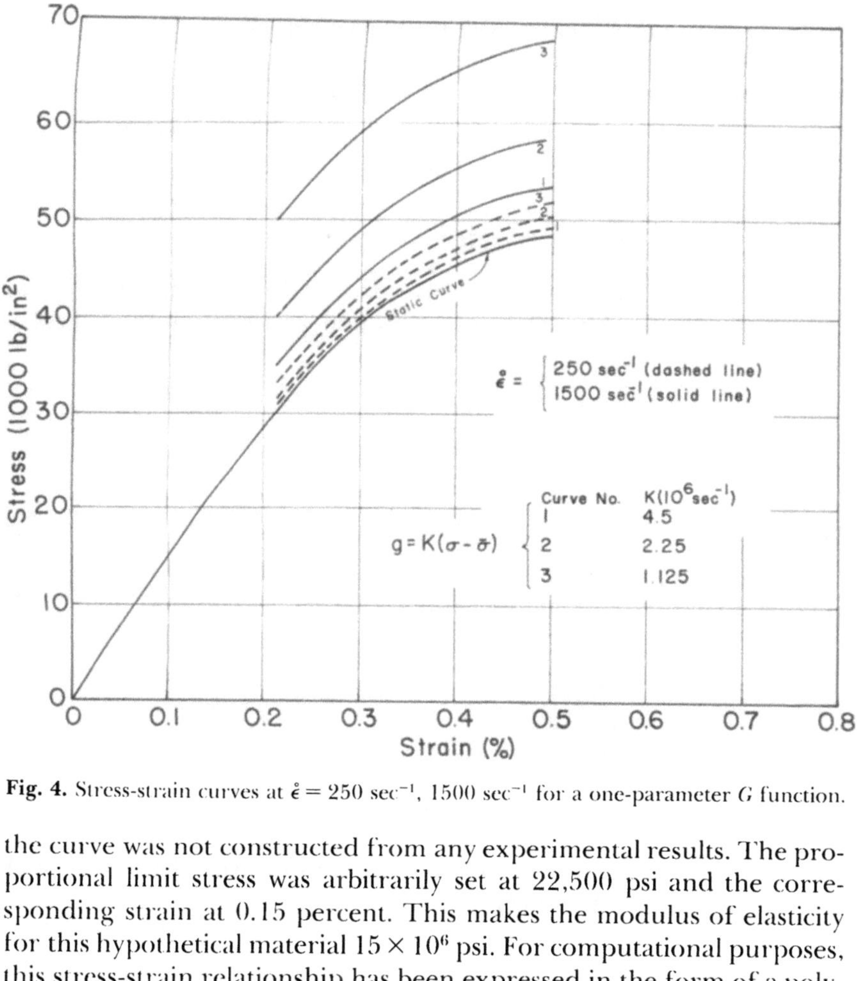

Fig. 4. Stress-strain curves at $\dot{\epsilon} = 250$ sec^{-1}, 1500 sec^{-1} for a one-parameter G function.

the curve was not constructed from any experimental results. The proportional limit stress was arbitrarily set at 22,500 psi and the corresponding strain at 0.15 percent. This makes the modulus of elasticity for this hypothetical material 15×10^6 psi. For computational purposes, this stress-strain relationship has been expressed in the form of a polynomial.

$$\sigma = f(\epsilon) = 15 \times 10^6 \epsilon \qquad \epsilon \leq 0.15\% \tag{7}$$

$$\sigma = f(\epsilon) = 15 \times 10^6[-139.260\epsilon^2 + 1.40667\epsilon - 0.000296] \qquad \epsilon > 0.15\%. \tag{8}$$

The problem considered is stated as follows: A semi-infinite cylindrical rod with a small diameter, made of a homogeneous material having the stress-strain characteristics indicated by (7) and (8) and shown in Fig. 1, is subjected to a step application of a uniform stress of 47,400 psi at the end of the rod.

The stresses and strains in this rod are computed as functions of time and the distance from the impact end. Thus, the stress and strain distributions along the axis of the bar at a given time can be compared, or the stress or strain variation with time at a given point on the bar can be compared. The method of characteristics was used for the computations when the constitutive relationship expressed by (1) was used. A simple finite difference routine was used when the constitutive relationship was given by (2).

COMPUTATIONAL PROCEDURE

For the propagation of plane waves in a uniform, homogeneous bar as assumed here, the equations of motion and continuity in the Lagrangian form are

$$\frac{\partial \sigma}{\partial x} = \rho \frac{\partial v}{\partial t} \tag{9}$$

$$\frac{\partial v}{\partial x} = \frac{\partial \epsilon}{\partial t}. \tag{10}$$

The symbols are defined as follows:

σ – engineering stress
ϵ – engineering strain
v – particle velocity
ρ – mass density
x – initial distance of a section of the bar from the origin
t – time

With these equations, one must have a constitutive relationship such as those given by (1) and (2).

Rate Independent Computations

For the rate independent material and a constant stress input, computation of the strain distribution in the bar at a given time is relatively simple. The slopes of the stress-strain curve are obtained by differentiating the polynomial given in (8). Then, if a density is assumed for the material, or, as was actually done, the velocity of elastic wave propagation is assumed, the velocity of propagation of any strain amplitude can be easily computed. With the velocities of the different strain amplitudes known, the position of each amplitude at a given time can be computed and the profile of the wave front established. This has been done and the results are shown in Fig. 5. These curves display all of the well known characteristics of a plastic wave: the initial step up to the yield strain, the sloping front from the yield strain to the input strain with the slope of this part decreasing with time, and the

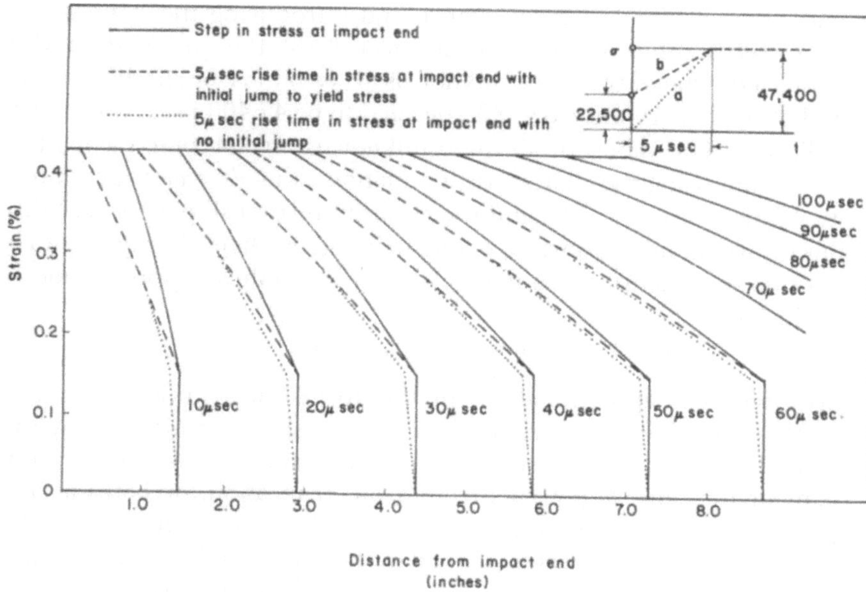

Fig. 5. Wave front profiles – strain-rate independent theory.

strain plateau which increases in length as time goes on. The dotted and dashed curves represent the wave front profiles for the input stresses shown in the inset in Fig. 5. These strain rate independent profiles will be compared later to corresponding profiles for rate dependent materials.

Rate Dependent Computations

When constitutive relationships of the type indicated by (1) are used to characterize the material, the basic equations are quasi-linear and hyperbolic. For equations of this type, the method of characteristics can be used to trace the propagation of plastic waves. This method, as it applies to plastic wave propagation, has been described in detail in numerous publications [7]. Consequently, only the form of the equations actually used in the computations will be given here. The ordinary differential equations in dimensionless form for propagation along the characteristic curves are

$$\begin{aligned} d\sigma - dv &= -g(\sigma, \epsilon)\, dt \qquad \text{along } dx = dt \\ d\sigma + dv &= -g(\sigma, \epsilon)\, dt \qquad \text{along } dx = -dt \qquad (11) \\ d\sigma - d\epsilon &= -g(\sigma, \epsilon)\, dt \qquad \text{along } dx = 0. \end{aligned}$$

By substituting one of the "g" functions given by (4), (5) or (6), changing (11) to a finite difference form, and introducing the static stress-strain relationship of Fig. 1 in the form of (8), the computations can be started. One of the three unknowns (σ, ϵ, v) must be specified at the end of the bar. In this case, σ is specified to be 47,400 psi as previously indicated. The values of all three unknowns are computed along the wave front using the characteristic equation along $dx = dt$, the continuity relationship $d\epsilon = -dv$, and the elasticity relationship $d\sigma = d\epsilon$. With this information as a starting point, an iterative scheme is used to "walk out into the interior mesh points." In this procedure, approximations of the form

$$d\sigma \simeq \Delta\sigma = \sigma_{i+1} - \sigma_i$$

are used. For approximations of this form, the truncation error is of the same order of magnitude as the mesh size multiplied by the second derivative. Furthermore, in initial value or "marching" problems of this sort, any errors (round off, truncation, or both) at any point are propagated along with the solution. Thus, the farther the solution goes, the greater the error becomes. It is extremely important, therefore, that errors be minimized near the beginning. Consequently, the computations reported here were started with a very small mesh size at the beginning of the propagation where the variables change very rapidly and second derivatives are large. Then, as the computations proceed, a shift is made to a coarser net to keep the computational time within acceptable bounds, to reduce storage requirements, and to reduce the total number of computations, thereby reducing the cumulative round-off error.

Mesh Changing Scheme and Stability and Accuracy Considerations

The computational scheme may be studied by first considering the $x - t$ plane (characteristic plane) as seen in Fig. 6. The order of the computation of the solution of σ, ϵ and v is indicated by a numbered sequence of points in this plane. In this sequence, computations begin with a small size mesh (mesh 1) and continue into the interior of the region of the x-t plane bounded by $x = t$ and $x = 0$. As the computations continue into this region, the mesh size is doubled from time to time (mesh 2, mesh 3) whenever the solution allows such a change, i.e., whenever second derivatives are small enough and convergence can be attained with the new enlarged mesh size. Note, however, that it is only possible to form some multiple of the last mesh size, i.e., mesh 2 is double the size of mesh 1. The reasons for this are, 1) previously computed values must be used to find new values, therefore, characteristic lines must pass through points at which the solution has been previously computed and 2) constancy of directions of the characteristic

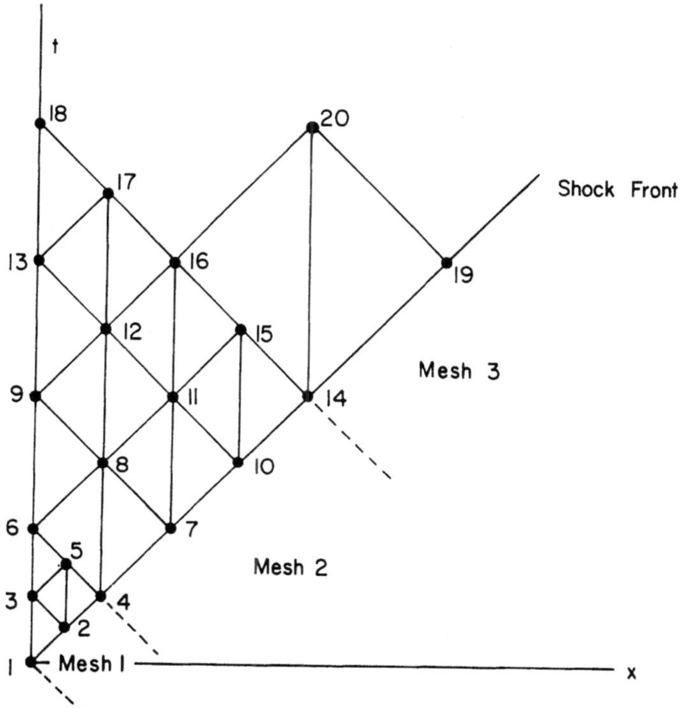

Fig. 6. x-t plane showing variable mesh size.

lines, i.e., $\Delta x = \Delta t$. Note also that the total number of mesh spaces in the last sequence (14–18) of mesh points just prior to changing mesh size must be an even number so that when the mesh size is doubled, the coarser mesh will be compatible with the finer mesh and fit into the pattern.

At points where a coarser mesh is compatible, the solution at point P (Fig. 7), the first interior point for the coarser mesh, is attempted after multiplying the value of the "g" function used in each characteristic difference equation by a factor of 10. This is equivalent to increasing the actual size of the truncation error, or the second derivatives, by an order of magnitude. If convergence can be reached for this extreme case, the mesh change is initiated and computations begin with the new, coarser mesh size. If at any later point in the computational sequence, convergence cannot be attained within a reasonable number of iterations, the computational procedure using that particular mesh size is aborted and computations begin anew with the previous mesh size from the point where the mesh size change was previously initiated. Whenever the mesh size is increased to the maximum desired size, no more changes in the mesh size are made and computations continue

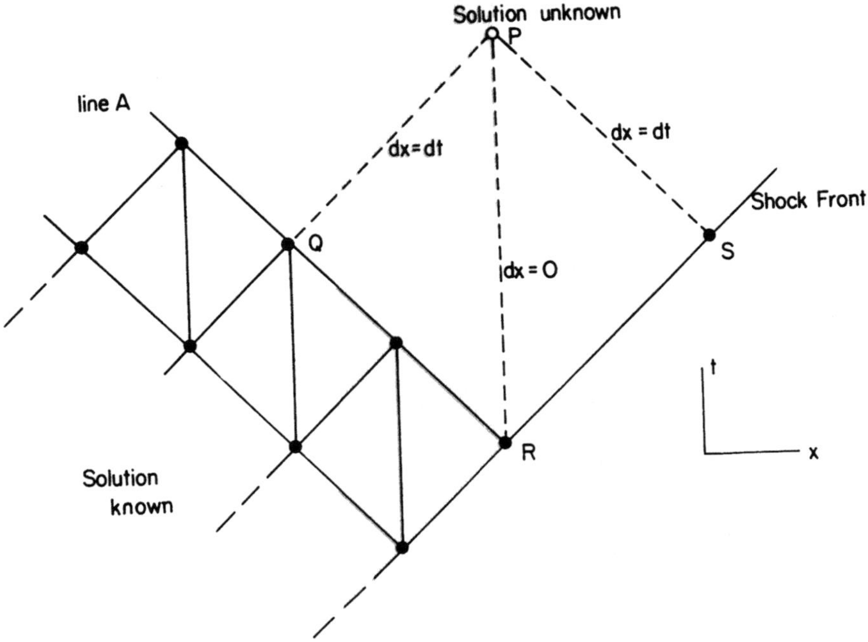

Fig. 7. Interior mesh point in x-t plane.

out to some desired value of time. On the average, 10–15 changes in mesh size are made during a single computation run.

A check on the truncation error was made for a single case by reducing the size of the final mesh by a factor of 2 and comparing the solution after many computations. The difference in the two solutions was in the fifth and sixth significant digits.

In order to minimize round-off error, the total number of computations was minimized by using the mesh changing scheme mentioned above. Also, the solution was carried along in 14 decimal digits. Convergence of the iterative scheme is assumed at each point whenever two successive trial solutions differ only in the twelfth digit. For convergence of this accuracy, about thirty iterations are needed at each point of the x, t plane to solve for the values of σ, ϵ and v. The final mesh size used in these computations is $\Delta x = \Delta t = 0.01$ (dimensionless) or $\Delta x = 0.05$ in., $\Delta t = 0.345$ μsec. Most of the computations were carried out to $t = 100$ μsec and $x = 14.5$ in.

No problems in stability were encountered in the computations reported here. Whenever convergence at a point was not obtained for one size of mesh, it was attained using a previous, finer mesh size. However, it was found in previous calculations for aluminum, with a step in velocity as the boundary condition, that the type of iterative scheme used determined whether or not convergence could be attained for t very near 100 μsec. The Gauss-Seidel method of iterating to a solution required a finer mesh size and required more iterations to converge than did the predictor-corrector method used here.

NUMERICAL RESULTS

The computed profiles of the wave front as it progresses along a bar with a "g" function of the form given by (4) are shown in Figs. 8 through 12. In Figs. 8, 9 and 10, the value of m is constant and τ is increasing. These curves have the well-known general features of a plastic wave front. There are, however, four features of these curves which should be especially noted. These are: 1) Although the strain at the impact end continues to increase after the impact, after 80 μsec, it has not reached the value shown in Fig. 5 for the strain rate independent propagation. 2) After 80 μsec, the strain is essentially constant from the impact end to about five inches from the end. For example, at five inches from the impact end, the strain is only 2.5 percent less than it is at the impact end for $\tau = 0.143 \times 10^{-6}$ sec and for $\tau = 8.0 \times 10^{-6}$ sec the strain at five inches is only 5.3 percent less than at the impact end. Attention is called to this point because one of the distinguishing differences frequently pointed out between propagation in strain rate sensitive and non-strain-rate sensitive materials has been the strain plateau which forms in the non-strain-rate materials [7]. It has been shown, it should also be noted, that

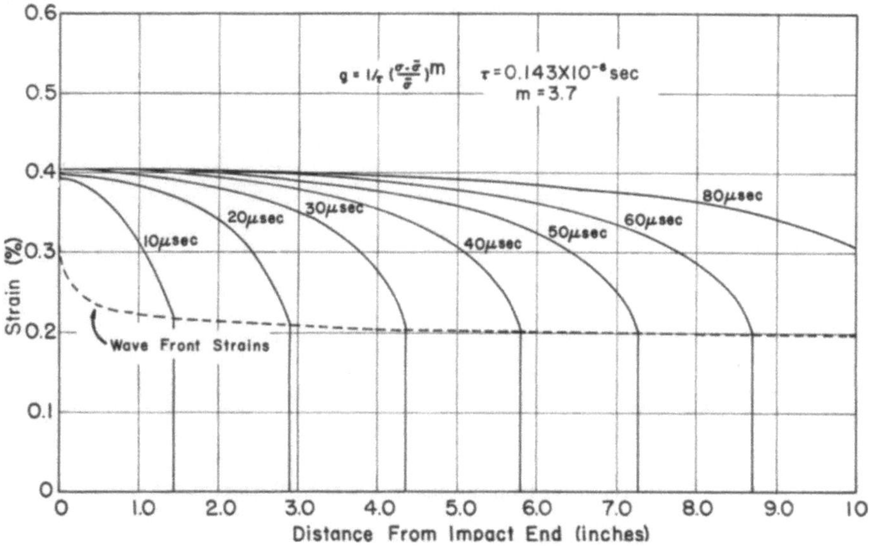

Fig. 8. Wave front profile, $\tau = 0.143 \times 10^{-6}$ sec, $m = 3.7$.

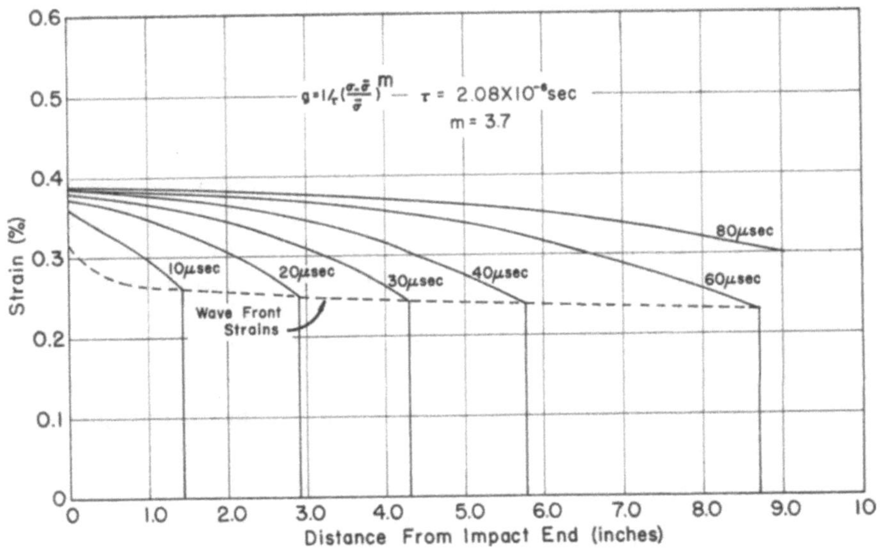

Fig. 9. Wave front profile, $\tau = 2.08 \times 10^{-6}$ sec, $m = 3.7$.

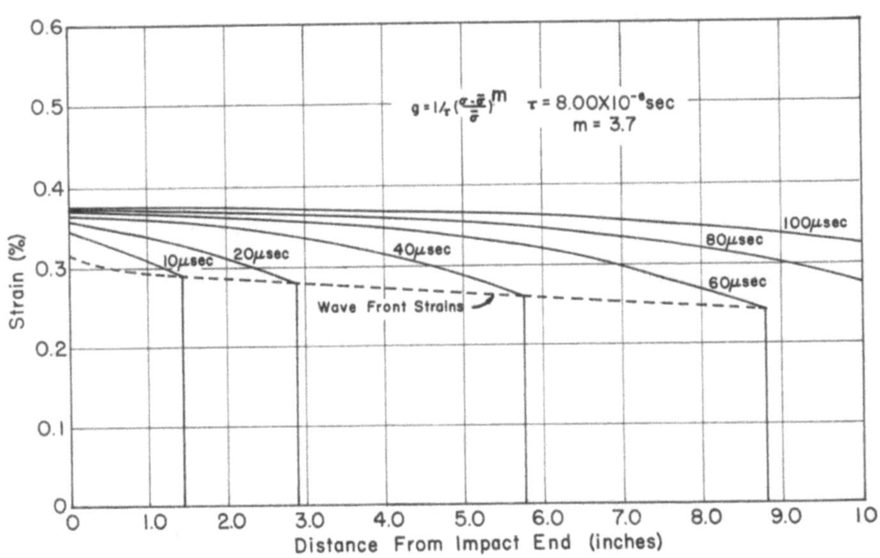

Fig. 10. Wave front profile, $\tau = 8.00 \times 10^{-6}$ sec, $m = 3.7$.

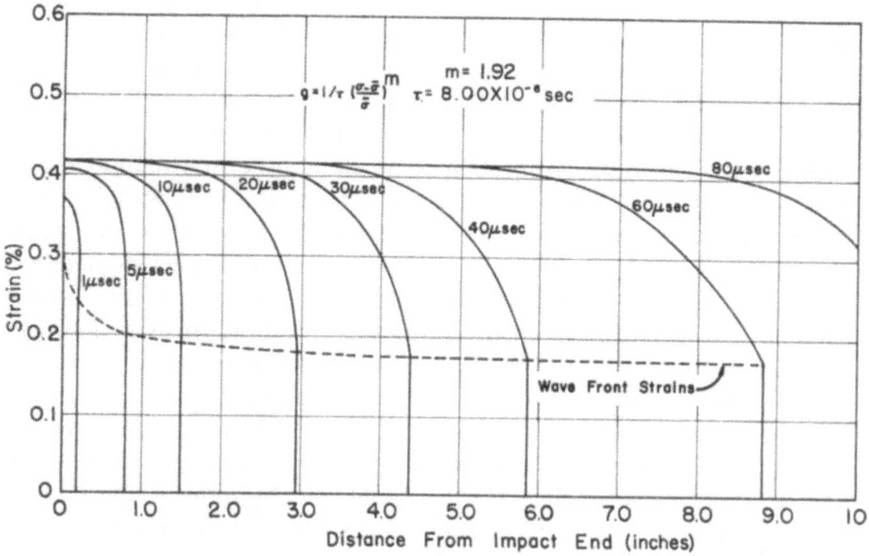

Fig. 11. Wave front profile, $\tau = 8.00 \times 10^{-6}$ sec, $m = 1.92$.

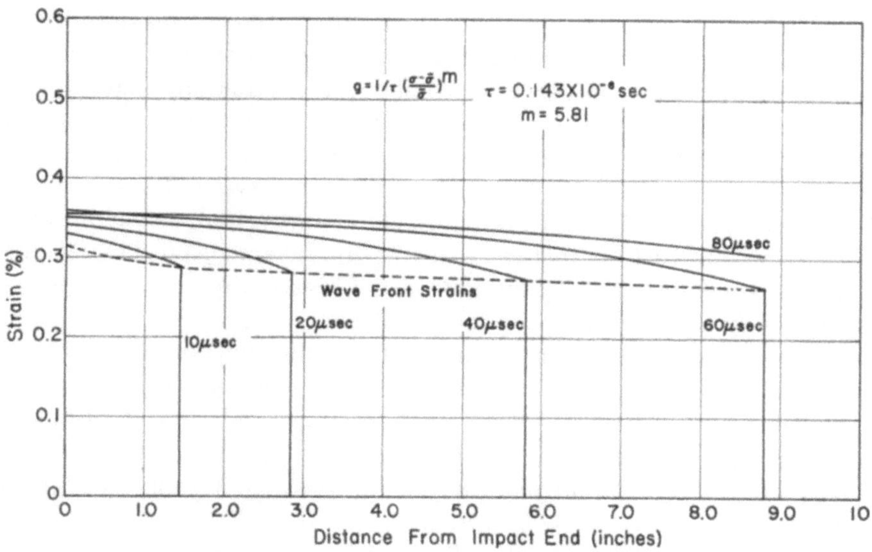

Fig. 12. Wave front profile, $\tau = 0.143 \times 10^{-6}$ sec, $m = 5.81$.

for a constant velocity impact as opposed to a constant stress impact, a constitutive equation such as the one used here results in no constant strain region, or strain plateau, developing in the first 140 μsecs after impact [16]. Malvern and Efron have presented computed results for aluminum which show that if a constant stress is applied at the end of the bar, the velocity very quickly approaches a constant value [7]. This implies that the wave profiles for a constant stress input should not differ much from those for a constant velocity input. The difference which apparently does exist indicates that the wave front profile is somewhat sensitive to variations in the input conditions. Under these circumstances, it would be difficult to determine in experimental measurements, in which the input conditions are not precisely known, whether or not a constant strain region should be expected. It would also be equally difficult to tell whether an observed constant strain region indicates a strain-rate independent material, or a rate dependent material. 3) The step amplitude at the wave front decays quickly in the first diameter or so of propagation, and then continues to decay, but much more slowly as propagation continues. For all three values of τ, the amplitude of the step exceeds appreciably the corresponding strain shown in Fig. 5 for the non-strain-rate material. 4) With the exception of strain levels slightly greater than the amplitude of the step at the wave front, the velocity of propagation of a given strain level is very nearly constant within the range covered by these computations even though the constant τ varies over nearly two orders of magnitude.

Since the velocity of propagation is essentially constant at a specified strain amplitude, the curve which shows the variation in strain with time at a given point, is the mirror image of the variation in strain along the bar at a time corresponding to the time of arrival of the wave front at the given point. This is shown by the curves in Fig. 13. These curves represent the strain variation at the impact end, and at one inch intervals out to four inches from the impact end. Note how all the curves appear to converge after 80 μsec. This is indicative of the formation of the strain plateau mentioned above.

Strain distributions for the linear strain-rate function $K(\sigma - \bar{\sigma})$ are shown in Figs. 14 and 15. These curves resemble rather closely those shown in Fig. 11 for $\tau = 8.0 \times 10^{-6}$ sec and $m = 1.92$ and all of the remarks made concerning those curves also apply to the linear strain rate curves. The strain plateau and the constant velocity at a given strain level are especially in evidence.

Profiles for the "g" function $K(\sigma - \sigma_y)$ are shown in Fig. 16. These curves differ markedly from the other profiles that are shown. The strain at the impact end increases very rapidly and in 30 μsec has reached a value more than double that reached in any of the other

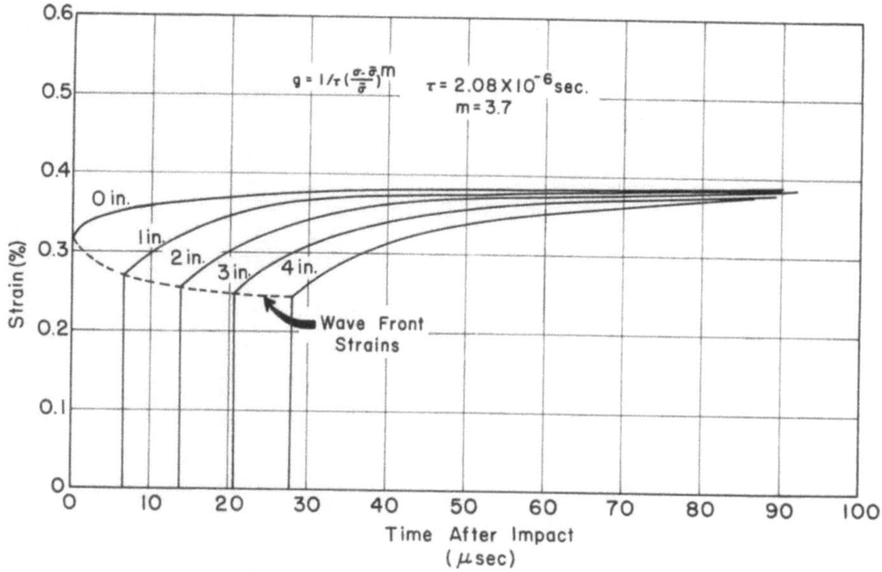

Fig. 13. Strain vs. time at $\tau = 2.08 \times 10^{-6}$ sec, $m = 3.7$.

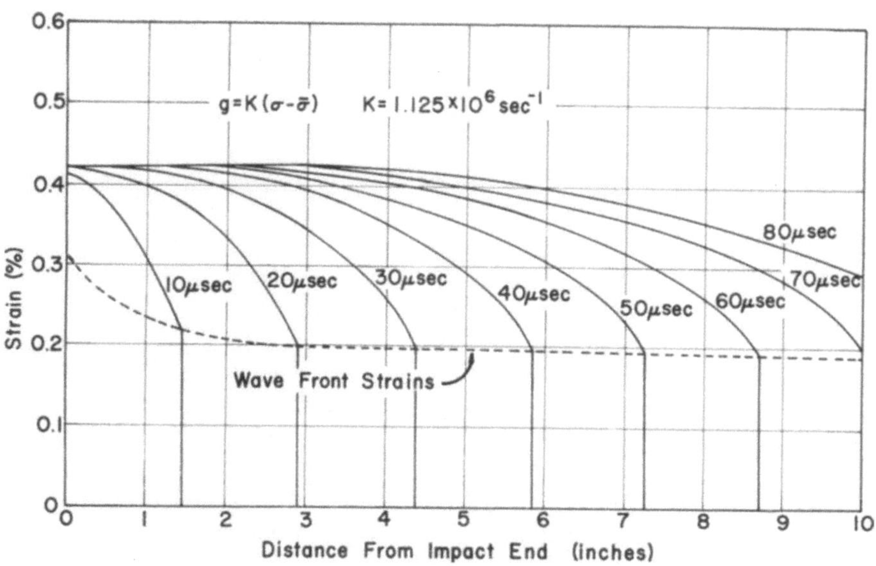

Fig. 14. Wave front profile, $K = 1.125 \times 10^{-6}$ sec^{-1}.

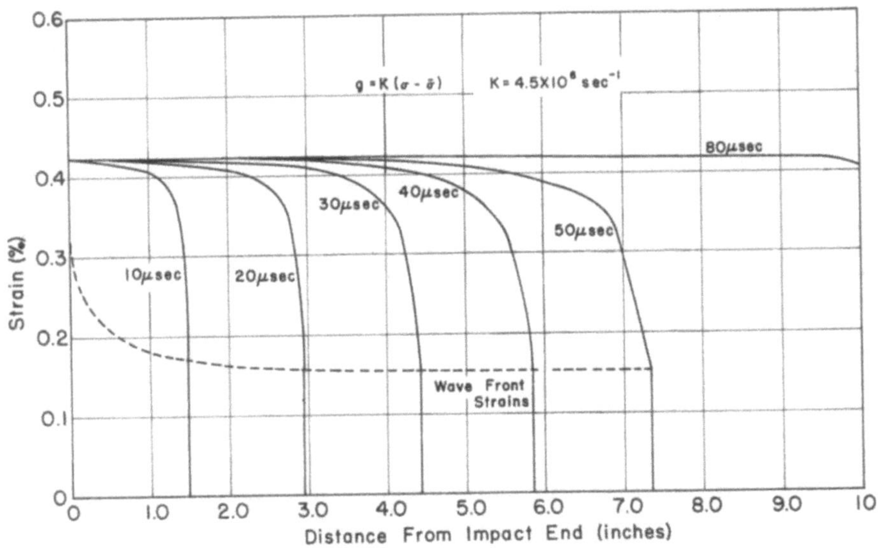

Fig. 15. Wave front profile, $K = 4.50 \times 10^6\ \mathrm{sec}^{-1}$.

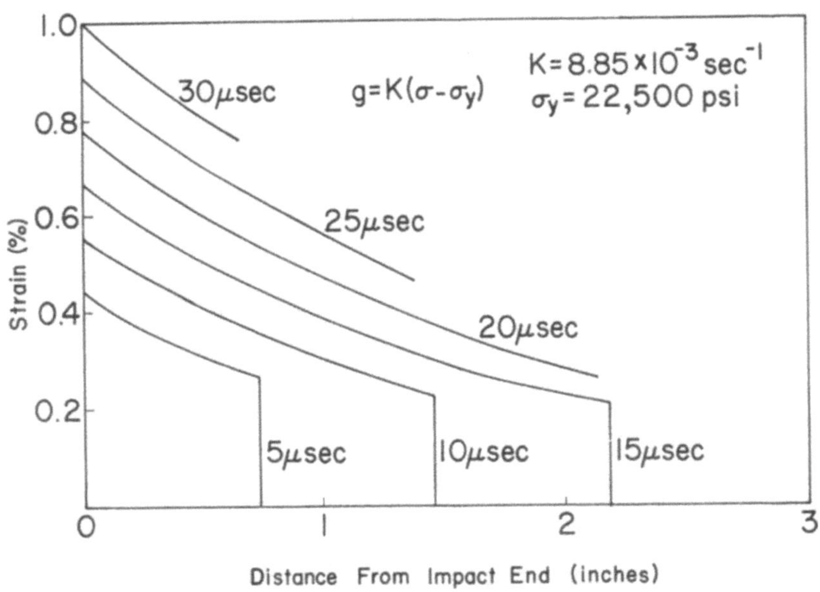

Fig. 16. Wave front profile, $g = K(\sigma - \sigma_y)$ for $K = 8.85 \times 10^{-3}\ \mathrm{sec}^{-1}$.

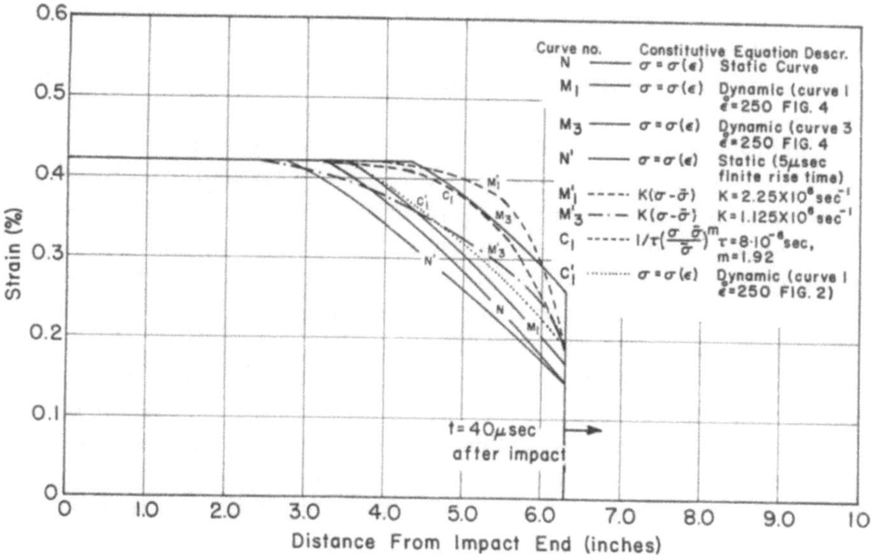

Fig. 17. Wave front profile at 40 μsec for several constitutive equations.

cases investigated. Obviously, if the input stress is constant, there will be a constant strain rate at the impact end and the strain will increase without limit. If K is chosen so as to make the strain rate at the end very small, then the rate of decay of the step at the wave front will also be very low. It appears, therefore, that this particular constitutive relationship is not applicable to the problem of plastic wave propagation in a rod.

For convenience in making comparisons, some of the profiles previously discussed are replotted in Fig. 17. These curves represent axial strain distributions 40 μsec after impact in semi-infinite bars, all of which have the same static stress-strain curve. In addition to the curves previously shown, some other profiles of interest are shown. These include 1) Curve N', which is the profile for a rate independent propagation with a stress input corresponding to curve b in the inset in Fig. 5. This is a stress which rises instantly to σ_y and then increases linearly until after 5 μsecs, 47,400 psi is reached. The stress then remains constant. 2) Curve M, which is the profile for a rate independent propagation assuming the dynamic stress-strain relationship shown by Curve 1 in Fig. 4. 3) Curve M_3, which is the profile for a rate independent propagation assuming Curve 3 in Fig. 4 as the stress-strain relationship. 4) Curve C', which is the profile for a rate independent propagation assuming Curve 1 of Fig. 2 as the stress-strain relationship.

The stress-strain relationships for Curves M_1, M_3 and C_1' are assumed to have the same slope in the elastic portion, as the static stress-strain curve. As a consequence, the yield strains for these curves are all somewhat higher than the yield strain shown by the static curve. This causes the step at the wave front to be somewhat higher for these curves than it is for Curve N. If this were not the case, the four profiles N, C_1', M_1 and M_3 would all be very close together, with M_1 agreeing most closely with N. The discrepancy between C_1' and N is so slight that it is extremely doubtful that any experimental measurement of the wave front profile would be accurate enough to justify any conclusion regarding which constitutive relationship to attribute to the material.

Finally, it might be noted that the differences between profiles N and N' are almost as great as the differences between N' and any of the other profiles. This is consistent with the remarks previously made concerning the sensitivity of the wave profile to the nature of the input stress to the bar. Since it is difficult to determine experimentally the precise form of the input, the validity of any form of the constitutive equation determined by wave profile measurements must be open to question.

CONCLUSIONS

1. The form of the constitutive equation influences the shape of the propagating wave front. The shape does not appear to be unique. By choosing different forms of the equation, proper combinations of constants or by slight variations in the input, one can apparently produce the same wave front profile. This is interpreted to mean that measurements of wave front shapes and strain propagation velocities are not reliable as indicators of the constitutive relationship.
2. The velocity of propagation of a given strain level is essentially constant for each of the constitutive relationships assumed. Thus, a constant velocity in itself provides no evidence at all as to rate sensitivity or lack of rate sensitivity.
3. The rate dependent constitutive relationships considered in this study, with the exception of $K(\sigma - \sigma_y)$, all lead to a constant or essentially constant strain region beginning at the end of the bar and progressing farther and farther along the bar as time increases. Thus, the strain plateau cannot be regarded as a distinguishing feature of non-strain-rate sensitive propagation when the load applied to the bar is a step function in stress.

ACKNOWLEDGMENTS

This work was supported in part by the Army Research Office (Durham), Contract DA 31-124-ARO-D-229.

References

1. J. E. Johnson, D. S. Wood and D. S. Clark, "Dynamic Stress-Strain Relation for Annealed 2S Aluminum Under Compression Impact," J. App. Mech., *20*, 523 (1953).
2. J. F. Bell, "Propagation of Large Amplitude Waves in Annealed Aluminum," J. App. Physics, *31*, 277 (1960).
3. J. Lubliner, "A Generalized Theory of Strain-Rate-Dependent Plastic Wave Propagation in Bars," J. Mech. Phys. Solids, *12*, 59 (1964).
4. N. Cristescu, "On the Propagation of Elastic-Plastic Waves in Metallic Rods," Bulletin De L'Academie Polonaise Des Sciences, Serie des Sciences Techniques, *11*, No. 4, 183 (1963).
5. T. von Karman and P. Duwez, "The Propagation of Plastic Deformations in Solids," J. of App. Phys., *21*, 987 (1950).
6. L. E. Malvern, "The Propagation of Longitudinal Waves of Plastic Deformation in a Bar of Material Exhibiting a Strain-Rate Effect," J. App. Mech., *18*, 203 (1951).
7. L. E. Malvern and L. Efron, Stress Wave Propagation and Dynamic Testing, Tech. Report No. 1, Grant G-24898, National Science Foundation.
8. J. F. Bell, "On the Direct Measurement of Very Large Strain at High Strain Rates," Experimental Mechanics, *7*, 8 (1967).
9. F. E. Hauser, J. A. Simmons and J. E. Dorn, "Strain Rate Effects in Plastic Wave Propagation," University of California MRL publication, Series No. 133, issue No. 3, June 1960.
10. J. D. Chalupnik, The Effect of a Hydrostatic State of Stress on the Propagation of Plastic Waves in Metals, Doctoral Dissertation, The University of Texas (1964).
11. P. Perzyna, "The Constitutive Equations for Rate Sensitive Plastic Materials," Quarterly of Applied Math., *20*, 321 (1963).
12. S. R. Bodner and P. S. Symonds, "Plastic Deformations in Impact and Impulsive Loading of Beams," Proc. Second Symposium on Naval Structural Mechanics, Pergamon Press, New York, 1960, p. 488.
13. V. V. Sokolovskii, "Propagation of Elastic-Visco-Plastic Waves in Bars" (in Russian), Prikl Mat Mekh, *12*, 261 (1948).
14. H. Watson, Jr., The Effects of Strain Rate and Temperature on Stress-Strain Characteristics of Copper and Iron, Doctoral Dissertation, The University of Texas (1967).
15. C. H. Karnes, Strain Rate Effects in Cold Worked High Purity Aluminum, Doctoral Dissertation, The University of Texas (1963).
16. E. A. Ripperger and H. Watson, Jr., "Strain Rate Effects and Plastic Wave Propagation in Aluminum Rods," Paper No. 350, Fifth National Congress for Applied Mechanics (1966).

PLANE-STRAIN PLASTIC WAVE PROPAGATION IN A DYNAMICALLY LOADED HOLLOW CYLINDER

IAN M. FYFE

University of Washington
Seattle, Washington

ABSTRACT

An experimental technique to create plane-strain plastic stress waves, propagating radially from the center of a hollow cylindrical specimen, is described. An optical system is used to measure the circumferential strain variation as a function of time, the propagation velocities, and, indirectly, the pressure loading required to produce these strains. A comparison is made between the experimental results and the theoretical predictions of a rate-independent plasticity theory for an aluminum alloy.

INTRODUCTION

It is readily appreciated that the design of experiments is greatly facilitated if the theory being verified is well founded. A fair illustration of this is the experimental work of B. Hopkinson [1] which occurred some forty years after the analysis of Pochhammer [2]. The extension of the Hopkinson bar experiments to plastic wave propagation did not have this advantage. The plastic bar experiments preceded not only the plastic analysis of the experiment, but even a well-founded theory on which to base this analysis. The simultaneous development of the experimental technique with the plastic theory may have proved to be advantageous to the theoretician, but it was most unfortunate from the experimentalist's point of view. Without the guidelines provided by the theory, the experimentalist was forced to assume that the bar experiments, so successful in elastic wave studies, would also prove to be so in the plastic case.

The observation of possible strain-rate effects, coupled with the uncertainties introduced by the assumption of negligible radial inertia, adds an almost insurmountable obstacle to the complete understanding of plastic wave propagation when the results from only one type of experiment are being utilized. The introduction of the uniaxial strain experiments of Rice, McQueen, and Walsh [3] in which a plane wave was generated in metal plates by explosives in contact with one surface, provided a suitable means for studying high intensity stress waves, but, as pointed out by Lee [4], the role of plasticity was relegated to a correction to hydrodynamic theory. This has been overcome to a certain extent in the plate impact experiments of Barker, Butcher and Karnes [5] which reduces the stress levels, and uses extremely accurate measuring techniques to observe the plastic behavior.

In an effort to increase the range of stress wave propagation experiments, work was undertaken to investigate the theoretical and experimental possibilities that might arise from the study of plastic stress waves propagating radially outward from the center of a cylindrical specimen. These waves were created by electrically exploding a copper wire along the axis. It was anticipated that experiments designed so much later in the evolution of dynamic plasticity, could incorporate a number of features that would both supplement earlier work, and also emphasize the plastic deformation aspects of the stress wave propagation. The preliminary results with regard to symmetry, the measurement of the loading, and response of the specimen to this loading are related elsewhere [6]. For the sake of continuity and to introduce more recent developments, a brief account only will be given here.

APPARATUS

The essential features of the loading mechanism are shown in Fig. 1. It consists of a high voltage (20 K.V.) capacitance system in which the stored electrical energy is suddenly released to the wire contained in the specimen. The success of this configuration in producing a simple deformation pattern depends on the character of the exploding wire. The very nature of an exploding wire would indicate that radial symmetry is to be expected. However, the two most desirable properties are the lack of variation along the length of the wire during and subsequent to the vaporization, and the resultant high pressures generated.

Due to the finite length of the wire and, hence, of the specimen, the period of observation is limited by end effects propagating into the center of the cylinder where the wave propagation characteristics are being measured. Assuming that the end effects propagate into the center section of the cylinder at elastic wave speed then, for the four inch long cylinder usually used, the measurement time is reduced to approximately eight microseconds.

To measure the response of the specimen to this type of loading, an

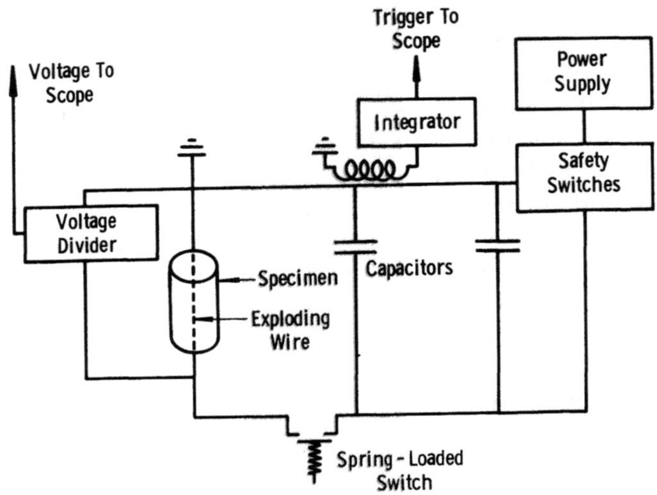

Fig. 1. Schematic of capacitor discharge system.

optical surface motion detection system was developed of the form shown in Fig. 2. In the present configuration of the system, a light beam from a laser is passed between a knife-edge and the specimen. The subsequent expansion of the specimen reduces the light intensity, as measured by a photomultiplier detector. To calibrate the system, the apparatus holding the specimen can be moved by micrometer to simulate its motion. As the calibration is a static process, considerable care must be taken to ensure that the detection system has a rise-time compatible with the dynamic event being measured.

The validity of this type of experimental system was checked by comparing the theoretical and measured responses of the outside of a specimen. The loading pressure acting on the inside surface of the cylindrical specimen was controlled to prevent loading beyond the elastic limit. This loading pressure history, acting on the inside surface, was required as a time-dependent boundary condition of the elastic theory, and was obtained by using the rather involved process described in [6]. Due to the simplicity of the strain-displacement relation in one-dimensional radial cylindrical motion ($\epsilon_\theta = u/r$) the response of the cylindrical specimen can be plotted as either the variation of the circumferential strain with time, or displacement with time at the free-surface. The results for aluminum 6061-T6 are presented in Fig. 3.

It is also possible, using the optical system, to measure the elastic wave speed from the oscilloscope trace of the free-surface displacement. As shown in the typical trace, Fig. 4, the time from the voltage peak to the indication of surface motion (A-B) can be measured for cylinders of different outside diameters. The slope produced by plot-

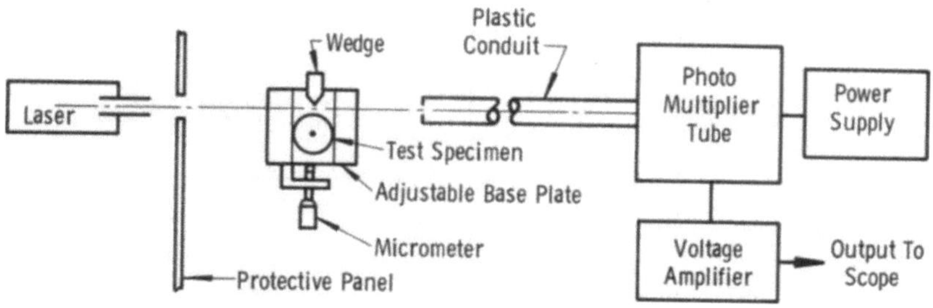

Fig. 2. Schematic of photomultiplier motion detector system.

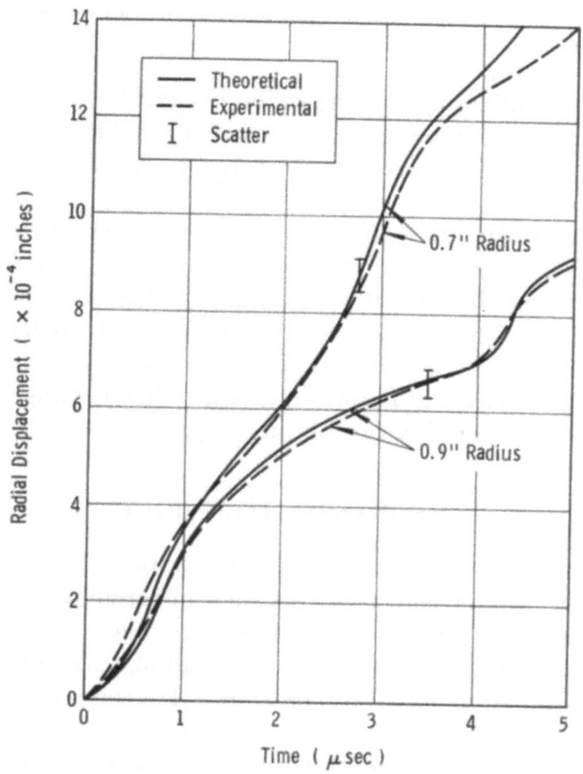

Fig. 3. Computed and measured radial surface displacements for elastically loaded aluminum (6061-T6) cylinders with an inner radius of 0.44 inches.

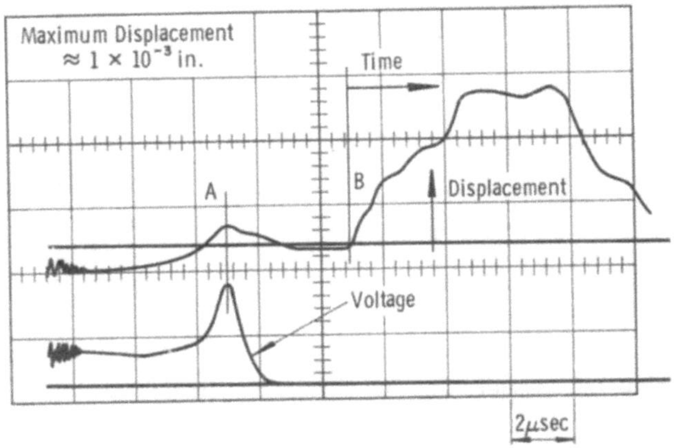

Fig. 4. Oscilloscope of free-surface displacement-time measurement with voltage peak timing point (*A*).

ting these time values against the different cylinder radii gives the required velocity. This type of measurement is possible because the peak voltage is a constant occurrence which represents the initial vaporization of the wire.

The testing procedures just described give a workable arrangement for producing one-dimensional radial deformations. The introduction of two strain components rather than one, makes it difficult to classify in terms of strain rate. However, although the stress levels may be varied quite readily by either changing the inside radius of the cylinder or the energy released to the wire, the shock-like loading restricts the application to the very high strain-rate regime.

THE INVERSE TECHNIQUE

Plastic waves in the plane-strain configuration can be readily obtained by the proper choice of specimen material and loading pressures. However, in the study of the deformations produced by the wave propagation it is important that the loading pressure-time history which produced these waves also be known. For this reason the experimental program can be considered in two parts: a) the determination of the loading pressure-time history, and b) the measurement of the corresponding free-surface displacements.

The useful range of the exploding-wire technique is dependent mainly on the methods available to obtain the values of the loading pressures. The initial method used in [6] was necessary in order to confirm the measuring processes used to obtain the free-surface motion, but a much simpler system is possible if use is made of the close agreement of elastic theory with its experimental counterpart.

The loading pressures acting on the cylindrical specimen are independent of the material itself, and are a function purely of the configuration and the surroundings of the exploding wire. The smaller the inside diameter of the specimen, the corresponding higher loading pressures produced. By introducing a material with a very high yield strength, for example 4130 steel, it is possible to load the specimen with relatively high pressures while the stresses still remain below the material yield point. The subsequent response of the specimen to this unknown pressure is that of an elastic material, and the elastic theory is therefore applicable.

In terms of the radial displacement u, the equation of motion for the elastic behavior is a second order partial differential equation. In the case of a dynamically loaded cylinder, the prescribed initial and boundary values required for a unique solution are:

$$u(r, o) = 0$$

$$\left.\frac{\partial u}{\partial t}\right|_{t=0} = 0$$

$$\sigma_r(a, t) = -p(t)$$

$$\sigma_r(b, t) = 0$$

where σ_r is the radial stress, u the radial displacement, $p(t)$ a prescribed pressure, a the inner radius and b the outer radius of the cylinder. However, by virtue of the theory related to the Cauchy problem, a unique solution for the second order partial differential equation can also be obtained, in a restricted region, by prescribing the values of u and $\partial u/\partial r$ along a non-characteristic radius. In the characteristic diagram, Fig. 5, this region is defined as ABC, with the line AB being the non-characteristic curve at $r = b$. Of the required values on the line AB, u is obtained directly from the measured response of the elastic cylinder, while the other quantity $\partial u/\partial r$ is a direct consequence of the stress-free boundary. Thus on AB we have

$$u(b, t) = u(t) - \text{measured}$$

and

$$\left.\frac{\partial u}{\partial r}\right|_{r=b} = -\frac{\nu}{1-\nu}\cdot\frac{u}{b}$$

where ν is Poisson's ratio for the elastic material. With this data it is possible to obtain the solution along the line AC. In the region aAC, which represents the initial propagation of the wave toward the outer surface, only simple waves exist. Using the properties inherent in sim-

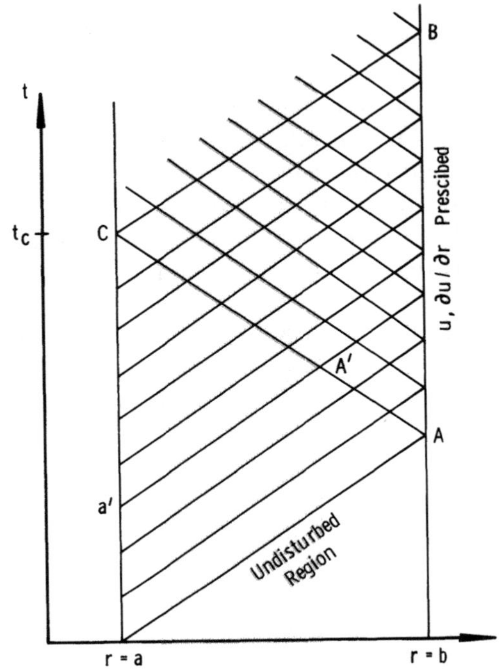

Fig. 5. Characteristic field of an elastically loaded cylinder of inner radius a and outer radius b.

ple waves, and the knowledge that aA borders an undisturbed region, it is possible to extend the solution at a point on AC to its corresponding point on aC (e.g., A^1 to a^1). Using this technique, it is then possible to obtain the value of σ_r anywhere on the line AC. However, on this line $\sigma_r = -p$, and hence we have the pressure-loading history acting on the inside diameter of the cylinder from time-zero to t_c. Keeping the energy released and the inner diameter of the specimens constant, the pressure loading is known regardless of the material used. The length of time over which this pressure history could be obtained was controlled by the size of the elastic cylinder.

Using some preliminary work in this area carried out by Anderson [7], a computer program was developed for this computation. Its validity was checked by calculating the pressure from the displacement supplied by a standard elastic analysis of the kind used to obtain the results given in Fig. 3. The pressure-loading history reproduced in this fashion agreed, within a few percent, with the original values. The mathematical procedures used in solving for the pressure with the free-surface displacement prescribed are identical with those used in solving for the free-surface displacement with the pressures prescribed, and is an inversion of the physical process rather than the mathematical. The

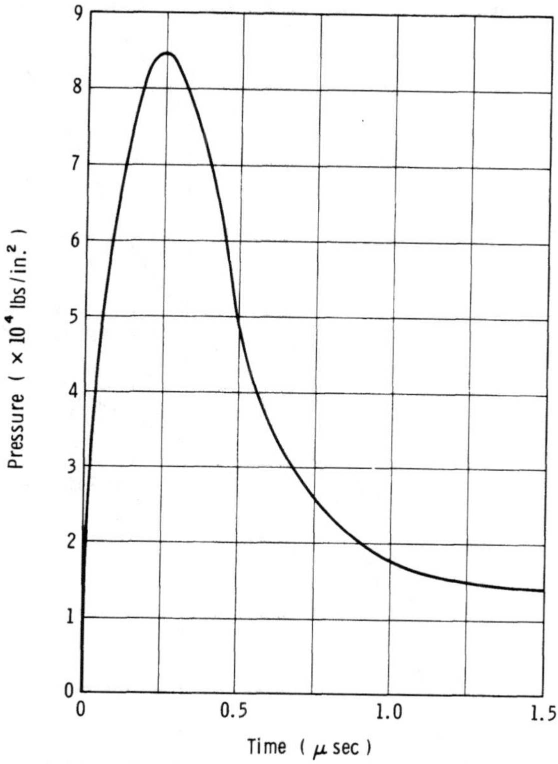

Fig. 6. Pressure-time variation calculated from the response of a steel (4130) cylinder. (Inner radius 0.252 inches, outer radius 0.425 inches.)

pressure as a function of time obtained by this method for a 4130 steel specimen with an inner radius of 0.252 inch is given in Fig. 6.

PLASTIC WAVE PROPAGATION EXPERIMENTS

In the initial plastic wave propagation studies, it was decided to eliminate the need to use the inversion method described above by using specimens identical in size, with regard to inside diameter and length, as those used in the initial elastic studies. The pressure loading as given in [6] thus applies to all cylinders with an inner radius of 0.44 inch. Further, to circumvent the problems of considering strain-rate effects in any theoretical development, the same material (aluminum 6061-T6) was used, as it is generally considered to be rate independent [8, 9]. The plastic flow was then obtained by annealing the specimen to reduce the yield stress to a value of 7000 lbs/ins.[2]. The variation of the radial displacement of the free-surface with time is given in Fig. 7

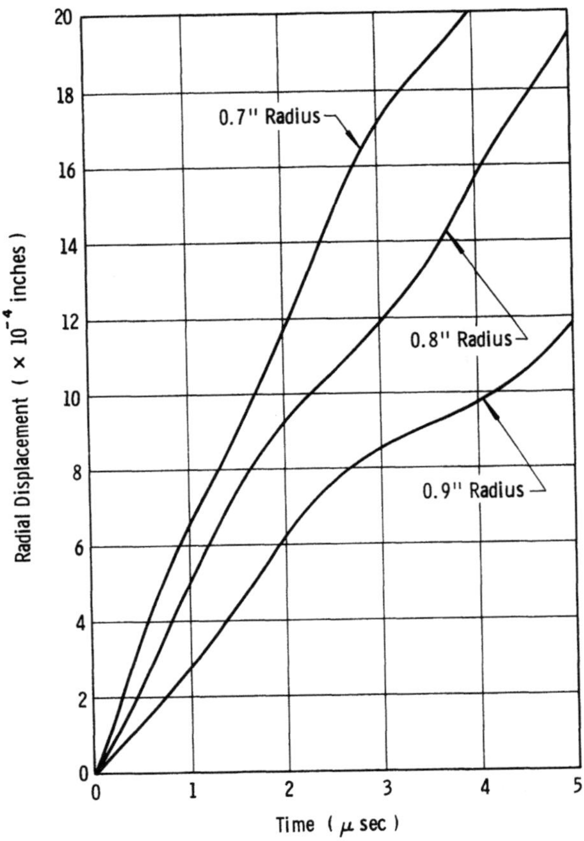

Fig. 7. Experimental values of the radial surface displacement for annealed aluminum (6061-T6) cylinders with an inner radius of 0.44 inches.

for the various outside diameters of these annealed specimens. It was observed that these strains were appreciably greater than those obtained under the identical loading conditions for the non-annealed aluminum specimens.

The rather unstable qualities of annealed aluminum may place unnecessarily severe restrictions on any theoretical development. To eliminate this problem, the specimen inner diameter was reduced with a subsequent increase in the loading. Using the inversion method with 4130 steel, the pressure loading predicted by this method indicated that the non-annealed aluminum 6061-T6 could be loaded into the plastic regime. This forecast was checked by observing small permanent deformation in the aluminum specimens. The results in Fig. 8 show the response of aluminum 6061-T6 due to a loading which induced plastic flow.

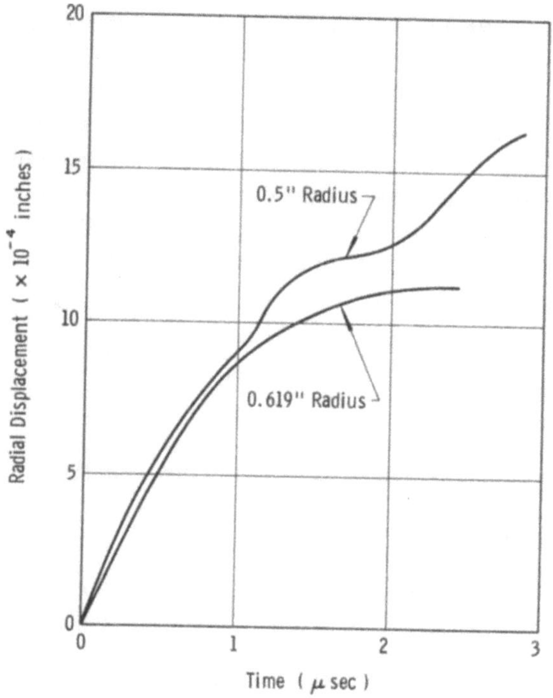

Fig. 8. Radial surface displacement of aluminum (6061-T6) cylinders subjected to the pressure loading given in Fig. 6 (inner radius 0.252").

THEORETICAL COMPARISON

The above results alone do not provide any particular insight as to the form that a constitutive equation for the plastic behavior of the material might take. However, the basic objective of this program is not to postulate new forms of constitutive equations, but rather to offer a measure against which any existing or proposed theory might be tested. It is to be expected that of the number of possible plasticity theories available, all should agree to some extent with these results, with the degree of agreement depending on the validity of the basic assumptions inherent in the theory. Thus, it is only by a comparison of different theories, in relation to the experiments, that the significance of any assumption can be isolated. The obvious theory on which to make a comparison should be an incremental flow rule. However, to avoid too cumbersome a computational problem, at this stage of development, it was decided to take advantage of a theory developed recently in a paper by Aggarwal, et al. [10]. In this paper it was suggested that despite the limitations of total-strain theory, a generalized form of the Koehler-Seitz bilinear model could be made applicable to dynamic problems. Additional incentives were: the equations appropriate to the problem being considered were available from [10] in a form compatible with the elastic solution already obtained, the tension test for the annealed aluminum readily fitted the bilinear model, and the desire to check a reasonable simple theory which might be useable for general analysis purposes.

The equations of motion as they apply to the different states of the material, and the yield criterion used are as follows:

Elastic

$$\frac{\partial^2 u}{\partial t^2} - c_1^2\left(\frac{\partial^2 u}{\partial r^2} + \frac{1}{r}\frac{\partial u}{\partial r} - \frac{u}{r^2}\right) = 0. \tag{1}$$

Von Mises Yield Criterion

$$6J_2^1 = (\sigma_r - \sigma_\theta)^2 + (\sigma_r - \sigma_z)^2 + (\sigma_\theta - \sigma_z)^2 = 2Y^2. \tag{2}$$

Plastic Loading

$$\frac{\partial^2 u}{\partial t^2} - c_2^2\left(\frac{\partial^2 u}{\partial r^2} + \frac{1}{r}\frac{\partial u}{\partial r} - \frac{u}{r^2}\right) = \frac{1}{\rho_0} G(\sigma_r^0, \sigma_\theta^0, r). \tag{3}$$

Plastic Unloading

$$\frac{\partial^2 u}{\partial t^2} - c_1^2\left(\frac{\partial^2 u}{\partial r^2} + \frac{1}{r}\frac{\partial u}{\partial r} - \frac{u}{r^2}\right) = \frac{1}{\rho_0} H(\epsilon_r^*, \epsilon_\theta^*, \epsilon_z^*, r). \tag{4}$$

where u is the radial displacement, ρ_0 the density, c_1 and c_2 the elastic and plastic wave velocities respectively, Y the yield stress in simple tension. The functions $G(\sigma_r^0, \sigma_\theta^0, r)$ and $H(\epsilon_r^*, \epsilon_\theta^*, \epsilon_z^*, r)$ depend respectively on the value of stress at the onset of yielding (see Fig. 9), and the strains at the onset of unloading. The quantities c_1, c_2, G, and H have the following forms:

$$c_1 = \left[\frac{E_1(1-\nu_1)}{\rho_0(1+\nu_1)(1-2\nu_1)}\right]^{1/2}$$

$$c_2 = \left[\frac{E_2(1-\nu_2)}{\rho_0(1+\nu_2)(1-2\nu_2)}\right]^{1/2}$$

$$G(\sigma_r^0, \sigma_\theta^0, r) = \left[1 - \frac{E_2}{E_1}\frac{(1+\nu_1)(1-2\nu_1)}{(1+\nu_2)(1-2\nu_2)}\right]\frac{\partial\sigma_r^0}{\partial r}$$

$$- \frac{E_2}{E_1}\left[\frac{(1+\nu_1)(\nu_2-\nu_1)}{(1+\nu_2)(1-2\nu_2)}\right]\frac{\partial(\sigma_\theta^0 - \sigma_r^0)}{\partial r}$$

$$- \frac{E_2}{1+\nu_2}\left[\frac{(1+\nu_2)}{E_2} - \frac{1+\nu_1}{E_1}\right]\frac{\sigma_\theta^0 - \sigma_r^0}{r}$$

$$H(\epsilon_r^*, \epsilon_\theta^*, \epsilon_z^*, r) = \frac{\nu_1}{1-\nu_1}\cdot\frac{\partial}{\partial r}(\epsilon_\theta^* - \epsilon_z^*) - \frac{\partial\epsilon_r^*}{\partial r} - \frac{1-2\nu_1}{1-\nu_1}\cdot\frac{\epsilon_r^* - \epsilon_\theta^*}{r}$$

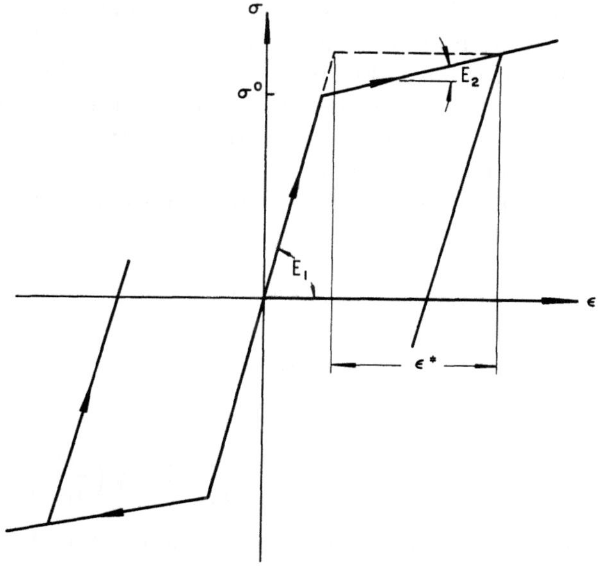

Fig. 9. Bilinear representation of a stress-strain curve in simple tension.

where E and ν are material constants with the subscripts 1 and 2 referring to the elastic and plastic regimes, respectively. It should also be noted at this point that the σ^0's and ϵ^*'s are functions of the spatial coordinates r, θ, z, and that the value of Y changes after initial yielding at each material point to allow for isotropic work-hardening. It has been shown [10] that a number of additional conditions must also be satisfied to meet the requirements of plasticity theory, these being,

$$0 < E_2 < E_1, \qquad \tfrac{1}{2} - (\tfrac{1}{2} - \nu_1)\frac{E_2}{E_1} \geqslant \nu_2 > \nu_1$$

and

$$J_2{}^1 \geqslant 0.$$

For the case of plastic incompressibility, this leads to a direct relationship for ν_2,

$$\nu_2 = \tfrac{1}{2} - (\tfrac{1}{2} - \nu_1)\frac{E_2}{E_1}. \qquad (5)$$

From a simple tension test of the annealed aluminum, it was possible to obtain the values for E_1, E_2 and Y. These values are 9×10^6 lbs/ins.^{2}, 4×10^5 lbs/ins.^{2} and 7×10^3 lbs/ins.^{2}, respectively. The value of ν_1 was obtained from an elastic wave experiment, and the value of ν_2 was calculated from (5) giving

$$\nu_1 = 0.351, \qquad \nu_2 = 0.493.$$

In the solution of the elastic equation of motion used in checking this technique, the method of characteristics was used. The additional equations (3) and (4) are also of the same form, and by adapting the method used for (1) to cover the additional material behavior of plastic loading and unloading, the solution for the plane-strain plastic wave propagation problem was obtained.

In characteristic form, these equations reduce to five first-order equations of the form

$$\frac{dr}{dt} = \pm c$$

$$du = p dt + q\, dr$$

and

$$dp \mp c dq = \frac{1}{r}\, q - \frac{u}{r^2} + \psi \left(\frac{G}{\rho_0 c_2{}^2}\right) + \gamma \left(\frac{H}{\rho_0 c_1{}^2}\right)$$

where the symbols p and q represent the standard mathematical notation for $\partial u/\partial t$ and $\partial u/\partial r$, respectively. These five equations are then solved numerically along the appropriate right- and left-running characteristics for the five unknowns u, $\partial u/\partial r$, $\partial u/\partial t$, r, t, and hence the stresses throughout the region. The symbols ψ and γ have the following values:

$\psi = 0$	$\gamma = 0$	elastic
$\psi = 1$	$\gamma = 0$	plastic loading
$\psi = 0$	$\gamma = 1$	plastic unloading

the combinations of ψ and γ being determined by (2) and the past history of the material point.

In this analysis, it was assumed that continuity of displacement, velocity, and the normal stress existed across the elastic-plastic interface. The radial displacements at the free-surface of the specimen predicted by this theory is shown in Fig. 10, together with the corresponding experimental values.

Although it is not the purpose of this paper either to defend or condemn the bilinear stress-strain law, a number of interesting observations can be made with regard to the results shown in Fig. 10. The displacements predicted by the bilinear theory differed very little from the results that could be obtained using classical elasticity theory, and indeed varying the initial yield stress term Y had no appreciable effect on the results. The very slight variation in the response produced by the plastic analysis appears to be a direct consequence of the similarity between the elastic equations (1) and the plastic loading equations (3); the left-hand sides of these two equations differ only by the wave speed

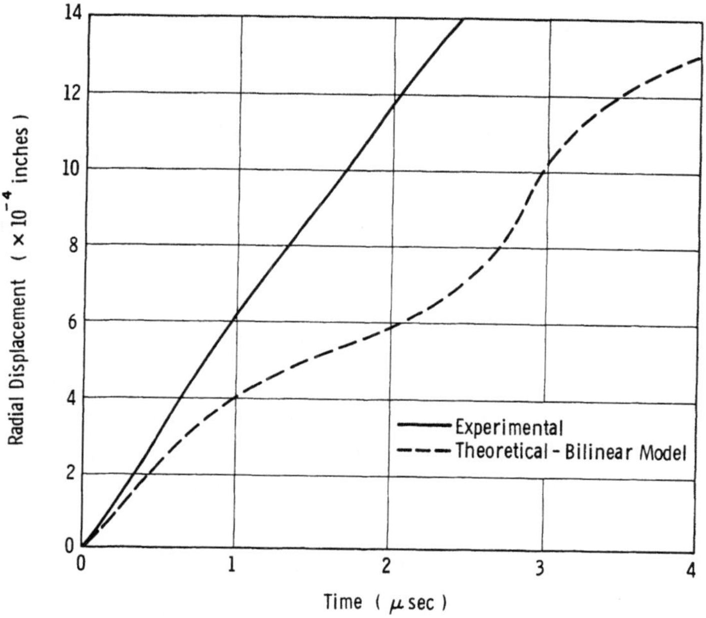

Fig. 10. Computed and measured radial surface displacements of an annealed aluminum (6061-T6) cylinder. (Inner radius 0.44 inches, outer radius 0.70 inches).

constants c_1 and c_2. If continuity is required at the elastic-plastic interface and the values of c_1 and c_2 are approximately equal, then the value of $G(\sigma_r^0, \sigma_\theta^0, r)$, which depends on the solution at the interface, may be small. As a consequence of this condition, the differential equation appropriate to both the elastic and plastic regimes will be essentially the same. It was noted in this case, where E_2 was considerably smaller than E_1 and plastic incompressibility was assumed, that c_2 was approximately equal to c_1. This limitation on plastic flow imposed when $E_2 \ll E_1$ seems a rather severe restriction on the bilinear model. The possibilities inherent in using a multilinear model or removing the incompressibility requirement was not considered, except it was noted that considerable plastic flow was predicted when the latter restriction was removed.

CONCLUSIONS

The introduction of a plane-strain configuration was dictated in part by the importance of considering the relative merits of dynamic plasticity theories in terms of their generalized formulation, rather than in an overly simple uniaxial form. Although the cylindrical geometry presents certain mathematical difficulties, as compared to the bar or plate impact experiments, the recent work of Perzyna and Bejda [11] indicates that these difficulties are by no means insurmountable. While at the same time consistency of the exploding-wire phenomena, and the pronounced behavior of the plastic flow is clearly advantageous.

The extension of these experiments to include materials with a rate-dependency does not require any change in technique. However, clearly the success of this experimental method depends largely on the use made of the results in relation to the theoretical analysis containing the constitutive equation in question.

ACKNOWLEDGMENTS

The author is indebted to R. Swift and F. A. Lee for their contributions to the experimental and theoretical aspects of the program. Also acknowledged is the support received by the U.S. Army Research Office, Durham, and the National Science Foundation during the course of this work.

References

1. B. Hopkinson, Phil. Trans. Roy. Soc. Lond., A *213*, 437 (1914).
2. L. Pochhammer, J. reine angew. Math. (Crelle), *81*, 324 (1876).
3. M. H. Rice, R. G. McQueen and J. M. Walsh, Solid State Physics, Eds. F. Seitz and D. Turnbull, Academic Press, Inc., *6* (1958).
4. E. H. Lee, Proc. 5th U.S. Congr. of Appl. Mech., 405 (1966).
5. L. M. Barker, B. M. Butcher, and C. H. Karnes, J. Appl. Physics, *37*, 1989 (1966).
6. R. R. Ensminger and I. M. Fyfe, J. Mech. Phys. Solids, *14*, 231 (1966).
7. D. A. Anderson, Private Communication (1966).
8. U. S. Lindholm, Behavior of Materials Under Dynamic Loading, Ed. N. J. Huffington, Jr., A.S.M.E. (1965).
9. C. J. Maiden and S. J. Green, J. Appl. Mech., *33*, 496 (1966).
10. H. R. Aggarwal, A. M. Soldate, J. F. Hook and J. Miklowitz, J. Appl. Mech., *31*, 181 (1964).
11. P. Perzyna and J. Bejda, Arch. Mech. Stos., *16*, 1215 (1964).

DYNAMIC PLASTICITY UNDER COMBINED STRESS

N. Cristescu

Mathematical Institute
Bucharest, Romania

ABSTRACT

The paper discusses the theory of propagation of elastic/viscoplastic waves in thin tubes subjected to combined tension and torsion. Several rate dependent constitutive equations are used. One compares the results obtained with various constitutive equations. In particular are discussed the coupling of plastic waves and the concepts of loading and unloading for viscoplastic constitutive equations.

1. FORMULATION OF THE PROBLEM

Let us consider a thin walled tubular specimen of initial length l_0. The end $x = l_0$ of the tube will be assumed to be fixed, while the other end $x = 0$ is put dynamically into a combined motion: a tension and a torsion (see Lindholm [1]). This motion of the end section $x = 0$ of the tube is transmitted along the tube by intermedia of waves. This mechanism of propagation is analyzed below for several kinds of constitutive equations, in order to give the possibility to choose the appropriate constitutive equation to be used for a certain material and a specific type of experiment.

We shall use cylindrical coordinates of reference x, r, θ, the $0x$ axis being directed along the symmetry axis of the tube. The components of the displacements in the axial and circumferential directions will be denoted by u and v, respectively. Due to the small thickness of the walls the components σ_r, σ_{rx}, $\sigma_{r\theta}$ will be assumed to be small and negligible with respect to σ_{xx} and $\sigma_{\theta x}$. On the other hand the whole problem is axi-symmetrical so that all derivatives with respect to θ will be considered zero $\left(\frac{\partial}{\partial\theta} = 0\right)$. Due to this last assumption and because the radial motion will be disregarded, the single coordinate which is involved in computation is the axial coordinate x. The rotation of various transverse sections of the tube will be obtained as a result of computation. This will be explained further.

2. THE EQUATIONS OF MOTION

Using the previous assumptions and taking into account that only two stress components are assumed to be different from zero (these will be denoted by $\sigma_{xx} = \sigma$ and $\sigma_{\theta x} = \tau$), the equations of motion are

$$\frac{\partial \sigma}{\partial x} + F_x = \rho \frac{\partial u_t}{\partial t}$$
$$\frac{\partial \tau}{\partial x} + F_\theta = \rho \frac{\partial v_t}{\partial t} \tag{2-1}$$

where F_x and F_θ are body forces components, ρ is the density and u_t and v_t are the components of the velocities in the axial and circumferential directions.

If we denote by u, v, w the displacements respectively in the x, θ and r directions, and we take into account the axial symmetry, then the strain components are

$$\epsilon_{rr} = \frac{\partial w}{\partial r}, \qquad \epsilon_{\theta\theta} = \frac{w}{r}, \qquad \epsilon_{xx} = \frac{\partial u}{\partial x} \; (= \epsilon)$$
$$\epsilon_{\theta x} = \frac{1}{2}\frac{\partial v}{\partial x} \; (= \gamma), \qquad \epsilon_{xr} = \frac{1}{2}\left(\frac{\partial u}{\partial r} + \frac{\partial w}{\partial x}\right), \qquad \epsilon_{r\theta} = \frac{1}{2}\left(\frac{\partial v}{\partial r} - \frac{v}{r}\right). \tag{2-2}$$

From all these strain components two components, namely ϵ_{xx} and $\epsilon_{\theta x}$ are dominant. These components will be further denoted by ϵ and γ, respectively.

The first invariant of the strain tensor is

$$\epsilon_m = \frac{\epsilon_{rr} + \epsilon_{\theta\theta} + \epsilon_{xx}}{3} \tag{2-3}$$

while the second invariant of the stress deviator is, for the case under consideration,

$$I_s^{(2)} = \frac{1}{2} s_{ij} s_{ij} = \frac{\sigma^2}{3} + \tau^2. \tag{2-4}$$

Two special motions corresponding to two special boundary conditions will be considered. The first corresponds to the case when everywhere $v = 0$. Then the system (2-1) reduces to a single equation of motion

$$\frac{\partial \sigma}{\partial x} + F_x = \rho \frac{\partial u_t}{\partial t} \tag{2-5}$$

and we have only a uniaxial longitudinal compressional motion. The other particular case corresponds to $u=0$; then instead of (2-1) we have

$$\frac{\partial \tau}{\partial x}+F_\theta=\rho\,\frac{\partial v_t}{\partial t} \tag{2-6}$$

which describes a uniaxial shearing motion. These two particular cases which occur when one or another component of the displacement is equal to zero, will be denoted below by the index 0 (zero).

3. THE CONSTITUTIVE EQUATIONS

In order to express the constitutive equations, we shall assume, as usual, that the strain-rate components may be decomposed in an elastic component and a plastic one

$$\dot{e}_{ij}=\dot{e}_{ij}^{E}+\dot{e}_{ij}^{P} \tag{3-1}$$

where e_{ij} are the rate of strain deviator components. The elastic part of the strain rate satisfies always the Hooke's law

$$2G\dot{e}_{ij}^{E}=\dot{s}_{ij} \tag{3-2}$$

and

$$\dot{\sigma}_m=3K\dot{\epsilon}_m \tag{3-3}$$

where s_{ij} are the components of the stress deviator, σ_m is the mean stress and G and K are the usual elastic constants.

What concerns the plastic component of the strain, having in mind to apply the theory to various kinds of materials (plastics, metals, etc.) we shall assume quite general constitutive equations able to emphasize not only plastic inviscid properties, but also viscoplastic effects. In a general form, such constitutive equations can be written as (Cristescu [2], Ch. X)

$$\dot{\epsilon}_{ij}^{P}=A_{ijkl}\dot{s}_{kl}+B_{ij}. \tag{3-4}$$

Giving various expressions for the coefficients A_{ijkl} and B_{ij} we can obtain several constitutive equations used in dynamic plasticity (see §4).

For the particular stress state under consideration, (3-4) becomes

$$\begin{aligned}\dot{\epsilon}_{xx}^{P}&=\varphi_{11}\dot{s}_{xx}+\varphi_{12}\dot{s}_{\theta x}+\psi_1\\ \dot{\epsilon}_{\theta x}^{P}&=\varphi_{21}\dot{s}_{xx}+\varphi_{22}\dot{s}_{\theta x}+\psi_2.\end{aligned} \tag{3-5}$$

Using (2-2, 3-1–3-3) and the simplified notation mentioned before (3-5) can be written as

$$\frac{\partial u_t}{\partial x} = \left(\frac{2}{3}\varphi_{11} + \frac{1}{E}\right)\dot{\sigma} + \varphi_{12}\dot{\tau} + \psi_1$$
$$\frac{\partial v_t}{\partial x} = \frac{4}{3}\varphi_{21}\dot{\sigma} + \left(2\varphi_{22} + \frac{1}{G}\right)\dot{\tau} + 2\psi_2 \tag{3-6}$$

taking into account also that $\sigma_m = \sigma/3, E = 9KG/(3K+G)$. For the sake of simplicity we shall use the notations

$$\alpha_{11} = \frac{1}{3}\left(2\varphi_{11} + \frac{1}{G} + \frac{1}{3K}\right), \qquad \alpha_{12} = \varphi_{12}, \qquad \beta_1 = \psi_1$$
$$\alpha_{21} = \frac{4}{3}\varphi_{21}, \qquad \alpha_{22} = 2\varphi_{22} + \frac{1}{G}, \qquad \beta_2 = 2\psi_2 \tag{3-7}$$

so that the constitutive equation (3-6) will be written in the simple form

$$\frac{\partial u_t}{\partial x} = \alpha_{11}\dot{\sigma} + \alpha_{12}\dot{\tau} + \beta_1$$
$$\frac{\partial v_t}{\partial x} = \alpha_{21}\dot{\sigma} + \alpha_{22}\dot{\tau} + \beta_2. \tag{3-8}$$

4. SPECIAL CASES OF THE CONSTITUTIVE EQUATIONS

Several special cases of constitutive equations (3-8) will be examined. These will be denoted by $P1$ to $P5$.

P1. The simplest possible case is the elastic one:

$$E\frac{\partial u_t}{\partial x} = \dot{\sigma}$$
$$G\frac{\partial v_t}{\partial x} = \dot{\tau} \tag{4-1}$$

which is obtained from (3-8) for

$$\varphi_{11} = \varphi_{12} = \varphi_{21} = \varphi_{22} = \psi_1 = \psi_2 = 0.$$

P2. A more complicated constitutive equation corresponds to

$$\varphi_{11} = \varphi_{12} = \varphi_{21} = \varphi_{22} = 0. \tag{4-2}$$

For instance the Hohenemser-Prager [3] constitutive equation is of this form

$$\dot{\epsilon} = \frac{1}{E}\dot{\sigma} + \frac{1}{3\eta}\left(1 - \frac{k}{\sqrt{I_s^{(2)}}}\right)\sigma$$
$$\dot{\gamma} = \frac{1}{2G}\dot{\tau} + \frac{1}{2\eta}\left(1 - \frac{k}{\sqrt{I_s^{(2)}}}\right)\tau. \tag{4-3}$$

There are many other constitutive equations used in the literature which can be obtained from (3-8) with the conditions (4-2), but with other expressions for the functions β_1 and β_2.

P3. The last particular case (4-2) can be considered at its turn as a special case of the more general case

$$\varphi_{12} = \varphi_{21} = 0. \tag{4-4}$$

The constitutive equations (3-8) now become

$$\begin{aligned} \frac{\partial u_t}{\partial x} &= \left(\frac{2}{3}\varphi_{11} + \frac{1}{E}\right)\dot{\sigma} + \psi_1 \\ \frac{\partial v_t}{\partial x} &= \left(2\varphi_{22} + \frac{1}{G}\right)\dot{\tau} + \psi_2. \end{aligned} \tag{4-5}$$

Such constitutive equations will be called quasi-linear uncoupled constitutive equations.

P4. If all the coefficients φ_{ij} are different from zero but

$$\psi_1 = \psi_2 = 0 \tag{4-6}$$

the constitutive equations will be called quasi-linear coupled constitutive equations. An example of such constitutive equations is the Prandtl-Reuss constitutive equation

$$2G\dot{e}_{ij} = \dot{s}_{ij} + \frac{2G}{H'}\frac{s_{kl}\dot{s}_{kl}}{s_{mn}s_{mn}}s_{ij} \tag{4-7}$$

where the work-hardening law was written in the form

$$s_{ij}s_{ij} = H\left(\int s_{kl}\, de_{kl}^{P}\right) \tag{4-8}$$

and H' is the derivative of H with respect to its argument.

For the case under consideration (4-7) becomes

$$\begin{aligned} \dot{\epsilon} &= \left[\frac{1}{E} + \frac{4}{27}\frac{1}{H'I_s^{(2)}}\sigma^2\right]\dot{\sigma} + \frac{2}{3}\frac{1}{H'I_s^{(2)}}\sigma\tau\dot{\tau} \\ \dot{\gamma} &= \frac{2}{9}\frac{1}{H'I_s^{(2)}}\sigma\tau\dot{\sigma} + \left[\frac{1}{2G} + \frac{1}{H'I_s^{(2)}}\tau^2\right]\dot{\tau}. \end{aligned} \tag{4-9}$$

The Hencky-Iliushin constitutive equation written in differential form can also be considered as a special case of (3-8) with (4-6) satisfied.

Generally the conditions (4-6) may characterize more general constitutive equations than the classic inviscid ones.

P5. Finally the last special case which will be considered is an example of the general constitutive equation of the type (3-2, 3-4), of the form

$$\dot{e}_{ij} = \frac{\dot{s}_{ij}}{2G} + \frac{1}{2\eta}\left(1 - \frac{k}{\sqrt{I_s^{(2)}}}\right) s_{ij} + \frac{\mathscr{F}(I_s^{(2)})}{2I_s^{(2)}} s_{kl}\dot{s}_{kl}s_{ij}. \tag{4-10}$$

This can be obtained easily (Cristescu [2]) assuming that the rate of strain component can be decomposed as

$$\dot{e}_{ij} = \dot{e}_{ij}^{E} + \dot{e}_{ij}^{VP} + \dot{e}_{ij}^{P} \tag{4-11}$$

where $\dot{e}_{ij}^{VP}$ is the viscoplastic rate of strain component and $\dot{e}_{ij}^{P}$ is the plastic inviscid rate of strain component. In (4-10) k is a plastic constant, η is the viscosity coefficient while the function $\mathscr{F}$ describes the work-hardening properties of the body, according to $s_{ij}\dot{e}_{ij}^{P} = \mathscr{F}(I_s^{(2)})s_{kl}\dot{s}_{kl}$.

In particular, for the problem under consideration, the equation (4-10) can be written as

$$\dot{\epsilon} = \left[\frac{1}{E} + \frac{4}{27}\frac{\mathscr{F}(I_s^{(2)})}{I_s^{(2)}}\sigma^2\right]\dot{\sigma} + \frac{2}{3}\frac{\mathscr{F}(I_s^{(2)})}{I_s^{(2)}}\sigma\tau\dot{\tau} + \frac{1}{3\eta}\left(1 - \frac{k}{\sqrt{I_s^{(2)}}}\right)\sigma$$

$$\dot{\gamma} = \frac{2}{9}\frac{\mathscr{F}(I_s^{(2)})}{I_s^{(2)}}\sigma\tau\dot{\sigma} + \left[\frac{1}{2G} + \frac{\mathscr{F}(I_s^{(2)})}{I_s^{(2)}}\tau^2\right]\dot{\tau} + \frac{1}{2\eta}\left(1 - \frac{k}{\sqrt{I_s^{(2)}}}\right)\tau. \tag{4-12}$$

Such constitutive equations describe the main properties emphasized by the general form of the constitutive equation (3-4). If necessary, other effective examples of the constitutive equations (3-4) can be imagined, which would emphasize other possible mechanical properties.

By putting $\mathscr{F} = 0$ in (4-12) we obtain (4-3) as a particular case.

The computations which follow were done mainly for the general form of the constitutive equation (3-8). However, several formulae will be specialized for the previous five particular cases; in these formulae indices from 1 to 5 will be used to indicate the corresponding particular case.

5. VELOCITIES OF PROPAGATION

If we compute the characteristic lines of the system (2-1, 3-8) we obtain four families of characteristic lines satisfying the differential equation

$$\rho^2(\alpha_{11}\alpha_{22} - \alpha_{12}\alpha_{21})(dx)^4 - \rho(\alpha_{11} + \alpha_{22})(dx)^2(dt)^2 + (dt)^4 = 0. \tag{5-1}$$

Thus we obtain two velocities of propagation which are furnished by

$$\left.\begin{matrix} c_{TL}^2 \\ c_{LT}^2 \end{matrix}\right\} = \frac{\alpha_{11} + \alpha_{22} \pm \sqrt{(\alpha_{11} - \alpha_{22})^2 + 4\alpha_{12}\alpha_{21}}}{2\rho(\alpha_{11}\alpha_{22} - \alpha_{12}\alpha_{21})}. \tag{5-2}$$

It is easy to see that both velocities of propagation are real, and that these are finite if

$$\begin{vmatrix} \alpha_{11} & \alpha_{12} \\ \alpha_{21} & \alpha_{22} \end{vmatrix} \neq 0.$$

The velocities (5-2) are generally distinct and $c_{TL} > c_{LT}$; they can coincide, generally, only in isolated points.

Therefore, for the general case (3-8) as well as for the particular cases $P4$ and $P5$ the velocities are variable and furnished by (5-2).

For the particular case $P3$ putting in (5-2) $\varphi_{12} = \varphi_{21} = 0$ we obtain

$$c_{L3}^2 = \frac{1}{\rho\alpha_{11}}, \qquad c_{T3}^2 = \frac{1}{\rho\alpha_{22}}. \tag{5-3}$$

These velocities are again variable, but furnished by much more simple formulae.

Finally, for the particular cases $P1$ and $P2$, by using (4-2) in (5-3) we obtain

$$c_{L1}^2 = c_{L2}^2 = \frac{E}{\rho}, \qquad c_{T1}^2 = c_{T2}^2 = \frac{G}{\rho}. \tag{5-4}$$

Now the meaning of the two indices L and T is clear: they mark the longitudinal and the transverse (shearing) waves, respectively. The meaning of these indices in (5-2) will be explained later on.

In the cases when a uniaxial motion occurs, the computation must be repeated from the beginning. If $v_t = 0$, attaching to (2-5) only the first equation (3-8) (with $\varphi_{12} = 0$) we obtain the velocity of propagation of the longitudinal waves

$$c_{L0}^2 = \frac{1}{\rho} \frac{3E}{2E\varphi_{11} + 3} \tag{5-5}$$

which for the particular cases $P1$ and $P2$ reduces to the first velocity (5-4). If $u_t = 0$, then attaching to (2-6) the second equation (3-8) (in which $\varphi_{21} = 0$) we obtain the velocity of propagation

$$c_{T0}^2 = \frac{1}{\rho} \frac{G}{2G\varphi_{22} + 1} \tag{5-6}$$

which reduces to the second velocity (5-4) in the particular cases $P1$ and $P2$.

6. DIFFERENTIAL RELATIONS SATISFIED ALONG THE CHARACTERISTIC LINES

In order to perform an integration scheme and to study the properties of the waves involved it is necessary to write the differential relations satisfied along the characteristic lines.

For the system (2-1, 3-8) the differential relations satisfied along the characteristic lines (5-1) can be found easily. These are

$$\mp\rho c(1-\rho\alpha_{22}c^2)du_t \mp \rho^2\alpha_{12}c^3\,dv_t + (1-\rho\alpha_{22}c^2)d\sigma$$
$$+\rho\alpha_{12}c^2\,d\tau + [\rho\beta_1 c^2(1-\rho\alpha_{22}c^2) + \beta_2\alpha_{12}\rho^2c^4$$
$$\mp c(1-\rho\alpha_{22}c^2)F_x \mp \rho c^3\alpha_{12}F_\theta]dt = 0 \quad (6\text{-}1)$$

where for c one of the two expressions (5-2) may be replaced (thus in (6-1) we have four distinct relations), while "d" stands for "interior derivative" along a characteristic line. It is sometimes useful to write the equations (6-1) in another form

$$\rho\alpha_{21}c^2\,du_t + (1-\rho\alpha_{11}c^2)dv_t \mp \alpha_{21}c\,d\sigma$$
$$+\frac{1-\rho\alpha_{11}c^2}{\mp\rho c}\,d\tau + \left[\mp\beta_1\rho\alpha_{21}c^3 \mp \beta_2 c(1-\rho\alpha_{11}c^2)\right.$$
$$\left.+\alpha_{21}c^2F_x + \frac{1-\rho\alpha_{11}c^2}{\rho}F_\theta\right]dt = 0 \quad (6\text{-}2)$$

which is certainly equivalent with (6-1).

A main conclusion is that all the unknown functions involved in the problem are present in all the differential relations (6-1), and thus all the four waves are at the same time shearing waves and longitudinal waves (this will be discussed in the next section). This conclusion holds for the general constitutive equation (3-8), but also for the particular cases $P4$ and $P5$. (We must replace only in (6-1) the expressions of α_{ij} by the corresponding values from $P4$ or $P5$.)

However, if a constitutive equation of the type $P3$ is used, then (6-1) reduces to

$$\mp\rho c_{L3}\,du_t + d\sigma + (\rho\beta_1 c_{L3}{}^2 \mp c_{L3}F_x)dt = 0 \quad (6\text{-}3)$$

because $1-\rho\alpha_{22}c_{L3}{}^2 \neq 0$. Similarly for $c = c_{T3}$ we obtain from (6-2).

$$\mp\rho c_{T3}\,dv_t + d\tau + (\rho\beta_2 c_{T3}{}^2 \mp c_{T3}F_\theta)dt = 0. \quad (6\text{-}4)$$

Thus constitutive equations of the type (4-5) have the property to separate the two previous coupled waves into two separated simple waves: one producing an axial compression, while the other a pure torsion. However, if β_1 is function of v_t and τ too, and β_2 is function of u_t and σ too, then the two kinds of waves are partially coupled, in the sense that the two kinds of waves influence each other only through the coefficients β_1 and β_2.

Certainly, the constitutive equations $P1$ and $P2$, which are special cases of $P3$, possess the same property. We easily obtain for $P2$

$$\mp\sqrt{\rho E}\, du_t + d\sigma + \left(\beta_1 E \mp \sqrt{\frac{E}{\rho}}\, F_x\right) dt = 0$$
$$\mp\sqrt{\rho G}\, dv_t + d\tau + \left(\beta_2 G \mp \sqrt{\frac{G}{\rho}}\, F_\theta\right) dt = 0 \tag{6-5}$$

and for $P1$

$$\mp\sqrt{\rho E}\, du_t + d\sigma \mp \sqrt{\frac{E}{\rho}}\, F_x\, dt = 0$$
$$\mp\sqrt{\rho G}\, dv_t + d\tau \mp \sqrt{\frac{G}{\rho}}\, F_\theta\, dt = 0. \tag{6-6}$$

Since in (6-6) the functions β are no more present, the two waves are completely separated.

In conclusion, there are two stages in simplifying the theory made for the quasi-linear coupled constitutive equation (3-8) (or $P5$ or $P4$ – the mathematical difficulties being of the same order). First, if the constitutive equation is uncoupled, of the type $P3$, then the two waves separate, but both velocities of propagation are variable. In the second stage when we pass to the constitutive equation $P2$ or $P1$, both velocities of propagation become constant.

7. THE COUPLING OF THE TWO KINDS OF WAVES

The possibility of coupling of the two kinds of waves was already discussed previously, but to make clear this subject we can consider it by using the jump conditions.

From (3-8) and (2-1) we obtain the dynamic jump conditions across a wave front

$$\left[\frac{\partial u_t}{\partial x}\right] = \alpha_{11}\left[\frac{\partial \sigma}{\partial t}\right] + \alpha_{12}\left[\frac{\partial \tau}{\partial t}\right]$$
$$\left[\frac{\partial v_t}{\partial x}\right] = \alpha_{21}\left[\frac{\partial \sigma}{\partial t}\right] + \alpha_{22}\left[\frac{\partial \tau}{\partial t}\right] \tag{7-1}$$

and

$$\rho\left[\frac{\partial u_t}{\partial t}\right] = \left[\frac{\partial \sigma}{\partial x}\right], \qquad \rho\left[\frac{\partial v_t}{\partial t}\right] = \left[\frac{\partial \tau}{\partial x}\right] \tag{7-2}$$

where $[\Phi]$ stands for the jump of the function Φ. The kinematic jump conditions are

$$\left[\frac{\partial \Psi}{\partial t}\right] = -c\left[\frac{\partial \Psi}{\partial x}\right] \quad \text{for} \quad \Psi = u_t, v_t, \sigma, \tau. \tag{7-3}$$

Introducing (7-3) into (7-2), and the result in (7-1) we obtain the following two equivalent conditions

$$\left[\frac{\partial u_t}{\partial x}\right](1-\rho\alpha_{11}c^2)=\rho\alpha_{12}c^2\left[\frac{\partial v_t}{\partial x}\right]$$
$$\left[\frac{\partial v_t}{\partial x}\right](1-\rho\alpha_{22}c^2)=\rho\alpha_{21}c^2\left[\frac{\partial u_t}{\partial x}\right] \tag{7-4}$$

From (7-4) it is evident that generally, for (3-8), $P5$ and $P4$, because $\alpha_{12} \neq 0$ and $\alpha_{21} \neq 0$, both kinds of waves are coupled: both produce a longitudinal and a circumferential motion. This is the reason why the velocities of propagation (5-2) were denoted by TL and LT.

It is only if $\alpha_{12}=\alpha_{21}=0$, i.e., for $P1$–$P3$, that the two waves separate, i.e., for boundary conditions which prescribe at the end of the tube a velocity parallel to the axis of the tube, and, simultaneously, an angular velocity about this axis, the two motions propagate along the tube by two separate kinds of waves. One is longitudinal (denoted by L) and the other shearing (denoted by T). In the special cases $P2$ and $P3$ the two kinds of waves can influence each other by the intermedia of the factors β_1 and β_2 (see eqs. 6-3–6-5), while in the special case $P1$ these are completely separated. This is the reason that constitutive equations of the form $P3$ were called partially coupled.

If we compare $P3$ and $P4$ we may observe that starting from the same Prandtl-Reuss form of constitutive equation

$$2G\dot{e}_{ij}=\dot{s}_{ij}+\lambda s_{ij} \tag{7-5}$$

we obtain either $P4$ if in order to give the explicit expression of λ we use a "global" yield condition (or work-hardening rule), or $P3$ if we use a "piecewise" yield condition (or work-hardening rule). Here "global" yield condition means a yield condition expressed by a single equation, while "piecewise" yield condition is a yield condition expressed by several distinct equations. For instance if the work-hardening properties are expressed by two equations

$$\sigma=\hat{\sigma}(\epsilon), \qquad \tau=\hat{\tau}(\gamma) \tag{7-6}$$

where $\hat{\sigma}$ and $\hat{\tau}$ are strictly increasing functions, then from (7-5) we obtain $P3$ and not $P4$. Since from the physical point of view "piecewise" yield conditions are either not acceptable, or the functions $\hat{\sigma}$ and $\hat{\tau}$ are to be linked in a functional form, it results that generally plastic waves are coupled. This must be checked experimentally.

Concerning the mechanism by which the two parts of the rate of strain components (elastic and plastic) are propagating, it is easy to show (in a similar manner as was done by Cristescu [4, 5]) that the elastic part of the rate of strain component propagates always by waves. The plastic (inviscid) component of the strain propagates always by waves (for $P3$, $P4$ and $P5$) while the viscoplastic components propagate in an intermediate manner (for $P2$, $P3$ and $P5$). The coupling concerns the plastic inviscid components, while the viscoplastic components can be at most partially coupled.

Returning to the relations (7-4) we can conclude that, if one of the three possibilities arises

$$c_{LT}^2 \lesseqqgtr \frac{1}{\rho(\alpha_{11} + \alpha_{12})} \tag{7-7}$$

we have

$$\left[\frac{\partial u_t}{\partial x}\right] \lesseqqgtr \left[\frac{\partial v_t}{\partial x}\right] \tag{7-8}$$

respectively. In the first case the shearing motion is dominant with respect to the longitudinal motion, while in the third case a reverse situation arises. The equality in (7-7) and (7-8) occurs in isolated points or for a particular type of (theoretical) loading when everywhere $\dot{\sigma} = \dot{\tau}$, which can be called "diagonal loading." Similarly we obtain

$$c_{TL}^2 \gtreqqless \frac{1}{\rho(\alpha_{21} + \alpha_{22})} \tag{7-9}$$

from which again yields (7-8) and similar conclusions.

8. INITIAL AND BOUNDARY CONDITIONS

In order to solve the problem, the initial and boundary conditions must be prescribed.

As for the initial conditions, we shall assume that at $t < 0$ the tube is at rest and that some constant uniformly distributed stress state is present, i.e.,

$$\text{for} \quad \left.\begin{matrix} 0 \leqslant x \leqslant l_0 \\ \\ t < 0 \end{matrix}\right\} \quad \begin{matrix} u_t = v_t = 0 \\ \\ \sigma = \sigma_0, \qquad \tau = \tau_0. \end{matrix} \tag{8-1}$$

Possibly $\sigma_0 = \tau_0 = 0$.

The boundary conditions will be prescribed as follows. The end $x = l_0$ of the tube is fixed:

$$\left.\begin{matrix} x = l_0 \\ t \geqslant 0 \end{matrix}\right\} \quad u_t = v_t = 0 \tag{8-2}$$

while the end $x = 0$ is put into a combined motion

$$\left.\begin{matrix} x = 0 \\ t \geqslant 0 \end{matrix}\right\} \quad u_t = U(0, t), \qquad v_t = V(0, t). \tag{8-3}$$

Using the conditions (8-1–8-3) we can integrate the system of equations (2-1, 3-8) (or one of the systems (4-1, 4-3, 4-5, 4-9, 4-12)), in order to obtain

$$u_t(x, t), \qquad v_t(x, t), \qquad \sigma(x, t), \qquad \tau(x, t).$$

Then the longitudinal and circumferential motions of the points lying on a certain circle $x = x^*$ can be obtained from

$$u(x^*, t) = \int_0^t u_t(x^*, \bar{t})d\bar{t}, \qquad \theta(x^*, t) = \frac{1}{r}\int_0^t v_t(x^*, \bar{t})d\bar{t}$$

where r is the radius of the tube. Summing up $u(x, t)$ along the tube (fixed t) we obtain the variation of the length of the tube, at the considered time t.

9. LOADING/UNLOADING CRITERIA

To make precise the cases of applicability of various constitutive equations under consideration, the loading/unloading criteria for these constitutive equations must be discussed. Several possibilities may arise, but always *loading* will be associated to the idea that a certain variation of the stress produces an increase of some of the two plastic strain components. Because the loading/unloading criteria for the constitutive equations $P1$–$P4$ are either well known or particular cases of $P5$, we shall discuss further (according to Cristescu [2]) only the loading/unloading criteria for the constitutive equation (4-10).

A variation of the stress state which produces an increase of both viscoplastic and plastic components of the strain will be named *total loading.* In this case the current stress state satisfies the conditions

$$I_s^{(2)} > k^2 \quad \text{and} \quad I_s^{(2)} = \mathscr{T}(W^P) \tag{9-1}$$

where $\mathscr{T}(W^P)$ stands for the value of $I_s^{(2)}$ at the current yield state and $dW^P = s_{ij}d\epsilon_{ij}^P$. $\mathscr{T}(W^P)$ is an isotropic work-hardening parameter, which can be considered either independent or dependent on the loading rate history. In (9-1) we have assumed that for any W^P we have $\mathscr{T}(W^P) > k^2$; if a contrary situation arises, the whole discussion of the loading/unloading criterion can be done in a similar manner. In the special case when $\mathscr{T}(W^P) < k^2$ and $I_s^{(2)} < k^2$, then $F(I_s^{(2)})$ is the function which defines the isotropic quasi-static work-hardening.

Due to the presence of the ideal (inviscid) plastic part of the model, the total loading occurs only if the stress and strain state satisfies (9-1), while the stress increments satisfy the condition

$$\dot{I}_s^{(2)} > 0. \tag{9-2}$$

If

$$I_s^{(2)} > k^2, \qquad I_s^{(2)} = \mathscr{T}(W^p), \quad \text{but} \quad \dot{I}_s^{(2)} = 0 \tag{9-3}$$

the corresponding stress variation will be named *partial neutral loading*.

If

$$I_s^{(2)} > k^2 \quad \text{and} \quad I_x^{(2)} < \mathscr{T}(W^p) \tag{9-4}$$

or

$$I_s^{(2)} > k^2, \qquad I_s^{(2)} = \mathscr{T}(W^p) \quad \text{and} \quad \dot{I}_s^{(2)} < 0 \tag{9-5}$$

are satisfied together with

$$\dot{I}_s^{(2)} + 2\frac{G}{\eta}\left(1 - \frac{k}{\sqrt{I_s^{(2)}}}\right) I_s^{(2)} > 0 \tag{9-6}$$

the stress variation will be named *partial loading* (viscoplastic only).

If (9-4) is satisfied, but instead of (9-6) we have

$$\dot{I}_s^{(2)} + 2\frac{G}{\eta}\left(1 - \frac{k}{\sqrt{I_s^{(2)}}}\right) I_s^{(2)} = 0 \tag{9-7}$$

then a *pure viscoplastic relaxation* (defined locally) occurs.

If (9-4) is satisfied but

$$\dot{I}_s^{(2)} + 2\frac{G}{\eta}\left(1 - \frac{k}{\sqrt{I_s^{(2)}}}\right) I_s^{(2)} < 0 \tag{9-8}$$

the corresponding stress variation will be named *quasi-unloading;* the plastic strains still increase but slower than in a relaxation process, i.e., this process must correspond to a certain unloading at the boundaries of the body.

Various kinds of loading (increase of plastic strains) may occur for any finite $\dot{I}_s^{(2)}$, either positive or negative, as long as $I_s^{(2)} > k^2$. When $\dot{I}_s^{(2)} \to -\infty$ the unloading becomes instantaneous and pure elastic. The instantaneous unloading must be used in order to determine at any moment the magnitude of the elastic components of the strain.

Finally, for $I_s^{(2)} < k^2$ the body behaves pure elastically.

If $\eta \to \infty$ from the previously discussed conditions we obtain the classic loading/unloading conditions of classic plasticity. When the viscosity increases, the influence of the partial loading regime on the total magnitude of the strain becomes negligible, and the relaxation time defined as

$$\tau = \frac{\eta_*}{G} = \frac{\eta}{G}\left(1 - \frac{k}{\sqrt{I_s^{(2)}}}\right)^{-1}$$

tends to infinity.

For the present model the relaxation is a viscoplastic phenomenon which occurs exactly as for the Hohenemser-Prager model, down to $I_s^{(2)} = k^2$ which are stable stress states. However, the present model shows more or less, either plastic or viscoplastic properties depending on the duration of the experiment and on the maximum stress reached. Thus, for a certain experiment, the dominant rheological behaviour can be made precise by comparing the time of experiment not only with the relaxation time, but also with a conventionally introduced "distinguishing time" (Cristescu [2], Ch. III).

10. SOME REMARKS

The previously mentioned theory can be used to study the dynamic plastic behaviour of various materials under combined loading (biaxial loading of tubes). The whole picture of wave propagation and reflection can be described by integrating the previously mentioned equations. The methods of integration are the well-known numerical methods (see Cristescu [2]) which for the constitutive equations $P1$–$P3$ can be applied very easily. It is only after more experimental facts would be available, that such an integration would be possible, and a full description of the wave propagation phenomenon could be given.

ACKNOWLEDGMENT

The author would like to thank J. D. Campbell for useful discussions and suggestions.

References

1. U. S. Lindholm, Dynamic deformation of metals, *In:* Behaviour of Materials Under Dynamic Loading, Amer. Soc. Mech. Engrs., 42 (1965).
2. N. Cristescu, *Dynamic Plasticity,* North Holland Publ. Company (1967).
3. K. Hohenemser and W. Prager, Z. Angew. Math. Mech., *12,* 216 (1932).
4. N. Cristescu, About the propagation of elastic/plastic waves in thin rods, *In:* Dynamika strojov, II, Proc. Conf. on the Dynamics of Machines, Prague 1963 (Vyd. Slov. Akad. Vied, Bratislava), 87 (1966).
5. N. Cristescu, Some dynamic problems of one-dimensional elastic/visco/plastic bodies, Proc. XIth Intern. Congr. of Applied Mechanics, München, 1964 (Springer-Verlag, Berlin-Heidelberg-New York), 390 (1966).

UNIFIED THEORY OF THERMOMECHANICAL BEHAVIOR OF VISCOELASTIC MATERIALS

K. C. Valanis

Iowa State University
Ames, Iowa

1. INTRODUCTION

Workers in the field of continuum mechanics have had occasion to witness, in recent times, a significant evolution in the theory of irreversible thermodynamics of viscoelastic materials. Following Onsager's work in the early 1930's there ensued an intense activity consisting in attempts to give a thermodynamic basis to the mechanical theory of *small* viscoelastic deformations which, of course, constitute processes that may be regarded as small deviations from an equilibrium state. Worthy of mention at this point is the work of Biot [1, 2] and other able researchers [3–6]. It is of historical interest that this activity left in its wake a divided opinion. Strong objections were voiced from the mathematical wing of "natural philosophers" who in a tours de force attacked such assumptions which, apparently, were arbitrary and, at most, of debatable validity.

These assumptions were often based on physical intuition and in some cases, they were unduly restrictive, but nonetheless they provided a basis from which viscoelastic behavior was better understood and moreover they led to the later development of mathematically more elegant expositions.

In 1964 Coleman [8] published his treatise on "Thermodynamics of Materials with Memory." This was a work of power and depth, such depth according to some people, that it gave "no status at all" to the Onsagerist theories. This view and the subsequent work of Coleman's prompted me to re-examine and re-evaluate critically the Onsagerist theories; the present work is the product of this re-evaluation.

The main results of the present paper may be summarized in terms of a general theorem and its corollary which establish the existence of a VISCOELASTIC POTENTIAL from which the stress tensor and the entropy density are derivable.

Theorem: The constitutive equations of a viscoelastic material with an initial elastic response and in the presence of large deformation and time varying, spatially inhomogeneous thermal field, are:

$$\tau^{ij} = 2\frac{\rho}{\rho_0}\frac{\partial\Psi}{\partial C_{ij}}, \qquad \eta = -\frac{\partial\Psi}{\partial\theta} \tag{1-1a, b}$$

$$\theta\dot{\gamma} = -\frac{\partial\Psi}{\partial q_\alpha}\dot{q}_\alpha - h^i\theta_{,i} \tag{1-2}$$

$$-\frac{\partial\Psi}{\partial q_\alpha}\dot{q}_\alpha \geqslant 0, -h^i\theta_{,i} \geqslant 0 \tag{1-3a, b}$$

*where τ^{ij} is the Piola stress tensor, C_{ij} is the right Cauchy-Green tensor, Ψ is the free energy per unit undeformed volume and η is the entropy per unit undeformed volume; θ is the absolute temperature, h_i is the heat flux vector and $\dot{\gamma}$ is the rate of irreversible entropy generation; finally q_α are hidden thermodynamic coordinates and ρ and ρ_0 are the current and reference density, respectively.**

The proof of this theorem is given in the later sections of the paper. The interesting aspect of this theorem is that the stress tensor and the entropy density are indeed derivable from a viscoelastic potential. In view of the fact that the deformation of a viscoelastic material is a dissipative process this is indeed a surprising result. However, the potential which in this case is the free energy density, is now a function not only of C_{ij} and θ but also of the thermodynamic coordinates q_α. This is what distinguishes the elastic from the viscoelastic potential.

The inequalities (1-3a, b) provide potent thermodynamic constraints on the form of the functions Ψ and h^i.

Corollary: The constitutive equations of a viscoelastic material with an initial elastic response and in the presence of *small* deformation and time-varying spatially inhomogeneous thermal fields are:

$$\sigma^{ij} = \frac{\partial\Psi}{\partial\epsilon_{ij}}, \qquad \eta = -\frac{\partial\Psi}{\partial\theta} \tag{1-4a, b}$$

$$\theta\dot{\gamma} = -\frac{\partial\Psi}{\partial q_\alpha}\dot{q}_\alpha - h^i\theta_{,i} \tag{1-5}$$

* In Equations (1-1–1-3) as well as in subsequent equations, a repeated suffix implies summation in the usual fashion, unless otherwise stated.

$$-\frac{\partial \Psi}{\partial q_\alpha}\dot{q}_\alpha \geqslant 0, \qquad -h^i\theta_{,i} \geqslant 0 \tag{1-6a, b}$$

where ϵ_{ij} is the small deformation strain tensor.

2. SOME GENERALLY ACCEPTED RELATIONS

To prepare the ground for the proof of this theorem, I begin by introducing some general thermodynamic relations which are valid* for small as well as large deformations, in the light of the following notation: x_i are the orthogonal set of material coordinates; y_i are the orthogonal set of spatial coordinates; t, ρ, ϵ and Q denote the time, density of a medium, internal energy per unit undeformed volume and rate of heat absorption per unit undeformed volume, in that order; τ_{ij}, $\dot{\gamma}_{ij}$ and h_i denote the stress tensor, the deformation rate tensor and the heat flux vector (per unit undeformed area) in that order, all defined in the material system; η denotes the entropy per unit undeformed volume and $\dot{\gamma}$ denotes the rate of irreversible entropy generation per unit undeformed volume; finally θ denotes the absolute temperature and $\theta_{,i}$ the temperature gradient in the material system. A dot over a quantity will denote its material derivative.

In the above notation the first law of thermodynamics in conjunction with the principle of conservation of momentum takes the form:

$$\dot{\epsilon} = \frac{\rho_0}{\rho}\tau^{ij}\dot{\gamma}_{ij} - h^i;_{\,i} + Q \tag{2-1}$$

where h_i pertains to heat entering the body and the semi-colon denotes covariant derivative in the material system; the rate of irreversible entropy generation is given by (2-2):

$$\theta\dot{\gamma} = \frac{\rho_0}{\rho}\tau^{ij}\dot{\gamma}_{ij} + \theta\dot{\eta} - \dot{\epsilon} - \frac{1}{\theta}h^i\theta_{,i}. \tag{2-2}$$

The Clausius-Duhem inequality (2-3) states that the rate of irreversible entropy generation $\dot{\gamma}$ must be either zero (for reversible processes) or positive (for irreversible processes). Thus

$$\dot{\gamma} \geqslant 0. \tag{2-3}$$

Evidently, since the above relations are valid for all materials, they cannot by themselves describe the constitutive behavior of a particular material.

For this to be accomplished, some other, additional, assumptions are necessary which are supported by experimental evidence. Conversely

* Valid in the sense that they are accepted by the majority of researchers in the field.

a postulated material behavior is inadmissible if it violates any one of the equations (2-1), (2-2) and (2-3).

3. RE-EVALUATION OF THERMODYNAMIC THEORY OF SMALL DEFORMATION OF VISCOELASTIC MATERIALS

The thermodynamic theory of small viscoelastic deformations was made possible through the introduction of the hidden coordinates. The physical significance of these coordinates was not immediately obvious.* However, it is known that an elastic material with small deformations is described by the following thermomechanical constitutive relations:

$$\eta = \eta(\epsilon, \epsilon_{ij}) \tag{3-1a}$$

$$\theta = \theta(\epsilon, \epsilon_{ij}) \tag{3-1b}$$

$$\Psi = \Psi(\theta, \epsilon_{ij}) \tag{3-1c}$$

where ϵ_{ij} is the small deformation strain tensor ($\epsilon_{ij} \approx \gamma_{ij}$) and Ψ is the free energy per unit undeformed volume, i.e.:

$$\Psi = \epsilon - \theta\eta. \tag{3-2}$$

In the above of thermodynamic theory ϵ_{ij} and θ (or ϵ) play the role of thermodynamic coordinates. It appears reasonable, therefore, that to describe non-elastic, i.e., irreversible, behavior additional coordinates have to be introduced which play the role of dissipation parameters. These were called hidden coordinates and are invariably denoted by q_α.

The introduction of the parameters q_α led to the following thermomechanical relations for viscoelastic materials, assuming time varying but spatially homogeneous temperature:

$$\eta = \eta(\epsilon, \epsilon_{ij}, q_\alpha) \tag{3-3a}$$

$$\theta = \theta(\epsilon, \epsilon_{ij}, q_\alpha) \tag{3-3b}$$

$$\Psi = (\epsilon, \epsilon_{ij}, q_\alpha). \tag{3-3c}$$

We pursue the evaluation of past theories (these have been restricted to homothermal fields, i.e., $\theta_{,i} = 0$). From (3-3a) and (2-2) it follows that

$$\theta\dot{\gamma} = \sigma_{ij}\dot{\epsilon}_{ij} + \theta\left[\frac{\partial\eta}{\partial\epsilon}\,\dot{\epsilon} + \frac{\partial\eta}{\partial\epsilon_{ij}}\,\dot{\epsilon}_{ij} + \frac{\partial\eta}{\partial q_\alpha}\,\dot{q}_\alpha\right] - \dot{\epsilon} \tag{3-4}$$

where σ_{ij} is the small deformation stress tensor defined in a cartesian

* The author showed in [9] that in the absence of thermal gradients, the hidden coordinates are functionals of the deformation history.

system, and $\rho \approx \rho_0$. It was now assumed, quite reasonably, that under homothermal conditions dissipation was a direct consequence of deformation, i.e., $\dot{\gamma} = 0$ and $\dot{q}_\alpha = 0$ when $\dot{\epsilon}_{ij} = 0$ whereupon as a consequence of (3-4):

$$\frac{1}{\theta} = \frac{\partial \eta}{\partial \epsilon}\bigg|_{\epsilon_{ij}, q_\alpha}. \tag{3-5}$$ *

If we now consider η, ϵ and Ψ as the dependent variables and upon introducing a new set of independent variables θ, ϵ_{ij} and q_α it follows easily from (3-3) that

$$\eta = \frac{\partial \Psi}{\partial \theta}, \quad \theta \frac{\partial \eta}{\partial \epsilon_{ij}} = -\frac{\partial \Psi}{\partial \epsilon_{ij}}, \quad \theta \frac{\partial \eta}{\partial q_\alpha} = -\frac{\partial \Psi}{\partial q_\alpha} \tag{3-6}$$

and thus in view of (3-5) and (3-6), (3-4) becomes

$$\theta \dot{\gamma} = \left(\sigma_{ij} - \frac{\partial \Psi}{\partial \epsilon_{ij}}\right) \dot{\epsilon}_{ij} - \frac{\partial \Psi}{\partial q_\alpha} \dot{q}_\alpha. \tag{3-7}$$

It now became logical to call X_{ij} and X_α where

$$X_{ij} = \sigma_{ij} - \frac{\partial \Psi}{\partial \epsilon_{ij}}, \quad X_\alpha = -\frac{\partial \Psi}{\partial q_\alpha} \tag{3-8}$$

irreversible forces, since these are indeed responsible for irreversible entropy generation, and to introduce the controversial Onsager's relations which say that X_{ij} and X_α on one hand and $\dot{\epsilon}_{ij}$ and $\dot{q}_\alpha$ on the other are linearly related and furthermore that these relations are symmetric in their coefficients. To be more precise,

$$X_{ij} = b_{ijkl}\dot{\epsilon}_{kl} + b_{ij\alpha}\dot{q}_\alpha \tag{3-9a}$$

$$X_\alpha = b_{ij\alpha}\dot{\epsilon}_{ij} + b_{\alpha\beta}\dot{q}_\beta \tag{3-9b}$$

where as a result of the assumed symmetry,

$$b_{ijkl} = b_{klij}, \; b_{\alpha\beta} = b_{\beta\alpha}. \tag{3-10}$$

The symmetry inherent in (3-9) has been the subject of much controversy; † this equation, however, contains an even more subtle feature which, so far, appears to have gone unnoticed. I shall return to this point shortly. Equations (2-3), (3-7) and (3-9) yield the result

$$0 \leq \theta\dot{\gamma} = b_{ijkl}\dot{\epsilon}_{ij}\dot{\epsilon}_{kl} + 2b_{ij\alpha}\dot{\epsilon}_{ij}\dot{q}_\alpha + b_{\alpha\beta}\dot{q}_\alpha\dot{q}_\beta. \tag{3-11}$$

* Though (3-5) is usually assumed to apply in the presence of thermal gradients, all its past derivations were obtained under the restriction of a homothermal field.

† It has been shown recently by the author [10] that the assumption of symmetry is not necessary for the development of the thermodynamic theory.

Also (3-8) in conjunction with (3-9) yields the relations

$$\frac{\partial \Psi}{\partial \epsilon_{ij}} + b_{ijkl}\dot{\epsilon}_{kl} + b_{ij\alpha}q_\alpha = \sigma_{ij} \tag{3-12a}$$

$$\frac{\partial \Psi}{\partial q_\alpha} + b_{ij\alpha}\dot{\epsilon}_{ij} + b_{\alpha\beta}\dot{q}_\beta = 0. \tag{3-12b}$$

Through the use of some quasi-physical arguments [7] Ψ is expressed, in the case of an isothermal process, as a quadratic in the strain components and the hidden coordinates, i.e.,

$$\Psi = \tfrac{1}{2}a_{ijkl}\epsilon_{ij}\epsilon_{kl} + a_{ij\alpha}\epsilon_{ij}q_\alpha + \tfrac{1}{2}a_{\alpha\beta}q_\alpha q_\beta \tag{3-13}$$

and thus (3-12a, b) becomes

$$a_{ijkl}\epsilon_{kl} + a_{ij\alpha}q_\alpha + b_{ijkl}\dot{\epsilon}_{kl} + b_{ij\alpha}\dot{q}_\alpha = \sigma_{ij} \tag{3-14a}$$

$$a_{ij\alpha}\epsilon_{ij} + a_{\alpha\beta}q_\beta + b_{\alpha\beta}\dot{q}_\beta + b_{ij\alpha}\dot{\epsilon}_{ij} = 0. \tag{3-14b}$$

Because $a_{\alpha\beta}$ and $b_{\alpha\beta}$ are both positive definite symmetric matrices [9] there exists an orthonormal matrix $Q_{\alpha\beta}$ which diagonalizes $a_{\alpha\beta}$ and simultaneously reduces $b_{\alpha\beta}$ to a unit matrix, i.e.,

$$Q_{\alpha\gamma}Q_{\beta\delta}a_{\gamma\delta} = a_\alpha\delta_{\alpha\beta} \quad (\alpha \text{ not summed}) \tag{3-15a}$$

$$Q_{\alpha\gamma}Q_{\beta\delta}b_{\gamma\delta} = \delta_{\alpha\beta}. \tag{3-15b}$$

Letting $\bar{q}_\alpha = Q_{\alpha\beta}q_\beta$, and making use of (3-15) in (3-14) and dropping bars, we obtain the "canonical" form of (3-14), i.e.,

$$a_{ijkl}\epsilon_{kl} + a_{ij\alpha}q_\alpha + b_{ijkl}\dot{\epsilon}_{kl} + b_{ijl}\dot{q}_l = \sigma_{ij} \tag{3-16a}$$

$$a_{ij\alpha}\epsilon_{ij} + a_\alpha q_\alpha + \dot{q}_\alpha + b_{kl}\dot{\epsilon}_{kl} = 0 \quad (\alpha \text{ not summed}). \tag{3-16b}$$

Without actually solving (3-16) explicitly for σ_{ij} we proceed to examine it more critically on the basis of its physical implications rather than its mathematical form. To keep the algebra simple consider the one-dimensional form of (3-16) and let $\alpha = 1$. Then (3-16) becomes

$$a_{11}\epsilon_1 + a_{12}q_1 + b_{11}\dot{\epsilon}_1 + b_{12}\dot{q}_1 = \sigma_1 \tag{3-17a}$$

$$a_{12}\epsilon_1 + a_{22}q_1 + \dot{q}_1 + b_{12}\dot{\epsilon}_1 = 0 \tag{3-17b}$$

as a result of (3-17b), (3-17a) can also be written as

$$\sigma_1 = (a_{11} - b_{12}a_{12})\epsilon_1 + (a_{12} - a_{22}b_{12})q_1 + (b_{11} - b_{12}b_{12})\dot{\epsilon}_1. \tag{3-17c}$$

The solution of (3-17b) is straightforward and in fact

$$q_1 = -\int_{-\infty}^{t} e^{-a_{22}(t-\tau)}\left(a_{12}\epsilon_1(\tau) + b_{12}\frac{\partial \epsilon_1}{\partial \tau}\right) d\tau. \tag{3-18}$$

Attached to the lower limit of integration is the implication that

$$q_1\Big|_{t=-\infty} = 0.$$

Letting now the deformation process start at time $t = 0$ (rather than $t = -\infty$), and in view of (3-18)

$$\dot{q}_1 = a_{22} \int_0^t e^{-a_{22}(t-\tau)} \left\{ a_{12}\epsilon_1(\tau) + b_{12}\frac{\partial \epsilon_1}{\partial \tau} \right\} d\tau - a_{12}\epsilon_1(t) - b_{12}\dot{\epsilon}_1(t). \quad (3\text{-}19)$$

Consider the viscoelastic materials which possess an initial elastic response (these being by far the most common), and consider a strain history which takes place isothermally over a very short period of time Δt but at an exceedingly high rate. Then from (3-19),

$$\dot{q}_1 \approx -b_{12}\dot{\epsilon}_1(t) \quad (3\text{-}20)$$

the integral, as a result of the smallness of Δt, being negligible since it is of the order $(\Delta t\ \dot{\epsilon}_1)$ and since ϵ_1 is small compared with $\dot{\epsilon}_1$.

It is a well-observed fact, that these materials under these conditions of loading behave almost elastically (reversibly), insofar as if the strain path is reversed the material will return to its initial configuration with a negligible amount of work having been performed over the cycle.

However, (3-20) is not in accord with this observation since now

$$\theta\dot{\gamma} = \tfrac{1}{2}(b_{11} - b_{12}^2)\dot{\epsilon}_1^2. \quad (3\text{-}21)$$

Thus contrary to expectation the rate of irreversible entropy production increases * with the square of the rate of strain according to this theory. Furthermore, the total irreversible entropy produced over such a cycle is of order $\dot{\gamma}\Delta t \approx \dfrac{1}{2\theta}(b_{11} - b_{12}^2)\epsilon_1\dot{\epsilon}_1$; this is clearly an inadmissible situation.

Thus, as far as these materials are concerned, if the theory is at all applicable it can apply only to processes that are associated with slow straining.

4. A NEW POINT OF VIEW

The only † possible way to resolve the previous paradox is to take

$$b_{11} = b_{12} = 0. \quad (4\text{-}1)$$

* $b_{41} - b_{12}^2$ is a positive quantity since b_{ij} is positive definite in view of (3-11).

† The apparent alternative $b_{11} = b_{12}^2$ leads to the same consequences, upon defining a new hidden coordinate p_1 such that $p_1 = q_1 - b_{12}\epsilon_1$.

Then indeed under high rates of strain

$$\dot{\gamma} \approx 0. \tag{4-2}$$

The general implication is that, in the case of viscoelastic materials with initially elastic response, $\dot{\gamma}$ does not depend explicitly on the rates of strain but only implicitly, through q_α. Thus in (3-11) $b_{ijkl} = 0$, $b_{ij\alpha} = 0$ and,

$$\theta\dot{\gamma} = b_{\alpha\beta}\dot{q}_\alpha\dot{q}_\beta \geqslant 0 \tag{4-3}$$

the equality being valid if and only if $\dot{q}_\alpha = 0$.

We shall see that this equation is of extreme significance in that a) it removes this inconsistency which existed in the theory of linear irreversible thermodynamics of viscoelastic materials and b) it gives rise to some elegant and far-reaching results. Thus, (3-12) now take the surprising form

$$\sigma_{ij} = \frac{\partial\bar{\Psi}}{\partial\epsilon_{ij}} \tag{4-4a}$$

$$\frac{\partial\Psi}{\partial q_\alpha} + b_{\alpha\beta}\dot{q}_\beta = 0. \tag{4-4b}$$

These are the isothermal constitutive equations of linear viscoelastic materials with an initial elastic response.

Equation (4-4a) is identical in form to the constitutive equation of an elastic material, only now Ψ is a function of ϵ_{ij} as well as q_α; also (4-4b) is identically satisfied for elastic materials since Ψ does not depend on q_α; furthermore, in this case $b_{\alpha\beta} = 0$.

Let's now re-examine the behavior of this type of viscoelastic material with one non-vanishing component of strain and one hidden coordinate, as before.

Recalling (3-17) and writing $\dot{\gamma}\theta = b\dot{q}_1^2$ one obtains the relations

$$\sigma_1 = a\epsilon_1 + dq_1 \tag{4-5}$$

$$0 = d\epsilon_1 + cq_1 + b\dot{q}_1 \tag{4-6}$$

where $a_{11} \equiv a$, $a_{12} \equiv d$, $a_{22} \equiv c$.

Integrating (4-6) one obtains

$$q_1 = -\frac{d}{c}\,\epsilon_1(t) + \int_{-\infty}^{t} e^{-(c/b)(t-\tau)}\,\frac{\partial\epsilon_1}{\partial\tau}\,d\tau. \tag{4-7a}$$

Note that in the equilibrium configuration $q_1 = -\dfrac{d}{c}\,\epsilon_1$.

Substitution of (4-7) in (4-5a) yields the result

$$\sigma_1 = \left(a - \frac{d^2}{c}\right)\epsilon_1(t) + \frac{d^2}{c}\int_{-\infty}^{t} e^{-(c/b)(t-\tau)}\,\frac{\partial\epsilon_1}{\partial\tau}\,d\tau. \tag{4-8}$$

Equation (4-8) represents a three-element linear viscoelastic model as shown below:

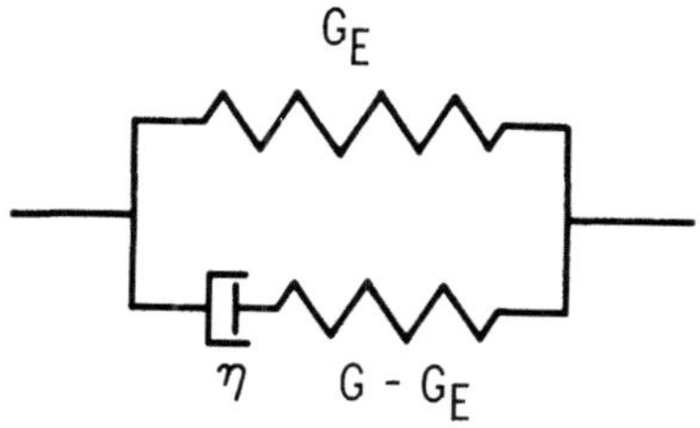

where

$$G_E = a - \frac{d^2}{c}, \qquad G - G_E = \frac{d^2}{c} \tag{4-9a}$$

$$G = a \tag{4-9b}$$

$$\eta = b. \tag{4-9c}$$

With a little algebra one can show that under all loading conditions, $q_1 = -\left(\frac{d}{c}\right) \times$ (strain of the dashpot), i.e., q_1 has a clearly understood significance. From a physical viewpoint, under high rates of loading the dashpot has little time to relax, $q_1 \approx 0$ and the model behaves like two springs in parallel. Thus, in this case $\gamma = 0$. For very slow loading the dashpot extends without offering any resistance $q_1 = \sigma_1/G_E$, but $\dot{q}_1 \approx 0$ and thus, again $\dot{\gamma} = 0$. The theory is now internally consistent.

It can also be shown that Ψ, where $\Psi = \frac{1}{2}(a_{11}\epsilon_1^2 + 2a_{12}\epsilon_1 q_1 + a_{22}q_1^2)$, is the strain energy stored in the springs. Indeed if ϵ_{12} and ϵ_{11} are respectively the strains in the spring and dashpot that are lying in series such that $\epsilon_{12} + \epsilon_{11} = \epsilon_1$, then

$$\begin{aligned} \Psi &= \frac{1}{2}\left(a_{11} - \frac{a_{12}^2}{a_{22}}\right)\epsilon_1^2 + \frac{1}{2}\frac{a_{12}^2}{a_{22}}\epsilon_{12}^2 \\ &= \tfrac{1}{2}G_E\epsilon_1^2 + \tfrac{1}{2}(G - G_E)\epsilon_{12}^2 \end{aligned} \tag{4-9d}$$

which proves the assertion.

Also the rate of entropy production can be expressed explicitly in terms of the properties of the model and, in fact,

$$\theta\dot{\gamma} = b\dot{q}_1\dot{q}_1 = \frac{a_{12}^2}{a_{22}}\dot{\epsilon}_{11}(\epsilon_1 - \epsilon_{11}) \tag{4-9e}$$

where use has been made of the relation $q_1 = -\frac{a_{12}}{a_{22}}\epsilon_{11}$ and (4-6). From (4-9e)

$$\dot{\gamma} = \frac{1}{\theta}\dot{\epsilon}_{11}\sigma_{11}. \tag{4-9f}$$

Thus, under isothermal conditions, the rate of irreversible entropy production is equal to the rate of work done in extending the dashpot divided by the absolute temperature at which the current extension is taking place. Equations (4-9d) and (4-9f) merely confirm the fact that under isothermal conditions

$$\dot{W} = \dot{\Psi} + \dot{\gamma}\theta. \tag{4-9g}$$

We now proceed to examine a more general model such as the one shown below:

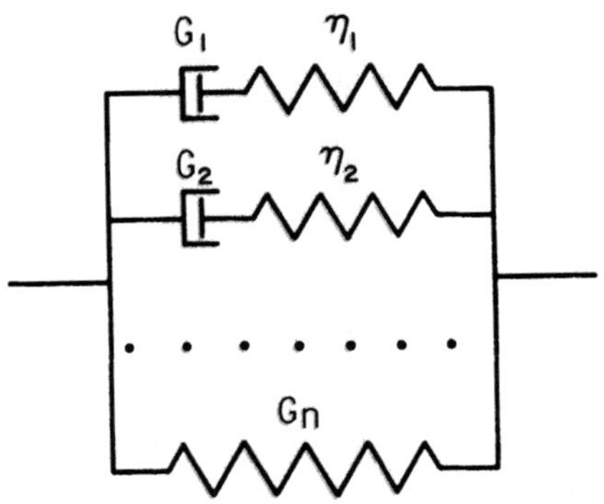

In this situation,

$$\theta\dot{\gamma} = b_{\alpha\beta}\dot{q}_\alpha\dot{q}_\beta \quad (\alpha, \beta = 1, 2 \cdots n-1). \tag{4-10}$$

Equation (4-10) is a positive definite quadratic form and can be reduced to the canonical form

$$\theta\dot{\gamma} = \sum_{r=1}^{n-1} \eta_r(\dot{q}_r)^2 \tag{4-11}$$

by a suitable linear transformation on the q_α's.

Equations (4-4a) and (4-4b) now become

$$\sigma_1 = \frac{\partial\Psi}{\partial\epsilon_1} \tag{4-12a}$$

$$\frac{\partial\Psi}{\partial q_\alpha} + \eta_\alpha\dot{q}_\alpha = 0 \quad (\alpha \text{ not summed}). \tag{4-12b}$$

Let Ψ be a quadratic, as previously, in q_α and ϵ, i.e.,

$$\Psi = \tfrac{1}{2}a_{11}\epsilon_1{}^2 + a_{1\alpha}\epsilon_1 q_\alpha + \tfrac{1}{2}a_{\alpha\beta}q_\alpha q_\beta. \tag{4-13}$$

The transformation which reduces (4-10) to the canonical form (4-11) may be chosen (without loss of generality) in such a way as to reduce (4-13) to the canonical form

$$\Psi = \tfrac{1}{2}a_{11}\epsilon_1{}^2 + a_{1\alpha}q_\alpha\epsilon_1 + \sum_\alpha \tfrac{1}{2}q_\alpha q_\alpha a_\alpha. \tag{4-13a}$$

Thus (4-12a) and (4-12b) now yield the result

$$\sigma_1 = a_{11}\epsilon_1 + a_{1\alpha}q_\alpha \tag{4-14a}$$

$$\epsilon_1 a_{1\alpha} + a_\alpha q_\alpha + \eta_\alpha \dot{q}_\alpha = 0 \quad (\alpha \text{ not summed}). \tag{4-14b}$$

From (4-14b) one can obtain q_α explicitly and in fact

$$q_\alpha = -\frac{a_{1\alpha}}{\eta_\alpha}\int_{-\infty}^{t} e^{-\rho_\alpha(t-\tau)}\epsilon_1(\tau)\, d\tau \tag{4-15}$$

where $\rho_\alpha = \left(\dfrac{a_\alpha}{\eta_\alpha}\right)$. After integration by parts one obtains an equation analogous to (3-18). i.e.,

$$q_\alpha = -\frac{a_{1\alpha}}{a_\alpha}\epsilon_1(t) + \frac{a_{1\alpha}}{a_\alpha}\int_{-\infty}^{t} e^{-\rho_\alpha(t-\tau)}\frac{\partial\epsilon_1}{\partial\tau}\, d\tau \tag{4-16}$$

and thus the stress may be obtained from (4-14a) in the form

$$\sigma_1 = \left(a_{11} - \sum_{\alpha=1}^{n-1}\frac{a_{1\alpha}{}^2}{a_\alpha}\right)\epsilon_1(t) + \sum_{\alpha=1}^{n-1}\frac{a_{1\alpha}{}^2}{a_\alpha}\int_{-\infty}^{t} e^{-\rho_\alpha(t-\tau)}\frac{\partial\epsilon_1}{\partial\tau}\, d\tau. \tag{4-17}$$

It is easily verified that η_α are indeed the viscosities of the dashpots, $a_{11} - \sum \dfrac{a_{1\alpha}{}^2}{a_\alpha}$ is the stiffness G_n of the lone spring and $\dfrac{a_{1\alpha}{}^2}{a_\alpha}$ is the stiffness G_α of the spring in series with dashpot α.

Note again *that the coefficients η_r entering in the expression for the generation of irreversible entropy production (4-11), are the viscosities of the dashpots.* To my knowledge this is a new result.

Generalization to Three Dimensions

The generalization to 3 dimensions is straightforward. The expression for $\dot{\gamma}$ is still (4-11) but now Ψ is expanded in the form

$$\Psi = \tfrac{1}{2}a_{ijkl}\epsilon_{ij}\epsilon_{kl} + a_{ij\alpha}\epsilon_{ij}q_\alpha + \sum_\alpha \tfrac{1}{2}a_\alpha q_\alpha q_\alpha. \tag{4-18}$$

Hence, in view of (4-12),

$$\sigma_{ij} = a_{ijkl}\epsilon_{kl} + a_{ij\alpha}q_\alpha \tag{4-19a}$$

$$0 = a_{ij\alpha}\epsilon_{ij} + a_\alpha q_\alpha + \eta_\alpha\dot{q}_\alpha \quad (\alpha \text{ not summed}). \tag{4-19b}$$

Equation (4-19b) yields immediately

$$q_\alpha = -\frac{a_{ij\alpha}}{\eta_\alpha}\int_{-\infty}^{t} e^{-\rho_\alpha(t-\tau)}\epsilon_{ij}(\tau)d\tau \quad (\alpha \text{ not summed}) \tag{4-20}$$

which after integration by parts simply becomes

$$q_\alpha = -\frac{a_{kl\alpha}}{a_\alpha}\left\{\epsilon_{kl}(t) - \int_{-\infty}^{t} e^{-\rho_\alpha(t-\tau)}\frac{\partial\epsilon_{kl}}{\partial\tau}\,d\tau\right\} \quad (\alpha \text{ not summed}). \tag{4-21}$$

Substitution of (4-21) in (4-19a) yields the result

$$\sigma_{ij} = \left(a_{ijkl} - \sum_\alpha \frac{a_{ij\alpha}a_{kl\alpha}}{a_\alpha}\right)\epsilon_{kl}(t) + \sum_\alpha \frac{a_{ij\alpha}a_{kl\alpha}}{a_\alpha}\int_{-\infty}^{t} e^{-\rho_\alpha(t-\tau)}\frac{\partial\epsilon_{kl}}{\partial\tau}\,d\tau \tag{4-22}$$

which in obvious notation may be written in the more compact form,

$$\sigma_{ij} = C^0_{ijkl}\epsilon_{kl}(t) + \int_{-\infty}^{t} C^1_{ijkl}(t-\tau)\frac{\partial\epsilon_{kl}}{\partial\tau}\,d\tau. \tag{4-23}$$

If the material is isotropic then

$$C^0{}_{ijkl} = \lambda^0\delta_{ij}\delta_{kl} + \mu^0(\delta_{ik}\delta_{jl} + \delta_{il}\delta_{jk}) \tag{4-24a}$$

$$C^1{}_{ijkl}(t) = \lambda^1(t)\delta_{ij}\delta_{kl} + \mu^1(t)(\delta_{ik}\delta_{jl} + \delta_{il}\delta_{jk}). \tag{4-24b}$$

Substitution of (4-24) in (4-23) yields the classical relation

$$\sigma_{ij}(t) = \lambda^0\epsilon_{kk}(t)\delta_{ij} + \delta_{ij}\int_{-\infty}^{t}\lambda^1(t-\tau)\frac{\partial\epsilon_{kk}}{\partial\tau}\,d\tau$$
$$+ 2\mu^0\epsilon_{ij}(t) + 2\int_{-\infty}^{t}\mu^1(t-\tau)\frac{\partial\epsilon_{ij}}{\partial\tau}\,d\tau \tag{4-25}$$

which can be reduced to a further more compact form by writing,

$$\lambda(t) = \lambda^0 H(t) + \lambda^1(t) \tag{4-26a}$$

$$\mu(t) = \mu^0 H(t) + \mu^1(t) \tag{4-26b}$$

where $H(t)$ is the Heaviside step function, in which case

$$\sigma_{ij}(t) = \delta_{ij}\int_{-\infty}^{t}\lambda(t-\tau)\frac{\partial\epsilon_{kk}}{\partial\tau}\,d\tau + 2\int_{-\infty}^{t}\mu(t-\tau)\frac{\partial\epsilon_{ij}}{\partial\tau}\,d\tau. \tag{4-27}$$

5. MATHEMATICAL FORMALIZATION

At this point we shall proceed with the fullest possible generality taking into consideration large deformation as well as thermal effects, i.e., large changes of temperature and thermal gradients.

Consider an infinite continuous medium. Given body forces g_i and heat source distribution Q the medium will experience deformation as well as thermal changes. The whole process is described by the following set of variables, sixteen in number:

$$\rho,\ \tau^{ij},\ \eta,\ \epsilon,\ h_j,\ y_i \text{ and } \theta$$

all of which have been defined. These unknowns are connected by the relations

$$\left|\frac{\partial y_i}{\partial x_j}\right| = \frac{\rho_0}{\rho} \tag{5-1}$$

$$\tau^{ij}{}_{;j} + \rho g^i = \rho \alpha^i \tag{5-2}$$

$$\dot{\epsilon} = \frac{\rho_0}{2\rho}\, \tau^{ij}\dot{C}_{ij} - h^i{}_{;i} + Q \tag{5-3}$$

$$\theta\dot{\gamma} = \frac{\rho_0}{2\rho}\, \tau^{ij}\dot{C}_{ij} + \theta\dot{\eta} - \dot{\epsilon} - \frac{1}{\theta}\, h^i\theta_{,i} \geqslant 0 \tag{5-4}$$

where (5-1) is the conservation of mass and (5-2) is the conservation of momentum; (5-3) and (5-4) have already been discussed. We have 16 unknowns and 5 equations in addition to the constraint (5-4), i.e., there is a deficiency of eleven relations. The implication is that the response of a particular material cannot be depicted from the field equations above but what is needed, in addition, is a set of eleven constitutive equations (equations of state) relating the state variables τ^{ij}, η, ϵ, h_i to the thermodynamic coordinates y_i (through C_{ij}) and θ.*

Equations of State for Elastic Materials

To fix ideas consider first an elastic material. By definition the state variables of such a material depend on the current values of the thermodynamic coordinates (possibly on their current gradients but not on their past values). Thus the equations of state will be of the form

$$\tau^{ij} = \tau^{ij}(C_{ij},\ \theta,\ \theta_{,i}) \tag{5-5a}$$

$$\eta = \eta(C_{ij},\ \theta,\ \theta_{,i}) \tag{5-5b}$$

$$\epsilon = \epsilon(C_{ij},\ \theta,\ \theta_{,i}) \tag{5-5c}$$

$$h_i = h_i(C_{ij},\ \theta,\ \theta_{,i}) \tag{5-5d}$$

where we have included dependence on $\theta_{,i}$ as a possibility.

Let the time range of interest be $(-\infty,\ t)$ and let τ be such that $-\infty < \tau \leqslant t$; we denote the history of a function $f(\ \ )$ by $f(\tau)$ and denote the current value $f(t)$ of the function by f. Following Coleman we ob-

* The reasons for the choice of C_{ij} as a tensorial thermodynamic variable have been discussed in the literature extensively and will not be elaborated upon here.

serve that the history of motion $y(x_k, \tau)$ and the history of temperature $\theta(x_k, \tau)$ (assumed to possess all requisite derivatives) completely determine the state variables through (5-5) as functions of x_k and τ for all x_k and τ. Hence, (5-2) and (5-3) determine g_i and Q uniquely. Furthermore, (5-4) being valid for all sets of g_i and Q, places important restrictions on the form of the equations of state (5-5). Thus, all histories of motion and temperature, with the requisite amount of smoothness imposed by (5-5), (5-1), (5-2) and (5-3), are "thermodynamically admissible."

At this point we introduce the free energy density Ψ in terms of which (5-4) becomes

$$\theta\dot{\gamma} = \frac{2\rho_0}{\rho}\tau^{ij}\dot{C}_{ij} - \dot{\Psi} - \eta\dot{\theta} - \frac{1}{\theta}h^i\theta_{,i} \geqslant 0. \tag{5-6}$$

Evidently since $\Psi = \epsilon - \eta\theta$ and in view of (5-5b, c)

$$\Psi = \Psi(C_{ij}, \theta, \theta_{,i}). \tag{5-5e}$$

As a consequence of (5-5e) we can rewrite (5-6) as follows:

$$0 \leqslant \theta\dot{\gamma} = \frac{2\rho_0}{2\rho}\tau^{ij}\dot{C}_{ij} - \frac{\partial\Psi}{\partial C_{ij}}\dot{C}_{ij} - \frac{\partial\Psi}{\partial\theta}\dot{\theta} - \frac{\partial\Psi}{\partial\theta_{,i}}\dot{\theta}_{,i} - \eta\dot{\theta} - \frac{1}{\theta}h^i\theta_{,i}. \tag{5-7}$$

Fixing C_{ij}, θ, $\theta_{,i}$ (at $\tau = t$), we fix τ^{ij}, η, ϵ and h_i (at $\tau = t$), as a result of (5-5). Furthermore, since any history $C_{ij}(\tau)$ and $\theta(\tau)$ is admissible then, obviously $\dot{C}_{ij}$, $\dot{\theta}$, $\dot{\theta}_i$ may be chosen at will. For all such choices inequality (5-7) must not be violated. Evidently this is only possible if and only if

$$\frac{\partial\Psi}{\partial\theta_{,i}} = 0 \tag{5-8a}$$

$$\tau^{ij} = \frac{2\rho}{\rho_0}\frac{\partial\Psi}{\partial C_{ij}} \tag{5-8b}$$

$$\eta = -\frac{\partial\Psi}{\partial\theta} \tag{5-8c}$$

$$-h^i\theta_{,i} \geqslant 0. \tag{5-9}$$

Thus, in all possible thermomechanical states *the stress tensor and entropy density of an elastic material are derivable from a potential which is the free energy density* Ψ such that

$$\Psi = \Psi(C_{ij}, \theta). \tag{5-10}$$

Also, for any such thermodynamic process the inequality (5-9) must hold and may be regarded as an equality only in the special case of $\theta_{,j} = 0$, or $h^i = 0$.

Equations of State for Viscoelastic Materials

Viscoelastic materials differ from elastic ones, insofar as while they are in a dynamic state, their state of stress is a function of the whole previous history of their deformation; on the other hand, the state of stress of these materials so long as they are in a static state, is a function of the static configuration only and is independent of the previous history of deformation. Therefore, equations of state of these materials must take this fact into account.

One way of doing this is to introduce further thermodynamic coordinates which play the role of dissipation parameters. These are none other than the intrinsic coordinates q_α introduced in the linear theory.

At this point we are not overly concerned with the explicit form of q_α provided that these are functionals of the deformation history $C_{ij}(\tau)$, so long as the material is in motion, and simply functions of the equilibrium deformation when the material is at rest.

With these thoughts in mind, the equations of state for viscoelastic materials will be given by (5-11).

$$\tau^{ij} = \tau^{ij}(C_{ij}, \theta, q_\alpha, \theta_{,i}) \tag{5-11a}$$

$$\eta = \eta(C_{ij}, \theta, q_\alpha, \theta_{,i}) \tag{5-11b}$$

$$\epsilon = \epsilon(C_{ij}, \theta, q_\alpha, \theta_{,i}) \tag{5-11c}$$

$$h_i = h_i(C_{ij}, \theta, q_\alpha, \theta_{,i}). \tag{5-11d}$$

In addition,

$$\Psi = \Psi(C_{ij}, \theta, q_\alpha, \theta_{,i}). \tag{5-11e}$$

The intrinsic coordinates q_α are now defined by a general set of functional relationships of the type: *

$$\dot{q}_\alpha = f_\alpha(C_{ij}, \theta, q_\alpha) \tag{5-12}$$

which are the non-linear counterparts of (4-19b) with thermal effects included, subject to the initial condition that $q_\alpha = 0$ prior to the initiation of a thermodynamic process.

Just as in the case of elastic materials given the functions $y_i(x_u, t)$ and $\theta(x_u, t)$, sufficiently smooth, q_α may be found by solving the set of first order differential equations (5-12) hence τ^{ij}, η, ϵ and h_i may be found from (5-11) and hence $q_i(x_u, t)$ and $Q(x_u, t)$ are determined uniquely from (5-2) and (5-3). On the other hand, (5-6) now assumes the form

* For materials with an initial elastic response. Note that (5-12) satisfies the stipulation on q_α stated previously.

$$\theta\dot{\gamma} = \left(\frac{2\rho_0}{\rho}\,\tau^{ij} - \frac{\partial\Psi}{\partial C_{ij}}\right)\dot{C}_{ij} - \left(\eta + \frac{\partial\Psi}{\partial\theta}\right)\dot{\theta} - \frac{\partial\Psi}{\partial q_\alpha}\,\dot{q}_\alpha - \frac{\partial\Psi}{\partial\theta_{ij}}\,\dot{\theta}_{,i} - \frac{1}{\theta}\,h^i\theta_{,i} \geqslant 0. \tag{5-13}$$

Repeating our previous argument, fixing C_{ij}, θ, $\theta_{,i}$ and q_α at time t fixes * q_α but leaves $\dot{C}_{ij}$, $\dot{\theta}$ and $\dot{\theta}_{,i}$ arbitrary. By varying one of the last three in turn while keeping the remaining constant, one arrives at the conclusion that inequality (5-13) will not be violated if and only if

$$\frac{\partial\Psi}{\partial\theta_{,i}} = 0 \tag{5-14}$$

$$\tau_{ij} = \frac{2\rho}{\rho_0}\,\frac{\partial\Psi}{\partial C_{ij}} \tag{5-15a}$$

$$\eta = -\frac{\partial\Psi}{\partial\theta} \tag{5-15b}$$

$$\theta\dot{\gamma} = -\frac{\partial\Psi}{\partial q_\alpha}\,\dot{q}_\alpha - \frac{1}{\theta}\,h^i\theta_{,i} \geqslant 0. \tag{5-15c}$$

Now prescribing C_{ij} and θ at a *specific* x_k for all values of τ in the interval $(-\infty, t)$ fixes the first term of inequality (5-15c). Furthermore, this term is independent of $\theta_{,i}$. Hence, for $\theta_{,i} = 0$ and in view of inequality (5-15c),

$$-\frac{\partial\Psi}{\partial q_\alpha}\,\dot{q}_\alpha \geqslant 0. \tag{5-16a}$$

Similarly, since we can arbitrarily vary the second term of the inequality by varying $\theta_{,i}$ for any fixed value of the first term, it follows that

$$-h^i\theta_{,i} \geqslant 0. \tag{5-16b}$$

Finally as a result of (5-15b) and the definition of free energy, ϵ and η are not functions of $\theta_{,i}$. Thus, the theorem stated in the introduction is proved.

A further relation can be derived at this point. Using the definition of free energy in conjunction with (5-15a, b) and the first law of thermodynamics, one finds that

$$h^i{}_{;i} = \theta\left(\frac{\partial\Psi}{\partial\theta}\right) - \frac{\partial\Psi}{\partial q_\alpha}\,\dot{q}_\alpha + Q. \tag{5-17}$$

* As a result of (5-12).

6. APPLICATIONS

I shall proceed with the application of the above theory with the object of developing explicit forms of constitutive equations of viscoelastic materials for two cases:

(i) Non-isothermal small deformation
(ii) Isothermal large deformation

Before this is done I would like to discuss at greater length the character of the intrinsic coordinates q_α and how they relate to the rate of irreversible entropy generation. It will be recalled that, where viscoelastic materials with an initial elastic response are concerned, $\dot{\gamma}$ does not depend on the strain rates explicitly, but only implicitly through $\dot{q}_\alpha$. Thus, in complete generality one can write, in the absence of heat conduction,

$$\theta\dot{\gamma} = \Gamma_0(\dot{q}_\alpha) \tag{6-1}$$

where Γ_0 is an unknown function and $\dot{q}_\alpha$ are given by (5-12).

The intrinsic coordinates q_α remain largely undefined quantities provided they satisfy (5-12) for any physically meaningful functions f_α. However, it is important to note that once Γ_0 and f_α have been given, then ψ can no longer be functionally independent but must satisfy (5-15c) which in the absence of heat conduction becomes

$$\theta\dot{\gamma} + \frac{\partial \Psi}{\partial q_\alpha}\dot{q}_\alpha = 0. \tag{6-2}$$

In the presence of heat conduction (6-1) must, of course, be modified since now $\dot{\gamma}$ must depend on $\theta_{,i}$ as well. Thus,

$$\theta\dot{\gamma} = \Gamma(\dot{q}_\alpha, \theta_{,i}). \tag{6-3}$$

Again, once f_α and Γ are given, Ψ is determined from (5-15c). Furthermore, since Ψ does not depend on $\theta_{,i}$ the part of $\theta\dot{\gamma}$ which does not contain $\theta_{,i}$ must be equated to $-\dfrac{\partial \Psi}{\partial q_\alpha}\dot{q}_\alpha$ whereas the remaining part must be equal to $\dfrac{1}{\theta} h^i\theta_{,i}$.

Let us consider the physical situations where q_α and $\theta_{,i}$ are small. Evidently, $\theta_{,i}$ imply small thermal gradients. Also, on the basis of our early discussion q_α will be small (a) if the strains are small, (b) if the strains are large but the time of interest lies within a small interval following the initiation of straining and (c) if the strains are large but the rate of straining has been slow in the recent past.*

* (b) and (c) are a direct consequence of the definition of q_α by (5-12). For a lengthy discussion of this point see [9].

In these situations, in the light of q_α and $\theta_{,i}$ being small and γ being positive for non-zero values of $\dot{q}_\alpha$ and $\theta_{,i}$, (6-3) admits the following Taylor series expansion for suitable choice of q_α:

$$\theta\dot{\gamma} = \sum_\alpha \eta_\alpha \dot{q}_\alpha \dot{q}_\alpha + b_{\alpha i}\dot{q}_\alpha \theta_{,i} + a_{ij}\theta_{,i}\theta_{,j} \tag{6-4}$$

where the coefficients η_α, $b_{\alpha i}$ and a_{ij} may depend explicitly on C_{ij} and θ. Equations (5-15c) and (6-4) will be compatible if

$$-\frac{\partial\psi}{\partial q_\alpha} = \eta_\alpha \dot{q}_\alpha \quad (\alpha \text{ not summed}) \tag{6-5a}$$

$$-h_i = a_{ij}\theta_{,j} + b_{\alpha i}\dot{q}_\alpha \tag{6-5b}$$

since Ψ does not depend on $\theta_{,i}$. Also h_i must vanish for vanishing $\theta_{,i}$ so, evidently, $b_{\alpha i}$ must be zero. Thus, in the presence of heat conduction and some special cases of large deformation *, we have the relations

$$\frac{\partial\psi}{\partial q_\alpha} + \eta_\alpha \dot{q}_\alpha = 0 \tag{6-6a}$$

$$-h_i = a_{ij}\theta_{,j}. \tag{6-6b}$$

With this elaboration we are now in a position to return to our task as set out at the beginning of Section 6. The fundamental equations which are applicable to all cases of heat conduction and strain are (5-15a), (5-15b), (5-15c) and (5-17). Equation (6-6a) applies when conduction is absent and (6-6a) and (6-6b) apply jointly under the conditions specified.

Case (i). Non-isothermal Small Deformation

For the sake of simplicity let us also assume that temporal variations in θ are also small, i.e., $\theta = \theta_0 + \zeta$ where θ_0 is the reference temperature. Let the reference state be defined by the condition $\sigma_{ij} = 0$, $q_\alpha = 0$, $\zeta = 0$, $\chi = 0$ where χ is the entropy increment associated with the above process such that

$$\eta = \eta_0 + \chi. \tag{6-7}$$

Then (5-15a, b, and c) become

$$\sigma_{ij} = \frac{\partial\Psi}{\partial\epsilon_{ij}} \tag{6-8a}$$

$$\chi = -\frac{\partial\psi}{\partial\zeta} \tag{6-8b}$$

* Such that q_α and $\theta_{,i}$ are small.

$$\theta_0 \dot{\gamma} = -\frac{\partial \Psi}{\partial q_\alpha} \dot{q}_\alpha - \frac{1}{\theta_0} h_i \theta_{,i} \tag{6-8c}$$

whereas (5-17) becomes

$$h_{i,i} = \theta_0 \left(\frac{\partial \Psi}{\partial \zeta}\right) - \frac{\partial \Psi}{\partial q_\alpha} \dot{q}_\alpha + Q. \tag{6-9}$$

In this event ψ can be expanded in a Taylor series and terms higher than the second can be ignored, whereas terms of the first order must vanish to satisfy the initial conditions. In the case of an isotropic material this expansion will take the form

$$\psi = \tfrac{1}{2}\mu \epsilon_{ij}\epsilon_{ij} + \tfrac{1}{2}\lambda \epsilon_{kk}\epsilon_{ii} + \hat{a}_\alpha q_\alpha \epsilon_{kk} + \tfrac{1}{2} a_\alpha q_\alpha q_\alpha + a' \epsilon_{kk}\zeta + a'_\alpha q_\alpha \zeta + \tfrac{1}{2}\beta \zeta^2 \tag{6-10}$$

where all repeated indices are summed. Substitution of (6-10) in (6-8) and (6-9) and after some routine manipulations yields the following expression for q_α (when $Q = 0$):

$$\eta_\alpha q_\alpha + a_\alpha q_\alpha + \hat{a}_\alpha \epsilon_{kk} + a'_\alpha \zeta = 0 \quad (\alpha, \text{ not summed}) \tag{6-11}$$

and the thermomechanical constitutive equations: *

$$\sigma_{ij} = \delta_{ij} \int_0^\xi \lambda(\xi - \xi') \frac{\partial \epsilon_{kk}}{\partial \xi'} d\xi' + \int_0^\xi \mu(\xi - \xi') \frac{\partial \epsilon_{ij}}{\partial \xi'} d\xi' + \delta_{ij} \int_0^\xi \alpha(\xi - \xi') \frac{\partial \zeta}{\partial \xi'} d\xi' \tag{6-12a}$$

$$a_T k \zeta_{,ii} = \int_0^\xi C_v(\xi - \xi') \frac{\partial^2 \zeta}{\partial \xi'^2} d\xi' - \theta_0 \int_0^\xi \alpha(\xi - \xi') \frac{\partial^2 \epsilon_{kk}}{\partial \xi'^2} d\xi' + \sum_\alpha \eta_\alpha \frac{\partial q_\alpha}{\partial \xi} \frac{\partial q_\alpha}{\partial \xi} \tag{6-12b}$$

where †

$$\eta_\alpha = \eta_{0\alpha} a_T\{\zeta(t)\} \tag{6-13}$$

k is the coefficient of heat conduction, and

$$\xi = \int_0^t \frac{d\tau}{a_T[\zeta(\tau)]}. \tag{6-14}$$

Evidently, the material which is described by (6-11) and (6-12) is thermorheologically simple in accordance with (6-13) where $a_T[\zeta(t)]$ is the "shift factor" of temperature.

The physical significance of the function $\alpha(\xi)$ is obvious from (6-11). That is, if $\epsilon_{ij} = 0$ and $\zeta(\xi) = H(\xi)$ where H is the Heaviside step function

* The algebraic details of the derivation of (6-11) and (6-12) are omitted: the functions $\lambda(\xi)$, $\mu(\xi)$, $\alpha(\xi)$, and $C_v(\xi)$ are finite sums of decaying exponentials.

† The dependence of η_α on temperature as given by (6-13) is of a special character typical of thermorheologically simple materials.

then $\sigma_{ij} = \delta_{ij}\alpha(\xi)$. In other words $\alpha(\xi)$ is the hydrostatic stress induced in a volume of isotropic viscoelastic material when strain is prevented and the initially uniform temperature is given a unit step jump and is kept constant thereafter.

The physical significance of $C_r(\xi)$ is found from (6-12b). In the absence of strain,

$$k\zeta_{,ii} = -h_{i,i} \tag{6-15}$$

where $-h_{i,i}$ is the rate of heat input per unit volume. Thus, ignoring second order terms, in $\dot{q}_\alpha$, in (6-12) we let $\dfrac{\partial\zeta}{\partial\xi} = H(\xi)$. Then,

$$C_v(\xi) = \frac{\text{Rate of heat input/unit volume}}{a_T}. \tag{6-16}$$

Note that (6-11) and (6-12) are analogous to the elastic thermomechanical equations where λ, μ, α and C_v are constants.

I am not aware of any experimental determinations of $C_v(\xi)$ and $\alpha(\xi)$ and as far as I know these functions are taken as constants in practical applications.

Case (ii). Large Isothermal Deformation

As mentioned in the opening paragraphs of Section 6, the choice of functions f_α is open. Then in the light of (6-11) we set

$$\dot{q}_\alpha + \rho_\alpha q_\alpha + \phi_\alpha(C_{ij}) = 0 \quad (\alpha \text{ not summed}) \tag{6-17}$$

where ρ_α and C_α are positive constants and ϕ_α are functions of C_{ij} only. The differential equation (6-17) can be solved for q_α in the light of the initial condition that at $t = -\infty$ the material is undisturbed. Thus,

$$q_\alpha = -\int_{-\infty}^{t} e^{-\rho_\alpha(t-\tau)}\phi_\alpha(C_{ij}(\tau))\,d\tau. \tag{6-18}$$

If (6-15) is integrated by parts and the assumption is made that $\dfrac{\partial\phi_\alpha}{\partial C_{ij}}$ depends on C_{ij} only for large values of α and further that $\rho_\alpha \to \infty$ as $\alpha \to \infty$ then (6-18) becomes [9]

$$q_\alpha = -\frac{1}{\rho_\alpha}\phi_\alpha(C_{ij}(t)) + \frac{1}{\rho_\alpha}\frac{\partial\phi_\alpha}{\partial C_{kl}}\int_{-\infty}^{t} e^{-\rho_\alpha(t-\tau)}\frac{\partial C_{kl}}{\partial\tau}\,d\tau. \tag{6-19}$$

Since in the case considered here $\dot{q}_\alpha$ are not necessarily small the expansion of $\Gamma(\dot{q}_\alpha)$ into a Taylor series will not terminate at the quadratic term but may include higher order terms. Thus,

$$\Gamma(\dot{q}_\alpha) = b_{\alpha\beta}\dot{q}_\alpha\dot{q}_\beta + b_{\alpha\beta\gamma}\dot{q}_\alpha\dot{q}_\beta\dot{q}_\gamma + \cdots \tag{6-20}$$

where $b_{\alpha\beta}$, $b_{\alpha\beta\gamma}$, etc., need not depend on C_{ij} since q_α depend on the deformation through (6-18).

For the sake of fixing ideas let $\Gamma(\dot{q}_\alpha)$ terminate at the quadratic term in (6-20). The equation (6-20) will be compatible with equation (5-15c) in the absence of thermal gradients if

$$b_{\alpha\beta}\dot{q}_\beta + \frac{\partial\Psi}{\partial q_\alpha} = 0. \tag{6-21}$$

Utilizing (6-17) and substituting for q_β in (6-21) one finds that

$$b_{\alpha\beta}(p_\beta q_\beta + \rho_\beta \phi_\beta) = \frac{\partial\Psi}{\partial q_\alpha} \tag{6-22}$$

whereupon integration with respect to q_α yields

$$\Psi = \tfrac{1}{2} b_{\alpha\beta}\rho_\beta q_\alpha q_\beta + b_{\alpha\beta}\phi_\beta q_\alpha + \Psi_0(C_{ij}). \tag{6-23}$$

At this point one may use (5-15a) to determine τ_{ij} and in fact

$$\tau_{ij} = \frac{2\rho}{\rho_0}\left\{\frac{\partial\psi_0}{\partial C_{ij}} + b_{\alpha\beta}q_\alpha \frac{\partial\phi_\beta}{\partial C_{ij}}\right\}. \tag{6-24}$$

At this point we can utilize (6-19) to substitute directly for q_α in (6-24) and obtain the constitutive equation

$$\tau_{ij} = \frac{2\rho}{\rho_0}\left\{\frac{\partial\psi_0}{\partial C_{ij}} - \frac{b_{\alpha\beta}}{\rho_\alpha}\phi_\alpha(C_{ij})\frac{\partial\phi_\beta}{\partial C_{ij}}\right\} + \frac{2\rho}{\rho_0}\frac{b_{\alpha\beta}}{\rho_\alpha}\frac{\partial\phi_\beta}{\partial C_{ij}}\frac{\partial\phi_\alpha}{\partial C_{kl}}\int_{-\infty}^{t} e^{-p_\alpha(t-\tau)}\frac{\partial C_{kl}}{\partial\tau}\,d\tau. \tag{6-25}$$

It is a well known experimental fact, that in the equilibrium configuration the stress tensor is derivable from a potential. A perusal of (6-25) shows that this is possible only if $b_{\alpha\beta}$ is expressible in the form:

$$b_{\alpha\beta} = \beta_\alpha \delta_{\alpha\beta} \quad (\alpha \text{ not summed}) \tag{6-26}$$

in this event

$$\tau_{ij} = \frac{2\rho}{\rho_0}\left\{\frac{\partial\phi_0}{\partial C_{ij}} + \sum_\alpha \frac{\beta_\alpha}{\rho_\alpha}\frac{\partial\phi_\alpha}{\partial C_{ij}}\frac{\partial\phi_\alpha}{\partial C_{kl}}\int_{-\infty}^{t} e^{-p_\alpha(t-\tau)}\frac{\partial C_{kl}}{\partial\tau}\,d\tau\right\} \tag{6-27}$$

where

$$\phi_0 = \psi_0 - \sum_\alpha \tfrac{1}{2}\phi_\alpha^2\left(\frac{\beta_\alpha}{\rho_\alpha}\right). \tag{6-28}$$

It is interesting that at very short times following the initiation of the deformation

$$\tau_{ij} = \frac{2\rho}{\rho_0} \frac{\partial \Psi_0}{\partial C_{ij}} \tag{6-29}$$

whereas at long times, i.e., as the material approaches an equilibrium state,

$$\tau_{ij} = \frac{2\rho}{\rho_0} \frac{\partial \phi_0}{\partial C_{ij}}. \tag{6-30}$$

In other words, ψ_0 and ϕ_0 are the instantaneous and equilibrium potentials, respectively.

Finally, if we retain third order terms in the expansion given by (6-20) and omitting details we find the following form of the constitutive equation:

$$\tau_{ij} = \frac{2\rho}{\rho_0} \left\{ \frac{\partial \phi}{\partial C_{ij}} + \int_{-\infty}^{t} \phi_{ijkl}(C_{rs}(t);\, t-\tau) \frac{\partial C_{kl}}{\partial \tau}\, d\tau \right.$$
$$\left. + \int_{-\infty}^{t} \int_{-\infty}^{t} \psi_{ijklmn}(C_{rs}(t);\, t-\tau_1-\tau_2) \frac{\partial C_{kl}}{\partial \tau_1} \frac{\partial C_{mn}}{\partial \tau_2}\, d\tau_1\, d\tau_2 \right\}. \tag{6-31}$$

ACKNOWLEDGMENT

This work was initiated at the Jet Propulsion Laboratory, Pasadena, California during the summer of 1966, and was completed under the partial support of the National Science Foundation. The author wishes to express his appreciation to these agencies.

References

1. M. A. Biot, J. Appl. Phys. *25*, 1385 (1954).
2. M. A. Biot, Proc. Third U.S. Nat. Cong. Appl. Mech., 1 (1958).
3. A. J. Staverman and F. Schwartzl, "Die Physik Der Hochpolymeren" Springer-Verlag, Berlin *IV* (1956).
4. A. J. Staverman, Proc. 2nd Int. Congr. Rheol., Academic Press (1954).
5. S. C. Hunter, J. Mech. Phys. Sol. *9*, 39 (1961).
6. A. C. Eringen, Phys. Rev. *117*, 1174 (1961).
7. R. A. Schapery, J. Appl. Phys. *35*, 1451 (1964).
8. B. D. Coleman, Arch. Rat. Mech. Anal., *17*, 1 (1964).
9. K. C. Valanis, J. of Math. and Phys., *45*, 197 (1966).
10. K. C. Valanis, J. of Math. and Phys., *46*, 164 (1967).

STATIC AND DYNAMIC MATERIAL BEHAVIOR OF SYNTACTIC FOAM

W. E. JAHSMAN

Lockheed Missiles & Space Company
Palo Alto, California

ABSTRACT

The use of foams or porous solids in a variety of applications such as energy absorbers, buoyancy materials, etc. has renewed interest in the accurate representation of the properties of these materials. In particular, a "compressive constitutive law" or equation of state is needed in the calculation of the dynamic response of the material to suddenly applied loads. Static testing to provide such data is appealing because of its simplicity; however, the importance of rate effects cannot be determined by this one method alone. Therefore, additional but numerically limited elevated strain-rate tests must be run for this purpose.

In the present paper, results of uniaxial strain static and gas gun compression tests on syntactic foam are reported. The foam is buoyant and is composed of hollow glass microspheres (average diameter 100 microns) embedded in an epoxy resin. Static testing consists of compressing a 0.25 cm × 2.5 cm dia. wafer between carefully aligned 2.5 cm dia. steel pistons. Lateral expansion of the wafer is suppressed by mounting it in a thick-walled (10 cm OD, 2.5 cm ID) cylinder. The degree of expansion is monitored by a circumferential strain gage mounted on the outside of the cylinder, coplanar with the foam wafer. Loading is increased until crushing of the microspheres, as indicated by a flattening of the stress-strain curve and by photomicrograph, is complete.

Dynamic testing is conducted using the LMSC plane wave generator ("gas gun"). Impact pressure and specimen particle velocity are recorded using a quartz transducer as impactor. Dimensions of the foam specimen for this test are equivalent to those of the static test. The data reduction technique for the conversion of stress-strain relation to impact stress-particle velocity behavior is based on an analysis of Courant and Friedrichs [1] which can be adapted to a rate-independent crushable-locking material (i.e., foam). Comparison between static and dynamic tests shows reasonable agreement at higher stress levels, with discrepancies at lower stress levels interpreted in terms of rate effects.

1. INTRODUCTION

Syntactic foam is a composite consisting of hollow sphere fillers incorporated in a resin matrix.* The fillers, usually glass micro-spheres

* The word "syntactic" is the adjective form of "syntax" (Greek *sun*, together, + *tassein*, to put in order), which means connected system or order; or orderly arrangement. Hence, syntactic foam is an orderly arrangement of hollow spheres in resin.

approximately 100 microns (~4 mils) in diameter, provide strong, impervious supports for otherwise weak, irregular voids. As a result, syntactic foam has attracted considerable attention both as a convenient and relatively light-weight buoyancy material [2] and as a porous solid with excellent shock attenuating characteristics [3]. The latter characteristic is achieved through crushing of the spheres and filling in the voids with resin.

Interest in the use of syntactic foam as a shock attenuator led to the present study of its static and dynamic mechanical properties. Particularly important is the influence of loading rate on stiffness and crushing strength, since oversensitivity of either of these parameters can complicate the prediction of the effectiveness of a foam system as an energy absorber.

Accordingly an experimental and analytical program was undertaken to establish the magnitude of the rate effect over the range of interest. From a materials testing standpoint, it is clear that the rate range is bounded from below by zero, which is well approximated by the so-called "static" testing machines, and from above by rates achieved under "instantaneous" impact, which are approached in gas gun tests. Assuming monotonicity of the material behavior over this extended rate range, one may argue that data from the "bounding" experiments will exhibit the maximum discrepancy and hence provide a gross measure of the material sensitivity to rate effects. The experiments conducted in this program are designed to follow this philosophy.

2. MATERIAL AND SPECIMEN PREPARATION

Seven 5-cm (2-in.) diameter syntactic foam cylinders were cast from a viscous mixture of hollow borosilicate spheres (31.2 weight %) and epoxy resin (68.8 weight %). Sphere diameters ranged from 0.3- to 1.25×10^{-2} cm. Air bubble entrapment in the mixture was minimized by using a hot casting technique. As a result density variation among the seven cylinders was less than 4%; this figure was reduced to 2% by selecting the most uniform five of the seven cylinders for actual testing. Table 2-1 lists, together with other properties, the as-cast dimensions and densities of the selected cylinders.

Determination of the sonic velocity (column 7) was made by dividing the cylinder length by the sonic transit time (column 6). This latter measurement was obtained by introducing a sharp (<1 μsec) pulse to one end of the cylinder by means of a piezoelectric ceramic disc. The disc was acoustically attached to one end of the cylinder by a heavy grease and an identical element was mounted on the other end of the

Table 2-1

SYNTACTIC FOAM CYLINDER AND SONIC TEST DATA

Cylinder No.	Length (cm)	Diameter (cm)	Weight (gm)	Density (gm/cm^3)	Sonic Transit Time (μsec)	Sonic Vel. ($cm/\mu sec$)	Dyn. Modulus ($kbar = dynes/cm^2 \times 10^9$)
1	15.75	5.176	285.4	0.8612	60.3	0.261	58.8
2	15.80	5.227	286.8	0.8460	60.8	0.260	57.3
3	15.48	5.258	286.3	0.8518	59.7	0.259	57.3
5	15.22	5.263	285.0	0.8608	58.4	0.260	58.3
6	15.88	5.232	290.7	0.8515	61.3	0.259	57.2

cylinder to receive the signal. The waveform received by the element was displayed on a Tektronix 535A oscilloscope and the sonic transit time was measured by a delayed sweep system. To insure accurate time measurement, this system was calibrated with a time mark generator. Comparison of sonic velocities measured by this technique for known reference materials (aluminum, lead) with published data [4] shows good agreement, with a maximum deviation of 2.3% for aluminum and 6.6% for lead.

From these four cylinders, disc specimens were machined for static and dynamic testing. Discs for static testing came from cylinders 1 and 3 and measured approximately 2.5 cm in diameter by 0.25 cm thick. Discs for dynamic testing were taken from all four cylinders and measured roughly 5 cm in diameter by 0.3 cm thick. All discs appeared quite uniform and a random check of density showed little variation between cylinder and the discs used in static testing. Some tendency toward increased density (by 3%) of the discs prepared for dynamic testing was noted (column 9, Table 4-1), suggesting that machining may have introduced a thin layer of higher density (2.13 gm/cm^3) glass chips at the surface although inspection at 100× was not conclusive.

A photomicrograph (also at 100×), showing the interior arrangement of the glass spheres in the epoxy resin matrix, is given in Fig. 2-1. The dark circles with the light spot in the center represent the spheres which were unbroken prior to sectioning, while the light circles are initially broken spheres which filled with epoxy during mixing. In some cases a dark circle may be found inside a light one, showing that smaller whole spheres may lodge within larger broken ones during the mixing process. In general, the size and spacing distribution of the spheres appears to be random, strongly suggesting that the assumption of statistical isotropy of the virgin material be used in the subsequent analysis. This has been done and more recent independent measurements of elastic properties at several orientations further substantiate this assumption.

3. STATIC TESTING

For reasons which will become apparent in the next section, the static tests were conducted using a fixture which approximated the "uniaxial strain" condition—i.e., complete lateral constraint with deformation limited to the direction of the maximum applied stress. Since a triaxial test apparatus was not available for these measurements, a far simpler piston and cylinder device (Fig. 3-1) was used. In this device a disc specimen (usually 0.002 cm oversize) was pressed into position at the midlength of the cylinder and the load was then applied to the sur-

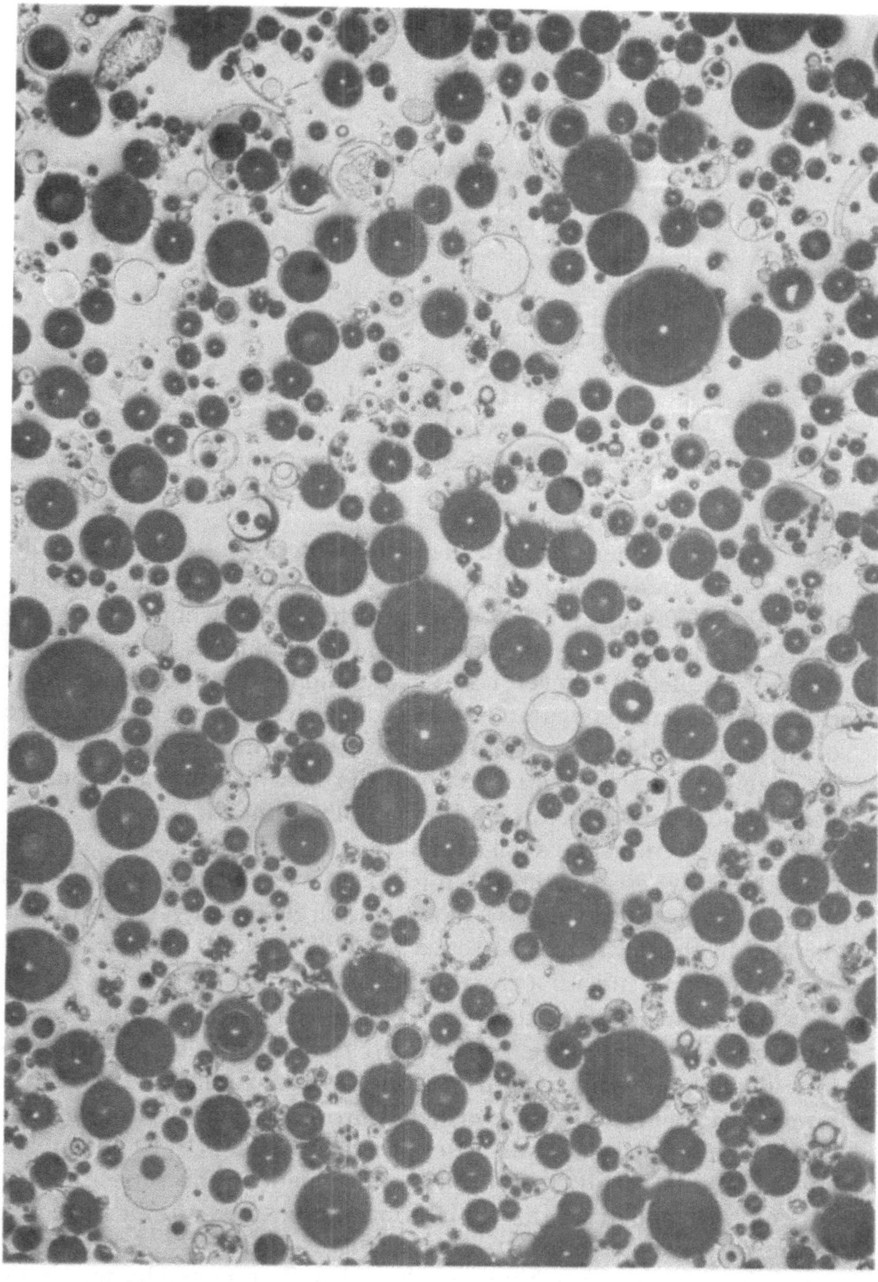

Fig. 2-1. Photomicrograph of as-received syntactic foam disc.

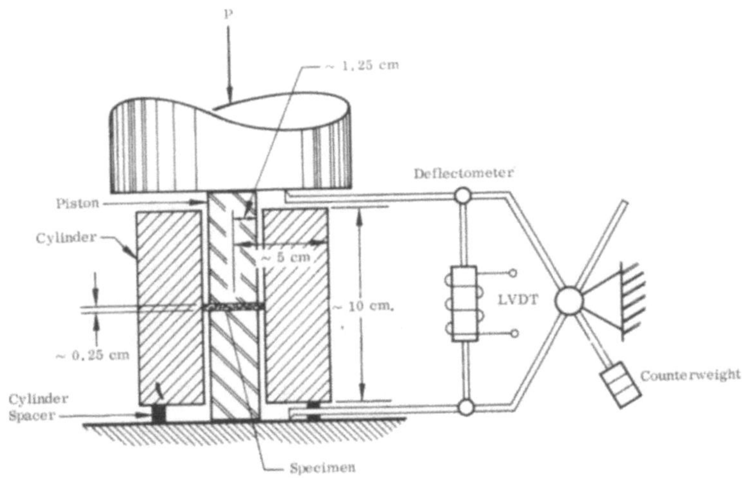

Fig. 3-1. Piston and cylinder device for approximating uniaxial strain conditions.

face of the specimen through the pistons. Because of the finite stiffness of the cylinder (30 × 10^6 psi (2 × 10^{12} dynes/cm^2) Young's modulus, with 180,000 psi (12.4 × 10^9 dynes/cm^2) yield strength Maraging steel—true also for the pistons), lateral constraint of the specimens was not complete. However, it was monitored in some tests by means of externally mounted circumferential strain gages. For most of the range of the test variables, the strain measurement, which could be used to estimate the circumferential strain of the specimen, was low; only at the high loads did it become substantial. Even then it was only a fraction of a percent of the axial strain.

In the early tests, a 200,000-lb (~100,000 kg) Tatnall machine was used to apply the load. Use of such a large capacity machine was dictated by the desire to achieve stresses well above 10 kbar (~145,000 psi) on a specimen with a cross-sectional area only slightly less than 1 in.^{2}. It was during these tests that the external circumferential strain measurements were made. Later a more conveniently located 120,000-lb (~60,000 kg) Tinius Olsen machine became available. Although a lower capacity machine, it nevertheless proved satisfactory for the load range of interest which was better specified by then (<10 kbars). No circumferential strain measurements were made in this series.

Length change in the direction of maximum stress was measured by means of the deflectometer shown in Fig. 3-1. Care was taken to account for both test machine stretch and deflectometer hysteresis by subtracting from the load-deflection curve obtained for a specimen the "pistons only" characteristic of the system. This characteristic was found to be quite repeatable so that after a few trial runs, a representa-

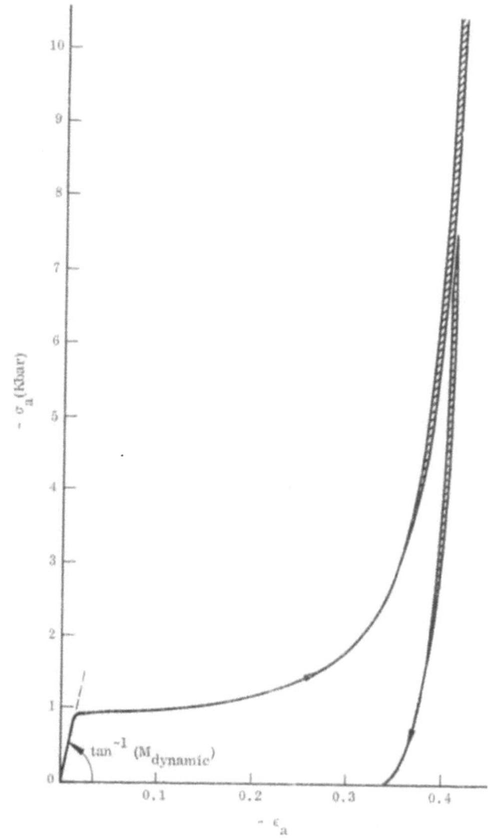

Fig. 3-2. Static axial stress-axial strain curve for syntactic foam. Cross hatching indicates experimental scatter.

tive trace was used for correcting all subsequent specimen load-deflection curves reaching a similar load.

At lower loads tilting of the loading heads gave rise to erroneous deflectometer readings. This difficulty was overcome by averaging two deflectometer readings, one taken at the front of the machine and the other taken 180° opposite. Thus, two specimens were expended for a single "averaged" curve. Each individual trace was repeatable, thereby demonstrating the small variability from specimen to specimen.

The final product of the data reduction described here is shown in Fig. 3-2, where axial stress $-\sigma_a$ (load ÷ original area, since the cylinder prevented changes of more than 0.4%) is plotted against axial strain $-\epsilon_a$ (length change ÷ original length). Estimates of specimen circumferential strain have been made * and found to be of order 0.2% at an axial stress of 10 kbar. Since axial strain at this stress level is about 40%, the

* See App. A for details.

condition of uniaxial strain is well approximated for syntactic foam using the piston and cylinder device. Figure 3-2 is a composite of curves obtained for four specimens, with the scatter band indicated by the cross hatching.

The crushing characteristic of the syntactic foam is quite apparent. Up to a stress of almost 1 kbar, the foam behaves linearly and reversibly; then with almost no increase in stress a dramatic reduction in volume (as measured by the axial strain) develops due to the breaking of the hollow glass spheres. As the resin flows into the now unsupported voids a volume reduction of about $\frac{1}{3}$ is observed. However, once the voids are filled, the foam "locks up" and behaves almost like pure epoxy. Figure 3-3 is a photomicrograph of such a specimen which has undergone an irreversible uniaxial strain change of about 30%. It is seen that with the exception of glass chips and a few oblong voids (with their short dimension in the direction of load), the specimen is composed almost entirely of epoxy resin (white). Inspection of Fig. 3-2 shows that the unloading curve is also quite nonlinear, with the modulus larger than that of the virgin composite. This behavior is to be expected since the original material has less volume % resin to resist the load.

Excellent agreement can be shown between the measured values of specimen density and those predicted using a void volume fraction of $\frac{1}{3}$. The prediction can be made from the law of mixtures:

$$\rho = \alpha\rho_v + \beta\rho_r \tag{3-1}$$

where α and β are the void and resin volume fractions ($\frac{1}{3}$ and $\frac{2}{3}$, respectively) and ρ_v and ρ_r are void and resin densities. Void density is due almost completely to the glass sphere

$$\rho_v = \rho_{gl}\frac{4\pi r^2 h}{\frac{4}{3}\pi r^3} = 3(h/r)\rho_{gl} \tag{3-2}$$

where h and r are the thickness and radius of the sphere. The value of $h/r = \frac{1}{25}$ has been reported in reference [2] and will be used here together with densities of 2.13 gm/cm^3 for glass and 1.16 gm/cm^3 for resin. Substitution of these values into (3-2) and (3-1) leads to a $\rho =$ 0.858 gm/cm^3, which is within 3% of any of the measured densities listed in Table 2-1. However, some discrepancy is noted between the weight fractions calculated using these figures – 10 wt. % for the glass and 90 wt. % for the resin – and those reported in the Material Preparation Section; therefore, additional measurements are required before the volume fractions proposed here can be accepted with confidence.

Shown in dotted lines near the origin of Fig. 3-2 is the dynamic modulus given for cylinder 1 in Table 2-1. It is indistinguishable from the slope of the static curve, and more careful calculations indicate that it differs from the static value by at most a few percent. Hence, there is additional reason to believe that conditions of uniaxial strain are closely met.

4. DYNAMIC TESTING

The LMSC light gas gun was used to obtain impact stress–particle velocity data for syntactic foam. The gun is based on the Sandia (6.37 cm) design. A schematic of the LMSC setup is shown in Fig. 4-1. Interest in this behavior is due to the facts that measurements are now relatively straightforward and certain correlations may be made with static data. These correlations will be carried out in the next section.

For use in the present study, the gun was operated in the specimen-in-piston mode; i.e., the foam disc was mounted on the front of the 5-cm diameter aluminum piston to impact on an X-cut quartz pressure transducer. The transducer, developed at Sandia [5], was located at the muzzle end of the 7.5 meter long gun; current output from the transducer was displayed on a Hewlett-Packard 175A oscilloscope. Impact stress was obtained by first determining the trace current i_p by scaling with a calibration current made prior to each shot and then by using the formulas [5]

$$-\sigma_i = (400/\pi)(G/KU_{sq}) \tag{4-1a}$$

where the pressure coefficient G depends linearly on i_p

$$G = (L/D)i_p \tag{4-1b}$$

and the current coefficient K and the quartz shock velocity U_{sq} are nearly constant but exhibit a weak G-dependence

$$K = 1.9853 + 0.0014661G - 0.0000011657G^2 \tag{4-1c}$$

$$U_{sq} = \begin{cases} 573.85 - 0.0085136G, & 0 \leq G \leq 245.5 \\ 555.56 - 0.065957G, & 245.5 \leq G. \end{cases} \tag{4-1d}$$

In (4-la–d), $-\sigma_i$ is measured in kbars, G in ma/cm, i_p in ma, K in 10^{-8} coulomb/cm^2 kbar and U_{sq} in cm/msec. L and D are the quartz transducer thickness and active electrode diameters in cm (0.637 cm and 1.275 cm, respectively).

Transducer mechanical impedance has been given by Sandia [6], from particle velocity at the surface of the quartz is found by the relation

$$u_q = 10^3(-\sigma_i)/\rho_q U_{sq} \tag{4-2}$$

Table 4-1

GAS GUN DATA

Specimen No.	i_p^* (ma)	$-\sigma_i$ (kbar)	u_q (cm/msec)	t_1 (μsec)	t_2 (μsec)	V (cm/msec)	u_i (cm/msec)	ρ (gm/cm^3)
1439	37.5	1.63	1.08	363.6	364.0	8.38	7.30	0.884
1013	50.5	2.19	1.43	200.8	200.0	15.2	13.8	0.871
1010	53.8	2.33	1.52	142.8	142.9	21.4	19.9	—
1440	58.8	2.53	1.67	115.2	115.3	26.4	24.7	0.876
1009	68.6	2.95	1.95	96.33	96.81	31.6	29.6	—
1441	90.0	3.84	2.55	78.73	78.86	38.7	36.1	—
1443	120.0	5.09	3.40	62.44	62.38	49.0	45.6	0.861
1444	178.6	7.52	4.96	48.05	48.27	63.4	58.4	0.858
1328	227.0	9.44	6.23	41.64	41.97	71.2	64.0	—
1011	238.0	9.78	6.48	40.79	40.59	74.8	68.3	—

* Plateau values (see traces in Fig. 4-2).

Fig. 3-3. Photomicrograph of syntactic foam after loading of 10 kilobars.

where ρ_q is the initial density of the quartz (2.65 gm/cm^3), $-\sigma_i$ is given in kbar and u_q and U_{sq} are given in cm/msec. Particle velocity in the syntactic foam is then found from the velocity continuity condition

$$u_i = V - u_q \tag{4-3}$$

where V is the velocity of the piston-plus-foam flyer just prior to impact. This velocity is an average of two velocities obtained from two measured transit times between three equispaced (0.305-cm separation) shorting pins located next to the muzzle (see Fig. 4-1). Since impact stress is the same for both quartz and foam, (4-1–4-3) are sufficient for the calculation of the pressure-particle velocity behavior of the syntactic foam. Table 4-1 provides supplementary as well as specific data for the $-\sigma_i$ vs u_i curve shown in Fig. 4-2. Experimental points are shown in circles, and a light line has been drawn through them to guide the eye. The basis for the solid line is described in the next section. Also included in Fig. 4-2 are oscilloscope traces of transducer output at low, intermediate and high impact stresses. Differences in the shapes of these traces will be discussed later.

5. ANALYSIS

A comparison of the static and dynamic behavior of syntactic foam as exhibited in Figs. 3-2 and 4-2 is not possible without some further information about the constitutive relation for the foam. One of the simplest forms for this relation assumes that the material making up the foam is rate insensitive; i.e., that the only significant variables entering into this relation are axial stress and axial strain (for uniaxial strain conditions).

Because of its simplicity, a rate-insensitive form based on Fig. 3-2 is adopted here with the view that validity of the assumption can be tested by comparison between predicted and measured impact stress-particle velocity curves. As shown in App. B, when the uniaxial strain constitutive relation during loading has the general monotonic form shown in Fig. 5-1, then an expression for impact particle velocity can be obtained in closed form. Subdivided by impact stress level into three regimes (dispersive wave, dispersive wave preceding a shock wave, and single shock wave), it reads

$$u_i = \begin{cases} \dfrac{1}{\rho_0}\displaystyle\int_0^{-\sigma_i} \dfrac{dp}{g(p)} & 0 \leqslant \sigma_i \leqslant p^* \\[2ex] u_1 + \left[\dfrac{-\sigma_i + \sigma_1}{\rho_0}\right] \left[\dfrac{1 - \eta(-\sigma_1)}{g(-\sigma_1) - u_1}\right], & p^* \leqslant \sigma_i \leqslant p^+ \\ & (0 \leqslant -\sigma_1 \leqslant p^*) \\[2ex] \sqrt{\dfrac{-\sigma_i}{\rho_0}}\,\eta(-\sigma_i) & p^+ \leqslant -\sigma_i. \end{cases} \tag{5-1a–c}$$

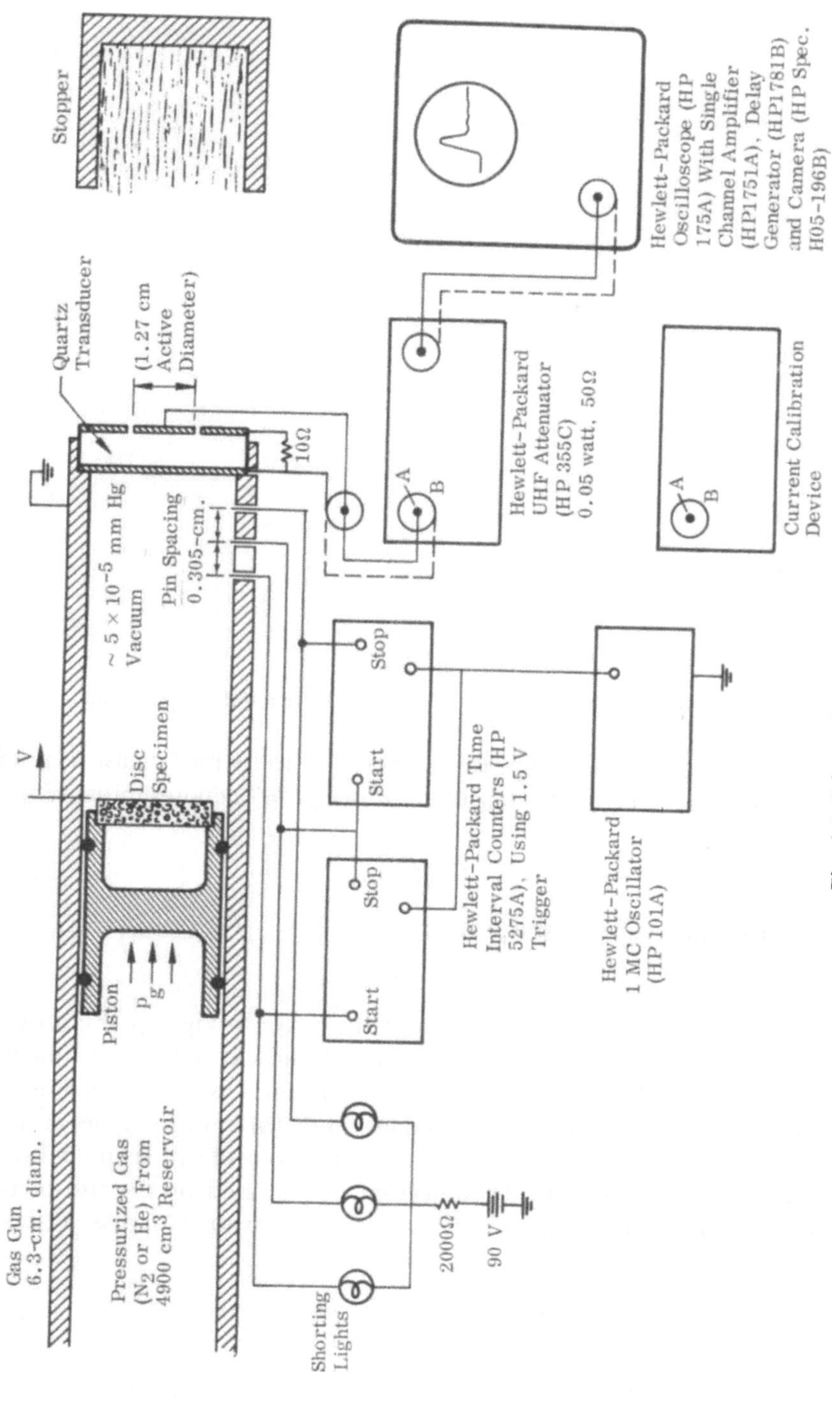

Fig. 4-1. Schematic of gas gun setup.

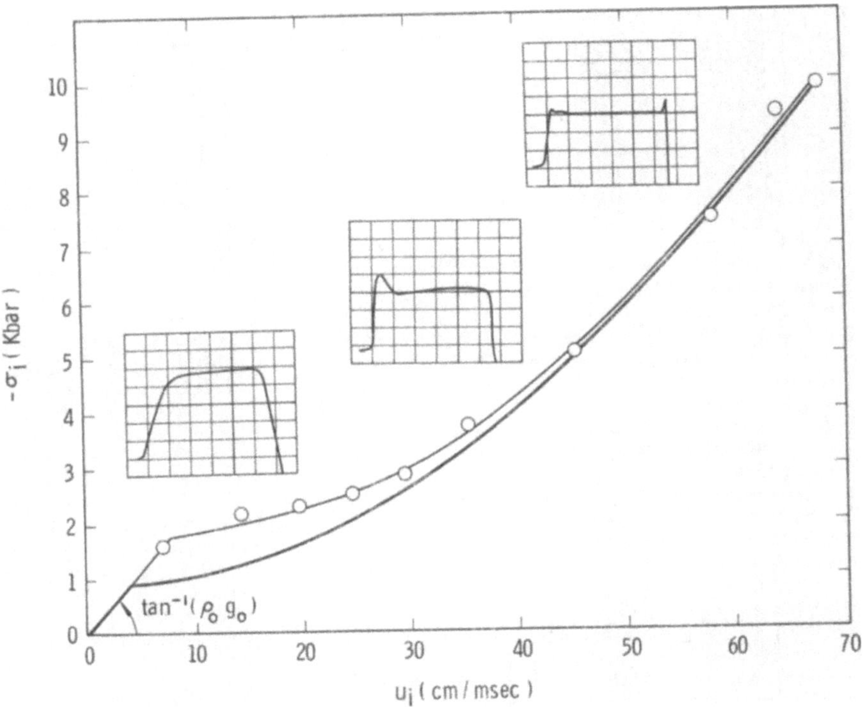

Fig. 4-2. Impact pressure-particle velocity behavior of syntactic foam.

The variables are compressive stress $-\sigma$ and strain measure η (cf. $-\sigma_a$ and $-\epsilon_a$ in Fig. 3-2); density ρ (subscript "0" denotes initial value), related to η by

$$\eta = 1 - \rho_0/\rho \tag{5-2}$$

and the characteristic shift rate g

$$g \equiv \sqrt{p'(\eta)/\rho_0} \tag{5-3}$$

where $p(\eta)$ is the functional form of the curve in Fig. 5-1. Because of the monotonicity of $p(\eta)$, one may invert and express η as a function of p (or $-\sigma$); hence, the shift rate in this analysis is regarded as dependent on compressive stress. Equation (5-1b) appears to depend on both $-\sigma_i$ and $-\sigma_1$ (note that u_1 is given in terms of $-\sigma_1$ from (5-1a) by replacing subscript i by 1); however, $-\sigma_1$ is fixed by the use of the Rankine-Hugoniot equation for mass conservation across a shock wave:

$$\frac{1 - \eta(-\sigma_i)}{g(-\sigma_1) - u_i} = \frac{1 - \eta(-\sigma_1)}{g(-\sigma_1) - u_1}. \tag{5-4}$$

Since (5-4) depends on u_i and $-\sigma_i$, iteration between (5-1b) and (5-4) is needed to find $u_i(-\sigma_i)$.

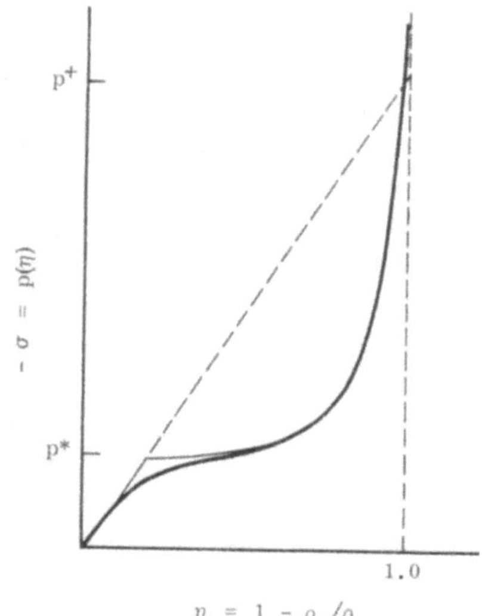

Fig. 5-1. Model of constitutive relation for syntactic foam. (Light line refers to eqs. 5.1a′, b′).

The locations of p^* and p^+ can be established by use of Fig. 5-1 and (5-1b–c), respectively. Up to a stress of p^*, a compressive pulse will tend to disperse as it propagates into the foam because the higher stresses travel with slower shift rates than the lower ones (see (5-3)). The reverse is true above p^* and a shock wave forms. As a result, the conditions ahead of and behind the shock are denoted by subscripts 1 and i. p^* is therefore located by the minimum shift rate or where $p''(\eta) = 0$; i.e., the inflection point in Fig. 5-1. The location of p^+ is determined by equating (5-1b) to (5-1c), with $-\sigma_1 = u_1 = \eta(-\sigma_1) = 0$ and $g(-\sigma_1) = \sqrt{M_0/\rho_0} = g_0$, where M_0 is the initial tangent modulus and g_0 is the initial elastic velocity. Then (5-1b) and (5-1c) combine to

$$\frac{p^+}{\eta(p^+)} = \rho_0 c_0^2 = M_0. \tag{5-5}$$

Since $\eta(p^+) \sim 0(1)$, extremely high impact stresses are necessary to produce a single shock wave because of the usually large initial tangent modulus. Reference to Fig. 3-2 shows that $-\sigma_i < p^+$ for the calculations of interest here and so (5-1c) will not be used.

Some simplification in (5-1a–b) results when the $-\sigma$ vs η curve is linear up to p^* (see light line in Fig. 5-1). These equations are replaced by

$$u_i = \begin{cases} \dfrac{-\sigma_i}{\rho_0 g_0} & 0 \leqslant -\sigma_i \leqslant p^* \\[2ex] \dfrac{p^*}{\rho_0 g_0} + \sqrt{\dfrac{-\sigma_i - p^*}{\rho_0}[\eta(-\sigma_i) - \eta^*]}, & p^* \leqslant -\sigma_i \leqslant p^+ \end{cases} \tag{5-1a′–b′}$$

where (5-2) and (5-4) have been used in eliminating g from (5-1b) and $\eta^* = \eta(p^*)$. In this form the equations can be used with the $-\sigma$ vs η curve exclusively to predict the impact stress-particle velocity relation for the material. Data from Fig. 3-2 have been used in this way with $p^* = 0.9$ kbar and $\eta^* = 0.0155$. The predicted curve is shown as the heavy solid line in Fig. 4-2. With the exception of the lower stress range the agreement with experimental data is quite good.

6. DISCUSSION AND CONCLUSIONS

Figure 4-2 provides a test of the hypothesis that syntactic foam is rate insensitive and that the static uniaxial strain stress-strain curve (Fig. 3-2) actually represents the general constitutive relation. Disagreement between the experimental data and the predicted behavior is greatest at low stresses (0(1 kbar)) where experimental stresses are about double those predicted analytically. The discrepancy decreases at the higher stress levels and virtually disappears at and beyond 7 kbar. This range of disagreement would extend somewhat further (to about 9 kbar) were the transient current readings rather than the plateau values used in the intermediate stress range. (The transient reading is defined as the initial overshoot seen in the center trace shown in Fig. 4-2; the plateau is the flat part of the trace following the overshoot.) The transient reading was discarded because it occurred in a time during which heterogeneity effects may be important. To assure a statistically homogeneous response, several "zones of heterogeneity" should be exposed to the stress wave. Taking the average diameter of a hollow sphere $2r$ as the dimension of a single such zone and the shift rate g at the stress level of interest wave speed, we arrive at a homogeneous state buildup time of

$$\Delta t = n(2r)/g \tag{6-1}$$

where n represents the number of zones exposed. For stresses just slightly above p^*, the shift rate g is considerably smaller than g_0, the elastic value (g_0 is approximately 0.26 cm/μsec, per Table 2-1). For a value of $g = 0.1$ cm/μsec, $2r = 10^{-2}$ cm and $n = 3$, (6-1) gives a buildup time of $\Delta t = 0.3$ μsec. Examination of the center trace in Fig. 4-2 shows that the transient is completed in approximately 0.4 sec (scale is 0.4 μsec/cm). As the stresses increase well above p^*, g increases toward g_0 (and beyond when the single shock wave develops) and the buildup time is correspondingly reduced.

A second effect—misalignment of specimen and transducer at impact—may mask the transient reading at low impact velocities because of the finite rise time associated with establishment of contact between specimen and gage from one edge to the other. If γ is the angle of misalignment, then the rise time is given by

$$\tau = D\gamma/V \tag{6-2}$$

where D is the active diameter of the gage. At a flyer velocity of 10 cm/msec, the rise time is $\tau = 0.2$ sec for an active gage diameter of $D =$ 1.275 cm and a misalignment angle of $\gamma = 5'$. As the flyer velocity increases, the misalignment effect ceases to be important; however, the much more gradual rise of the lower trace in Fig. 4-2 is quite apparent and is very probably due to this effect.

The influence of specimen heterogeneity and misalignment on the tracings have not been offered to explain the discrepancy between the curves in Fig. 4-2. Rather, it is believed that this difference can only be due to a rate or other material effect which must be included in the constitutive relation. Even in some of the static tests a small "delayed yield" type of behavior was observed at the knee of the stress-strain curve.

Because of the greater discrepancy at lower stress levels, a rate-dependent constitutive relation of the Malvern type may prove useful. Incorporation of such a relation in a further analysis will provide improved prediction capability; in addition further stress-particle velocity data near p^* will assist in filling out the curve in this important regime. Together the combined analytical-experimental approach makes a powerful tool in the study of dynamic behavior of materials.

References

1. R. Courant and K. O. Friedrichs, *Supersonic Flow and Shock Waves*, Interscience, p. 235 (1948).
2. H. Freed, "Buoyancy Material Development for Deep Submersibles," LMSC-806619, 15 Jan. 1967.
3. R. K. Linde and D. N. Schmidt, "Shock Propagation in Nonreactive Porous Solids," J. Appl. Phys., *37*, 3259 (1966).
4. *Handbook of Chemistry and Physics*, 47th Edition, The Chemical Rubber Company (1966–1967).
5. W. J. Halpin, O. E. Jones and R. A. Graham, "Submicrosecond Technique for Simultaneous Observation of Input and Propagated Impact Stresses," Proceedings, ASTM Meeting on Dynamic Behavior of Materials, Albuquerque, New Mexico (1962).
6. R. A. Graham, F. W. Neilson and W. B. Benedick, "Piezoelectric Current from Shock-Loaded Quartz—A Submicrosecond Stress Gauge," Sandia Corporation Report SC-R-65-899, Aug. 1965.
7. C. J. Tranter, "On the Elastic Distortion of a Cylindrical Hole by a Localized Hydrostatic Pressure," Quart. Appl. Math., *4*, 298 (1946).
8. S. P. Timoshenko and J. N. Goodier, *Theory of Elasticity*, McGraw-Hill Book Company, p. 58 (1951).

ACKNOWLEDGMENT

Among the many important contributors to this study, the author wishes to express special thanks to Dr. J. A. DeRuntz for his efforts in the material and specimen preparation and in the static experiments and to Mr. R. O. Doverspike for his supervision of the dynamic experiments. Dr. O. Hoffman has also made many helpful suggestions during long discussions devoted to this topic. This investigation was carried out under the Lockheed Independent Research Program and Navy Contract NOsp 65023.

APPENDIX A. ESTIMATE OF SPECIMEN CIRCUMFERENTIAL STRAIN

Desired is an estimate of the midlength circumferential strain at the inner radius of a thick-walled cylinder subjected to a localized internal pressure. The pressure is applied to a narrow band centered at the midlength of the cylinder and represents the surface load exerted by the disc specimen due to lateral expansion. Because of continuity of displacement between the disc specimen and the cylinder at the wall, cylinder strain at that point will equal the disc specimen strain. Available is a measurement of the midlength circumferential strain at the outer radius. The boundary value problem described here is sketched in Fig. A-1.

No published closed-form solutions for this problem have been found; however, solutions to related problems are available giving strain formulas which are believed to bracket the actual behavior. When the internal pressure p_i is uniformly distributed over the length $2c$, the results of Tranter [7] can be used to obtain a lower bound on the ratio $\epsilon_\theta(b)/\epsilon_\theta(a)$. In this reference, expressions are derived for

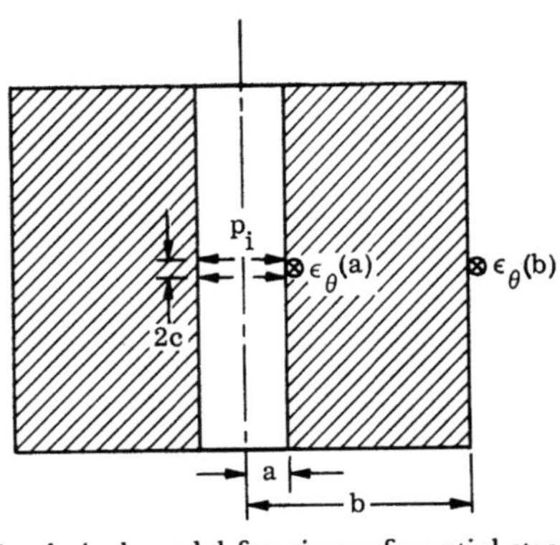

Fig. A-1. Analytical model for circumferential strain estimate.

stresses and displacements in an infinite medium surrounding a hole which is subjected to localized pressure. Pertinent to the problem here is the formula for the circumferential strain $\epsilon_\theta(r, 0)$ in a plane perpendicular to the axis of the hole and containing the center of the pressure band:

$$\epsilon_\theta(r, 0) = -\frac{2(1+\nu)}{\pi}\frac{p_i}{E}\frac{a}{r}\int_0^\infty \left[\alpha K_0(\alpha)K_1\left(\frac{r}{a}\alpha\right) - \frac{r}{a}\alpha K_0\left(\frac{r}{a}\alpha\right)K_1(\alpha) - 2(1-\nu)K_1(\alpha)K_1\left(\frac{r}{a}\alpha\right)\right]\frac{\sin\frac{c}{a}\alpha\, d\alpha}{\alpha D(\alpha)} \quad \text{(A-1a)}$$

where

$$D(\alpha) \equiv [\alpha^2 + 2(1-\nu)]K_1^2(\alpha) - \alpha^2 K_0^2(\alpha) \quad \text{(A-1b)}$$

and the K_i's are modified Bessel functions of the second kind of order i. E and ν are Young's modulus and Poisson's ratio for the medium and other dimensional variables are defined through Fig. A-1. The presence of the medium outside the radius $r = b$ acts to limit the radial displacement and thus the radial strain; hence the ratio

$$\frac{\epsilon_\theta(b, 0)}{\epsilon_\theta(a, 0)} = \frac{a}{b}\left\{\int_0^\infty \left[\alpha K_0(\alpha)K_1\left(\frac{b}{a}\alpha\right) K_0\left(\frac{b}{a}\alpha\right)K_1(\alpha) - 2(1-\nu)K_1(\alpha)K_1\left(\frac{b}{a}\alpha\right)\right]\frac{\sin\frac{c}{a}\alpha\, d\alpha}{\alpha D(\alpha)}\right\} \div \left\{\int_0^\infty \frac{K_1^2(\alpha)}{D(\alpha)}\frac{\sin\frac{c}{a}\alpha}{\alpha}\, d\alpha\right\} \quad \text{(A-2)}$$

is regarded as a lower bound to $\epsilon_\theta(b)/\epsilon_\theta(a)$.

Computations on (A-2) were carried out for the parameter values $\nu = 0.3$, $c/a = 0.07$ (approximately the thickness to diameter ratio of the compressed disc) and $b/a = 2, 3, 4$ and 5. At $b/a = 4$, corresponding to the radius ratio of the cylinder used in the experiments,

$$\left[\frac{\epsilon_\theta(b)}{\epsilon_\theta(a)}\right]_{\text{L.B.}} = 0.0119. \quad \text{(A-3)}$$

For the other extreme the Lamé formulas for internally pressurized thick-walled cylinders were used with the idea that axial as well as external constraint beyond $r = b$ has been removed. From [7] it is easily found that

$$\epsilon_\theta^*(r) = \frac{1}{E}\left[-(1+\nu)\frac{A}{r^2} + 2C(1-\nu)\right] \quad \text{(A-4a)}$$

where

$$-A = b^2(2C) = b^2[a^2 p_i/(b^2 - a^2)]. \quad \text{(A-4b, c)}$$

The ratio

$$\frac{\epsilon_\theta(b)}{\epsilon_\theta^*(a)} = \frac{2}{(1+\nu)(b/a)^2 + (1-\nu)} \tag{A-5}$$

gives an upper bound of

$$\left[\frac{\epsilon_\theta(b)}{\epsilon_\theta(a)}\right]_{\text{U.B.}} = 0.0931 \tag{A-6}$$

when the same parameters are inserted in (A-5) as in (A-2).

Inspection of (A-3) and (A-6) shows that while the bounds are not exactly narrow, a reasonable order of magnitude estimate for $\epsilon_\theta(a)$ in terms of $\epsilon_\theta(b)$ is

$$\epsilon_\theta(a) \sim 20\epsilon_\theta(b). \tag{A-7}$$

This figure, coupled with the recorded value of about 10^{-4} for $\epsilon(b)$ at a stress level of 10 kbar, yields on estimate of

$$\epsilon_\theta(a) \sim 2 \times 10^{-3} \tag{A-8}$$

for the maximum circumferential strain.* Since the corresponding axial strain is 4×10^{-1}, or 200 times as large, a uniaxial strain condition is apparently very nearly achieved through the test.

APPENDIX B. WAVE PROPAGATION ANALYSIS

The analysis presented herein is an elaboration of the derivation given in [1] and is included to make the paper self-contained. Attention is limited to one-dimensional wave propagation in a homogeneous isotropic medium. Motion perpendicular to the direction of wave propagation is prevented (the "uniaxial strain" condition). The equation of motion (conservation of momentum) is given by

$$\frac{\partial \sigma}{\partial a} = \rho_0 \frac{\partial u}{\partial t} \tag{B-1}$$

where σ is the normal stress acting in the direction of wave propagation and positive in tension, u is the particle velocity, a is the Lagrangian spatial coordinate, t is time, and ρ_0 is the initial material density. The Eulerian coordinate $x(a, t)$ may be found from the definition of particle velocity

$$u = \frac{\partial x}{\partial t}. \tag{B-2}$$

* This figure suggests the possibility of a small amount of plastic flow in the zone of contact between disc and cylinder. Repeated visual inspections and micrometer measurements showed no dimensional change in the interior so that (A-7) may well be an overestimate of the internal circumferential strain.

From conservation of mass, one has

$$v \equiv \frac{\rho_0}{\rho} = \frac{\partial x}{\partial a} \tag{B-3}$$

where v is the specific volume ratio and ρ is the instantaneous density. Equations (B-2) and (B-3) may now be combined to read

$$\frac{\partial v}{\partial t} = \frac{\partial u}{\partial a}. \tag{B-4}$$

To (B-1) and (B-4) for the unknowns σ, u and v must be added a third expression to complete the mathematical description. This expression is the constitutive relation $p(\eta)$ shown graphically in Fig. 5-1 (recall $\eta \equiv 1 - v$ from (5-2)).

The specific volume ratio may be eliminated by writing

$$\frac{\partial \sigma}{\partial t} = -p'(\eta)\frac{\partial \eta}{\partial t} = p'(\eta)\frac{\partial v}{\partial t} = \rho_0 g^2 \frac{\partial v}{\partial t} \tag{B-5}$$

where $g \equiv \sqrt{p'(\eta)/\rho_0}$ is referred to in [1] as the characteristic shift rate. Combination of (B-4) and (B-5) yields

$$\frac{\partial \sigma}{\partial t} = \rho_0 g^2 \frac{\partial u}{\partial a}. \tag{B-6}$$

Equations (B-1) and (B-6) constitute the governing equations for the unknown $\sigma = -p$ and u (g is regarded as a function of p; ρ_0).

It is convenient to transform (B-1) and (B-6) to characteristic coordinates described by

$$da = \pm g\ dt. \tag{B-7}$$

In this new coordinate system they become

$$-dp = \pm \rho_0 g\ du. \tag{B-8}$$

Prior to the formation of shock waves (i.e., $-\sigma \leq p^*$), use of the characteristics as coordinates is particularly attractive since integration may be carried out directly. For example, following impact a simple centered wave develops. Near the origin in t-a space the characteristics and stress profile have the appearance shown in Fig. B-1.

The relation between impact stress and particle velocity is obtained by integrating (B-8) along the negative characteristic (minus sign) from the undisturbed region below $da_0 = +g_0\ dt$ to the impacted interface. This integral may be written as

$$u_i = \frac{1}{\rho_0}\int_0^{-\sigma_i} \frac{dp}{g(p)} \tag{B-9}$$

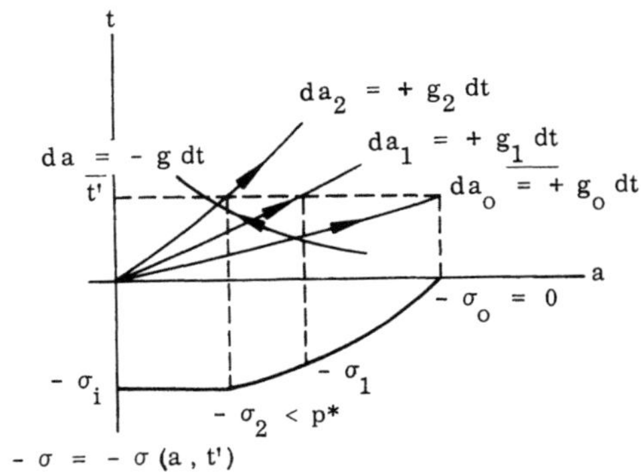

Fig. B-1. Characteristic network and stress profile for impact stresses below p^*.

since $-\sigma = u = 0$ in the undisturbed region (subscript i denotes interface value). For stresses in excess of p^* a shock wave will form and follow the dispersive precursor described by (B-9). In this case, Fig. B-1 should be replaced by Fig. B-2.

Conditions below $da_1 = +g_1\, dt (0 \leqslant -\sigma \leqslant p^*)$ are already described by (B-1–B-9); to relate variables across the shock front (state 1 to state i), the Rankine-Hugoniot equations for mass and momentum conservation across the shock are used:

$$\rho_i(c_s - u_i) = \rho_1(c_s - u_1) \tag{B-10}$$

$$-\sigma_i + \sigma_1 = \rho_1(c_s - u_1)(u_i - u_1). \tag{B-11}$$

All variables except c_s, the shock velocity, have been defined previously. For the shock front to remain stable (i.e., neither overtake nor fall behind the dispersive precursor) its velocity must be identical to the

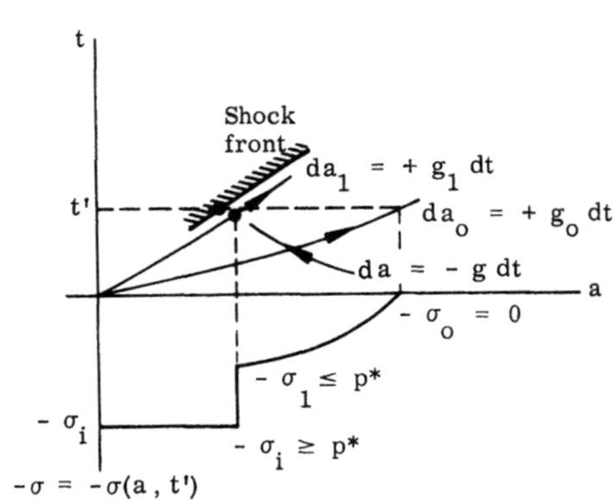

Fig. B-2. Characteristic network, shock front and stress profile for impact stresses above p^* (and less than p^+).

shift rate associated with the conditions just preceding the shock

$$c_s = g(-\sigma_1) \tag{B-12}$$

Use of (B-10–B-12) and (5-2) enables one to arrive at (5-1a–c).

Above p^+, (B-10) and (B-11) are simplified by setting u_1 and σ_1 equal to zero and (B-12) no longer applies. Shock velocity may be eliminated between the two and once again (5-2) is used to arrive at (5-1c).

THE EFFECT OF PRESSURE, TEMPERATURE, AND LOADING RATE ON THE MECHANICAL PROPERTIES OF ROCKS

J. B. CHEATHAM, JR.

Rice University
Houston, Texas

ABSTRACT

The influence of confining pressure, pore-fluid pressure, temperature, and rate of loading on the strength and ductility of rocks is reviewed in this paper. Effects of pressure and temperature on the mechanical behavior of a large variety of rock types have been studied by means of triaxial compression tests, and indentation experiments have been used to evaluate the influence of loading rates at ambient temperature and pressure. Data relating the combined effects of all of these variables over a wide range of values are not available; therefore. the effects of independent variations of pressure, temperature, and loading rate are considered in detail and interactions between these parameters are evaluated under limited conditions.

INTRODUCTION

The dependence of the mechanical behavior of rocks on pressure and temperature has been studied extensively by geologists and geophysicists, and the effect of loading rate has been evaluated by mining and petroleum engineers. Pressures as great as 300,000 psi and temperatures to 1500° F in conjunction with extremely low rates of loading are of interest in the geological studies. Laboratory experiments have been conducted to approximate conditions in the earth's crust to depths of 20 km for studies of faulting and folding. In mining and petroleum engineering problems pressures and temperatures rarely exceed 20,000 psi and 300° F, respectively; however, dynamic loading is an important consideration in blasting and percussive drilling.

An abundance of data have been published relating rock properties to the effects of pressure and temperature at low strain rates and to the effect of loading rates at atmospheric conditions. However, relatively few experiments have been performed in which pressure, temperature, and loading rate have been varied simultaneously over a large range of values of each of these variables. The purpose of this paper is to review the current state of knowledge of rock mechanics as related to the influence of pressure, temperature, and loading rate on the mechanical behavior of rock. Strength and ductility, the two mechanical properties of most interest, will be the primary concern in this survey. Other properties frequently evaluated are density, static and dynamic elastic constants, hardness, porosity, and fluid permeability.

EFFECTS OF CONFINING PRESSURE

An interest in the behavior of rock at great depth in the earth led early investigators to study the dependence of ductility on pressure. Adams and Nicolson [1] evaluated the ductility of rocks as a function of confining pressure as early as 1901. Later in 1911, von Karman [2] observed that a jacketed sample of marble confined in a high fluid-pressure environment could flow without fracturing. This testing procedure, known as the "triaxial test," has been the basis for many experiments by other investigators.

Typical stress-strain curves for the triaxial compression of rocks are illustrated in Fig. 1. Generally, the strength and ductility of rocks increase simultaneously as the confining pressure is increased. Differential axial stress, defined as the axial stress in excess of the confining fluid pressure, is plotted as a function of percent axial strain for the triaxial stress-strain curves.

The change in mode of failure with increasing confining pressure is also illustrated in Fig. 1. At atmospheric pressure, uniaxial compres-

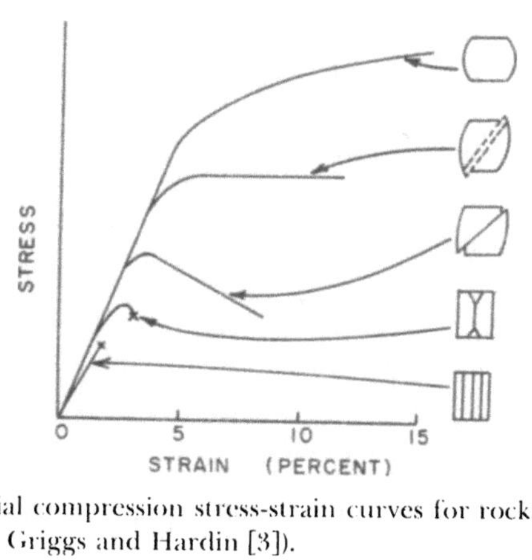

Fig. 1. Typical triaxial compression stress-strain curves for rocks and corresponding modes of failure (after Griggs and Hardin [3]).

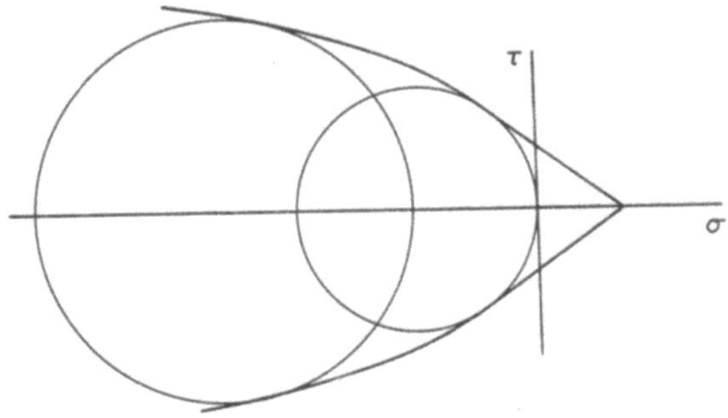

Fig. 2. Illustration of Mohr theory of failure of rocks.

sion normally results in axial tensile fractures. As the confining pressure is increased, the failure mode changes from brittle fracture to ductile flow. The stress-strain curve at an intermediate confining pressure indicates that the stress reaches a maximum value and then decreases for larger values of axial strain. Although the sample is partially fractured, the confining pressure acting on the jacket holds the fractured portions together and friction between the fractured surface continues to support an axial load. At higher confining pressures the samples deform without loss of cohesion. Griggs and Handin [3] have given a detailed discussion of the failure modes that are observed in triaxial tests.

Rock strength as a function of mean stress can be represented by Mohr circles as shown in Fig. 2. A nonlinear envelope can usually be fitted to these circles. The Mohr theory of failure predicts that fracture or flow will occur on a plane represented by the point of tangency of

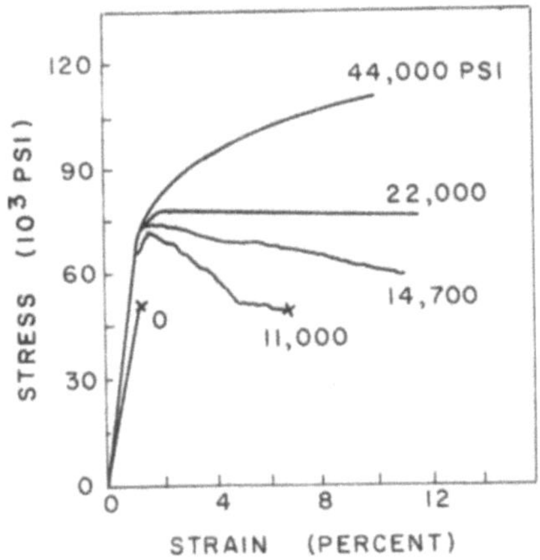

Fig. 3. Influence of confining pressure on strength of dry Solenhofen limestone at room temperature (after Heard [4]).

the envelope to a particular Mohr circle. The actual fracture direction does agree fairly well with the prediction of the Mohr theory in some cases [4].

The results of a comprehensive study of the ductility of Solenhofen limestone have been reported by Heard [4]. This limestone has a small grain size, is homogeneous and isotropic to a high degree, and is also chemically stable within a wide range of temperatures and pressures. Data for dry samples at room temperature and a strain rate of approximately 2×10^{-4} in./in. per second are given in Fig. 3. It can be seen that these results are similar to the idealized curves shown in Fig. 1. Experimental results for a number of different rock types have been published by Griggs, et al. [5] and Handin, et al. [6]; their data illustrate that both strength and ductility increase with pressure.

EFFECTS OF PORE FLUID

Interstitial fluid can exert a mechanical effect on rock due to the pore pressure as well as chemical effects related to the chemical nature of the fluid and the rock. Figure 4 shows data by Robinson [7] for water-saturated Indiana limestone at constant confining pressure and decreasing pore pressure. As the pore-fluid pressure decreases, the strength and ductility increase. Handin, et al. [6] have shown that the stress-strain curves nearly coincide when both the confining pressure and the pore pressure are varied simultaneously so as to maintain a constant differential pressure. It can therefore be concluded that the effective stress concept of soil mechanics also applies to rocks. For example, a compression test conducted at 10,000 psi confining pressure

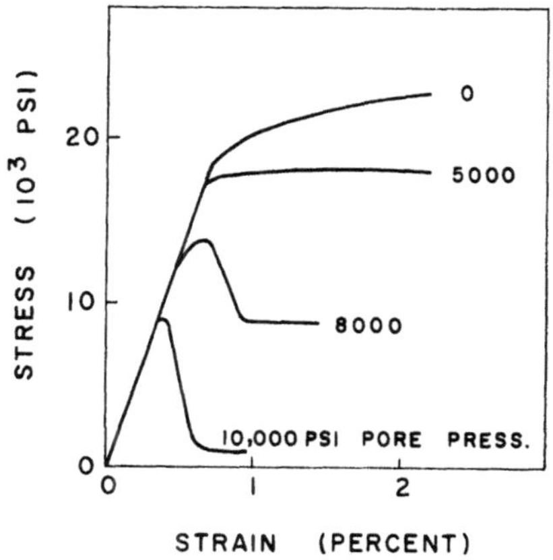

Fig. 4. Effect of pore pressure on behavior of Indiana limestone at a constant confining pressure of 10,000 psi (after Robinson [7]).

and 5000 psi pore pressure should give quantitatively the same result as a compression test on an identical sample of rock at 5000 psi confining pressure and atmospheric pore pressure.

Boozer, et al. [8] have shown that some fluids are not chemically inert when in contact with certain rock types. For example, the ductile flow of Indiana limestone is a thermally activated process in which the activation energy is proportional to the free surface energy. Its strength is reduced by adsorption resulting from ionic bonding on calcite surfaces because the adsorption causes a reduction of free surface energy. These authors based their selection of interstitial fluids for experiments on sandstone on the results of preliminary tests which showed that the tensile strength of quartz glass rods was lowered nearly 50 percent when the rods were immersed in water rather than in *n*-hexadecane. Oleic acids produced an effect intermediate between that of water and *n*-hexadecane. Hardness measurements on calcite crystals immersed in the above fluids were used to obtain preliminary data for experiments on limestone. The crystal hardness was found to be much lower in oleic acid and saturated calcium carbonate water solution than in *n*-hexadecane. However, as shown in Fig. 5, the strength of Indiana limestone was not decreased appreciably in the presence of these fluids, although the mode of failure of the sample saturated with oleic acid apparently changed.

EFFECTS OF TEMPERATURE

The combined effects of temperature and pressure on the strength and ductility of Solenhofen limestone are indicated by the data of Heard [4] shown in Fig. 6. At a constant confining pressure of 44,000 psi an increase in temperature from 68° F to 750° F decreases the ulti-

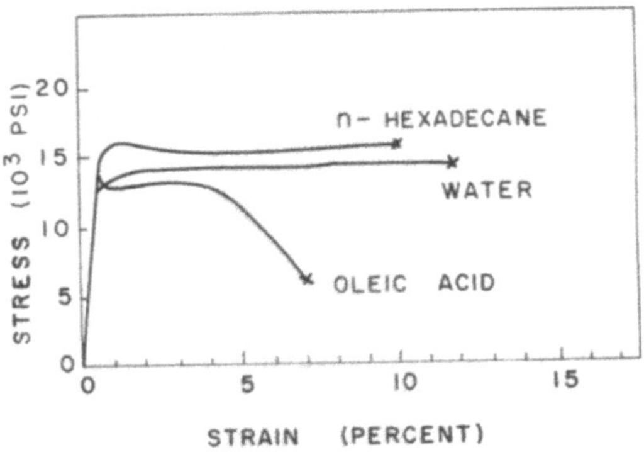

Fig. 5. Effect of various interstitial fluids on the behavior of Indiana limestone at 5000 psi confining pressure, room temperature and 0.07% sec strain rate (after Boozer, et. al. [8]).

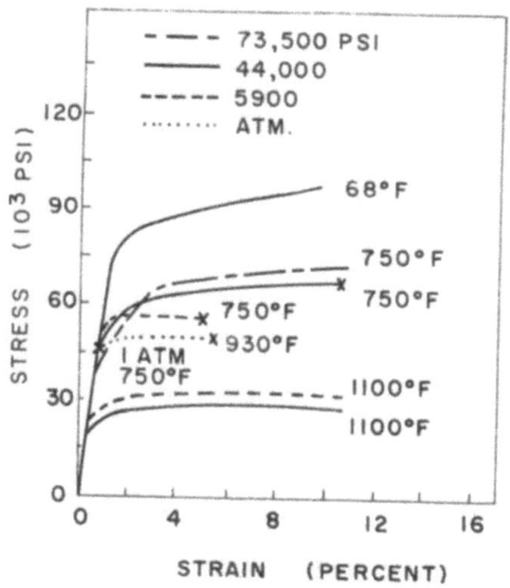

Fig. 6. Effect of temperature on the strength of dry Solenhofen limestone at various confining pressures (after Heard [4]).

mate strength by about 30%. A further increase of temperature to 1100° F causes an additional reduction in strength to about one-third of that measured at room temperature. For temperatures and pressures at which ductile failure occurs, a temperature change of a few hundred degrees has a much greater influence on strength than a confining pressure change of several thousand psi. For Solenhofen limestone the transition from brittle to ductile behavior occurs at a pressure of 15,000 psi for tests at room temperature and at a temperature of about 900° F for tests conducted at atmospheric pressure. Similar results have been obtained by Griggs, et al. [5] and Handin, et al. [6] for a variety of rock types.

EFFECTS OF RATE OF LOAD APPLICATION

The strain-rate sensitivity of rocks has been demonstrated by Robertson [9] in creep tests. Most of the experiments involving combined variations of rate of load application, temperature, and pressure have been conducted over a range of low loading rates. Indentation tests have been made under both static and dynamic conditions, primarily at ambient pressure and temperature. A few impact tests have been performed on rock subjected to pressure, but these were done at room temperature. Impact experiments will be discussed in a subsequent section.

Boozer, et al. [8] varied strain rates from 0.06 to 13 percent per second in their studies of the effects of fluid composition on rock properties. Strain rates were varied from 4×10^{-3} to 80 percent per

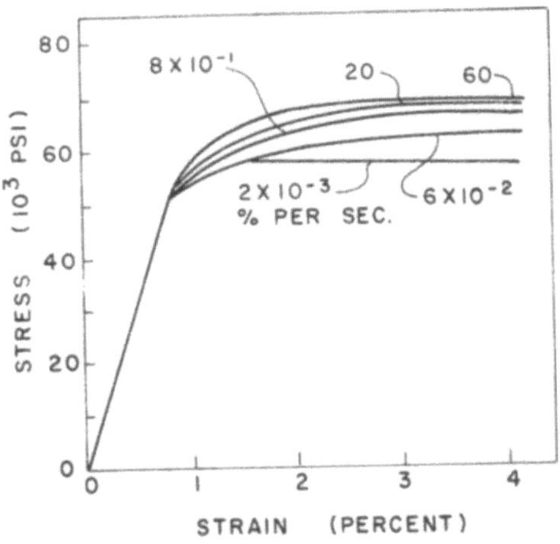

Fig. 7. Strain-rate effects for mineral-oil saturated Solenhofen limestone at 20,000 psi confining pressure and room temperature (after Serdengecti and Boozer [10]).

second while confining pressure and temperature ranged to 20,000 psi and 300° F, respectively, in results reported by Serdengecti and Boozer [10]. The above investigations of the interrelationships between these variables, although somewhat limited in variations of

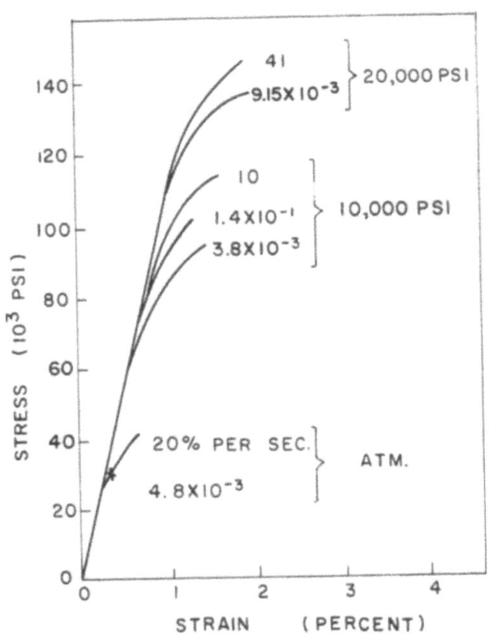

Fig. 8. Strain-rate effects for Pala Gabbro at room temperature and confining pressures of atmospheric, 10,000 psi and 20,000 psi (after Serdengecti and Boozer [10]).

pressure and particularly temperature, represent about the only such data available.

The effect of strain rate is illustrated in Fig. 7 for oil-saturated Solenhofen limestone at 20,000 psi and room temperature. The rock strength increases as the strain rate is increased from 2×10^{-3} to 60 percent per second. A similar effect of strain rate as well as the effect of confining pressure can be seen for Palo gabbro in the data shown in Fig. 8. While these data do not illustrate the influence of strain rate on ductility, other tests by these investigators show that as the strain rate is increased the failure becomes more brittle for constant temperature and confining pressure.

MECHANICAL EQUATION OF STATE

An equivalent effect of strain rate and temperature on the strength of Solenhofen limestone was observed by Serdengecti and Boozer [10]. This equivalence is similar to the effect reported by Zener and Holomon [11] for metals. In the range of strain rates from 10^{-3} to 100 percent and temperatures from 78° F to 300° F, they found that the equivalence relation is of the form

$$\sigma_u \propto \{\dot{\epsilon} e^{Q/RT}\}^r = \xi^r$$

where σ_u = ultimate strength
$\dot{\epsilon}$ = strain rate
Q = heat of activation
R = universal gas constant
T = absolute temperature
ξ = equivalence parameter
r = function of stress state

From experimental data, the heat of activation is found to be 24,800 calories per gram-mole and the exponent, r, is approximately proportional to the confining pressure. At 5000 psi confining pressure, a value of 0.0046 is reported for r, while at 20,000 psi, the value is 0.0087. In the above relation a decrease in temperature from 300° F to 78° F is equivalent to increasing the strain rate by a factor of about 10^5.

Using the equivalence parameter relating strain rate to temperature effects and experimental data at confining pressures of 5000 to 20,000 psi, Sergendecti and Boozer derive the following mechanical equation of state:

$$\sigma_u = K_0\{e^{\Delta p/k_1 p_0}\}\xi^r$$

p_0 = reference confining pressure

$\Delta p = (p - p_0)$ difference between actual confining pressure and reference pressure

$$r = r_0 + \frac{\Delta p}{k_2 p_0}$$

$k_1, k_2 =$ material properties.

Assigning the reference state as 5000 psi, they found the data to fit an equation of the following form:

$$\log_{10} \sigma_u = \log_{10} K_0 + r_0 \log_{10} \xi$$

for values of

$$\log_{10} K_0 = 4.664 \text{ and } r_0 = 0.0046.$$

From experiments at 20,000 psi they also evaluated the material constants to be $k_1 = 40$ and $k_2 = 725$. Using these values, the equation of state predicts values of ultimate strength σ_u which are in good agreement with experimental results obtained at confining pressures of 10,000 and 15,000 psi. In addition to strength as a function of the equivalence parameter, ξ, as represented by the solid lines in Fig. 9, modes of failure can be assigned to regions on the graph as shown by the dashed lines.

The effects on the ultimate strength of Berea sandstone and Pala gabbro in the temperature range below 300° F are small; thus, the equivalence parameter is primarily a function of strain rate for these rocks. Ultimate strength data at 10,000 psi confining pressure plots as a straight line on a graph of σ_u versus log ξ. Although the ultimate

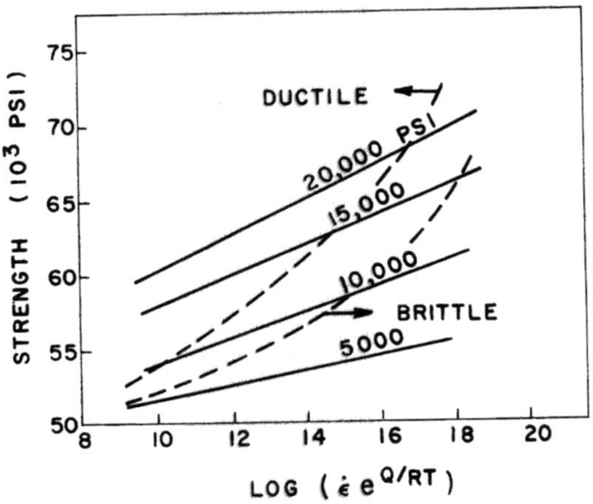

Fig. 9. Mechanical equation of state for Solenhofen limestone. Dashed lines indicate regions of graph representing points where brittle, transitional or ductile failure occurs (after Serdengecti and Boozer [10]).

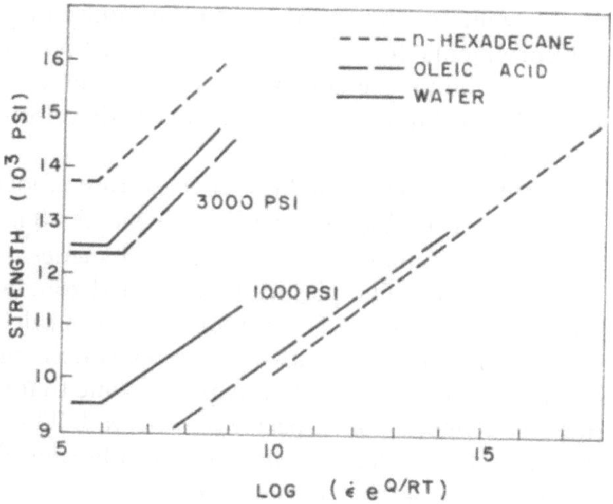

Fig. 10. Strength of Indiana limestone as a function of the logarithm of the strain-rate temperature equivalence parameter (after Boozer, et. al. [8]).

strength is approximately independent of temperature for these two rock types in the temperature range studied, the mode of failure does depend on temperature. Caution should therefore be observed in applying the equivalence concept to rocks in general.

The strength of Indiana limestone as a function of confining pressure, saturation fluid, and equivalence parameter are shown in Fig. 10 from data of Boozer, et al. [8]. The pore pressure was maintained at atmospheric in these tests; temperatures were varied from 78° F to 300° F and strain rates from 10^{-3} to 13 percent per second. Strength is a linear function of the equivalence parameter on the semi-logarithmic plot for each combination of confining pressure and saturation fluid.

RESULTS OF IMPACT TESTS

Maurer [12] reports that the energy per unit volume of crater is constant for impact of spherical bullets into sandstone and granite at velocities from 10 to 8000 ft/sec. He also found that crater volume is proportional to energy of impact and inversely proportional to the rock strength squared in this range of impact velocities:

$$V_c \propto \frac{E_c - E_0}{\sigma_u^2}$$

where V_c = crater volume
E_c = total energy
E_0 = threshold energy
σ_u = ultimate compressive strength of rock.

In addition, the volume is proportional to the square of the depth of the crater. Experiments reported by Vanzant [13] also showed that crater depth and volume are constant at constant impact energy for marble, although the bullet mass was varied with velocities ranging from static to 2000 ft/sec.

Pavlova and Shriner [14] studied the effect of rate of loading on the plastic deformation of marble under a circular die. A 10 mm diameter die was struck by a pneumatic drop hammer at velocities ranging from static to about 160 ft/sec. The plastically deformed zone was detected by means of gamma radiation. Translation gliding in calcite crystals is not visible; however, when an irradiated sample is deformed the gliding causes a visible color change [15]. A plastic zone that decreased in size with increased velocity of impact was observed. The dynamically deformed samples had a denser color, apparently caused by greater distortion of the crystal lattice at higher velocities.

Slow and rapid loading tests have been compared by Maurer [16] for blunt wedges forced into Indiana limestone under confining pressure. In his "static" tests craters were formed in 0.1 to 5 seconds, while in the rapid loading tests craters were formed in 0.001 to 0.005 seconds. Drilling mud acting on the horizontal rock surface provided a membrane with a low fluid permeability in the form of a filter cake. Force-displacement curves for both loading rates are similar and of the same form as the dynamic loading curves shown in Fig. 11; however, the threshold force required to cause cratering is greater for rapid loading at a given differential pressure. Results of other tests in which water was the confining fluid are shown in Fig. 11. In this case the pore pres-

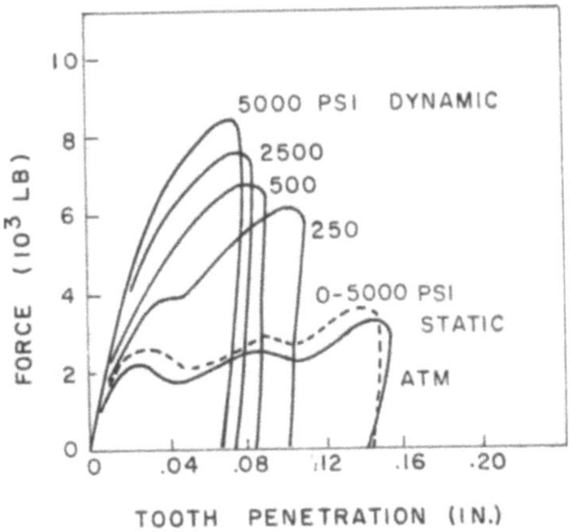

Fig. 11. Force penetration curves for static and dynamic loading of water-saturated Indiana limestone confined in water (equal pore pressure and confining pressure)—loaded by 45° wedge with 1/32 × 1/2 inch flat apex (after Maurer [16]).

sure is equal to the confining pressure. At low rates of loading all tests from atmospheric pressure to 5000 psi yield approximately the same force-displacement curve. With increased rate of crater formation, however, the water does not have sufficient time to flow into the fractures. This time delay allows a differential pressure to act on the chip during its formation, and conditions are therefore much the same as in the tests with drilling mud. Brittle fractures were produced at all pressures during the static tests with water, whereas a brittle to ductile transition occurs during the dynamic tests at a pressure between 250 and 300 psi. In the drilling mud tests, brittle to ductile transition also occurs at 250 to 500 psi for dynamic loading, but the transition takes place at 1250 to 1500 psi mud pressure with static loading. It should be emphasized that these phenomena are mechanical effects of the inert fluid and are not inherent properties of the constitution of the rock.

CONCLUDING REMARKS

The mechanical behavior of rock is influenced by the confining pressure, the pore-fluid pressure, the temperature, and the rate of load application. Triaxial test data indicate that both strength and ductility of rocks are increased as the effective confining pressure is increased. At a given confining pressure, an increase of temperature generally results in a decrease of rock strength and an increase of ductility. Increasing strain rate has qualitatively the same effect as decreasing temperature; thus, rock strength is increased and ductility is decreased as the strain rate is increased at constant pressure and temperature.

A mechanical equation of state has been derived for certain varieties of rock. This equation relates the compressive strength of a rock to the effects of confining pressure, temperature, and strain rate. The chemical composition of the interstitial fluid can also be related to the mechanical behavior of various rocks.

Over a wide range of impact velocities the volume of a crater formed in dry rock at ambient conditions has been observed to correlate with the impact energy. The time dependency of crater formation by impact in liquid-saturated rocks is influenced by the pressure interaction between the external confining fluid and the internal pore fluid.

Currently, most rock mechanics research efforts involve either static or creep experiments at high pressures and temperatures or dynamic loading of dry samples at ambient conditions. From the limited data available, it can be concluded that dynamic loading tests at ambient conditions cannot reproduce conditions in the earth where significant pressure and temperature variations exists.

ACKNOWLEDGMENT

The author wishes to express his appreciation to Dr. Paul Gnirk for his assistance during the preparation of this paper and to the American Petroleum Institute for support under Project API 67F.

References

1. F. D. Adams and J. R. Nicolson, "An Experimental Investigation into the Flow of Marble," Trans. Royal Soc. London, Ser. A, *195*, 363 (1901).
2. T. von Karman, "Festigkeitsversuche unter allseitigem Druck," Zeitschr. Ver Deutsch. Ingenieure, *55*, 1749 (1911).
3. D. T. Griggs and J. Handin, "Observations on Fracture and on Hypothesis of Earthquakes," *Rock Deformation,* Geol. Soc. Amer. Mem. 79, 347 (1960).
4. H. C. Heard, "Transition from Brittle to Ductile Flow in Solenhofen Limestone as a Function of Temperature, Confining Pressure, and Interstitial Fluid Pressure," *Rock Deformation,* Geol. Soc. Amer. Mem. 79, 193 (1960).
5. D. T. Griggs, F. L. Turner and H. C. Heard, "Deformation of Rocks at 500° to 800° C," *Rock Deformation,* Geol. Soc. Amer. Mem. 79, 39 (1960).
6. J. Handin, R. V. Hager, M. Friedman and J. N. Feather, "Experimental Deformation of Sedimentary Rocks under Confining Pressure: Pore Pressure Tests," Amer. Assoc. Pet. Geol. Bull., *47*, No. 5 (1963).
7. L. H. Robinson, "Effects of Bore and Confining Pressures on Failure Characteristics of Sedimentary Rocks," Trans. A.I.M.E., *213* (1958).
8. G. D. Boozer, K. H. Hiller and S. Serdengecti, "Effects of Pore Fluids on the Deformation Behavior of Rocks Subjected to Triaxial Compression," *Rock Mechanics,* Pergamon Press, 579 (1963).
9. E. C. Robertson, "Experimental Study of the Strength of Rocks," Bull. Geol. Soc. Amer., *66*, 1275 (1955).
10. S. Serdengecti and G. D. Boozer, "The Effects of Strain Rate and Temperature on the Behavior of Rocks Subjected to Triaxial Compression," Proc. Fourth Symp. Rock Mech., Penn State U., 83 (1961).
11. C. Zener and J. H. Holomon, "Plastic Flow and Rupture of Metals," Trans. A.S.M., *33*, 163 (1944).
12. W. C. Maurer, "The 'Perfect-Cleaning' Theory of Rotary Drilling," Trans. A.I.M.E., *225*, I-1270 (1965).
13. S. W. Vanzant, "Dynamic Rock Penetration Tests at Atmospheric Pressure," *Rock Mechanics,* Pergamon Press, 61 (1963).
14. N. N. Pavolva and L. A. Shreiner, "Effect of Loading Speed on Plasticity of Marble Under Indentation," Soviet Physic-Doklady, *6*, No. 9, 224 (1961).

15. J. Handin, D. V. Higgs, D. R. Lewis and P. K. Weyl, "Effects of Gamma Radiation on the Experimental Deformation of Calcite and Certain Rocks," Geol. Soc. Amer. Bull., *68*, 1203 (1957).
16. W. C. Maurer, "Bit-Tooth Penetration under Simulated Borehole Conditions," Trans. A.I.M.E., *234*, I-1433 (1965).

APPENDIX

Closing Comments by Session Chairmen

Session Ib. W. PRAGER,
University of California at San Diego

As I understand it, my assignment is to comment on classical continuum theory in general, and on its relations to the papers presented in the first session in particular, and, last but by no means least, to do this in less than ten minutes.

The common problem at the core of these and many other papers presented at this Symposium is the choice of a *constitutive equation*. This is a problem that has intrigued me ever since the late 1920's. If you asked me to try to distill the experience of these forty years into a couple of sentences, I should say that two facts appear to be important to me in this area. Firstly, if a constitutive equation is to be useful in the solution of practical problems, it can, at best, apply only to *limited ranges* of strain, strain rate, temperature, etc. Secondly, even though a constitutive equation is an essential ingredient of a continuum theory, it may well contain functions that are *not continuous* or *continuously differentiable.*

Since, from the practical point of view, there is no such thing as a *universal* constitutive equation, the *explicit* statement of a range of validity is an integral part of any constitutive equation. We must resist the temptations of trying to apply a constitutive equation beyond this range, or even of trying to derive from a constitutive equation the limits of its range of validity. Accordingly, it seems futile to me to expect that the constitutive equation of, say, an *elastic* solid should *by implication* furnish a yield limit. This will always have to be stated explicitly as a limit of its range of validity. It is, of course, not necessary for this purpose directly to give the yield stress. One may instead state that yielding sets in, say, when the work expended per unit of volume on the elastic change of shape reaches a critical value. This was done by Hencky [ZaMM, 4(1924) 323–334], who showed that the yield condition of von Mises could be derived in this manner. Dr. Reiner uses a failure criterion of a similar nature: the work expended in the change of shape is partly stored in the elastic elements of his model, and partly dissipated into heat in the viscous elements. Dr. Reiner assumes that failure occurs when the first part reaches a critical value. Now, Dr. Reiner's model contains only elastic and viscous elements, and is therefore unable to represent time-independent plastic effects. To incorporate these, we may use elements with dry friction that, up to a certain limit, can support stress without deforming but slip when a critical stress is reached. If we wanted to construct a simple failure criterion for a model of this kind, it would perhaps be more reasonable to assume that failure occurs when the work done in overcoming friction reaches a critical value, because this work may be taken as an indication of the extent to which the original structure of the solid has been permanently damaged. I mention this alternative failure criterion not to cast doubts on Dr. Reiner's work but only to underscore the first point I made at the start of these remarks. There is no universal constitutive equation, and an assumption that may be eminently reasonable in one set of circumstances may be obviously wrong in others.

I believe that the papers by Drs. Bell and Dillon nicely illustrate my second point: a continuum theory may well involve functions that are not continuous or continuously differentiable. There may be physical reasons for introducing discontinuities of this kind. For example, the slip theory of Batdorf and Budiansky taught an important lesson along these lines by showing that a simple physical model made it necessary to abandon the conventional idea of a yield locus with continuously turning tangent. Even though the virgin yield locus might be smooth, strain-hardening would promptly produce vertices which would profoundly affect subsequent plastic behavior.

Alternatively, discontinuities may be introduced for the sake of mathematical convenience. If this assertion appears paradoxical to you, let me remind you of the manner in which the use of a piecewise linear yield locus often enables us to treat problems that could not be handled if we insisted on a yield locus with continuously turning tangent.

Dr. Bell chooses to represent his experimental results by a polygon rather than a curve with continuously turning tangent, not for the sake of mathematical convenience, but because he believes that this sequence of straight segments (in the $\sigma^2-\epsilon-$plane) represents physical reality. It seems to me that this raises important questions: what is the physical phenomenon behind the sudden change from one value of his deformation mode index to the next, and what causes the quantization of this index? I must confess that these questions bother me more than the prospect of eventually having to use this sort of piecewise linear relationship in the solution of practical problems.

A similar remark may be made regarding the stepwise yielding observed by Dr. Dillon. Regardless of whether we are inclined to accept this as a physical characteristic of the material or as the result of the interaction of test specimen and testing machine, we might well explore its use as a mathematical device. Consider, for instance, the behavior of a work-hardening plastic beam under slowly increasing loads when the relation between bending moment and curvature is represented by a stairlike line. At any stage of the loading process, the beam may be divided into a number of segments, throughout any one of which the curvature has the value corresponding to one of the risers of the stairs, while at the junction of two segments the bending moment has the value corresponding to one of the treads of the stairs. This kind of treatment may in fact be simpler mathematically than a treatment based on a continuously rising diagram of bending moment versus curvature.

I have singled out two facts that appeared to me to be important in the choice of constitutive equations for plastic solids. If you permitted me to add a third such fact, it would be this: contrary to what one would be inclined to expect, thermodynamics of reversible or irreversible processes has so far failed to make a significant contribution to this field. Arguments that appear to be of thermodynamic nature are unavoidable in problems that involve the interaction of stress, strain, and temperature. On closer inspection, however, these arguments turn out to be only quasi-thermodynamical in the sense that they use the vocabulary rather than the principles of thermodynamics. Dr. Drucker, who is the next speaker, is well known for his use of quasi-thermodynamical arguments of this kind, and I hope that he will elaborate on my remark.

Session II. D. C. DRUCKER,
Brown University, Providence, Rhode Island

I have known Professor Prager a long time and shall start by saying that in essence I agree with almost all that he has said in his concluding remarks. The only disagreement in principle is on our feeling about the future success of the thermodynamic approach. A good deal of my time these days is spent in thinking about thermodynamics and plastic or inelastic behavior in general.* It makes me sad to report that my latest hope of combining elastic and perfectly plastic elements in a clever manner has just been deflated by one of my students, Dr. David Rubin. He has written a very nice thesis on how a thermodynamic

* Sponsored by the Office of Naval Research under Contract Nonr 562(20) with Brown University.

treatment can be based upon any such explicit state description. He shows, however, that while everything is clear, explicit, and self-consistent, the result is physically incorrect for a metal no matter how complex the assemblage of elements. Nevertheless, hope does spring eternal.

As I did not know what Professor Prager was going to say, I had planned to start by noting that Professor Perzyna, in the introduction to his "Thermodynamics of Viscoplasticity" brought out both the primitive and the controversial status of this subject. He deals with both extremes of the research spectrum. One is the very sensible pragmatic approach of choosing the simplest of all representations which can give a reasonable approximate solution for a given problem of practical importance. If you want to solve boundary value problems, you must idealize your constitutive relations or your thermodynamics drastically. On the other extreme there is the effort to which he and a number of us are devoting our time (represented in viscoelasticity by Professor Valanis' presentation), aimed at a rather general thermodynamic approach to the inelastic behavior of materials.

Perhaps a reduction to time-independent plastic relations can illustrate best the confusion which results from a mixture of the two extremes. Most elaborate forms used at present in a thermodynamic treatment (they look very elaborate indeed in the large-strain general-tensorial formulation) are really at the same level of representation as perfect plasticity with a simple yield criterion, or they go one step beyond to include kinematic hardening. This, of course, as Professor Prager emphasizes, is not in any way a real description of the physical behavior of material. No one takes any such drastic idealization as representing the actual response of real metals. It is simply the least you can put down and hope to get an answer which has engineering importance. The details of stress-strain behavior of a structural metal are enormously more elaborate as can be and often has been demonstrated by experiments in combined tension and shear. Once we leave the elastic range, all hope disappears of finding both a simple and a complete description of the mechanical behavior of metals. Obviously, then, we cannot hope for simplicity and completeness in a thermodynamic description.

When solving problems, simplicity comes first; but what should we do when attempting to describe physical behavior in its essential detail? One very reasonable choice is to uncouple the thermal and the mechanical aspects except for the influence of temperature on the properties of the material which is taken into account only approximately. This is one approach, but it is not thermodynamics; it is the antithesis of thermodynamics. Most speakers at this Symposium have wisely followed this path.

What must we do if we wish to do thermodynamics? Clearly, we must satisfy the first and second laws and build a mathematically consistent theory. The main point I wish to make, however, is that mathematical consistency is not the area in which our major problem lies. A complex mathematical development of complete generality, which satisfies the first and second laws, may look impressive but is of little if any use. Complete generality is chaos.

How can I be so positive in making so negative a statement? Let us go back again to time-independent plasticity. If you use just the first and second laws, nothing comes out that you don't already know. Nothing comes out which is of the essence of plasticity theory. Consequently, you cannot expect that in a more general treatment anything will come out which is really of the essence of a coupled thermodynamic time- and temperature-dependent plasticity theory. When you look at the details of the most complex or the most elementary of the modern mathematical treatments which produce definite results, you find implicitly or explicitly an arbitrary choice of state variables or a very strong assumption about the state variables. Instead of illuminating the physical nature of the assumptions about the material, an elaborate mathematical superstructure obscures the essential step which is taken. For example, often only plastic strain is added as a state variable to those variables appropriate for purely elastic materials. Perhaps because there are nine new state variables in tensorial space, some readers get the impression that a general thermodynamic theory is proposed. First of all, of course, plastic strain is clearly not a state variable as can easily be seen from the simple example of compressing a bar plastically and then pulling it back to its original dimensions. In the limiting case of isothermal time-independent plasticity theory, this is just the kinematic hardening assumption. It is a great way of solving particular problems with rather small plastic strains, but it is not more than the crudest beginning of a real thermodynamic description of metal behavior. Certainly it is not enough of a start to warrant very learned discussions about second order and higher order effects.

Possibly because the mathematics is so opaque, assumptions are made which are so drastic as to distort the essence of the theory. Again let me revert to isothermal time-independent theory to describe what can happen. Suppose we do as we once did, before we learned better, and say that "obviously" the rate of plastic work is positive; $\sigma_{ij}d\epsilon_{ij}^p$ or $\sigma_{ij}\dot{\epsilon}_{ij}^p > 0$. We then concentrate on plastic work and the second law of thermodynamics, when in fact it is not $\sigma_{ij}\dot{\epsilon}_{ij}^p > 0$ but $(\sigma_{ij} - \sigma_{ij}^0)\epsilon_{ij}^p \geqslant 0$ which gives plasticity theory. As shown in the figure on p. 408, the yield locus need not, and with sufficient Bauschinger effect does not, enclose the origin. The rate of doing plastic work $\sigma_{ij}\dot{\epsilon}_{ij}^p$ then can be negative as illustrated.

If in a thermodynamic theory, we generalize but keep the assumption that plastic work is positive, clearly we settle on the wrong physical basis even though we speak in terms of entropy production.

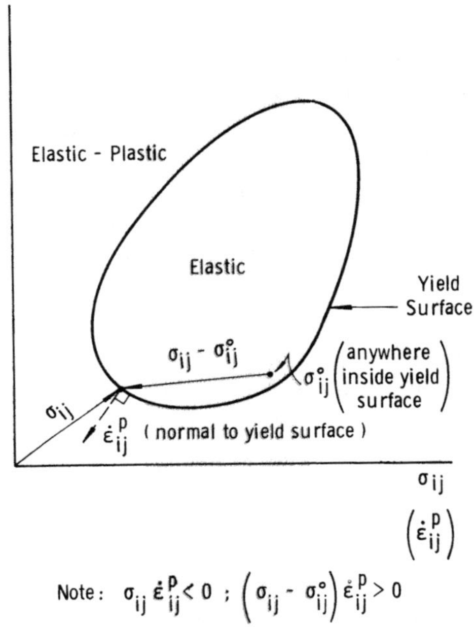

These pitfalls are mentioned, not because many people have fallen into error but to emphasize the need to concentrate on a different level of endeavor. Somehow we must elucidate the essence of physical behavior first and later cast it into mathematical form. Professor Bodner phrased it well when he said that we must look for a basic physical law not just a mathematical convenience.

When a drastic simplification is made, it should shine out clearly in the mathematics so everyone can see it and everyone can evaluate it. This Professor Perzyna does when he solves problems as did Dr. Lindholm in the analysis of his experimental data. Dr. Hahn and his coauthors also were clear in the exposition of their basic variables, as was Dr. Krafft in his treatment of the plane-strain crack. In this group of papers, only Dr. Krafft paid detailed attention to the microscale and its effect on the macroscale. He emphasized the role of inclusions and their spacing in determining the spread of the plastic zone and the final separation of material.

The difficulty with the assemblage-of-element thermodynamic approach described in the opening paragraph of this closing discussion does not, I believe, lie at this microstructural level but rather at the much finer dislocation level. Some of the aspects of dislocation behavior can be modeled by assemblages of elastic and perfectly plastic elements of different degrees of complexity. However, the mechanisms of importance in the range of 1% or more of plastic strain in a polycrystalline structural metal are not represented well by such models. The basic difficulty is that the models have too great a degree of mechanical restorability. They can by purely mechanical means be brought back to a reference or annealed state no matter how elaborate the prior path of loading. Their virtue is their drawback. The easy definability of state and the fixed set of state variables is a gross oversimplification of the behavior of real metals. An annealed metal which is deformed plastically cannot be brought back to the annealed state by purely mechanical working. The lack of a loading path from the deformed to the annealed state reflects those rearrangements on the microscale which do in fact account for work-hardening and many of the other properties of single crystals and polycrystals.

Perhaps in thermodynamic terms the mechanically irrestorable (as distinguished from irreversible) changes on the microscale correspond to continual changes in phase. This certainly is true in the sense that the changing properties of a plastically deforming metal are not contained within any of the classes of models or assemblages we use to simulate metals. Are assemblages of model elements, each with a well-defined state, inadequate because they correspond to a fixed phase? If so, the point once more is that emphasis must be placed on the basic physical aspects in simple mathematical terms rather than on the simple physical idealizations in complex mathematical form. Again, full generality is not the goal; full generality is complete chaos and contains no information. Unifying instructive classifications of physical behavior are required which contain the essence of the macroscopic problem. Such ideas as using plastic work or plastic strain as a state variable won't do because they are trivially oversimple. On the other extreme, a continual change of phase without thermodynamic prescription is impossibly general even if true in the strictest physical sense. Some concept between an assemblage of well-defined states and a continual unspecified irrestorability or phase change is required.

Let me then repeat in conclusion that what I believe should be done is to worry far less about general mathematical consistency and to concentrate on the physical aspects. These must be put down in clear and reasonably acceptable form before it is worthwhile subjecting them to formal mathematical treatment.

Session III. J. D. CAMPBELL,*
University of Oxford, England

It has become increasingly clear during the last two decades that the rate dependence of the mechanical behavior of solids is a key factor in understanding the fundamental processes involved in anelastic deformation. The realization that this is true for metals as for non-metals was delayed for two reasons: firstly, many common alloys are relatively rate-insensitive at normal rates of strain, so that it is necessary to use apparatus capable of measuring stress and strain accurately over very wide ranges of strain rate; secondly, the fundamental processes of plastic flow in metals could not be investigated experimentally until adequate techniques were developed for observing dislocations and their properties.

With the advent of electronic methods of control and measurement of stress, strain and strain rate, and the development of techniques for high-speed testing, the accurate determination of rate effects in metals became possible; and some ten years ago the direct measurement of mean dislocation velocities was achieved by the double-etch method of Gilman and Johnston. The results of those measurements showed that, in certain materials at least, very large increases in the mean dislocation velocity correspond to quite moderate increases in the applied stress. This explains the small rate-sensitivity of such materials. It has been shown by electron microscopy that dislocations multiply very rapidly as plastic straining takes place. If the dislocation density increases in this way, a given strain rate may be maintained with a reduced mean dislocation velocity; this may even lead to a decrease in the flow stress, so that a yield drop is observed. It is possible that, at a given plastic strain, the density of *mobile* dislocations increases with the applied stress; experimental observations on this point are not yet available. Any such increase in the number of dislocations participating in the deformation will reduce the dislocation velocity and thus tend to mask the rate effect.

Experiments such as those referred to in Dr. Gillich's paper indicate that certain annealed pure metals are effectively rate-insensitive over fairly wide ranges of rate. Another group of rate-insensitive materials consists of high-strength alloys in which the plastic flow is controlled by hard second-phase particles or other relatively large obstacles. In materials such as those of either group, the behavior may be adequately represented by a simple rate-independent stress-strain curve, for uni-

* On leave at Brown University.

axial stressing. For general three-dimensional stress, the laws of classical plasticity theory describe the deformation.

The rate dependence of the flow stress, even if it is small, may have important consequences. For example, as Bodner and Rosen have shown, if the flow stress decreases with increasing strain rate, the specimen may show discontinuous straining represented by a serrated stress-strain curve. It must be noted, however, that this behavior depends on the characteristics of the machine in which the deformation is carried out, as well as on the properties of the material. This is a very important point which concerns all types of material testing: it is in general necessary to insure that the interaction between the material and test apparatus is taken into account, if the basic properties of the material are to be determined. Another example in which this interaction is of importance is provided by the discontinuous yielding observed in mild steel and other alloys. Here the deformation becomes highly heterogeneous and Lüders-band formation occurs. A further example, also involving non-uniform flow, is the phenomenon of necking in the tensile test. In such cases, the determination of the basic material behavior becomes extremely difficult.

All these phenomena can only be properly understood by taking account of two factors: the intrinsic rate dependence of the material itself, and the heterogeneous nature of the deformation and of the original state of the material. Under conditions of non-uniform flow, initial irregularities of geometry or material state may become magnified during the deformation, and it is then very hard to deduce the basic material properties from macroscopic measurements; however, some progress has been made in developing phenomenological theories to describe these types of heterogeneous deformation, using empirical laws for the material behavior.

The direct measurement of dislocation velocities was a major step towards the goal of relating macroscopically observable quantities—stress, strain, strain rate—to the basic deformation mechanisms in metals. In addition to the results obtained by the double-etch method, dislocation mobilities have been derived indirectly from measurements of ultrasonic attenuation. These two techniques involve different ranges of stress and strain, and in general they do not give the same behavior. I think it is important to distinguish between the mean dislocation velocity for motion over relatively large distances, and the local velocity attained by dislocations moving over distances which are small compared with the spacing of obstacles in the lattice. There has been some success in interpreting measurements of both types, in terms of various theoretical models for the rate-controlling processes. Examples of these are the overcoming of point defects, phonon viscosity, and the motion of kinks along the dislocation line; some of these models were discussed in Professor Gilman's paper. It is clear that there is a great variety of possible rate-controlling processes, and it seems that even in a given material more than one of these may be significant under given conditions. Thus a constitutive relationship derived by considering a single rate process is likely to be valid only within a small range of the variables concerned. I would therefore emphasize the importance of the point made by Professor Prager: that any constitutive law must include a statement of the ranges of variables for which it is valid, if it is to be useful.

A further complication in attempting to deduce the macroscopic behavior from models of the basic deformation mechanisms is the heterogeneous nature of the deformation on the microscale even when it is uniform on the macroscale. This is well illustrated in the results given by Dr. Edington in his paper; these showed that even in a single crystal of a pure metal, highly complex non-uniform distributions of dislocations develop during plastic flow. In such circumstances, the concept of mean dislocation density is of doubtful value. In alloys of practical importance, heterogeneity on a larger scale occurs because of the polycrystalline nature of the material, variations in grain size and orientation, second phase particles, inclusions, residual stresses and other factors.

In view of the complexity of the deformation processes in common alloys, therefore, it seems unlikely that it will prove possible to deduce general constitutive laws from a knowledge of the basic flow mechanisms. Thus, as remarked by Professor Gilman in his paper, it is at present more realistic to be satisfied with phenomenological descriptions. In formulating these descriptions, however, we should still be guided as far as possible by general considerations based on dislocation theory. We start with the basic flow relation for plastic strain due to the motion of dislocations, which states that the shear strain rate is proportional to the product of the density of mobile dislocation ρ_m and their mean velocity v. We then try to obtain some indication as to how each of these two variables depends on macroscopic quantities such as the applied stress and the plastic strain. This seems to be the most hopeful way in which to try to formulate constitutive relationships which will be physically reasonable, whose validity can be tested by experiments.

As I have already mentioned, the mobile dislocation density is not a very well understood quantity. All that can be measured at present is the total dislocation density, and results such as those given in Dr. Edington's paper show that it increases approximately linearly with plastic strain. It is customary to assume that the mobile number is a constant fraction of the total; I believe this may be an oversimplification, and it is to be hoped that experimental techniques will be developed to throw light on the problem of the influence of the applied stress. The mean velocity at low dislocation densities can be measured by the double-etch method, but this technique cannot be used at densities corresponding to even moderate plastic strains; at such strains interactions between dislocations become important and the mean velocity decreases. The mobile fraction may also decrease as a consequence of dislocation interaction.

It seems that neither mobile dislocation density nor mean dislocation velocity can at present be determined experimentally, except under severely limiting restrictions. Various semi-empirical expressions have been proposed for the two quantities, and it has been shown by Gilman, Hahn and others that many of the observed phenomena of dynamic straining can be qualitatively explained by combining the basic flow equations with an equation describing the elastic behavior of the specimen and testing machine.

In a more general approach, we may assume that both ρ_m and v depend on the applied stress and the dislocation sub-structure of the material. Dr. Edington's paper showed clearly that this sub-structure is a function not only of the current plastic strain but also of the history of straining. As a first approximation, however, we may neglect the history and simply take the plastic strain as defining the dislocation density. This approach leads to the formulation put forward by Malvern, in which the plastic strain rate is taken to be a function of the applied stress and the plastic strain. As is well known, this has proved to explain some of the observed wave effects, and it has also been shown to give a reasonable description of some of the data relating to rate-dependent behavior at medium strain rates. A more specific form of Malvern's equation is based on the overstress hypothesis, in which the plastic strain rate is taken to be a function of the amount by which the applied stress exceeds a work-hardening stress corresponding to static conditions. However, as Professor Bodner pointed out in his paper, this approach cannot be applied to materials showing rate dependence at low rates, since no static stress-strain relation can then be defined.

In certain metals it may be possible to characterize the deformation in terms of a single stress-dependent activation energy, over limited ranges of temperature and strain rate. The most widely used model of this type is Seeger's intersection model, according to which the flow stress for a given dislocation sub-structure is a linear function of the logarithm of the strain rate. This model has been used very successfully by Dorn, Lindholm, Maiden and others, in interpreting experimental data obtained for several face-centered cubic metals.

At very high stresses dislocations are able to overcome the barriers which oppose their motion, without the aid of thermal activation. Under these conditions it appears that the flow stress increases linearly with the strain rate, rather than with its logarithm; this corresponds to a very large increase in the rate sensitivity. It is thought that in this region the rate is controlled by damping forces of a viscous nature which act on dislocations as they move through the lattice. The rates at which this behavior occurs are at the limit of the experimentally attainable range and therefore I think that data in this region are less well established than those obtained at lower rates. Certainly more experiments at very high rates would be most valuable, and I have no doubt that improved techniques will enable the behavior in this range to be explored further within the next few years.

Nearly all the existing data on dynamic plasticity relate to simple states of stress or strain. The experimental difficulties of extending the work to more general stress systems are very considerable, especially at the highest rates, which are the ones of most interest. The experiments at medium rates described in Dr. Lindholm's paper are a very promising start in this direction, and clearly we may expect further interesting studies of this nature to be undertaken in the future. Only by such work can empirical three-dimensional constitutive laws be established for rate-dependent materials.

Several hypotheses have been put forward as to the form these laws might take: Dr. Perzyna has generalized Malvern's overstress hypothesis, while Dr. Cristescu has proposed an equation based on decomposing the strain rate into elastic, plastic and visco-plastic components. At the present meeting, Dr. Bodner has put forward a law in which the second invariant of the plastic deformation rate is taken to be a function of the second invariant of the deviatoric stress. This law, if modified to take work-hardening into account, would correspond to a generalization of Malvern's basic rate equation. These are interesting proposals which must be tested against experimental results obtained for materials subjected to dynamic combined stresses.

In summary, it seems that detailed studies of dislocation mechanisms, while of great intrinsic interest, are not likely to provide in the foreseeable future the information needed for the calculation of macroscopic behavior under dynamic loading. It is therefore necessary to employ semi-empirical laws, based as far as possible on existing knowledge of the basic deformation processes. Some success has been achieved along these lines, but consideration has so far been almost entirely limited to simple states of stress. Much further experimentation remains to be done, especially at the highest rates and under general stress states, in order to determine the range of validity of the various forms of constitutive law.

Session IV. C. D. LUNDERGAN, Sandia Corporation, Albuquerque, New Mexico

The papers presented in the fourth session of the symposium can be related in the context of extending and unifying the techniques used to determine the dynamic mechanical properties of materials. These papers are concerned with that portion of the state of the art represented by one-dimensional stress, both short and long bar, one-dimensional strain and the cylindrical configuration resulting in plane strain. The various recording techniques were well represented as mechanical, optical, electrical and electro-mechanical devices were used to record the dynamic response.

Dr. Karnes observed in his presentation that to obtain dynamic mechanical material properties for the time intervals of the order of microseconds, it is necessary to place the specimen in a defined state of stress from the outset of the observation and to maintain this mathematically amenable condition throughout the time of observation. The strain rates under the condition of one-dimensional strain were limited only by the wave propagation characteristics of the material itself. He also noted the assiduousness that must be employed in the preparation and conduction of an experiment; the indirect method that must be used in relating the observations to the desired properties; and the complexity of the analysis which must follow the multitude of stress wave interactions.

He also presented comparisons of the results of various kinds of instrumentation such as quartz gage, the capacitor and the interferometer; all of which were favorable. Reference was made to the use of the Gillman-Johnston dislocation model used by Taylor to relate the attenuation of peak stress with distance. Again the results were favorable. The intent being to show the internal consistency of the information obtained in this relatively new procedure for investigating dynamic behavior.

Dislocation dynamics was also called into play by Dr. Maiden in his representations of the work done at General Motors. Agreement, or at least limited explanation of some of the strain-rate observations, was found. They observed the strain-rate sensitivity of aluminum as a function of alloying, temperature, and hardness. This was all done in the one-dimensional stress configuration. Using this information, they predicted what might well be observed in aluminum in the one-dimensional strain configuration.

Professor Fyfe recognized some years ago that a further extension of experimental techniques is required to provide a more stringent test of the intrinsical nature of the dynamic properties generated by the conditions of one-dimensional stress and one-dimensional strain and to this end he has made considerable progress in the development of a technique which subjects the specimen to plane strain and is still amenable to mathematical description. The compatibility of the observed behavior in the elastic regime and the mathematical model was established and used to validate the experimental procedure and to define initial conditions. The resolution of the descrepancies in the plastic regime will achieve the desired end of discerning which of the models are valid and the extent to which the material properties are intrinsic. So again, the intent is to extend the methods of investigation and provide a means of unifying the field of study.

Unification was also the intent in part of the paper of Professor Ripperger. Using a general form of the constitutive equation and by introducing large variations of the parameters involved, he obtained the stress-strain relations for various constant strain rates. In turn he obtained a wave profile in a semi-infinite bar for both the rate independent model and for several variations of the rate dependent model. He concluded the establishment of constitutive equations from the observation of wave front shapes and strain propagation velocity under one-dimensional stress conditions was ambiguous and the deductions on the existence of rate sensitivity or the lack of it in materials using this technique is questionable.

Now as to the expected course of the subject of dynamic mechanical properties of materials—the demands of the industrial and military complexes will insure an increase of activities for the reasons noted by the introductory speaker. The larger part of this activity will be the generation of the properties of engineering materials. There will, as there is now, be some difficulties resulting from the acquisition, analysis, and interpretation of the data. The source of this difficulty is the relative newness of the study, it has not as yet reached the point in development that tests can be conducted in a routine manner. Another difficulty is that generalizations of several classes of materials, as for example the polymers, have not been completed in the research laboratories. While the demand for properties exists now, there are not as yet a sufficient number of engineering laboratories producing reliable dynamic data.

The activity in the research laboratory will correspondingly increase. This activity will be in the form of a more intensive effort to describe additional classes of materials, such as the polymers, the ceramics, and the distended materials. More conditions will be explored, such as stress relaxation, multiaxial stress configurations and the effects of phase transitions. The environments will be extended as well into the fast heating rates and the larger temperature ranges.

In summary, the design needs which control the purse strings will provide impetus to this field of study of wave propagation of materials in the microsecond and submicrosecond regime.

Session V. D. M. FORNEY, JR., Wright-Patterson Air Force Base, Dayton, Ohio

When I noted my position on the program, it occurred to me that my prepared remarks about the presented material might be redundant since, by then, my predecessors would surely have adequately covered the essence of the important points. I decided, therefore, to confine my remarks, and make them brief, to the problems confronting us in the application of fundamental results of the type discussed in this conference to practical engineering situations.

As already pointed out, we have a long way to go in the development of comprehensive analytical techniques to solve engineering problems of materials dynamics based on solid theoretical grounds. This is not particularly surprising when one considers that such techniques for complex static cases, as well as low-frequency dynamic cases involving negligible strain rate effects (e.g., vibrations), are still only marginally adequate themselves. This, on the face of it, could be a very discouraging point of view; however, in light of the progress that has been made in the understanding of fundamental aspects of materials behavior, even in the last ten years, one then has reason to be very hopeful. We must realize, nevertheless, that theoretically justified techniques for accurately defining the high strain rate dynamic behavior of materials are largely nonexistent and the interim methods of analysis are indeed meager.

One of the most critical problems we face today in engineering applications work related to the subject of our conference is the lack of a fully descriptive theory of wave propagation in non-homogeneous or multiphase materials. Without such a theory, it is not possible to predict the response of such a composite system to impulsive loading, even under the most ideal conditions, nor to theoretically design a candidate system for test. This fact is understandable, though, when we realize that there is even notable disagreement today in what constitutes the logical approach to the formulation of the problem of one ideal solid dispersed as a phase in a second one, and so on. One-, two- and three-dimensionally reinforced composites are being utilized today in many roles involving high rate loading and improvements in the theoretical bases for the configurations of these systems are critical requirements.

Another problem area of real consequence is the lack of descriptive dynamic fracture criteria for composite materials. The problem appears especially critical when we realize that knowledge of fracture mechanisms in homogeneous materials under dynamic conditions is still largely undeveloped, while fracture behavior under *static* conditions is still to be fully resolved.

In the absence of well-developed analyses of both dynamic response and fracture properties, it is necessary to rely on approximations from which it is possible to draw only few generalizations. For example, a fictitious equation of state must be assumed for a composite material based on a theory for homogeneous solids, and a threshold fracture resistance can be defined only in terms of a set of standard test conditions and not in terms of definitive stress and strain fields. Still one step further is the important need for an understanding of dynamic behavior at elevated temperatures.

An additional serious complicating factor in the understanding of dynamic materials behavior is the difficulty of experimentation, not only to measure quantities that are required for use in constitutive relations describing materials response as well as to validate new theories, but also to compare the dynamic response characteristics of components with those predicted in design. It appears that, even though impressive advances in experimental skill have occurred, we have a long way to go in this field; we have seen during the course of this conference that there is a significant disagreement among many experimenters on techniques. It will be an important task to develop experimental techniques which introduce more realistic environments; not just simply more realistic environments one at a time, but rather in their proper combination. It is a well-known fact that environmental effects are not simply additive, but are synergistic in nature. Examples of environmental conditions under which dynamic response measurements are presently needed for current engineering applications are: up to very high temperatures, high heat fluxes, high pressures, corrosive conditions, and others. Examples of engineering applications of current interest and importance are: the design of protective systems such as armor and sacrificial shock absorbing systems; re-entry bodies such as nose cones and space capsules; design of structures to withstand earthquake and other ground-transmitted shocks, sonic boom overpressures, air blast loads, underwater detonations; design of oil well drill rigs; the application of explosive forming manufacturing techniques; and many others.

A problem common to all fields of science and engineering, and certainly in substantial evidence in the dynamic materials behavior area, is the very large separation in thinking, point of view and orientation between the fundamentalists in materials science and the design engineers who apply engineering data and analytical tools to practical engineering situations. Because of the many restraints of time, simplicity and practicality of a design procedure on the one hand, and the number of analytical complexities and idealizations attributable to fundamental analyses on the other, much of the scientific knowledge developed over the years in very careful research and in very fundamental work is lost to the engineering community. Obviously, the more advanced our technology becomes, the more imperative it is to reduce, or hopefully eliminate this separation. Unfortunately, there is no sufficiently organized class of research engineers properly oriented to specifically reconcile the differences and convert fundamental research results into a form and language suitable for adoption in the usual design situation. I think each of us, whether we be materials fundamentalists or materials-design engineers, needs to improve on our efforts to reconcile this gap.

MIX
Papier aus verantwortungsvollen Quellen
Paper from responsible sources
FSC® C105338

If you have any concerns about our products,
you can contact us on
ProductSafety@springernature.com

In case Publisher is established outside the EU,
the EU authorized representative is:
Springer Nature Customer Service Center GmbH
Europaplatz 3, 69115 Heidelberg, Germany

Printed by Libri Plureos GmbH
in Hamburg, Germany